Introduction to Modern Mathematics

Introduction to Modern Mathematics
second edition

DORA McFARLAND
EUNICE M. LEWIS

University of Oklahoma

D. C. HEATH AND COMPANY
Lexington, Massachusetts Toronto London

Published simultaneously in Canada.

Printed in the United States of America.

International Standard Book Number: 0-669-62562-0

Library of Congress Catalog Card Number: 72-4465

Preface

The mathematics used today in our elementary schools is a matter of interest to many people in various walks of life. Among those having a particular interest in today's school mathematics are those preparing to teach in the elementary schools, those already teaching, and those who realize that some knowledge of mathematics is an essential part of their general education. There is still another group made up of parents, relatives, and friends of the children attending school. They are interested in knowing about the mathematics being currently taught to these children.

The selection of the content for the first edition of *Introduction to Modern Mathematics* was designed to meet the needs of these various groups. Extensive testing in college classes by the authors and their colleagues supported the selection and organization of the content, as well as the level of rigor used. Because of the varying backgrounds of those who would study this textbook, the presentation of the material is somewhat informal in approach. Concepts are developed inductively by means of examples designed so that the reader through his investigation of them will be led to make his own generalizations. These he may verify with those presented in the textbook.

The theme of this textbook is the study of the real number system and its subsystems, as suggested for the first six hours of the Level I Recommendations of the Committee on the Undergraduate Program in Mathematics (CUPM). Other topics consist of geometry (metric and nonmetric), numeration systems, and relations and functions.

With constant change in the elementary school mathematics program brought about by the influence of the continuing mathematics revolution, a *second edition* was deemed advisable. This edition continues with the basic theme, the development of the real number system. Extensions have been made to topics in the first edition, and there are additions of new ones.

The original chapter on *sets,* as preparation for their use in the development of the basic operations on the natural and whole numbers, now includes a study of sets and operations on them as an operational system having certain properties. Mapping, basic to the function concept, is introduced in this chapter. It is used here in the development of cardinal numbers. Mapping is used in appropriate places throughout the textbook, finally appearing in the chapter *Relations and Functions.*

The chapter on *sets of points* now includes an introduction to some transformational geometry of the plane, convex and nonconvex sets, and some of the standard simple closed curves (polygons). The chapter on *metric geometry* now includes the study of perimeters, areas, and volumes of some of the standard figures. An innovative approach is the use of models by which the student discovers area, perimeter, and volume formulas.

The use of flow charts in problem solving in business and industry has caused their increased appearance in elementary forms in school mathematics textbooks. This seemed to be sufficient evidence of the need of some knowledge of such charts. These are introduced in the final discussion of algorithms in Chapter 7.

A new chapter, *Probability,* has been added. The reason for including this topic is due to the increased interest in it among all segments of our society. This interest has been felt in the elementary school mathematics curriculum.

Comments from those who have taught the first edition as a three-hour terminal course indicate concern that these students have no contact with the "fractions" taught in the elementary school program. To meet this need, a new chapter, *Nonnegative Rationals,* was written and appears in this, the second edition.

This brings a suggestion for the structuring of a three-hour terminal course. It is recommended that for such a course, Chapters 1 through 7 be taught in the order given in the textbook, concluding with Chapter 10, *Nonnegative Rationals.* This can be done without breaking the continuity of the development of the topics. Notice that in Chapter 10, frequent references are made to Chapter 9, *The System of Rational Numbers,* for further rationalization. These may or may not be used, depending on the level of rigor one wishes to achieve in this chapter.

Chapter 5, *The System of Natural Numbers,* and Chapter 6, *The System of Whole Numbers,* introduce the student to a study of two of the subsystems of the real number system. With this as a start, he should begin to see the pattern that leads to the development of the real number system.

This book has been designed to be taught as a five-hour semester course or two three-hour semester courses. It assumes very little previous mathematics training and should lead the student to a confident feeling about the mathematics current in our schools today.

We wish to thank a number of our colleagues from the Mathematics Department of the University of Oklahoma for their suggestions and contributions to this revision.

D. McF.
E. M. L.

Contents

4 SETS OF POINTS

5 THE SYSTEM OF NATURAL NUMBERS

6 THE SYSTEM OF WHOLE NUMBERS

7 ARITHMETIC OF THE WHOLE NUMBERS

8 THE SYSTEM OF INTEGERS

9 THE SYSTEM OF RATIONAL NUMBERS

10 ARITHMETIC OF NONNEGATIVE RATIONAL NUMBERS

Introduction to Modern Mathematics

1

A Brief Look to the Future

1.1 A WORD TO ALL READERS

Before beginning the study of a new subject or the further investigation of a familiar one, it is always well to survey the situation. One needs to know something of the circumstances that led up to the present study and to have at least a glimpse of what the future holds in store. Mathematics, our subject under discussion, has a long and interesting history. Besides being a challenging study in itself, it has grown and developed as a companion service for other areas of knowledge. Science and technology come immediately to mind as endeavors where mathematics contributed skills that made progress possible. Naturally, with new concepts becoming a requirement in these developing areas, it was necessary for the mathematics program to change, too, if it was to be useful.

It is our purpose in this chapter to suggest to you some of the changes that have taken place—some in the content, but more in the method of approach. There was a time when elementary and secondary school instruction in mathematics emphasized facts and routine computation with very little development of understanding. Simple notions of functions, sets, numeration systems, and reasons for operational procedures were reserved until advanced college study. Much of this has been changed since the early 1950's. Sputnik really shook us up. Experimental programs, with highly skilled mathematicians leading the way, sprang up in the United States, in England, and on the continent of Europe. The letters SMSG (School Mathematics Study Group) and the movement they represent had a profound influence on school textbooks and consequently on instruction. Students no longer learned "how" with little attention to "why." There were many other experimental programs and some still continue, but the reasons for change are less obvious and the new material and methods have found their way to a greater or less degree into the school textbooks. That which was called "the new math" was never really "new." It was mathematics approached in a more meaningful way. Now that these interpretations are incorporated into most elementary and secondary school textbooks, it is fair to say that "the new math" is no longer even thought to be "new."

Modern mathematics as it is to be treated here makes no special prior demands on the reader. No mathematical background is assumed beyond one or two years of high school mathematics. There will be emphasis on structure. Notice how the number system is developed through natural, whole, integral, rational, and real systems in five separate chapters. These systems are built, each one upon the preceding one, as the properties that each system adds are developed. The structure that is evident helps to develop clearer meaning and hence greater understanding and skill. Elementary set theory is used at the very beginning (Chapter 2) as a method to develop and give meaning to the simple laws (rules) of arithmetic. The study of man's struggle to record the results of his counting makes an interesting story. It also serves to increase our appreciation of the decimal system, which is often accepted very casually. Geometry will have its share of attention, too; first the number line and its application for simple arithmetic; then sets of points—their union and interesection. This will play a part in the development of measurement on the line, in the plane, and in space. None of this resembles the formal proofs of Euclidean geometry.

If you read further, you will find an introduction to the simpler aspects of mathematical systems—for example, "clock arithmetic." This leads naturally into modular systems and from there we finally arrive at a definition for a *field*—a term used earlier in connection with the system of rational numbers. The notion of function is central through much of mathematics and a chapter on "Relations and Functions" treats this subject in a correct, but not too sophisticated, manner. If you want a taste of probability and statistics, there is a chapter that gives some of the fundamental notions without the discussion becoming too complicated.

We hope this has been helpful and has served to whet your appetite for what lies ahead.

1.2 A SPECIAL WORD TO THOSE WHO PLAN TO TEACH

This is not a book on methods of teaching, but one on mathematics. Moreover, it is not intended that the mathematics presented here be taught to children, although you should watch for situations in which the topics presented here may be related to classroom activities. This material is more concentrated than that presented in the elementary school textbooks, and yet it will not be especially difficult for you if you are willing to look at familiar mathematical ideas in a new setting and to examine and accept unfamiliar ones as they are introduced.

Every teacher of elementary mathematics should become affiliated with the National Council of Teachers of Mathematics (NCTM). This is the acknowledged leader of professional organizations involving teachers of elementary mathematics. A journal, *The Arithmetic Teacher,* is published by NCTM and is provided as part of the membership benefits. It is addressed to the elementary school teacher. In it you will find suggestions for teaching certain topics, enrichment materials, ideas for club programs, construction and use of visual materials. Included also will be articles to challenge the teacher to an understanding of unfamiliar topics that may be relevant to current thinking in the mathematics for the elementary school. The NCTM is an active organization at the local, state, and national levels. The meetings provide great opportunities for information and inspiration—opportunities that will be invaluable for your professional growth.

In addition to membership in professional organizations, each teacher should begin the building of a personal library. The part devoted to mathematics might include your college textbooks and a file of the issues of *The Arithmetic Teacher.* The list of publications that follows is by no means complete. You must be alert to add up-to-date references. Only by constant vigilance can you build a library that will sustain a creditable professional growth.

1.3 A SUGGESTED PROFESSIONAL LIBRARY FOR ELEMENTARY SCHOOL TEACHERS

The Arithmetic Teacher, a journal. Washington, D.C.: National Council of Teachers of Mathematics. An official journal of the National Council of Teachers of Mathematics devoted to the interests of those concerned with elementary school mathematics.

BIGGS, EDITH E., and JAMES R. MACLEAN. *Freedom to Learn.* Menlo Park, Calif.: Addison-Wesley, 1969. This book embodies the philosophy of the London Schools Council. An approach to mathematics through active learning.

BRUNER, JEROME S. *The Process of Education.* Cambridge, Mass.: Harvard University Press, 1962. A landmark effort in mid-twentieth century education sponsored by the National Academy of Science was the Woods Hole Conference in 1959. This book resulted from the discussions at this conference.

The Continuing Revolution in Mathematics. Washington, D.C.: The National Council of Teachers of Mathematics, 1968. A joint effort of the National Council of Teachers of Mathematics and the National Association of Secondary School Principals in presenting a progress report on the reform movement. A discussion of what is included in the expression "the new math," and a description of several of the innovative elementary school mathematics programs.

COPELAND, RICHARD W. *How Children Learn Mathematics.* New York: The Macmillan Company, 1970. An interpretation of Piaget's research in arithmetic. Emphasis is based on how children learn mathematics rather than on the techniques used in teaching them.

Dienes, Zoltan P. *Mathematics in the Primary School.* New York: St. Martin's Press, 1964. Innovative ideas about content and method with many examples showing how mathematical concepts are taught through games and concrete experiences.

Elementary School Mathematics: New Directions. Washington, D.C.: U.S. Department of Health, Education, and Welfare, 1963. Information concerning early efforts made to improve the elementary school mathematics program. A description of several of the programs of the study groups is given.

Enrichment Mathematics for the Grades, 27th Yearbook. Washington, D.C.: National Council of Teachers of Mathematics, 1963. Material suitable for providing enriching experience to students. Bibliography.

Evaluation in Mathematics, 26th Yearbook. Washington D.C.: National Council of Teachers of Mathematics, 1961. Evaluation of instruction and learning provides valuable information to the teacher and the learner. The yearbook stresses the general theory of evaluation, techniques of evaluation for mathematics teachers, and evaluation principles applied to classroom problems.

Furth, Hans G. *Piaget for Teachers.* Englewood Cliffs, N.J.: Prentice-Hall, 1970. The author interprets what Piaget's discoveries can mean to our schools through a series of letters to teachers.

Glennon, Vincent J., and Leroy G. Callahan. *Elementary School Mathematics: A Guide to Research.* Washington, D.C.: Association for Supervision and Curriculum Development, National Education Association, 1968. Includes brief reports on selected research relating to questions asked about the teaching of elementary school mathematics.

Goals for School Mathematics: The Report of the Cambridge Conference on School Mathematics. Boston, Mass.: Houghton Mifflin, 1963. This report presents the goals for mathematics curriculum, K–12. A section on Curriculum for Elementary School (K–6) gives in detail those concepts which should be included in the future elementary school mathematics program.

The Growth of Mathematical Ideas: Grades K–12, 24th Yearbook. Washington, D.C.: National Council of Teachers of Mathematics, 1959. Points out basic concepts which should be continuously developed throughout the mathematics program; illustrates methods for use in the classroom; suggests a sequence for overall themes and their topics.

Instruction in Arithmetic, 25th Yearbook. Washington, D.C.: National Council of Teachers of Mathematics, 1960. Arithmetic in the elementary school is the focus of this yearbook. Part I deals with the cultural value of arithmetic, its structure, and its content. Part II deals with the learner and factors affecting him in learning arithmetic.

Mathematics for Elementary School Teachers. Washington, D.C.: National Council of Teachers of Mathematics, 1966. This textbook was prepared for use in preservice and inservice classes. It is easily read and gives a background for the teaching of a modern oriented elementary school mathematics program.

Mathematics in Primary Schools: Curriculum No. 1. London, England: The London Schools Council, 1966. Presents the philosophy of the Council concerning changes and suggests help for the teachers.

More Topics in Mathematics for Elementary School Teachers, 30th Yearbook. Washington, D.C.: National Council of Teachers of Mathematics, 1969. This is an extension of the subject matter presented in the 29th yearbook.

Notes on Mathematics in Primary Schools. New York: Cambridge University Press, New York Branch, 1968. Compiled by members of the Association of Teachers of Mathematics are notes on experiences for children as they participate in creative learning.

PAGE, DAVID A. *Number Lines, Functions, and Fundamental Topics.* New York: The Macmillan Company, 1964. This book presents a variety of mathematical concepts in a unique and fascinating manner.

PAPY, GEORGES, and FREDERIQUE PAPY. *Modern Mathematics,* Vols. 1 and 2. New York: The Macmillan Company, 1968. A textbook developed from the work done at the Belgian Center of Mathematical Pedagogy, by its director Professor G. Papy, University of Brussels, and his wife, Frederique. The Belgian mathematics reform movement is reflected in the content and methodology presented in these two volumes.

The Revolution in School Mathematics. Washington, D.C.: National Council of Teachers of Mathematics, 1961. This booklet explains the origins of the contemporary revolution in mathematics, how change is being implemented, and what administrative decisions are necessary in the local school system.

Studies in Mathematics. New Haven Conn.: Yale University Press, 1960–1961.

Vol. V, *Concepts of Informal Geometry*

Vol. VI, *Number Systems*

Vol. VII, *Intuitive Geometry*

Vol. IX, *A Brief Course in Mathematics for the Elementary Teacher*

These were written for School Mathematics Study Group (SMSG) under a National Science Foundation grant. They are designed to help elementary school teachers in developing a sufficient subject matter competence in the mathematics of the modern oriented elementary school program.

Teaching Mathematics in the Elementary School. Washington, D.C.: National Association of Elementary School Principals, National Education Association; National Council of Teachers of Mathematics, 1970. This contains articles dealing with good mathematics programs for children, as well as articles dealing with the many factors that influence the effectiveness of the elementary school mathematics program.

Topics in Mathematics for Elementary School Teachers, 29th Yearbook. Washington, D.C.: National Council of Teachers of Mathematics, 1964. Originally, this was a series of eight booklets presenting topics taught in a modern oriented elementary school mathematics program. Later, these were compiled and published as the 29th Yearbook. The material is written for the teacher, not the pupil. The material is easily understood.

2

An Introduction to Sets and Numbers

2.1 THE MEANING AND USE OF *SETS*

When thinking about various objects in our world, we often sort them into collections or classes. Although a philatelist may speak of his *collection* of stamps, often special words are used to identify particular kinds of collections. For example, a collection of stars is called a *constellation,* a coach boasts of his winning *team,* while a farmer is proud of his *herd* of Black Angus cattle. In mathematics we deal with various collections such as numbers, points, and lines. Instead of using a special word for each of these collections, mathematicians use a general term, *set.*

The development of the theory of sets is attributed to Georg Cantor, a German mathematician, who lived from 1845 to 1918. However, it was not until the middle of the twentieth century that this concept was adopted as a fundamental unifying principle in school mathematics beginning with the primary grades. The use of sets to unify and clarify mathematical concepts for children as well as for adults is both natural and appropriate. Historically, many mathematical concepts were discovered by early man through his manipulation of sets (groups) of things. Pairing members of sets, joining sets, and subdividing them gave him an intuitive notion of some of the fundamental concepts basic to mathematics. Today's school mathematics uses sets of points to introduce the student to the space in which he lives. The set language also provides for him a basis for an accurate and precise means of communicating mathematical ideas. Due to the important role of sets in the conceptual development of mathematics, some knowledge of elementary set theory seems appropriate.

A set may be described by a phrase or by listing its elements. The phrase "the students seated on the front row of this class" describes a set. The list "John, Mary, Sue, Bill" may be used to describe the same set in a different way, provided that these are the names of the students sitting on the front row of this class. The *set* is the *collection* of students. Each of these students is an *element* of this set. For example, Bill is an element of, or *belongs to,* this set. A set may have any number of elements. It may contain many elements, only a few, or none at all.

A set must be described so completely that there is no doubt as to whether an element belongs to the set or does not. Describing a set of books in the library as "the mathematics books" is sufficient for one to know that all the mathematics books in the library are included in this set, and that books in the field of home economics, for instance, are not included. Since we are able to tell whether or not a book or any other object in the library belongs to the set of mathematics books, we say that the set is *well defined.* The distinctive property that determines the inclusion or the exclusion of an element is called the *defining property* of the set. All sets must have such a property.

Would the phrase "the five most beautiful girls on the campus" describe a set? Is it well defined? Would everyone agree as to which girls would be included and which would not? Let us look at some other examples.

Example 1 Suppose that a child collects all the red blocks in the toy box. It certainly is not difficult to tell which blocks "belong" and which blocks "do not belong" to the collection. Thus, we say that the collection of the child's red blocks is a set, since it is well defined.

Example 2 Does the phrase "the best-built cars in the United States" describe a set? Since it is a matter of opinion whether or not the cars of a particular manufacturer fall into this category, the collection mentioned is not well defined and hence is not a set.

From the discussion thus far, one might be led to believe that the elements of a set are necessarily related in some way—that is, that they must fall into some obvious category. This is not the case, however. For instance, one set may consist of a ball, a bat, and a glove, equipment used in playing baseball, while another set may consist of such unrelated items as a pencil, a car, and a shoe. It is important, even though difficult, to identify the latter set with a descriptive phrase that identifies those elements to be included and those to be excluded. Probably the simplest way would be to point to each item and say, "My set consists of this car, this pencil, and this shoe."

It is also possible to have a set with elements that are themselves sets—for example, the set of football teams in the Big Eight conference. This set has eight elements and each element is a set of players.

2.2 THE ROSTER NOTATION

When possible, the most satisfactory method of describing a set is to display its elements. For instance, a child might describe his set of red blocks by merely pointing to them. With this method, there is no doubt as to which blocks belong in his set and which do not.

In many cases, however, it is not possible to display the elements of a set as the child did his red blocks. So we use a method of representing the set instead. Pictures or names are selected to represent the elements. Then these are enclosed within braces, { }. For example, suppose that it is not convenient to display the members of the set consisting of a ball, a bat, and a glove. Using symbols to represent these elements, we may denote the set by

{  } or {ball, bat, glove}.

This method of listing the elements (that is, representations of the elements) will be called the *roster notation.* This notation completely describes or defines the set, since there is no doubt concerning the membership. It should be noted that the sequence of listing the elements is immaterial; the set above might equally well have been described as

{bat, glove, ball}.

Also, an element is represented *only once.* Two different symbols would not be used to represent the same element, nor would one symbol representing an element be used more than once.

When there is a large number of elements in a set, it is inconvenient to list them. Sometimes the roster notation may be modified to describe such sets. For example, the set of all letters of our alphabet may be represented by

$$\{a, b, c, \ldots, z\},$$

where a few of the elements are listed and the symbol ". . ." is used to stand for the missing elements in the established pattern. Since the last element is known, it is listed last.

A certain set of birds may be described in roster notation as

$$\{h, w, r\}$$

where the lowercase letters h, w, and r represent a hummingbird, a woodpecker, and a robin, respectively. Often a set is named with an uppercase letter. This set of birds might be named set B. It is written

$$B = \{h, w, r\}$$

and read, "B is the set consisting of the elements h, w, and r."

Each of the elements h, w, and r belongs to set B. Using the symbol, $\in$, to denote membership in a set,

$$h \in B$$

is read, "h is an element of set B" or "h belongs to set B." Since a starling, s, does not belong to this set B,

$$s \notin B$$

denotes this. Observe that *is an element of* is negated by making a slash through $\in$.

In Section 2.4 another way of describing a set will be introduced. Instead of listing the elements, the set is described by stating its defining property. When the elements are listed as in the roster notation, the defining property is evident.

EXERCISE 2.2

1. Name five groups or collections of things employing the word customarily used to designate each. Example: A *herd* of cattle.
2. Since a set must be well defined, which of the following expressions describe sets? Explain.
   (a) All the bottles on my kitchen shelf.
   (b) All of the brothers of my Uncle John.
   (c) The ten most honest men in the U.S.A.
   (d) All of the names of the days of the week.
   (e) The first three letters of the alphabet.
   (f) The five best teachers in your school.
   (g) All the baseball teams in the National League.
   (h) All of the months of the year whose names begin with the letter J.
3. Each of the following statements defines a set. Explain. Use the roster notation to describe each.
   (a) The set of the last three letters of the alphabet.
   (b) The set of all 3-letter English words which can be constructed from the letters *a*, *t*, *r*. (Use each letter only once in a word.)
   (c) The set of symbols used in denoting twenty-four in Roman numerals.
   (d) The set of the squares of all counting numbers between 5 and 13. (HINT: The numbers *between* 5 and 13 do not include 5 and 13.)
   (e) The set of all the capital cities in your state.
   (f) The letters used in spelling your family name.
   (g) The names of the days of the week which begin with the letter M.
4. The letters of the alphabet which are *always* considered vowels form a set.
   (a) Describe this set using the roster notation.
   (b) Write, "The letter *e* is an element of the set of vowels, *V*," using the symbol indicating membership in a set.
   (c) Use the "does not belong to" symbol with this set *V*.
5. Since it is impractical to list all of the elements of the following, show a method of using the roster notations for
   (a) The set of counting numbers between 67 and 3000.
   (b) The set of counting numbers greater than 5.
   (c) The set of counting numbers greater than 15 and less than 100.

2.3 SETS AND SUBSETS

Let X represent the set of letters used in spelling the word "care," and let Y represent the set of letters used in spelling the word "are." They may be described in the following manner:

$$X = \{c, r, e, a\}, \qquad Y = \{e, r, a\}.$$

The first is read, "The set named by X *is the same as* (or *is equal to*) the set named by $\{c, r, e, a\}$." The second is read similarly. The use of the symbol, $=$, merely means that X and $\{c, r, e, a\}$ name the same set.

We find it natural to think of set Y as being *included in* set X, since each element of set Y also belongs to set X. A more emphatic statement would be that there is no element of Y that is not in X. This discussion leads to the following definition.

Definition 2.3(a) Set Y is a *subset* of set X if every element of set Y is an element of set X. This may be written as

$$Y \subset X.$$

Does each element of set X belong to set Y? Since the letter c found in set X is not in set Y, set X is *not* a subset of set Y. This is written

$$X \not\subset Y.$$

Since $Y \subset X$ and $X \not\subset Y$, then Y is said to be a *proper subset* of X. The following statement is used to define a proper subset:

Definition 2.3(b) If each element of set Y is an element of set X, but not every element of set X is an element of set Y, then set Y is a *proper subset* of set X.

Consider the set of letters used to spell "bat" and the set of letters used to spell "tab". Let

$$A = \{b, a, t\}, \qquad B = \{t, a, b\}.$$

Set A is a subset of set B, since every element of set A is included in set B. Furthermore, every element of set B is included in set A. Consequently, set B is a subset of set A. Neither of these is a proper subset of the other since

$$A \subset B \quad \text{and} \quad B \subset A.$$

Sets A and B have exactly the same elements and are called *identical sets.* The equals relation may be used,

$$A = B,$$

to say that set A "is the same as" set B.

From the preceding discussion, one should conclude that every set is a subset of itself, but not a proper subset.

Each of the sets A, B, X, and Y described above is a subset of the set of all the letters of the alphabet. Such an overall, inclusive set for a particular discussion is called the *universal set* for that discussion. It may be designated as U, and so we have here

$$U = \{a, b, c, \ldots, z\}.$$

It is customary to identify the universal set for each discussion. Of course, the universal set is not the same for all discussions. The set of all clocks may be the universal set for the set of alarm clocks, while the set of all school teachers may be the universal set for a discussion involving the set of elementary teachers. Is every set a subset of its universal set?

The concept of the *inclusion relation,* $\subset$, may be illustrated quite clearly by using *Venn diagrams,* named for an English logician, John Venn (1834–1923).

For example, the set of football players, *F*, is pictured thus. Each player is within the closed curve.

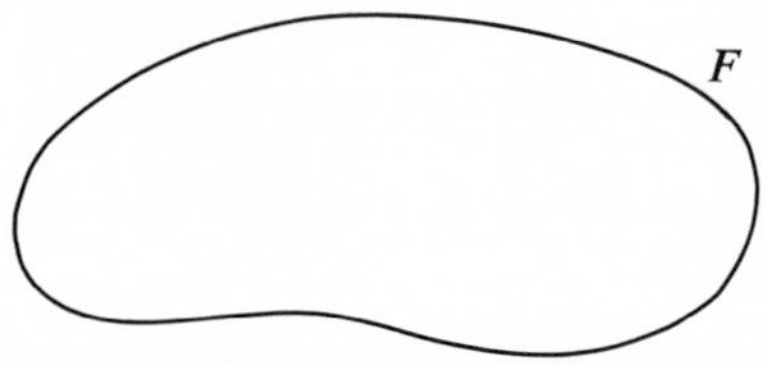

Locating the set of quarterbacks, *Q*, which is included in the set of football players, the diagram takes on this appearance.

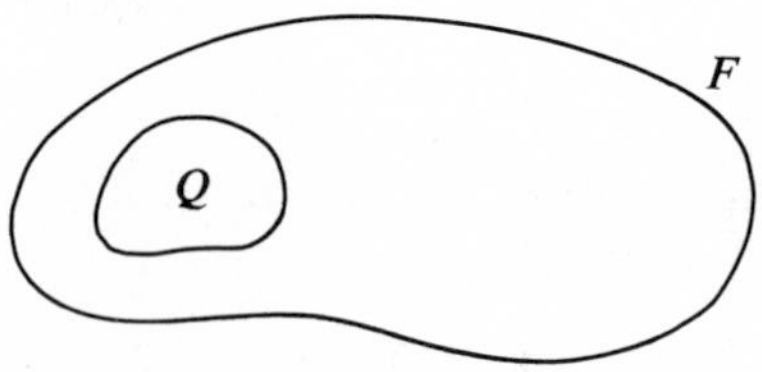

Where would the set of ends, *E*, be located in the diagram? Obviously, within this set of football players, but not within any part of the set of quarterbacks (unless some members of the team can play either position).

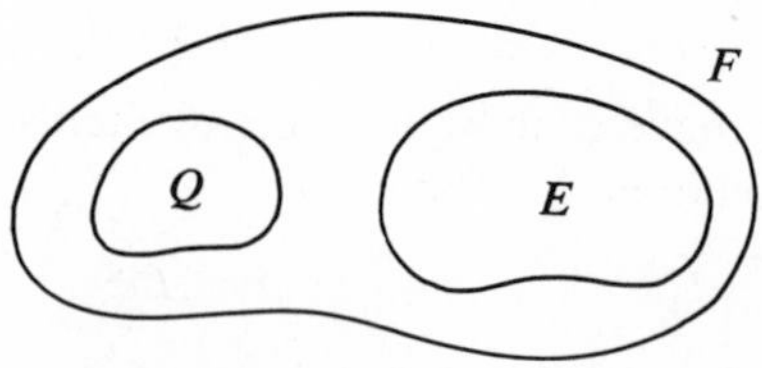

Where would the guards, the centers, the linebackers, the tackles, and all other players be located?

The all-inclusive set, *F*, is the *universal set*. Is the set of quarterbacks a subset of set *F*? Is it a proper subset?

Suppose that some of the quarterbacks also play end. Such players as these are represented by the shaded portion in the diagram. Is $Q \not\subset E$ or $E \not\subset Q$?

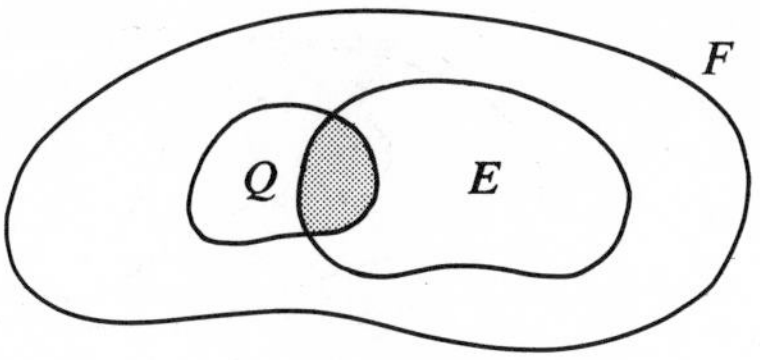

The following diagram shows a situation in which all quarterbacks also play end, but all ends do not play quarterback. That is, $Q \subset E$ and $E \not\subset Q$.

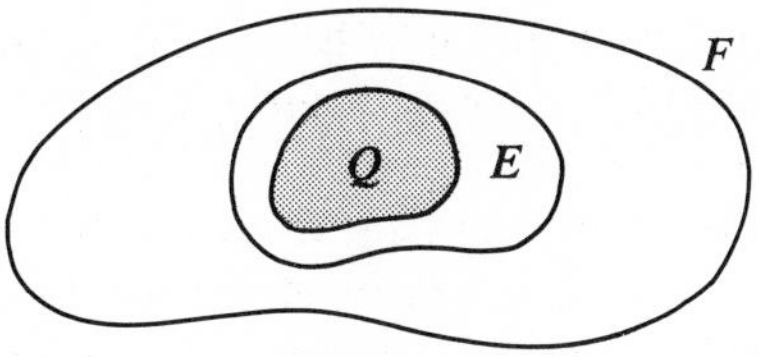

Another situation is represented next. In this, $Q \subset E$ and $E \subset Q$. What can be said of sets Q and E?

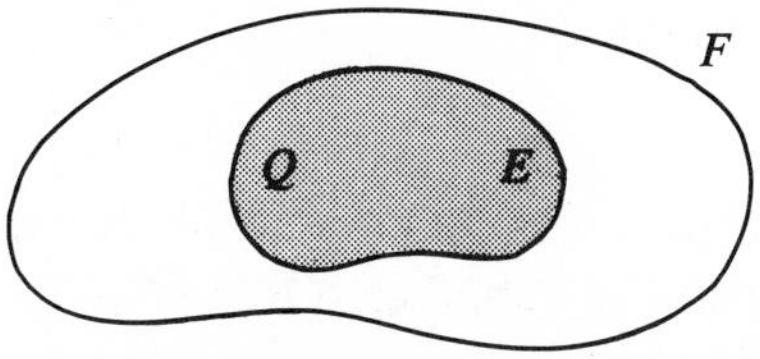

The relatedness of sets has been demonstrated in the preceding examples by means of Venn diagrams.

Is it possible to use a Venn diagram to represent the set of six-month-old infants who are quarterbacks? Obviously, since there are no infants who are quarterbacks the set is empty. Furthermore, the importance of such a set in the continued study of sets and their use makes it necessary that it be named. It is called *the empty set*. Some prefer to call it *the null set*. It is denoted by { } with no elements within the braces, or by ∅. Mathematicians commonly use ∅, while { } is more effective in teaching young children. The latter symbol emphasizes the empty condition of the set. In case the symbol { } is used, one should be warned not to place the symbol ∅ between the braces, also. By so doing the set would no longer be empty.

Since each set may be a subset of some set and a set is a subset of itself, the question arises: Is the empty set a subset of some set? First, consider the universal set, $U = \{a, b, c, \ldots, z\}$ and the sets

$$B = \{a, b\},$$
$$C = \{a, b, d, x\},$$

and the Venn diagram showing their relatedness:

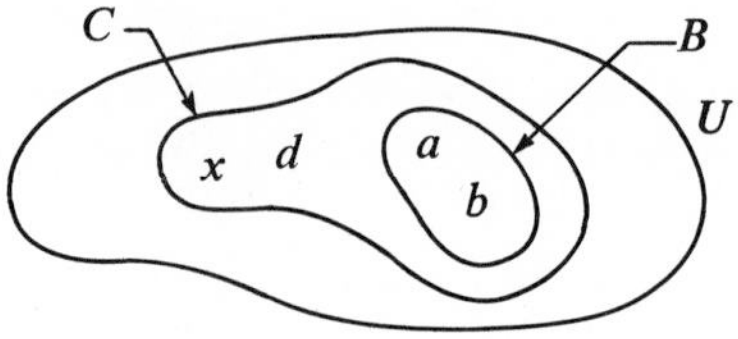

By Definition 2.3(a), $B \subset C$. Is set C a subset of set B? Obviously not, since set C has elements d and x that do not belong to set B. Set B, however, does not contain any elements that are not in set C. It is now possible to consider whether the empty set is a subset of any set.

Let set Z represent any set whatever. One must agree that either the empty set *is a subset* of set Z, or *it is not a subset* of set Z. That is, if one of these statements is accepted, the other one must be rejected. Since there is no element in the empty set that does not belong to set Z, the empty set is not disqualified as a subset of Z. Hence, $\varnothing \subset Z$ is accepted and $\varnothing \not\subset Z$ is rejected. Since Z represents any set, *the empty set is a subset of every set.* It follows that *the empty set is a subset of itself.* The empty set is a proper subset of every set except itself.

Thus far, we have limited our discussion to *a* subset of a set. May a set have more than one subset? You will find the answer in the discussion that follows. Given the set

$$R = \{a, b, c\},$$

it is possible to construct a subset consisting of only the elements a and c. Can you construct another subset of R? Is the following a complete list of the subsets of set R?

$$\begin{array}{lll} A = \{a, b\} & D = \{a\} & G = \varnothing \\ B = \{a, c\} & E = \{b\} & R = \{a, b, c\} \\ C = \{b, c\} & F = \{c\} & \end{array}$$

Justify the inclusion of sets G and R in this list. Set A is a proper subset of set R. Name a set that is not a proper subset of R.

EXERCISE 2.3

1. (a) Make a complete list of the subsets which may be formed from each of the following sets:
   (1) {Jo, Mary}
   (2) {Jo, Mary, Jane}
   (3) {Jo, Mary, Jane, Bill}

   (b) How many subsets did you make from each of the sets in part (a)?
   (c) Are you able to discover a pattern?
   (d) How many subsets may be formed from a set having
   (1) Five elements?
   (2) Eight elements?
   (3) Twelve elements?
   (4) Four elements?
   (5) One element?
   (6) No elements?

2. Let A equal the set of letters of the alphabet used to spell the word "team" and B be the set of letters used to spell "meat."
   (a) Use the roster notation to describe these sets.
   (b) Is A a proper subset of B? Illustrate with a Venn diagram, using
   $$U = \{a, b, c, \ldots, z\}.$$
   (c) Use the concept of subsets to explain why A and B are identical sets.
3. Tell which of the following are true and which are false. Be able to explain the answer you give.
   (a) $\{0\} = \varnothing$ (d) $\varnothing = \varnothing$ (g) $\{\ \} \subset \{\ \}$
   (b) $\{\varnothing\} = \varnothing$ (e) $\varnothing \subset$ set X (h) $\{\varnothing\} \subset \varnothing$
   (c) $0 \in \varnothing$ (f) set $R \subset$ set R (i) $\varnothing \subset \{\varnothing\}$
4. Let the universal set U be the letters of the English alphabet. Construct sets A, B, and C each a subset of U, such that
   (a) $A \subset B$ and $B \subset C$. (c) $A \not\subset B$ and $B \subset C$.
   (b) $A \subset B$ and $B \not\subset C$. (d) $A \not\subset B$ and $B \not\subset C$.
   Make a Venn diagram for each of the preceding conditions.
5. Make a Venn diagram for each pair of sets. Identify a universal set for each diagram.
   (a) The set of Plymouth cars and the set of Chrysler cars.
   (b) The set of Plymouths and the set of Fords.
   (c) $\{a, b, c, d\}$ and $\{e, f, g, h, x\}$.
   (d) $\{a, b, c, d\}$ and $\{b, d, x, y\}$.
   (e) $\{a, b, c\}$ and $\{c, b, a\}$.
   (f) $\{x, y, z, w\}$ and $\{z, x\}$.
   Select those pairs of sets for which the subset, $\subset$, relation holds.
6. Let $S = \{1, 2, 3, 4, 5\}$, $P = \{1, 3, 5\}$, and $Q = \{\ \}$. Which of the following statements are true?
   (a) $S \subset P$. (d) P is a proper subset of S.
   (b) $P \subset S$. (e) Q is a proper subset of S.
   (c) $Q \subset S$. (f) $S \subset Q$.

2.4 THE SET BUILDER NOTATION

You have now learned that a set must be well defined; that is, by means of a defining property one must be able to tell whether an element belongs or does not belong to the set. When a set is described by the roster notation, there is no question as to the existence of a defining property. However, this form of describing a set is not always practical. For example, how would you use the roster notation to describe the set of all the wild animals in Africa? To demand the listing of the elements of this set would be unreasonable. So, some form suitable for describing sets of all kinds is needed.

The *set builder notation* is a standard form which may be used to describe any set. It is convenient and also has the advantage over the roster notation of identifying the universal set as well as the defining property. This form and its use are best explained by examples.

Example 1 To describe the set, W, of all the wild animals in Africa, we shall first identify the universal set, U, as the set of all African animals. The set builder notation for the set W is then

$$\{x \in U : x \text{ is a wild animal}\}.$$

This is read, "The set of all elements x belonging to U, such that x is a wild animal." The statement of the defining property follows the symbol, :, in the set builder notation. The defining property gives the condition for membership in the set. It is also called the *set selector*. The following Venn diagram illustrates the set described.

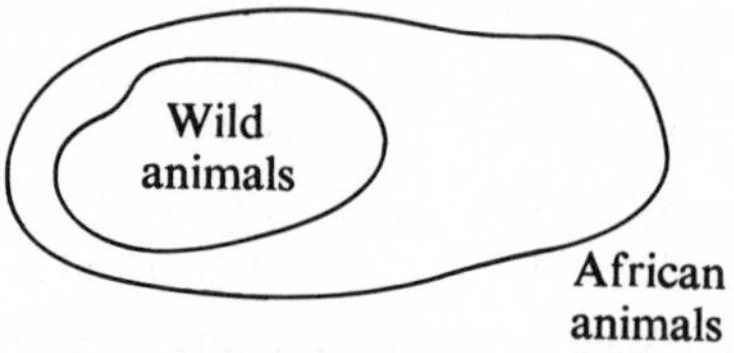

Example 2 Describe the following set using the set builder notation and illustrate it with a Venn diagram: The set, S, of all citizens of the United States who were born before 1900. Let A represent the set of all citizens of the United States. Set A is the universal set. Then the set S may be described by

$$\{x \in A : x \text{ was born before 1900}\}.$$

Would it be practical to attempt to list the members of this set? The Venn diagram is:

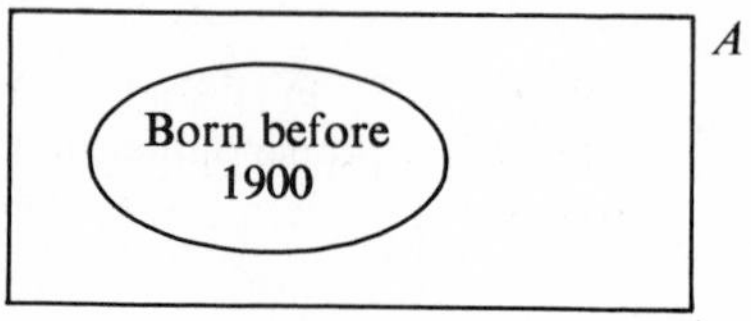

The set builder notation describes a set completely; however, the elements of the set are not on display as in the roster notation. If a listing of elements is desired, the set builder may be used to select them. Consider the role of the set builder in selecting the elements of the set in the following examples.

Example 3 Describe set G of all the people who are wearing glasses and are enrolled in this class. Let U represent all people enrolled in this class. Then, the description of this set is

$$G = \{x \in U : x \text{ is a person wearing glasses}\}.$$

A frame, $\square$, may be used instead of the letter, x. Then, the description appears like this:

$$G = \{\square \in U : \square \text{ is a person wearing glasses}\}.$$

Observe how the frame is helpful in understanding how names are selected and tested. A name is selected from the universal set and placed in the frame. Suppose Mary belongs to set U. Her name is placed in the frame as shown.

$$G = \{\boxed{\text{Mary}} \in U : \boxed{\text{Mary}} \text{ is a person wearing glasses}\}.$$

The test is to decide whether she qualifies for membership. The set selector, "$\square$ is a person wearing glasses," will either accept or reject her for membership. If Mary is wearing glasses, then she is accepted; otherwise she is rejected. As candidates are repeatedly chosen from U and are tested, the elements of the set are determined. If the elements selected are Mary, Susan, Jane, and Betty, the roster notation may be used to describe this set.

$$G = \{\text{Mary, Susan, Jane, Betty}\}$$

The Venn diagram shows the elements accepted by the set selector from set U.

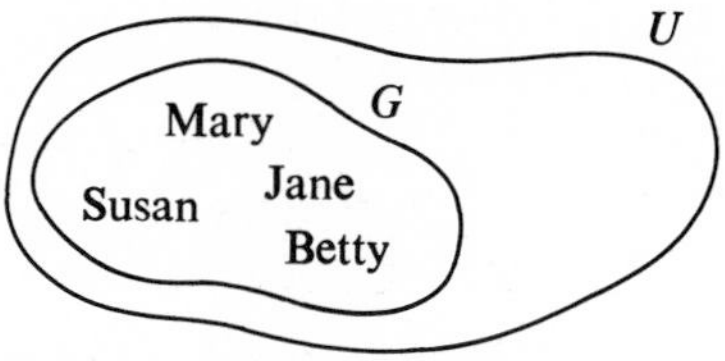

Example 4 Describe the set N of all counting numbers greater than 5. The universal set of counting numbers are those numbers used in counting, such as,

$$U = \{1, 2, 3, 4, 5, \ldots\}.$$

The set description in set builder notation is:

$$N = \{x \in U : x \text{ is greater than } 5\}.$$

Replacements for x are chosen from set U. The set selector, "x is greater than 5," tests each element presented for membership. Every number from U, except 1, 2, 3, 4, and 5, passes the test. So, described in roster notation,

$$N = \{6, 7, 8, 9, 10, \ldots\}.$$

The following Venn diagram illustrates the same set.

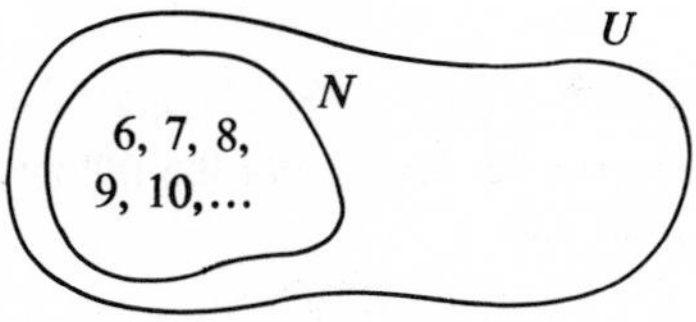

Example 5 Describe the set R of all whole numbers such that if each is added to 2, the sum is less than 6. The universal set of whole numbers is

$$U = \{0, 1, 2, 3, 4, \ldots\}.$$

Observe the difference in the set of whole numbers and the set of counting numbers. The description of set R is

$$R = \{x \in U : 2 + x \text{ is less than } 6\}.$$

Identify the set selector. What numbers of U does it accept? Make a Venn diagram showing the elements of W accepted by the set selector for membership in R.

Example 6 Describe the set S of all whole numbers consisting of 2. Let W represent the universal set of whole numbers. The set description is

$$S = \{x \in W : x \text{ is } 2\}.$$

The set selector may be written $x = 2$ instead of x is 2. Listing the elements of this set is quite simple, as 2 is the only number accepted. The roster notation describes the set as

$$S = \{2\}.$$

The Venn diagram illustrates the set.

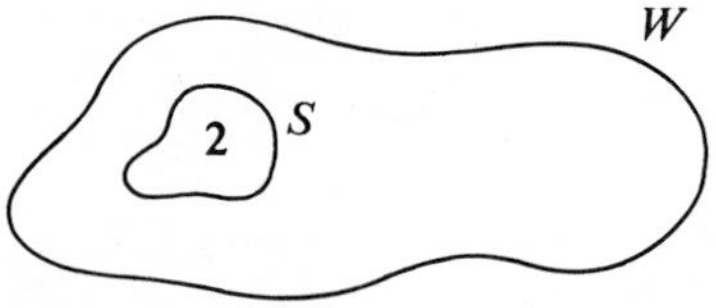

Since the set consists of only the number 2, no other numerals appear within the inner closed curve.

Example 7 Describe the set B of teachers in your school who have two heads. Let T represent the universal set.

$$B = \{y \in T : y \text{ has two heads}\}$$

describes (we hope) the empty set, since the set selector, "y has two heads," rejects each teacher submitted from T.

$$B = \{\ \ \}$$

Some prefer to describe the set as

$$B = \varnothing.$$

EXERCISE 2.4

1. Identify a universal set for each of the following and use the set builder notation to describe each one:
   (a) The set of all odd numbers.
   (b) The set of red dresses in Jane's closet.
   (c) The set of mathematics teachers in your school.
   (d) The set of toys belonging to Jim.
   (e) The set of national banks in Chicago.
   (f) The set of brick houses in your block.

(g) The set of rosebushes in Paul's flower garden.
(h) The set of secretaries who are married women.

In Problems 2 through 5 display the required sets using roster notation.

2. Given the universal set as the states of the U.S.A.:
   (a) $W_1 = \{x \in U : x$ has its name beginning with the letter A or $C\}$
   (b) $W_2 = \{y \in W_1 : y$ is located east of the Mississippi river$\}$
   (c) $W_3 = \{z \in W_1 : z$ is located south of the equator$\}$
3. Given the universal set $U = \{$airplane, sports car, horse and buggy, tricycle, kiddycar, roller skates, lumber wagon$\}$:
   (a) $W_4 = \{z \in U : z$ is suitable for use by a child less than 6 years of age$\}$
   (b) $W_5 = \{z \in U : z$ is suitable for a quick trip (6 days or less) from Maine to California$\}$
4. The universal set for the following problem is $U = \{$quarter, dime, dollar, half dollar, penny, nickel$\}$.
   (a) $W_6 = \{x \in U : 5x \in U\}$ (Note: $5x$ means 5 times x)
   (b) $W_7 = \{x \in U : 100x \in U\}$
   (c) $W_8 = \{x \in U : 500x \in U\}$
5. Given $U = \{1, 2, 3, 4, 5, 6, 7, 8, 9, 10, 11, 12\}$:
   (a) $W_9 = \{x \in U : 3x \text{ is in } U\}$
   (b) $W_{10} = \{x \in U : 10x \text{ is in } U\}$
   (c) $W_{11} = \{x \in U : 20x \text{ is in } U\}$
6. Describe the following sets using set builder notation.
   (a) The set of all counting numbers greater than 2.
   (b) The set of all counting numbers less than 7.
   (c) The set of all counting numbers that when added to 1 give a sum less than 8.
   (d) The set of all counting numbers that when added to 6 give a sum less than 2.

   What is the universal set in each case? Use the set selector of each to determine the elements of each set and then display them using roster notation and also Venn diagrams.

2.5 OPERATIONS ON SETS—INTERSECTION, UNION, AND COMPLEMENTATION

In the diagram, cars A, B, and C are represented on two city streets. Find a car that is on Main Street but not on University Boulevard. Find a car that is on University Boulevard but not on Main Street. Next find a car that is on both streets. It is located in the *intersection* of the two streets.

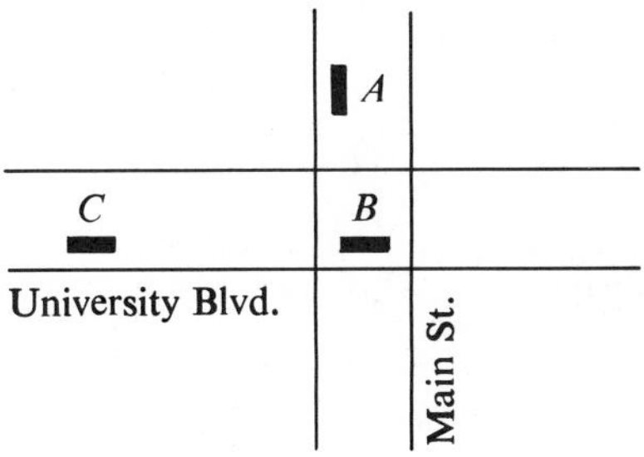

Let U represent all cars on all streets of this city. The set of cars on University Blvd. and the set of those on Main St. are each a subset of U. The Venn diagram is used to show this and to locate car B.

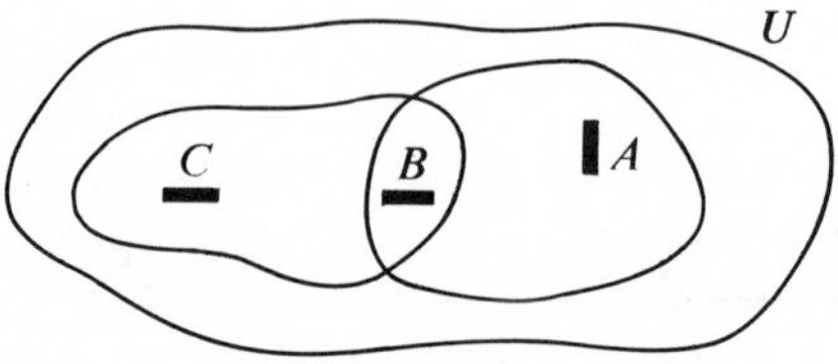

In thinking of all the houses in a town, one might recall that some were constructed of brick, while some had white roofs. It is quite possible that some of the brick houses also had white roofs. Where would these houses be located in the Venn diagram?

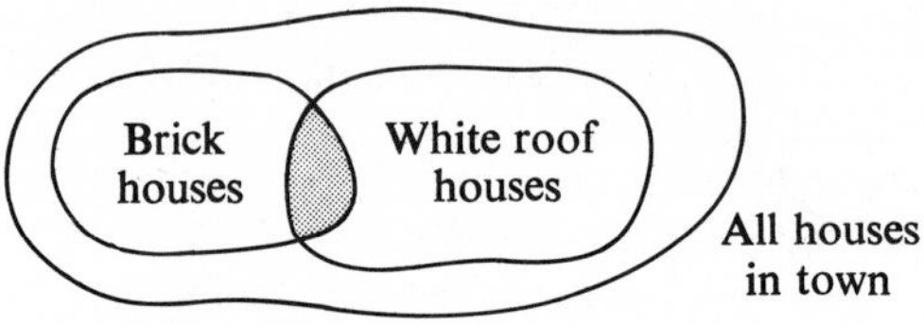

The part that is shaded in this diagram represents the houses that are brick and also have white roofs. How would the picture be drawn if there were no brick houses that had white roofs?

The set consisting of the elements (houses) that are in *both* the set of brick houses *and* the set of houses with white roofs forms a new set called the *intersection set*. This intersection set is assigned to a pair of sets. Since each pair of sets will have a unique (one and only one) intersection set assigned, this is a *binary operation*. It is called *intersection*. The symbol for this operation is $\cap$. This discussion leads to the following definition.

Definition 2.5(a) The *intersection* of set A and set B is the unique set, $A \cap B$, whose elements belong to both set A *and* set B.

Example 1 What is the intersection of sets A and B if

$$U = \{a, b, c, \ldots z\}$$
$$A = \{t, e, a, m\},$$
$$B = \{b, e, a, t\}?$$
$$A \cap B = \{t, e, a, m\} \cap \{b, e, a, t\} = \{t, e, a\}$$

Does each element in $A \cap B$ belong both to set A and to set B? Is $A \cap B$ a proper subset of set A? Is it a proper subset of set B?

$$B \cap A = \{b, e, a, t\} \cap \{m, e, a, t\} = \{e, a, t\}$$

Are the sets $A \cap B$ and $B \cap A$ the same? The accompanying Venn diagram shows that the elements t, e, and a are in both set A and set B; hence, they belong to the intersection of these two sets.

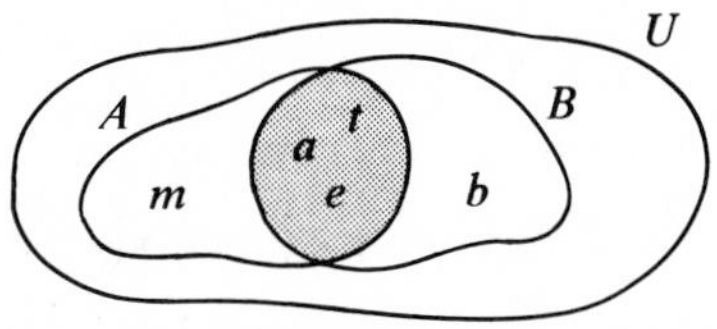

Example 2 Let U be the set of all animals on a farm,

$$P = \{\text{duck, goose, chicken}\}$$

and $$Q = \{\text{horse, mule}\}.$$

Then $$P \cap Q = \varnothing.$$

Explain why the intersection is the empty set. The Venn diagram illustrates that there are no elements in both P and Q.

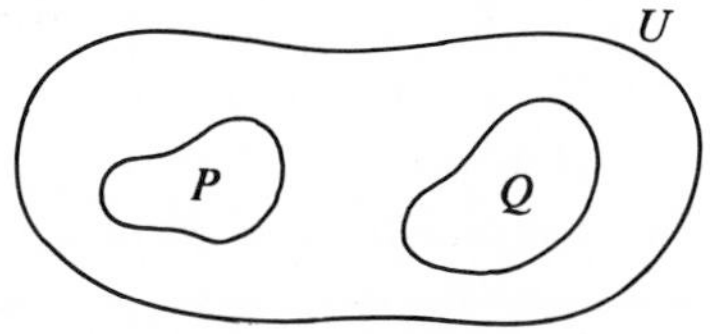

Since $P \cap Q = \varnothing$, P and Q are called *disjoint sets*.

Example 3 The intersection of two sets may be represented in set builder notation in the following manner:

$$\{x \in N : x \text{ is greater than } 5\} \cap \{x \in N : x \text{ is less than } 8\},$$
$$N = \{1, 2, 3, 4, \ldots\}.$$

Applying the set selector of each set, the description may be given in roster notation as

$$\{6, 7, 8, 9, \ldots\} \cap \{7, 6, 5, 4, 3, 2, 1\}.$$

What are the elements of the intersection set? The elements of the intersection set are shown in the Venn diagram.

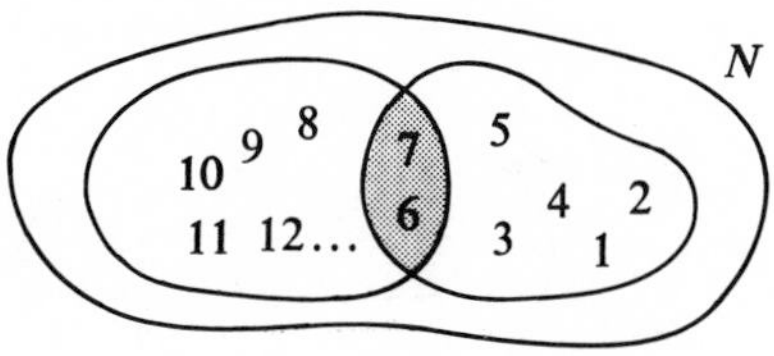

That is,

$$\{x \in N : x \text{ is greater than } 5\} \cap \{x \in N : x \text{ is less than } 8\} = \{6, 7\}$$

or

$$\{6, 7, 8, 9, \ldots\} \cap \{7, 6, 5, 4, 3, 2, 1\} = \{6, 7\}.$$

The intersection may be described in the following manner also:

$$\{x \in N : x \text{ is greater than 5 } \textit{and} \text{ less than } 8\}.$$

The conjunction *and* placed between the set selectors of the two sets forms the set selector of the intersection set. This means that the elements of the intersection set must belong to *both* set A *and* set B. See Definition 2.5(a).

Two sisters, Betty and Jane, are preparing an invitation list for their party. Each submits a list of her friends.

Betty's friends: Joe, Jerry, Marcia, Judy
Jane's friends: Jerry, Frances, Ruth, Joe

The following is the invitation list that the girls prepared:

Marcia, Jerry, Joe, Judy, Frances, Ruth.

Which names are on *either* Betty's list *or* on Jane's list? Are there any names on both of their lists? Will the girls send two invitations to each of those friends they have in common? No, as you would expect, the names of Jerry and Joe, friends of both girls, appear only once on the invitation list; hence, each will receive only one invitation just as the others.

The shaded portion of the Venn diagram represents all who are to receive invitations. Each receiving an invitation is either a friend of Betty or of Jane, or of both. Where would Jerry and Joe be located in the diagram? Are the two sets of friends disjoint? Identify an appropriate universal set.

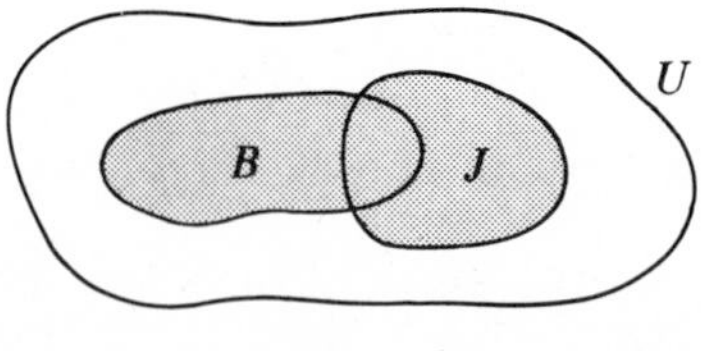

Suppose that Jerry and Joe are traveling in Europe when the invitation list is prepared. Betty and Jane agree to exclude them from their lists. Observe the change in the Venn diagram.

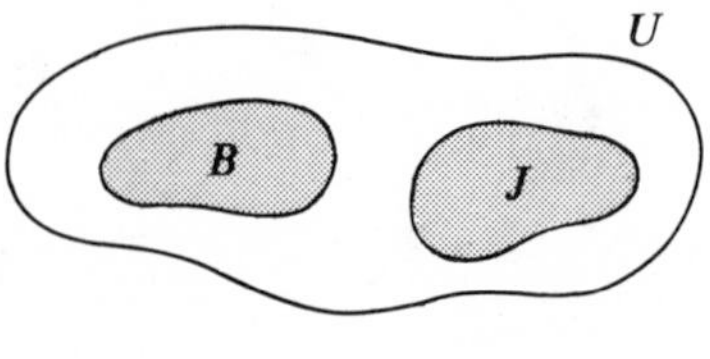

Why are the shaded portions not overlapping as in the preceding diagram?

The set consisting of the elements (names) that are on *either* Betty's list *or* on Jane's list, or on both, forms a new set called the *union set*. This union set is assigned to a pair of sets. Since each pair of sets will have a unique union set assigned, this is a *binary operation*. It is called *union*. The symbol for this is $\cup$.

Definition 2.5(b) The *union* of set A and set B is the unique set, $A \cup B$, whose elements are all the elements belonging to *either* set A *or* set B, or *both*. (Here *or* is used in the inclusive sense, meaning "either or both.")

Example 4 Consider a set consisting of all elements belonging to either set $V = \{a, e, i, o, u\}$ or to set $W = \{x, y, z\}$. Let $U = \{a, b, c, \ldots, x, y, z\}$. Then,

$$V \cup W = \{a, e, i, o, u, x, y, z\}.$$

The shaded portions of the Venn diagram represent this union set.

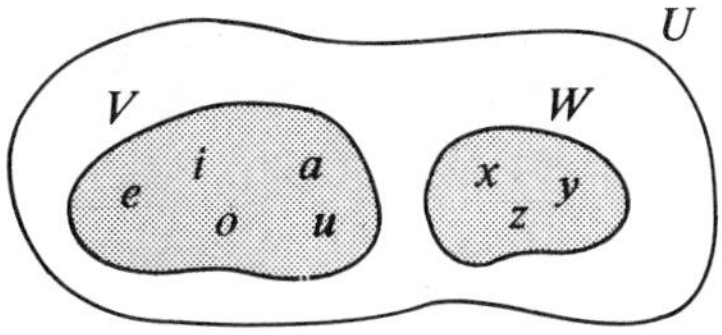

Are sets V and W disjoint?

If set W is changed to set $X = \{a, u, x, y, z\}$, then

$$V \cup X = \{a, e, i, o, u, x, y, z\}.$$

Observe that $V \cup W = V \cup X$. Why? (See Definition 2.5(b).) The shaded portion of the following Venn diagram represents $V \cup X$.

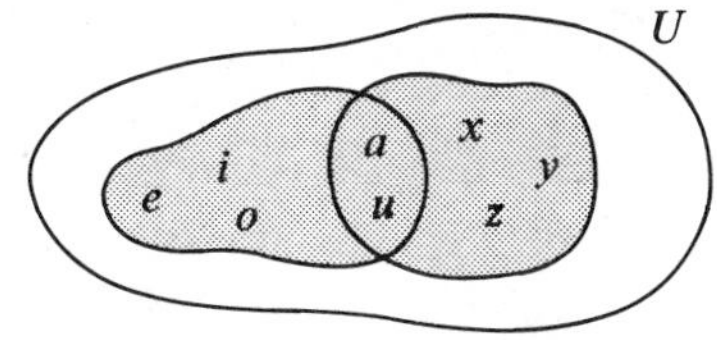

Observe that V and X are not disjoint sets. Elements a and u belong to both sets, but are listed only once in the union set, $V \cup X$.

Example 5 What is the union of set A and set B if

$$U = \{a, b, c, \ldots, x, y, z\},$$
$$A = \{a, c, t\},$$
$$B = \{x, e, g, k\}?$$
$$A \cup B = \{a, c, t\} \cup \{x, e, g, k\} = \{a, c, t, x, e, g, k\}.$$

Do the elements of $A \cup B$ belong either to set A or to set B (or to both)?

Does $A \cup B$ include all the elements of A and B and no others? Is $B \cup A$ the same as $A \cup B$?

$$B \cup A = \{x, e, g, k\} \cup \{a, c, t\}$$

Use the roster notation to describe $B \cup A$. Is it the same as the set $A \cup B$?

Example 6 What is the union of set X and set Y if

$$\begin{aligned} U &= \{a, b, c, \ldots, x, y, z\}, \\ X &= \{a, b, t\}, \\ Y &= \{r, b, c, q\}? \\ X \cup Y &= \{a, b, t, r, c, q\}. \end{aligned}$$

Observe that each element in $X \cup Y$ belongs either to set X or to set Y. In fact, element "b" belongs to both X and Y. Make a Venn diagram to illustrate this.

Example 7 Let the universal set, all cars, be represented by A. Use the set builder notation to describe the set consisting of either General Motors cars, G, or Chrysler cars, C. Thus:

$$G \cup C = \{x \in A : x \text{ is built by General Motors}\} \cup \{x \in A : x \text{ is built by Chrysler}\}$$

or

$$G \cup C = \{x \in A : x \text{ is built either by General Motors or by Chrysler}\}.$$

The following Venn diagram pictures this union:

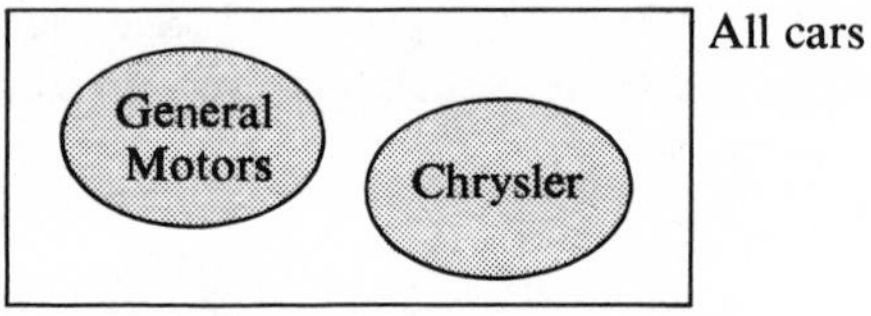

Where would the set of Ford cars fit into this diagram? Other makes of cars? Is $G \cup C$ a proper subset of A? Explain.

Example 8 Consider the following union, where set A represents the set of all cars:

$$\{x \in A : x \text{ is built by Ford}\} \cup \{x \in A : x \text{ is air-conditioned}\}.$$

A more concise way to describe this union in set builder notation is

$$\{x \in A : x \text{ is built by Ford } or \text{ is air-conditioned}\}.$$

The conjunction *or* placed between the set selectors of the two sets forms the set selector of the union set. This means that the elements of the union set must belong to *either* or *both* of the sets. See Definition 2.5(b).

In the next example, both binary operations ($\cap$, $\cup$) are performed on sets A and B. Notice the use of the conjunction *and* in the set builder notation of $A \cap B$, and the conjunction *or* in the set builder notation of $A \cup B$.

Example 9 Let W be the set of all women in Oklahoma, and

$$A = \{g \in W : g \text{ is a college freshman})$$

and

$$B = \{g \in W : g \text{ has red hair}\}.$$

Then

$$A \cap B = \{g \in W : g \text{ is a college freshman } \textit{and } g \text{ has red hair}\}.$$

That is, $A \cap B$ is the set consisting of redheaded college freshman women in Oklahoma.

$$A \cup B = \{g \in W : g \text{ is a college freshman } \textit{or } g \text{ has red hair}\}.$$

That is, $A \cup B$ is the set consisting of all college freshman women in Oklahoma along with all redheaded women in Oklahoma.

The Venn diagram for each operation is:

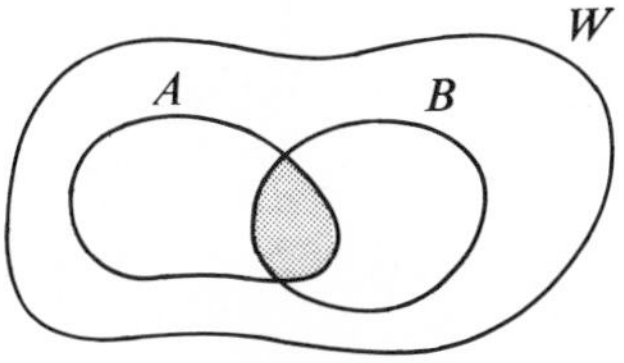

$A \cap B$

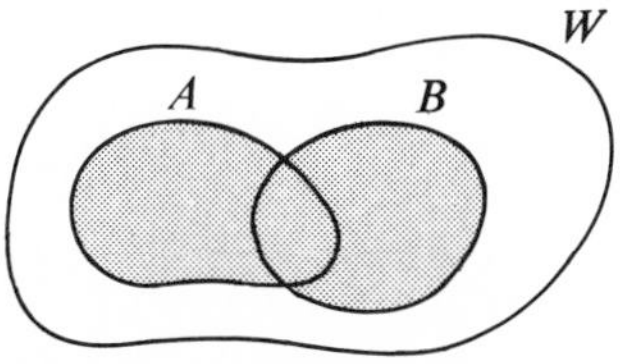

$A \cup B$

Each of the operations, intersection and union, with a set of sets forms a *binary operational system*. Using S to represent a set of sets, these operational systems may be designated by

$$(S, \cap) \quad \text{and} \quad (S, \cup).$$

Definition 2.5(c) If S is a set of elements and $*$ is an operation on set S, then $(S, *)$ is an *operational system*.

An operational system requires only one operation and a set of elements. The operations, intersection and union, are each binary. Is it possible to have an operation in which a *single* element has one and only one element of the set assigned to it?

A child has a deck (set) of alphabet cards. He draws from this deck the last twenty letters of the alphabet. What letters remain in the deck?

Let the deck of alphabet cards be represented by set $A = \{a, b, c, \ldots, x, y, z\}$ and D be the set of letters drawn from the deck. The remaining set, $\{a, b, c, d, e, f\}$, is called the *complement of D*. It is the set of elements of the universal set A which do not belong to set D. Since for each set of a universal set there is assigned a unique complement set, this is a *unary operation*. This is an operation on a *single set* where intersection and union are operations on *two sets*, hence the use of unary and binary as names to distinguish the two types of operations.

The symbol for the complement operation is $\sim$. To designate the complement of set D, the symbol for this operation is placed above the letter D, as $\tilde{D}$.

In summary, if

$$A = \{a, b, c, \ldots, x, y, z\},$$

the universal set, and

$$D = \{g, h, i, \ldots, x, y, z\},$$

then

$$\tilde{D} = \{a, b, c, d, e, f\}.$$

The shaded portion of the Venn diagram represents $\tilde{D}$.

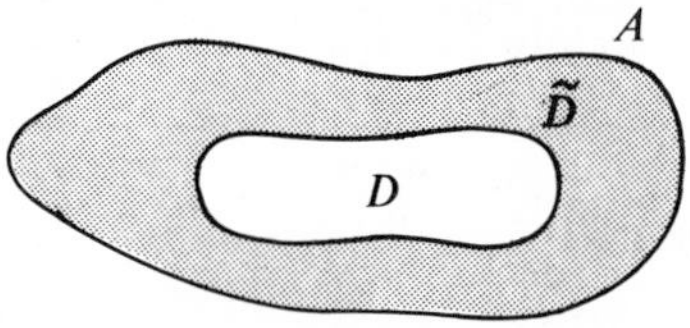

Where are the elements of D located in the diagram? The elements of A?

Definition 2.5(d) *Complementation* is a *unary operation* that assigns to each set D a unique set $\tilde{D}$ consisting of all elements in the universal set U that do not belong to D.

Example 10 Let $U = \{0, 1, 2, 3, 4, 5, 6, 7, 8, 9\}$, the universal set, and $B = \{0, 1, 9\}$. Find $\tilde{B}$.

This requires a look at the universal set and at set B. All of the elements of U that do not belong to B determine $\tilde{B}$. That is,

$$\tilde{B} = \{2, 3, 4, 5, 6, 7, 8\}.$$

The process of finding the complement of set B may be likened to removing set B from the universal set. The remaining set is called the complement of B. The complement of B is indicated by the shaded region.

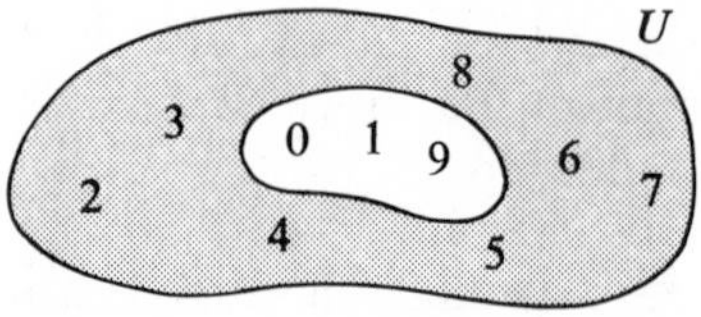

Example 11 The set of children living in my block consists of Jim, Mary, Susan, Bill, and Joe. If Susan moves away, what is the remaining set called? Let $U = \{\text{Jim, Mary, Susan, Bill, Joe}\}$ and $A = \{\text{Susan}\}$. If Susan moves away, the remaining set is called the complement of A. That is,

$$\tilde{A} = \{\text{Jim, Mary, Bill, Joe}\}.$$

Example 12 Given the universal set $U = \{1, 2, 3, 4, 5, 6, 7\}$ and $A = \{x \in U: x \text{ is an even number}\}$, describe $\tilde{A}$ in set builder notation.

Since the set selector of A chooses the elements of this set to be 2, 4, 6, then $\tilde{A}$ consists of all the remaining elements of U. That is,

$$\tilde{A} = \{1, 3, 5, 7\}.$$

Written in set builder notation,

$$\tilde{A} = \{x \in U : x \text{ is an odd number}\}.$$

Another way to express A in set builder notation is to place the complement symbol, $\sim$, directly over the set selector of set A, as

$$\tilde{A} = \{x \in U : \widetilde{x \text{ is an even number}}\}.$$

Make a Venn diagram to illustrate this example.

EXERCISE 2.5

1. Given $X = \{1, 2, 3\}$ and $Y = \{p, q, r\}$:
   (a) Find $X \cup Y$.
   (b) Find $Y \cup X$.
   (c) What can you say about $X \cup Y$ and $Y \cup X$?
2. Given the sets $R = \{a, c, d, f\}$ and $S = \{a, b, c, d\}$:
   (a) Find $R \cap S$.
   (b) Find $S \cap R$.
   (c) Is it true for all sets R and S that $R \cap S = S \cap R$?
   (d) Use a Venn diagram to illustrate the intersection of sets R and S.
3. Given $A = \{a, c, d, f\}$, $B = \{d, e, a\}$, and $C = \{a, b, c, d\}$:
   (a) Show that $(A \cup B) \cup C = A \cup (B \cup C)$.
   NOTE: The symbols "( )" are called *parentheses*. As used here, they indicate that the operation enclosed by the parentheses is to be performed first.
   (b) Show that $(A \cap B) \cap C = A \cap (B \cap C)$.
   (c) Show that $A \cup (B \cap C) = (A \cup B) \cap (A \cup C)$.
   (d) Show that $A \cap (B \cup C) = (A \cap B) \cup (A \cap C)$.
4. What is the intersection of the set of students in the freshman class and the set of students in the senior class?
5. Given $P = \{\text{car, driver, sunglasses}\}$ and $\varnothing$:
   (a) Find $P \cup \varnothing$.
   (b) Find $P \cap \varnothing$.
6. Given $A = \{1, 3, 4, 5\}$ and $B = \{2, 4, 6, 8, 10\}$:
   (a) Find $A \cap B$.

(b) Are A and B disjoint sets?
(c) Find $A \cup B$.
(d) Use Venn diagrams to illustrate $A \cap B$ and $A \cup B$.
(e) Is $A \subset B$? Is $B \subset A$?
(f) Is $3 \in A$? Is $3 \in B$?

7. Let $A = \{a, b, c, d, e\}$, $B = \{d, e, f, g, h, i\}$, $C = \{\ \}$:
   (a) Show that $(A \cap B) \cup C = A \cap B$.
   (b) Show that $(A \cup B) \cup C = A \cup B$.
8. Given the sets:

$$A = \{r, s, t\} \qquad C = \{s\} \qquad E = \{r, s\} \qquad G = \{s, t\}$$
$$B = \{r\} \qquad D = \{t\} \qquad F = \{r, t\} \qquad H = \{\ \}$$

   (a) Copy the grids below and place the correct element of $X = \{A, B, C, D, E, F, G, H\}$ in the spaces. (HINT: The operation used in the first grid is union; hence $A \cup A = A$, and A is to be placed in the first square as shown. The second grid is for the operation intersection, with $C \cap E = C$.)
   (b) Discuss the membership of set X.

$\cup$	A	B	C	D	E	F	G	H
A	A							
B				F				
C								
D								
E								
F								
G								
H								

$\cap$	A	B	C	D	E	F	G	H
A								
B								
C					C			
D								
E								
F							D	
G								
H								

9. Below are given grids for the operations of union and intersection on the null set, $\varnothing$, and any set X. Fill in the union grid and the intersection grid with the names for the correct sets. The union grid is begun with $\varnothing \cup \varnothing$, by matching $\varnothing$ (which begins the second row) with $\varnothing$ (which begins the second column). The answer, $\varnothing$, is shown in the proper box. Fill in the other spaces by matching $\varnothing$ with X, then X with $\varnothing$, and X with X. Continue in a similar manner to fill in the intersection grid.

$\cup$	$\varnothing$	X
$\varnothing$	$\varnothing$	
X		

$\cap$	$\varnothing$	X
$\varnothing$		
X		

10. Locate on the Venn diagrams for Example 9 of Section 2.5 the following:
    (a) Susan, a redheaded college senior in Oklahoma.
    (b) Mary, a blond college freshman in Oklahoma.
    (c) Mary's mother, who has red hair and lives in Tulsa, Oklahoma.
    (d) Susan's grandmother who lives in Ponca City, Oklahoma, and whose hair is white.
    (e) Mary's aunt who lives in New York City.
11. Use the set builder notation to describe the set of the letters of the alphabet used in spelling your first name and those used in spelling your mother's first name. Use the set builder notation to express the intersection of these sets, and then use any other notation to describe this intersection. Illustrate with a Venn diagram.
12. Let M be the set of all men teachers in your school system and W be the set of all women teachers in your school system. Use a Venn diagram to show $M \cap W$. What may be used as the universal set?
13. Let U be all of the University of Illinois students, E be all of the engineering students there, and S be all of the seniors there. Use a Venn diagram to show $E \cap S$ and $E \cup S$.
14. Given the universal set $U = \{1, 2, 3, 4, 5, 6, 7, 8\}$, let

    $$A = \{1, 3, 5, 7\}, \quad B = \{2, 4, 6, 8\}, \quad C = \{4, 5, 6, 7\}.$$

    Use roster notation to describe each of the following:
    (a) $\widetilde{A}, \widetilde{B}, \widetilde{C}, \widetilde{A \cup B}$.
    (b) $\widetilde{A \cup C}, \widetilde{A} \cup \widetilde{C}, \widetilde{\widetilde{A} \cup \widetilde{C}}, \widetilde{\widetilde{\widetilde{A} \cup \widetilde{C}}}$.
    (c) $B \cap C, \widetilde{B} \cap \widetilde{C}, \widetilde{\widetilde{B} \cap \widetilde{C}}, \widetilde{B \cap C}$.
15. The universal set of counting numbers less than 10 is described as $U = \{1, 2, 3, 4, 5, 6, 7, 8, 9\}$. If $E = \{n \in U : n \text{ is even}\}$,
    (a) Interpret $\{n \in U : \widetilde{n \text{ is even}}\}$.
    (b) Use roster notation to describe $\widetilde{E}$.
    (c) Make a Venn diagram, representing $\widetilde{E}$ by shading.
16. (a) Name three operational systems on sets.
    (b) Identify the binary and the unary ones.
    (c) Discuss in what ways they qualify as operational systems.

2.6 PROPERTIES OF INTERSECTION, UNION, AND COMPLEMENTATION

The instances cited in Exercise 2.5 lead one to suspect that there are certain properties (laws) that hold for all sets under the operations intersection, union, and complementation. This section will be devoted to the identification of these important properties.

From the results of Problems 1 and 2 in Exercise 2.5, $X \cup Y = Y \cup X$ and $R \cap S = S \cap R$. It is also quite evident that if any pair of sets were chosen, such as P and Q, the results would be similar. This is illustrated in the two Venn diagrams.

Is it possible to tell whether the shaded region in Figure 2–1 represents $P \cap Q$ or $Q \cap P$? Does the shaded region in Figure 2–2 represent $P \cup Q$ or

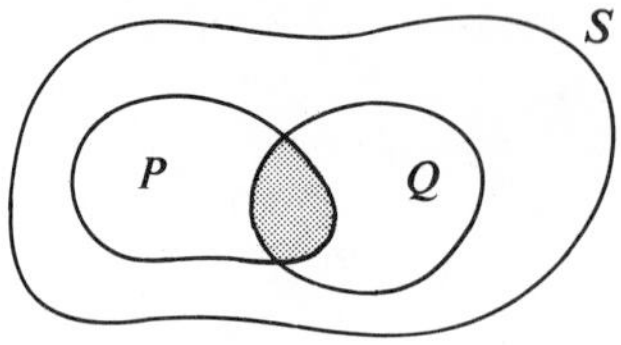

Figure 2–1

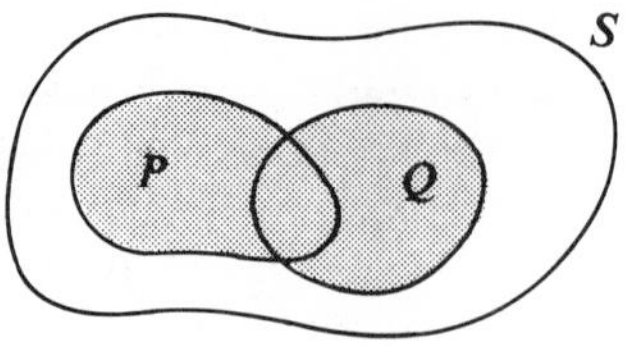

Figure 2–2

$Q \cup P$? Hence, $P \cap Q = Q \cap P$ and $P \cup Q = Q \cup P$. Since P and Q were selected as any pair of sets, the preceding statements apply to *every* pair of sets, and a property is evident. It is called the *commutative property* for (i) intersection and for (ii) union. Using S to represent the universal set of sets, the property is stated as:

Commutative Property

For each $P, Q \in S$,

(i) $$P \cap Q = Q \cap P,$$

(ii) $$P \cup Q = Q \cup P.$$

In Problems 3(a) and 3(b), Exercise 2.5, it was seen that, for particular sets A, B, and C,

$$(A \cap B) \cap C = A \cap (B \cap C)$$

and

$$(A \cup B) \cup C = A \cup (B \cup C).$$

Do these results hold for *all* sets?

Let P, Q, R be any three sets of the universal set S. Since these are any sets, any conclusions drawn concerning them will be true for every three sets of S. Is it true for these sets that $(P \cap Q) \cap R = P \cap (Q \cap R)$? The parentheses are used to indicate that the operation enclosed in the parentheses is to be performed first.

A Venn diagram is used to represent $(P \cap Q) \cap R$, and another to represent $P \cup (Q \cup R)$. Observe the stages necessary in developing the complete diagram for each.

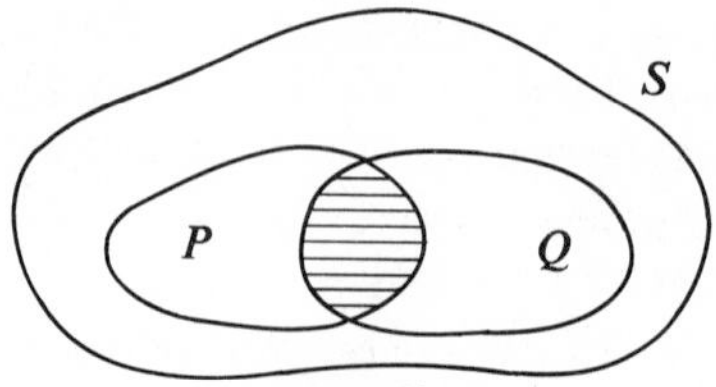

$(P \cap Q)$

Suppose an overlay containing set R is placed on the preceding diagram; the result would be as shown in this diagram.

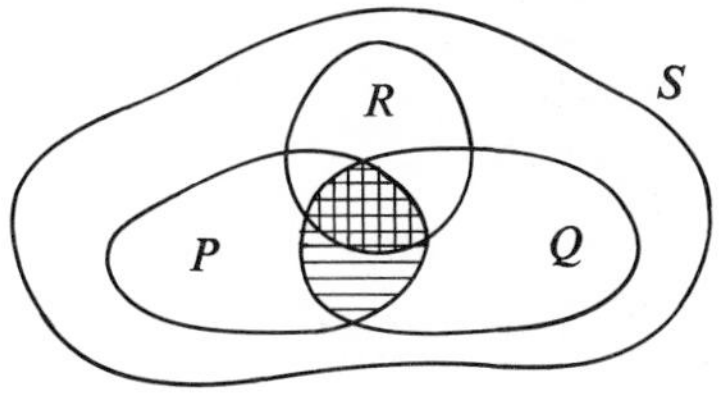

The crosshatch represents $(P \cap Q) \cap R$.

The other series of diagrams illustrates the construction of $P \cap (Q \cap R)$. The set of parentheses enclosing Q and R is a signal to perform the intersection on these first. This is shown in the following diagram.

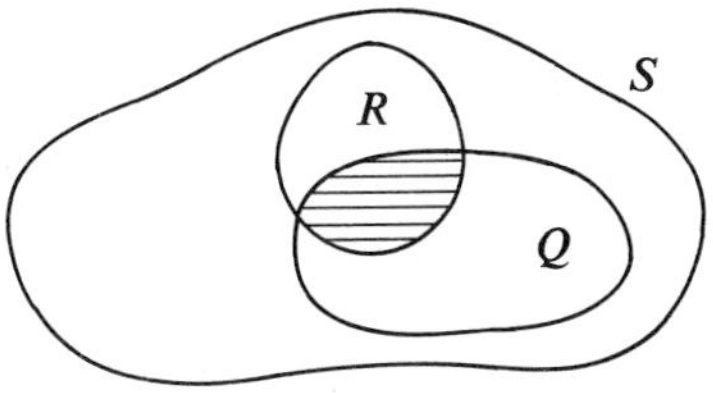

$(Q \cap R)$

Using an overlay containing set P, the resulting diagram becomes

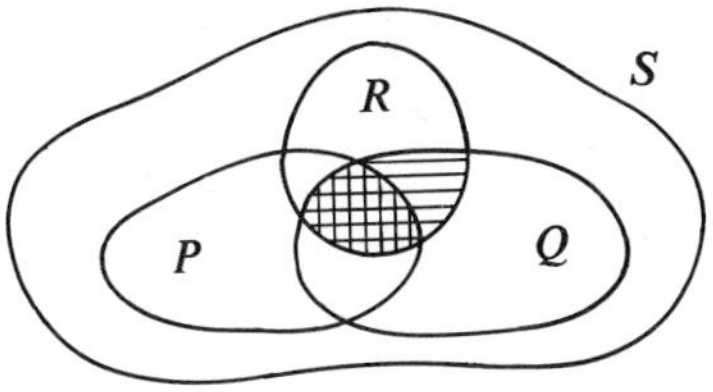

The crosshatch represents $P \cap (Q \cap R)$.

Is it possible to distinguish between the final results? Is the crosshatch the same for $(P \cap Q) \cap R$ and $P \cap (Q \cap R)$?

In a similar manner, Venn diagrams are constructed to represent $(P \cup Q) \cup R$ and $P \cup (Q \cup R)$.

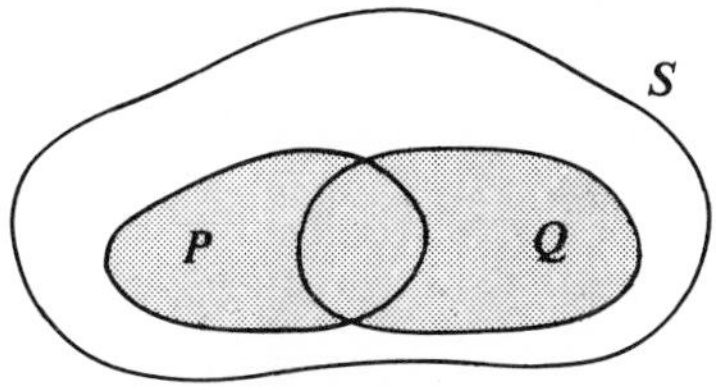

$(P \cup Q)$

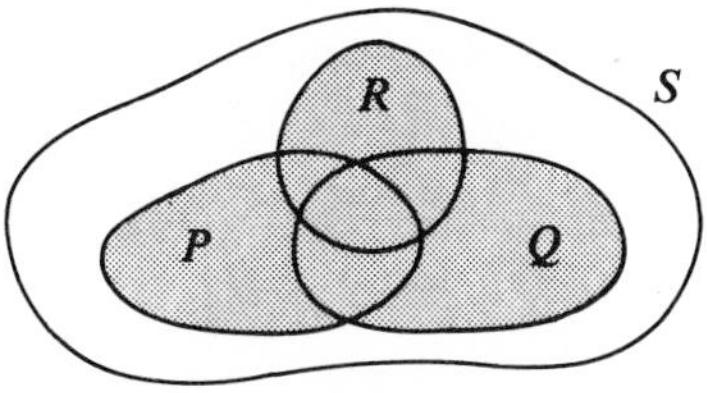

$(P \cup Q) \cup R$

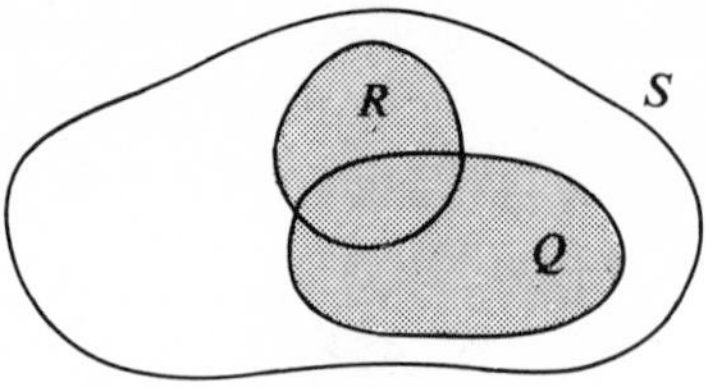

$(Q \cup R)$

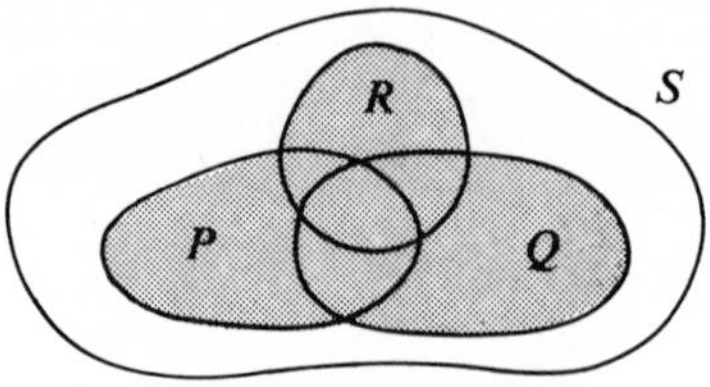

$P \cup (Q \cup R)$

Since for any sets P, Q, and R in S, $(P \cap Q) \cap R = P \cap (Q \cap R)$ and $(P \cup Q) \cup R = P \cup (Q \cup R)$, a property is evident. It is called the *associative property* for (i) intersection and (ii) union.

Associative Property

For each $P, Q, R \in S$,

(i) $$(P \cap Q) \cap R = P \cap (Q \cap R),$$
(ii) $$(P \cup Q) \cup R = P \cup (Q \cup R).$$

Using P, Q, and R any sets of the universal set S, make a Venn diagram to represent $P \cap (Q \cup R)$ and one to represent $(P \cap Q) \cup (P \cap R)$. Do your Venn diagrams look like these?

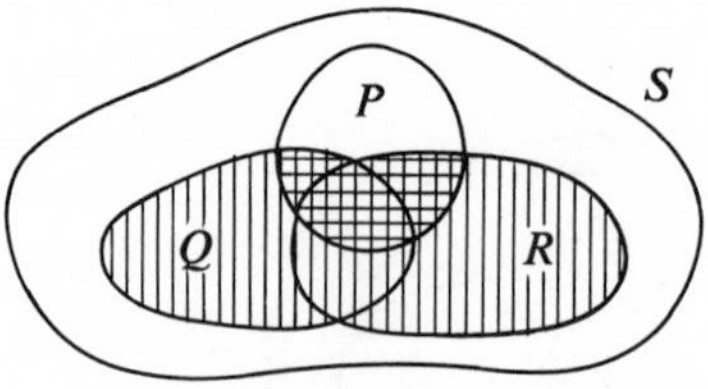

$P \cap (Q \cup R)$

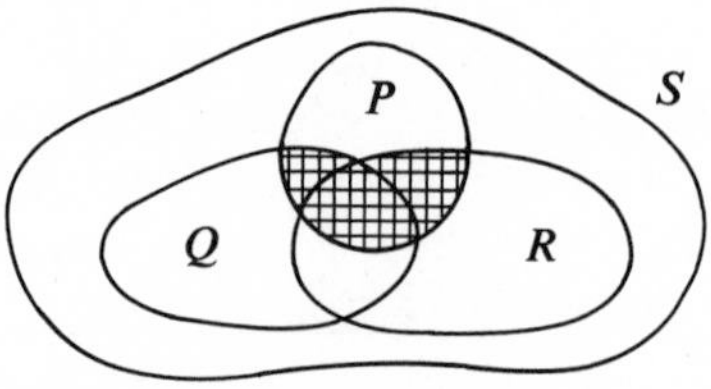

$(P \cap Q) \cup (P \cap R)$

Observe that the crosshatched region is the same in both diagrams. Hence, $P \cap (Q \cup R) = (P \cap Q) \cup (P \cap R)$.

Similarly it is possible to demonstrate with Venn diagrams that $P \cup (Q \cap R) = (P \cup Q) \cap (P \cup R)$. It is suggested that you make diagrams to test this statement.

These two statements bring together or connect the two operations, intersection and union. When stated in this way, they demonstrate the *distributive property* for (i) intersection over union and (ii) union over intersection.

Distributive Property

For each $P, Q, R \in S$,

(i) $$P \cap (Q \cup R) = (P \cap Q) \cup (P \cap R),$$
(ii) $$P \cup (Q \cap R) = (P \cup Q) \cap (P \cup R).$$

Combining intersection and union with complementation produces interesting results. Study the two Venn diagrams that follow. If $P \in S$, explain why $P \cap \tilde{P} = \varnothing$ and $P \cup \tilde{P} = S$.

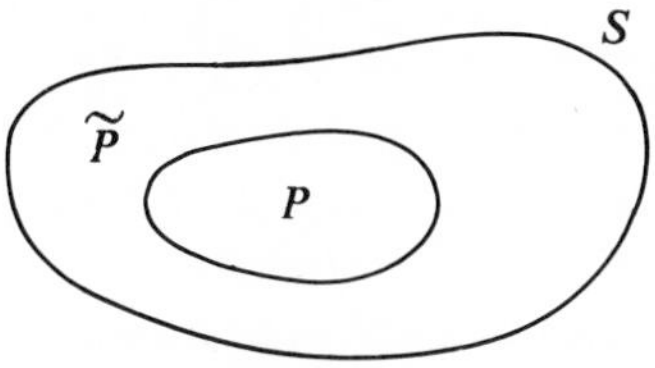

$P \cap \tilde{P} = \phi$

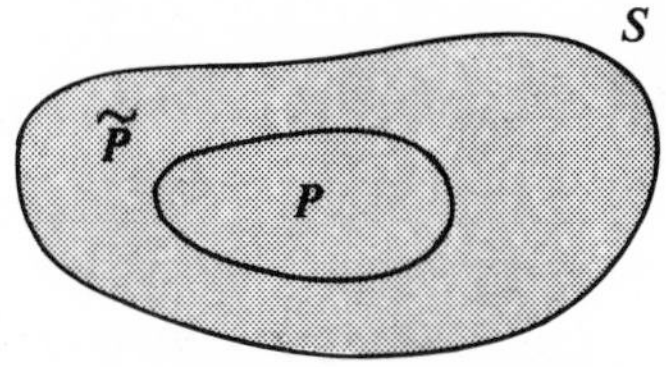

$P \cup \tilde{P} = S$

Construct a Venn diagram for $\widetilde{(P \cup Q)}$ and one for $\tilde{P} \cap \tilde{Q}$. Let $P, Q \in S$.

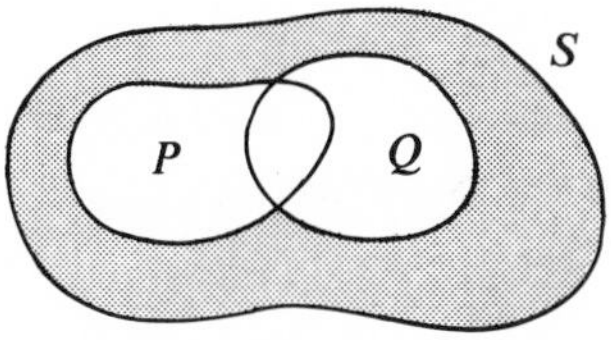

$\widetilde{(P \cup Q)}$

Figure 2–3

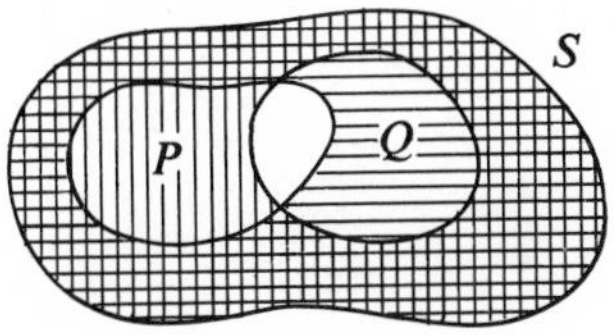

$\tilde{P} \cap \tilde{Q}$

Figure 2–4

Compare your diagrams with Figures 2–3 and 2–4. Locate $(P \cup Q)$ in Figure 2–3. Do you see that the shaded region represents $\widetilde{(P \cup Q)}$? In Figure 2–4, the horizontal marks indicate the region representing $\tilde{P}$. How is $\tilde{Q}$ represented? What does the crosshatched region represent? Compare this region with the dotted region in Figure 2–3. With P and Q any sets of S, these two Venn Diagrams lead one to conclude that

$$\widetilde{(P \cup Q)} = \tilde{P} \cap \tilde{Q}.$$

In a similar manner show that

$$\widetilde{(P \cap Q)} = \tilde{P} \cup \tilde{Q}.$$

Compare the Venn diagrams that you construct with Figures 2–5 and 2–6. The shaded region in Figure 2–5 represents $\widetilde{(P \cap Q)}$. Describe the region in this

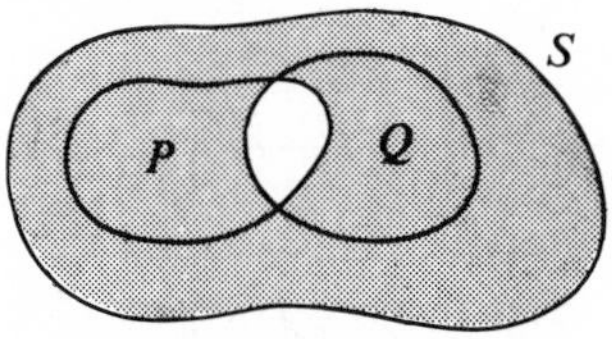

$\widetilde{(P \cap Q)}$

Figure 2–5

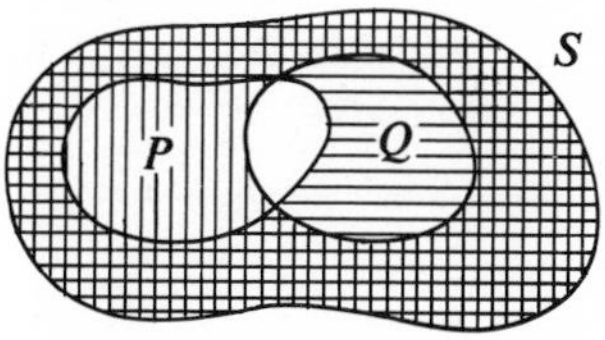

$\tilde{P} \cup \tilde{Q}$

Figure 2–6

figure that is not shaded. Since $\tilde{P} \cup \tilde{Q}$ is that region which is *either not P or not Q*, it is located in Figure 2–6 by the region having either vertical or horizontal marks or both. Is the "shaded" region in Figure 2–5 the same as that in Figure 2–6?

2.7 OPERATIONAL SYSTEMS AND PROPERTIES

The meaning of an operational system was introduced in Definition 2.5(c). This definition did not require the system to possess any properties. In the preceding section several properties were identified and named for the binary operations, intersection and union, on sets.

It is now possible to look at the two binary operational systems, $(S, \cap)$ and $(S, \cup)$, with accompanying properties. One might regard the inclusion of the properties as providing more power to these operational systems. A summary is given in the following diagram.

Binary
Operational Systems

$(S, \cap)$ ↙ ↘ $(S, \cup)$

↓ Properties ↓

Commutative for $\cap$	Commutative for $\cup$
Associative for $\cap$	Associative for $\cup$

↘ ↙

Distributive for
(i) $\cap$ over $\cup$
(ii) $\cup$ over $\cap$

EXERCISE 2.7

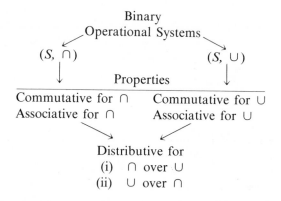

1. Use a Venn diagram to illustrate each of the following:
   (a) If P is a proper subset of Q, then
   (i) $P \cup Q = Q$ (ii) $P \cap Q = P$
   (b) If U is the universal set, then
   (i) $P \cup U = U$ (ii) $P \cap U = P$
   (c) $P \cup (P \cap Q) = P$
   (d) $P \cap (P \cup Q) = P$
2. Make a Venn diagram and indicate the result of each of the following with shading. Then write another expression that yields the same result. Name the property that each pair illustrates.

EXAMPLE:

$A \cap B$

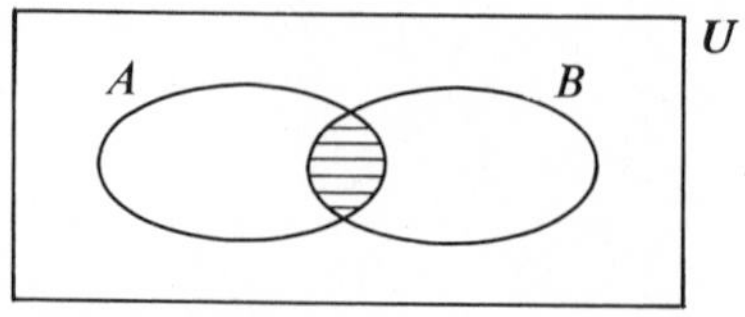

$B \cap A$

Property: Commutative for Intersection

(a) $A \cup B$ (d) $A \cap (B \cup C)$
(b) $(A \cap B) \cap C$ (e) $A \cup (B \cap C)$
(c) $(A \cup B) \cup C$

3. Explain why the following statements are true:
   (a) $P \cap \widetilde{P} = \varnothing$ (c) $\widetilde{\varnothing} = U$
   (b) $P \cup \widetilde{P} = U$ (d) $\widetilde{U} = \varnothing$

   NOTE: U is the universal set.

4. Make two Venn diagrams, one representing each statement. Show that the two results are the same. Let U be the universal set.
   (a) $\widetilde{(X \cup Y)}$; $\widetilde{X} \cap \widetilde{Y}$ (d) $\widetilde{(X \cup \widetilde{Y})} \cap Y$; $\widetilde{X} \cap Y$
   (b) $\widetilde{(X \cap Y)}$; $\widetilde{X} \cup \widetilde{Y}$ (e) $\widetilde{\widetilde{X}}$; X
   (c) $\widetilde{X} \cup (X \cap Y)$; $\widetilde{X} \cup Y$

2.8 MATHEMATICAL MAPPINGS

A prehistoric shepherd could have kept an account of his flock by placing a stone in a pouch of skin as each sheep left for pasture. As each of his sheep returned to the fold, he would remove a stone. If there were no extra stones and no extra sheep, the shepherd would know without counting that the size of his flock had not changed. Without being aware of it, he was employing a very important mathematical concept called *mapping*. He used this concept when he matched sheep and stones.

A child uses this matching process to determine whether there are enough birthday cookies for the children in her room. Janie, who is preparing to leave on a vacation, also demonstrates an intuitive knowledge of mapping. She asks her friends Faith, Tiffany, and Christy to take care of her pets consisting of a dog, a cat, and a lamb. Will each friend have a pet, and will each pet have a caretaker?

Some signficant types of mappings will be discussed in the examples that follow. In each, set P may be thought of as a set of sheep and set Q as a set of stones. The matching of the elements will be indicated in the diagrams with arrows, as in Figures 2–7 and 2–8. These diagrams show that

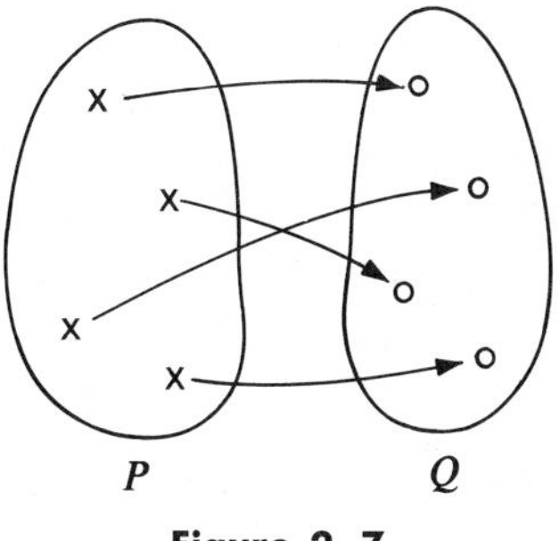

Figure 2–7

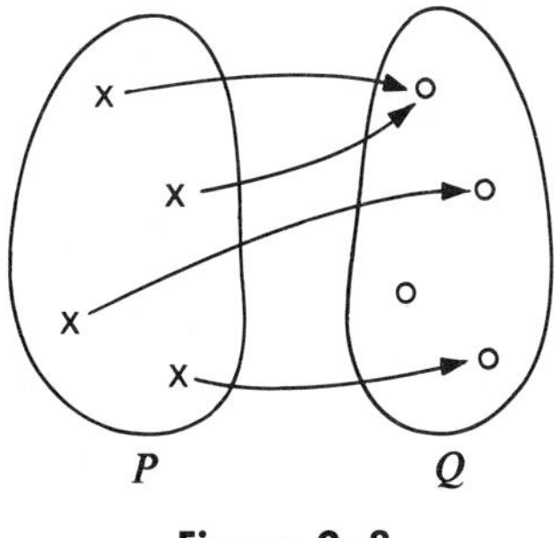

Figure 2–8

exactly one (at least one, and only one) arrow originates at each element of set P and points to an element of set Q. In Figure 2–7, each element of Q has only one arrow pointing to it. In Figure 2–8, one element of Q has two arrows pointing to it, while another element has no arrows. Each of these are mappings. Both have the requirement necessary for a mapping—that is,

(i) to each element of set P is assigned *at least one* element of set Q, and
(ii) to each element of set P is assigned *only one* element of set Q.

The direction of the arrows is significant. In both figures the arrows originate with elements of set P and point toward the elements of set Q. This indicates that the mapping is from P to Q, not from Q to P. The element at the end of an arrow is an *assignment* or *image*. The set P is called the *domain* and the set of images is called the *range*.

Example 1 Since the arrows in this diagram are coming from the elements of set P and pointing to elements of set Q, the mapping is from P to Q. Observe that each element of P has *one and only one* arrow originating from it. However, each element of Q does not have an arrow pointing to it. All of the elements of Q have not been selected as images. That is, in making the assignments, Q has not been exhausted. This is called an *into mapping* of P to Q. It may be spoken of as a *mapping of P into Q*.

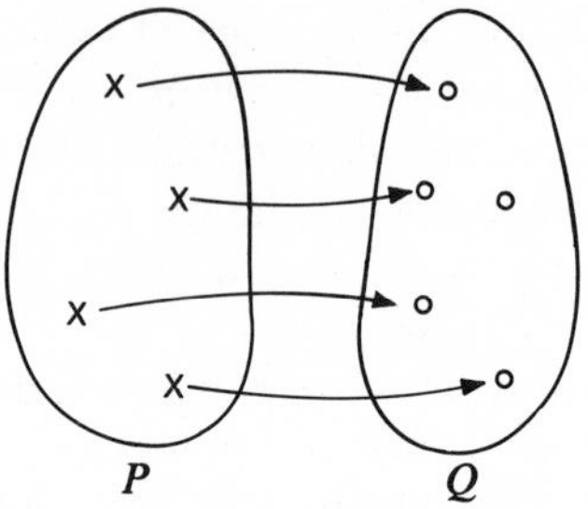

Example 2 How does one know that the next diagram illustrates a mapping of P to Q? Is there one and only one arrow originating from each element of P? Does an arrow point to each element of Q? Since *each* element of Q is an image of an element of P, this is a *mapping of P onto Q*.

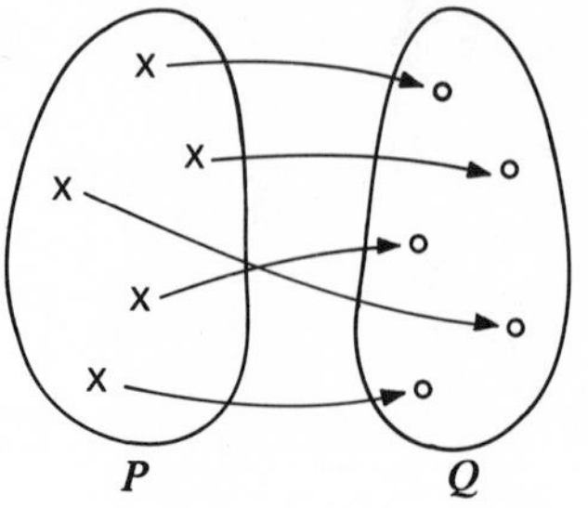

Example 3 This diagram illustrates an onto mapping of P to Q. It is possible for an onto mapping to have exactly one arrow originating from each element of P and more than one arrow pointing to one or more elements of Q. Make a diagram to illustrate this. However, in the diagram below, each element of Q has exactly one arrow pointing to it.

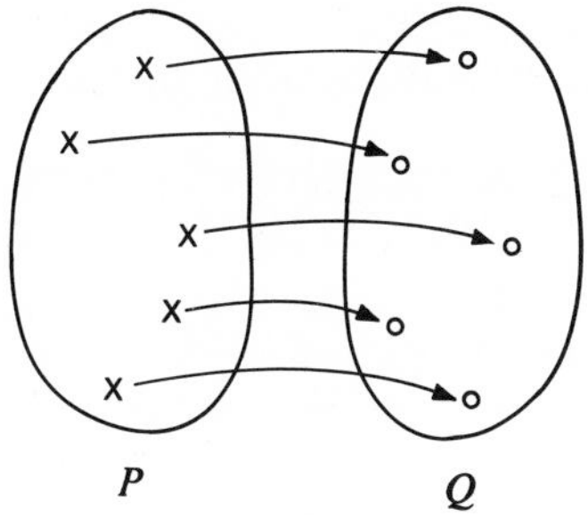

Now, suppose that these arrows are reversed as shown. The arrows in

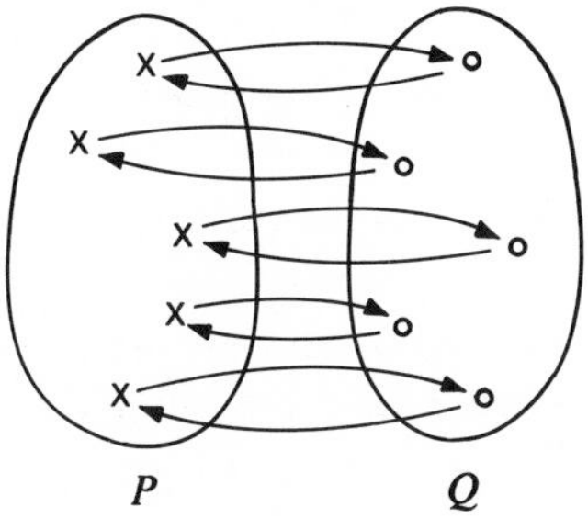

reverse illustrate a mapping of Q onto P. Explain. The reversed arrows have produced a "two-way road" effect. Consider the outcome if the arrows on your diagram had been reversed? Would the result have qualified for a mapping? Why?

Since P is mapped onto Q and it is reversible, this is called a *one-to-one mapping*. It is often called a *one-to-one correspondence* or a *one-for-one.*

This is the type of mapping used by the shepherd of ancient times as he sought to determine whether his flock was intact as they returned to the fold. It was also used by the birthday child to determine whether the cookies and children "came out even," and by Janie to tell whether her friends and pets "matched up."

The preceding examples give direction for the following definitions:

Definition 2.8(a) If to each element of set P there is assigned one and only one element of set Q, called the assignment or image, and

(i) if all elements of Q are not assigned, then this is a mapping of *P into Q*.

(ii) if all elements of Q are assigned, then this is a mapping of P *onto* Q.

(iii) if the mapping of P to Q is onto and reversible, then this is a *one-to-one mapping*.

One-to-one mapping is important in developing the concept of equivalent sets. The following examples are designed to point out the role of this mapping in determining whether sets are equivalent or not.

Example 4 Given $A = \{a, b, y, z\}$ and $B = \{1, 2, 25, 26\}$. A mapping of A onto B is shown. This is then followed by showing the mapping reversed.

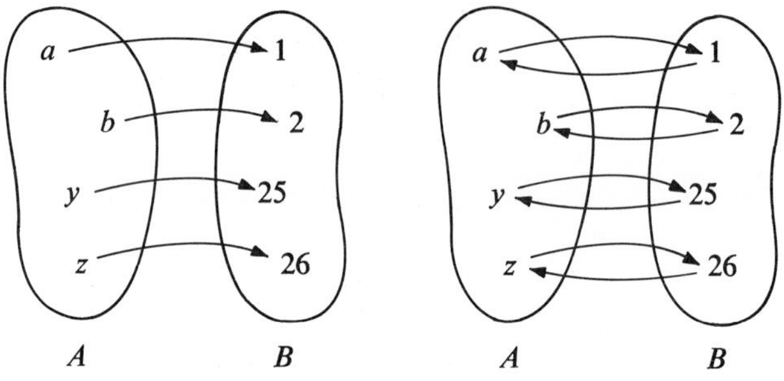

Mapping of A onto B One-to-one mapping

Since A is mapped onto B and the mapping is reversible, A and B are said to be *equivalent sets*. Using the symbol for equivalent, it is written

$$A \simeq B.$$

Example 5 Is set C equivalent to set B? The mapping of C onto B is shown in the diagram.

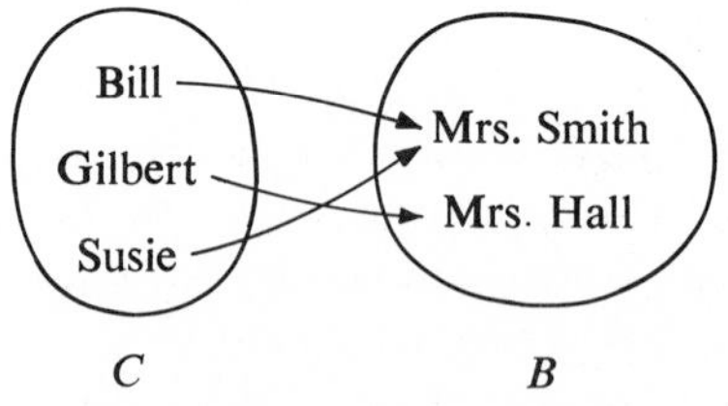

Examine the diagram to see if you think that this is an onto mapping of C to B. Is this mapping reversible? By reversing the arrows, more than one arrow will originate from Mrs. Smith. So, there can be no mapping from B to C. That is, reversibility does not exist in this case. Consequently, set C is not equivalent to set B. Using the symbol for *nonequivalent*, this is written

$$C \not\simeq B.$$

Example 6 Angela wishes to know whether Jimmy has enough horses for each of her friends to ride when they come to see her. Angela's friends are Susan, Mary, Jane, and Ginny. Jimmy saddles his horses Ginger, Nancy, and Lazy-Bones. He then assigns to each horse a rider.

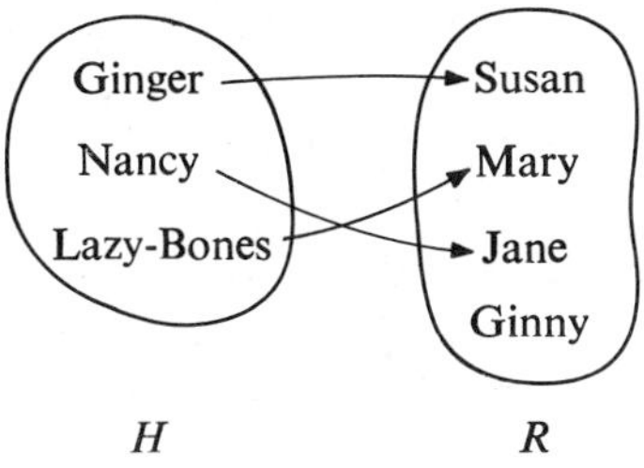

Explain why this is a mapping. Is this an into or an onto mapping? By reversing the arrows, is there a mapping of R to H? Explain.

Does each horse have a rider? Does each one of Angela's friends have a horse to ride? Obviously not. This means that the two sets are not equivalent.

From these examples, one conclues that if sets are to be equivalent it must be possible to establish an onto mapping that is reversible. The definition follows.

Definition 2.8(b) If a one-to-one mapping can be established between set X and set Y, then X is *equivalent* to Y.

EXERCISE 2.8

1. A teacher assigns boys in a gym class to activities as shown.

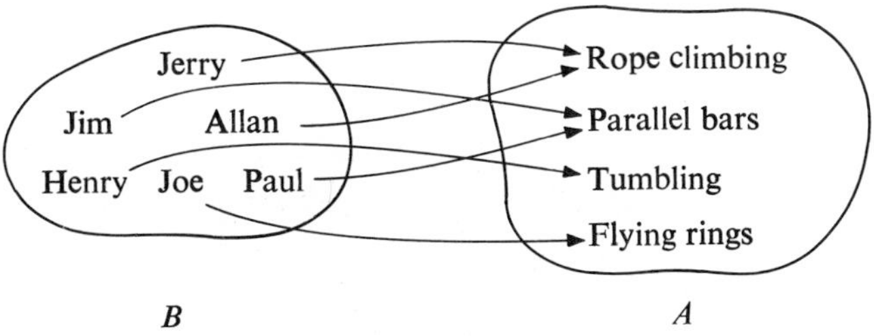

(a) Explain why this is a mapping of B to A.
(b) Is this an onto or an into mapping?
(c) Is this mapping reversible?

2. Given the following arrow diagram:

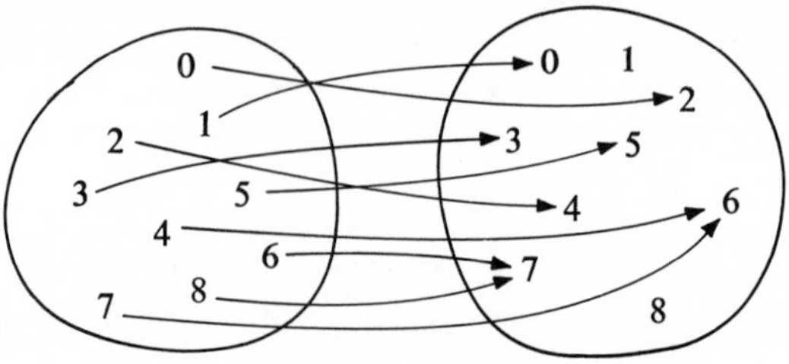

(a) Is this a mapping? (b) What is the domain? (c) What is the range?

3. The following arrow diagram is not a mapping because two requirements have not been fulfilled. Give these deficiencies.

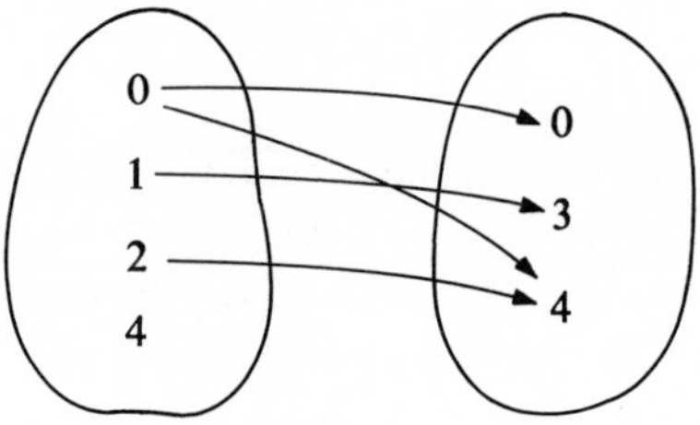

4. Use the accompanying arrow diagrams of mappings to determine the following:
   (a) The domain and range of each
   (b) Which are into and which are onto mappings
   (c) Which illustrate a one-to-one correspondence
   (d) Which represent a pair of equivalent sets

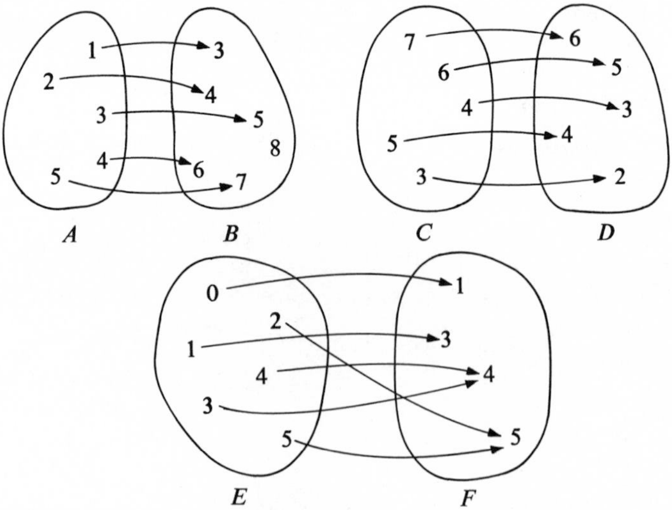

5. Given $X = \{2, 4, 5\}$ and $Y = \{s, a, c\}$:
   (a) Illustrate one way in which the elements may be paired to show that a one-to-one correspondence exists between set X and set Y.

(b) Are these sets equivalent or nonequivalent?

6. Given $P = \{15, 3, 5, 21\}$ and $Q = \{10, 11, 12\}$:
   (a) Is it possible to establish a one-to-one correspondence between set P and set Q?
   (b) Are these sets equivalent or nonequivalent?
7. At a class party, the committee bought vanilla, chocolate, and toffee ice cream. They planned to serve the ice cream in white, pink, and yellow bowls.
   (a) Has the committee provided for a one-to-one correspondence between the flavors of ice cream and the colors of the bowls? Illustrate one of the possible pairings.
   (b) Illustrate another one. How many pairings are possible?

2.9 CONCEPT OF NUMBER

The next step after knowing whether he had the same size flock of sheep, early man needed to be able to tell *how many* sheep he had, how many wives he owned, or how many animals he had killed during a hunt. Even though it is a very simple matter for us today to tell how many things we have in a set, the methods for doing this were developed slowly throughout prehistoric times. However man has, no doubt, always had some intuitive notion about number.

This leads one to inquire, "What is number?" A formal definition of number will not be attempted, but the concept of number will be developed from the concept of equivalent sets. Evidently, man very early became aware that all equivalent sets have a property of being alike in some way. That is, they have something in common. When attention is directed to the following sets, one is immediately aware of this "alikeness" property:

This property of "alikeness" of equivalent sets shall be identified as *number*.

Number is an abstract concept. It cannot be seen or pointed to any more than can *time* or *bravery*. Time passes and we use it, and we admire a brave man. Yet, we cannot see or point to time or bravery. We may point to a clock or to a brave man, but these are only representations of the ideas, not the ideas themselves. Likewise, number cannot be seen or pointed to. We have ways of representing number, which have been among man's most significant inventions. These inventions will be studied in much detail in the chapter which follows.

To continue the study of number, consider the sets shown above, now arranged in one-to-one matchings as indicated below:

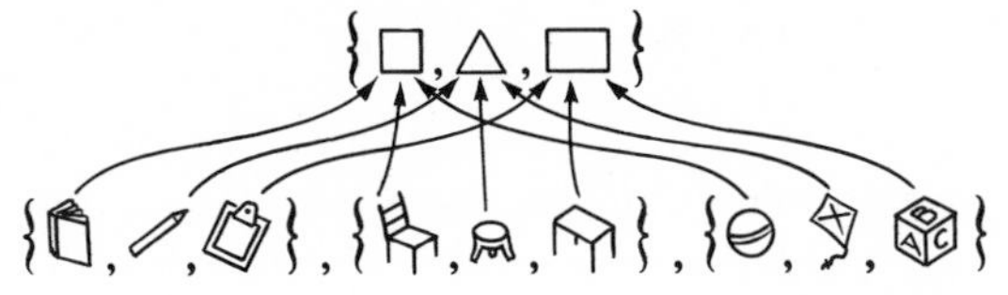

Since a one-to-one correspondence exists between each pair of these sets, they are equivalent. We say that *there are as many elements in one of these sets as there are in any one of the others.* Since each of the elements of the set of a student's supplies, the set of furniture, and the set of toys has been paired with each of the elements of the set

{□, △, ▭},

this shall arbitrarily be called the *standard set.* All sets equivalent to this set have *the same number property*—that is, *the same cardinality.*

The number which we shall assign to the standard set will be the same for all sets equivalent to it. A special *number notation,* the N-*notation,*

N{□, △, ▭},

will be used to represent this number. It is convenient to have a simpler name for this number. A symbol, *, is arbitrarily selected and is called blob. Since N{□, △, ▭} and * are names for the same number, this may be expressed as,

N{□, △, ▭} = *.

This is read, "The number assigned to the set consisting of □, △, and ▭ is blob." The equals symbol, =, is used here to mean *is the same as.* That is, both representations stand for the same number.

Since each of the following sets is equivalent to the standard set above, what number is assigned to each? The sets and their assigned numbers are represented below:

$A = \{r, s, t\}$	$B = \{a, b, c\}$	$C = \{x, y, z\}$
$N\{r, s, t\} = *$	$N\{a, b, c\} = *$	$N\{x, y, z\} = *$

The number notation assigning a number to each of the above sets may also be written in this way:

$$N(A) = * \qquad N(B) = * \qquad N(C) = *$$

"$N(A) = *$" is read, "The number assigned to set A is blob." Children will say, "The number of members (elements) in set A is blob."

In a similar manner, a number is assigned to every set. Since the choice of a symbol for a number is arbitrary, we shall now select the conventional symbols (with the usual names) as displayed below:

$N\{a, b\} = 2$ (*two*)	$N\{r\} = 1$ (*one*)
$N\{c, d, s, y\} = 4$ (*four*)	$N\{x, y, z\} = 3$ (*three*)

As you no doubt have observed, the number for the empty set has not yet been assigned. To the set with no elements shall be assigned the number 0 (*zero*); that is,

$$N\{\ \} = 0 \qquad \text{or} \qquad N(\varnothing) = 0.$$

This is read, "The number assigned to the empty set is 0," or more commonly, "The number of elements in the empty set is 0." When the N-notation is used with the roster notation, care must be taken not to place the symbol for zero, 0, between the braces. The number assigned to $\{0\}$ is the same as that assigned to $\{r\}$—that is, 1, not 0. In symbols, $N\{0\} \neq 0$, but $N\{0\} = 1$.

2.10 ORDER OF THE CARDINAL NUMBERS

Consider the following nonequivalent sets:

$$A = \{a, b\} \qquad C = \{r\}$$
$$B = \{c, d, s, y\} \qquad D = \{x, y, z\} \qquad E = \{\ \ \}$$

On attempting to make a one-to-one matching between set A and set B, it is found that set A can be matched with only a proper subset of set B:

$$\begin{array}{c} \{a, b\} \\ \updownarrow\ \updownarrow \\ \{c, d, s, y\} \end{array}$$

Thus, set B contains *more* elements than set A, and set A has *fewer* elements than set B. We shall use the words "larger" and "smaller" when talking about two sets to mean the following: If set B contains *more elements* than set A, then set B is larger than set A; if set A has *fewer elements* than set B, then set A is *smaller* than set B. That is, when "larger than" or "smaller than" is used in making a comparison of two sets, only the *number* of the elements is compared, not the size of the elements.

Which of the sets B, C, D, E in the preceding group are larger than set A? Which are smaller than set A? Let us arrange the sets A, B, C, D, and E in a sequence with the smallest set listed first and the largest last:

$$\begin{aligned} E &= \{\ \ \} \\ C &= \{r\} \\ A &= \{a, b\} \\ D &= \{x, y, z\} \\ B &= \{c, d, s, y\} \end{aligned}$$

Next we represent the number associated with each set in the preceding arrangement according to the assignments made in Section 2.8:

$$\begin{aligned} N(E) &= 0 \\ N(C) &= 1 \\ N(A) &= 2 \\ N(D) &= 3 \\ N(B) &= 4 \end{aligned}$$

Since there are more elements in set B than in set A, the *number* assigned to set B is said to be *greater than* the *number* assigned to set A. The symbol $>$ represents *is greater than* and we write

$$N(B) > N(A).$$

This is read, "The number of elements in set B is greater than the number of elements in set A," or "The number assigned to set B is greater than the number assigned to set A." Here another name for $N(B)$ is 4, while another name for $N(A)$ is 2, and so $N(B) > N(A)$ can be written

$$4 > 2$$

and read, "Four is greater than 2."

It is also the case that set A contains fewer elements than set B, and we shall say that the *number* assigned to set A is *less than* the *number* assigned to set B. The symbol for "is less than" is $<$, and we write

$$N(A) < N(B),$$

which is read, "The number of elements in set A is less than the number of elements in set B," or "The number assigned to set A is less than the number assigned to set B." Using the other names for these numbers in this case, we may write

$$2 < 4,$$

which is read, "Two is less than four." This leads us to say that if $4 > 2$, then $2 < 4$.

The order symbols, $>$, $<$, represent the *order relations—is greater than* and *is less than*. The symbols are used only to say things about numbers, never about sets. That is, it is correct to write $N(B) > N(A)$, but not $B > A$.

Using the order symbols with the symbols for the numbers assigned to the sets E, C, A, D, and B, we can write

$$0 < 1 < 2 < 3 < 4.$$

This is read, "Zero is less than one, is less than two, and so on." This could also be written

$$4 > 3 > 2 > 1 > 0$$

and read, "Four is greater than three, is greater than two, and so on."

Later, when a study of the different systems of numbers is made, a formal definition will be given to establish the *order* of numbers. For the present, when we speak of the order of numbers, we merely think of them as being "lined up" as

$$0, 1, 2, 3, 4, \ldots,$$

where the three dots mean that the numbers go on without end.

We now note that every pair of numbers a and b (where a and b are symbols which represent any number) are related in one, and only one, of the following ways:

$$a = b \quad \text{or} \quad a \neq b.$$

If $a \neq b$, then

$$a > b \quad \text{or} \quad a < b.$$

From this we have:

The Trichotomy Property

For any two numbers a and b, exactly one of the following statements may be made:

$$a = b, \quad \text{or} \quad a > b, \quad \text{or} \quad a < b.$$

This property says that at least one and only one of the three relations, $=$, $>$, $<$, exists between every pair of numbers. Suppose the first number of a pair *is greater than* the second; then the other two relations, *is equal to* and *is less than,* are eliminated from consideration. In other words, *only one* of the three relations holds.

When it is necessary to refer to the position or rank of an item in any "lineup" or sequence, the following names are used:

first, second, third, fourth, fifth, etc.

Thus, in the list 0, 1, 2, 3, 4, . . . , 0 is the first, or "number one" in the sequence, 1 is second, or "number two," and so on. These names designate *ordinal numbers.*

For example, an ordinal number is used to give the rank of a basketball team at the end of a tournament. The team that wins the tournament is *first,* or we may say that it is "number 1." The remaining contenders are ranked second, third, fourth, etc., or as number 2, number 3, number 4, etc.

On the other hand, the numbers used to specify *how many* elements there are in a set are called *cardinal numbers.* Thus, the basketball teams playing in the tournament are equivalent sets and all have the cardinal number five assigned to them, indicating that there are five players on each team.

It is easy to see that the ordinal number giving the tournament ranking is quite different from the cardinal number giving the number of players on each team. In general, a cardinal number is used to designate *how many,* while an ordinal number is used to designate *which one in a sequence.*

When sets are arranged in a sequence so that each succeeding set has one more element than the preceding set, and cardinal numbers are assigned to each, we have the following:

$\{\ \}$	0
$\{a\}$	1
$\{a, b\}$	2
$\{a, b, c\}$	3
$\{a, b, c, d\}$	4
$\{a, b, c, d, e\}$	5
$\vdots$	$\vdots$

Notice that the number 0 is assigned to the set with no elements (the empty set). Since each succeeding set in this arrangement has one more element than the preceding set, then each succeeding cardinal number is *one greater than* the preceding cardinal number, except in the case of 0. Why is 0 the exception?

When the cardinal numbers have been arranged in the order indicated above, they may be used for *counting*—that is, the process of determining how many elements there are in a set. It is customary to use the sequence beginning with 1:

$$1, 2, 3, 4, 5, \ldots$$

To count the elements in the set $\{\bigcirc, \triangle, \square\}$, we match each of these elements with the numbers in the above sequence:

$$\begin{array}{cccc} \{\bigcirc, & \triangle, & \square\} & \\ \updownarrow & \updownarrow & \updownarrow & \\ 1, & 2, & 3, & 4, \ldots \end{array}$$

Here the last element to be matched is the *third* element and the cardinal number of the set is *three.*

EXERCISE 2.10

1. Given $A = \{1, 2, 3\}$, $B = \{0, 2, 4, 6, 8\}$, $C = \{a, b, c\}$, $D = \{7, 9, 4\}$.
   (a) Which of the given sets are equivalent? Set up a mapping between these sets that establishes a one-to-one correspondence.
   (b) Indicate the cardinality of set A by using both the N-notation and the standard numeral.
   (c) Apply the trichotomy property to the numbers assigned to sets A and B; to sets C and D.
2. Consider the sets:

$$\begin{array}{ll} A = \{1, 3, 5, 7, 9, 11, 0\} & E = \{7, 9, 5\} \\ B = \{3, 7, 11\} & F = \{5, 7, 9\} \\ C = \{2, 3, 7\} & G = \{7, 11\} \\ D = \{9, 15, 7, 1\} & H = \{\ \ \} \end{array}$$

   (a) The following statements are made about the above sets. Some of the statements are true; some are false. Identify the statements as such.

(1) $3 \in A$	(11) $5 \notin A$
(2) $E \subset D$	(12) $N(C) \neq N(G)$
(3) $H \subset B$	(13) $0 \in H$
(4) $E \subset F$ and $F \subset E$	(14) $N(B) < N(A)$
(5) B is equivalent to C.	(15) $C \neq F$
(6) $G = 2$	(16) $G \subset B \subset A$
(7) $8 \in A$	(17) $0 \in N(A)$
(8) $N(B) = N(C)$	(18) $N(C) = F$
(9) $A > D$	(19) $H = 0$
(10) $E = F$	(20) $N(D) = 4$

   (b) Arrange the sets A, B, D, G, and H in order beginning with the set having the fewest elements.
   (c) Indicate the order of the numbers associated with the sets A, B, D, G, and H by using the proper order symbol.
   (d) Why is it incorrect to write $C = B$, but correct to write $N(C) = N(B)$?

(e) Explain the significance in the use of the symbol "=" in these two statements:

(1) $E = F$ (2) $N(B) = N(E)$

3. There are 43 people living in my block. All 43 are either Michigan-born or attend school, or both. Twenty-seven were born in Michigan and 18 attend school. Use a Venn diagram to show how many Michigan-born people are attending school. What may be used as the universal set?

4. In the following sentences, indicate each number as cardinal or ordinal.
   (a) Turn to page 95.
   (b) My address is 452 West Main Street.
   (c) John came in number 3 in the 50-yard dash.
   (d) The textbook used in Mathematics 70 has 250 pages.
   (e) The 2 states of greatest area are Alaska and Texas, with Texas number 2.
   (f) In 3 days the calendar will indicate month 4.
   (g) Bill is number 4 of 5 brothers.
   (h) I have gained 10 pounds within the past 2 months.
   (i) I was number 1 in line to buy ten 5-cent stamps.
   (j) My golf score was 18 at hole 3.

3

Systems of Numeration

3.1 EARLY RECORDING OF NUMBERS

Before we make any systematic study of our present-day method of recording the numbers we discussed in Chapter 2, we shall see how this method developed historically. It is clear from early records that people counted long before they recorded the results of their counting. Today, many primitive tribes have a spoken language but do not have a written language. (One of the tasks of the Linguistics Institute, held every summer on the campus of The University of Oklahoma, is to teach missionaries the basic methods of organizing into a written language any tribal dialect with which they may be confronted.) Notches on a stick, knots in a rope, and scratches on a stone were crude methods of recording the results of counting. Early peoples' simple kind of barter needed nothing beyond this.

As civilization developed, more detailed records were necessary, such as a calendar to indicate the proper time for planting crops, a boastful record of numbers of slaves captured in war, a survey to establish boundary lines when the yearly flooding of the Nile destroyed the temporary marks, reports of business transactions, and computations connected with the study of astronomy. All these needs stimulated efforts to develop some kind of systematic record keeping. Historians and archeologists estimate that the earliest evidences of such recording of numbers can be established as occurring about five thousand years ago.

The Egyptians in the valley of the Nile made records on stone or wrote on a sort of paper, called papyrus, made from a plant of that name. The symbols they used for writing are called *hieroglyphics*. An early record of Egyptian mathematics is the "Ahmes Papyrus," sometimes called the "Rhind Papyrus" from the name of the man who discovered it. This is written in hieratic script, the cursive form of hieroglyphic writing. This papyrus is supposed to have been copied about 1700 B.C. by the scribe Ahmes from an earlier work dated about 2000 B.C. It is a mathematical handbook, containing the solutions of eighty-five specific problems. This papyrus is now preserved in the British Museum.

The Babylonians, who lived in Mesopotamia in the valleys of the Tigris and Euphrates rivers, kept their records on clay tablets. A sharp instrument, or stylus, was used to press symbols on soft clay. These tablets were then baked and preserved as records. Archeologists have excavated them from the ruins of ancient cities, and careful study has led to the understanding of the method of writing that the Babylonian people used. Earliest dates are placed at about 3000 B.C. The symbols used by the Babylonians are called *cuneiform* (wedge-shaped) symbols in contrast to the *hieroglyphic* (picture) symbols of the Egyptians.

The symbols used to represent numbers are called *numerals* and the systems of writing them are called *numeration systems.* We shall examine the numeration systems of the Egyptians and the Babylonians, as well as a third one, that of the Romans, for two reasons. First, we shall thus become more fully aware of how our present (decimal) system has developed as the culmination of a long and extremely laborious struggle to find an accurate, compact system that lends itself readily to both recording and computation. Second, we shall study the characteristics of each system, some of which have been incorporated into our own, in order to understand and appreciate more fully the system we have inherited. No attempt has been made to present a complete history of numeration.

3.2 THE EGYPTIAN SYSTEM OF NUMERATION

The Egyptians recorded the numbers from one to nine in hieroglyphics as follows:

$$\begin{array}{ccccccccc} &&&&&&&& |\,|\,| \\ &&&& |\,|\,| & |\,|\,| & |\,|\,|\,| & |\,|\,|\,| & |\,|\,| \\ |, & |\,|, & |\,|\,|, & |\,|\,|\,|, & |\,|, & |\,|\,|, & |\,|\,|, & |\,|\,|\,|, & |\,|\,|. \end{array}$$

Then the next number, ten, was given a new symbol, $\cap$, and the pattern was repeated for

$$\begin{array}{rl} \text{twenty,} & \cap\cap, \\ \text{thirty,} & \cap\cap\cap, \\ & \vdots \\ & \cap\cap\cap \\ \text{ninety,} & \cap\cap\cap. \\ & \cap\cap\cap \end{array}$$

The Egyptians used a different picture, or symbol, for each power of 10. For example, 10 is the first *power* of ten; the second power of 10 is 10×10

or 10^2, written with the *exponent* 2; the third power of 10 is $10 \times 10 \times 10$ or 10^3; and so on. For 10^2 the scroll symbol is used. Table 3.1 gives several successive powers of ten.

Table 3.1 Egyptian Hieroglyphic Numerals

Our Numeral	*Egyptian Numeral*	*Descriptive Name*
1	𓏺	Stroke
10	𓎆	Heel bone
100	𓍢	Scroll
1,000	𓆼	Lotus flower
10,000	𓂭	Pointing finger
100,000	𓆐	Polywog
1,000,000	𓁨	Astonished man

The use of Hieroglyphic numerals is illustrated in the following examples.

Example 1.  means 21,492.

Example 2.  means 100,153.

The symbols represent powers of ten. Symbols are repeated until nine of them have been used. When a tenth one is needed, the whole group is replaced by the next power of ten. In Example 1, if 10 were added, another heel bone would be needed, and the whole group of ten heel bones would be replaced by a scroll.

To express the number in Example 2 in our notation we have:

$$100{,}000 + 100 + 10 + 10 + 10 + 10 + 10 + 1 + 1 + 1.$$

The symbols are usually written in order either from left to right, as shown in the examples, or from right to left (with the unsymmetrical symbols facing the other way), but the arrangement of the symbols has no effect at all on the number represented.

In short, the Egyptian system can be characterized in the following ways:

(a) It is based on powers of ten with a different symbol for each power.
(b) It is repetitive.
(c) It is additive.
(d) Position of symbols does not affect value.

There are several distinctive features of Egyptian arithmetic. It was concerned wholly with the solving of problems. The Ahmes Papyrus simply proposes a problem and then shows the solution, with no reasons for the steps or any indication of a general method of approach. Computation with whole numbers was rather direct and simple with these numerals. Addition could be accomplished by counting and replacing by the next power of ten when necessary, and subtraction could be done by reversing the process. Multiplication was done by an ingenious method of doubling and summing.

EXERCISE 3.2

1. Write 5362 in Egyptian notation.
2. Express in our system: 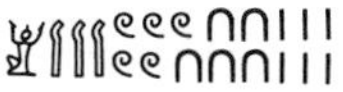
3. Express in Egyptian notation 1,001.
4. Express in Egyptian notation the number which must be added to

 to make .

5. Perform the following operations, using only Egyptian hieroglyphics and explaining the steps required.

(a)    (b) 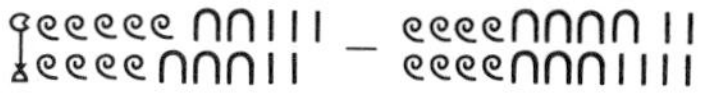

(c) How many years have elapsed since the discovery of America? Write the problem and the answer in Egyptian notation.

3.3 THE BABYLONIAN SYSTEM OF NUMERATION

Unlike the Egyptians with their large number of symbols, the Babylonians used only two, which we represent as follows: 𒁹, standing for one, and 𒌋, standing for ten. With these, they recorded numbers less than 60 by applying the additive principle used by the Egyptians:

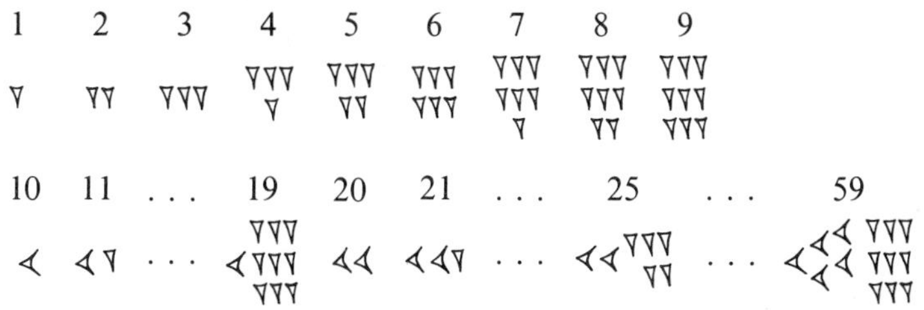

Observe that when both symbols are used, those for ten appear at the left of those for one, as

for 25.

For larger numbers they used a *place value system* based on powers of 60. For example,

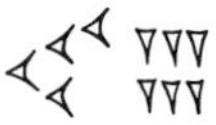

represents 46. By simply placing a 𒁹 to the left of the numeral for 46, we change the number represented to $(1 \times 60) + 46$:

𒁹𒌋𒌋𒌋𒌋𒁹𒁹𒁹𒁹𒁹𒁹

Thus, the symbol 𒁹 in this case, *because of its position*, has the value (1×60) instead of 1. As another example, the numeral

𒁹𒁹𒌋𒌋𒌋𒁹𒁹𒌋𒌋𒁹𒁹𒁹

may be thought of as grouped thus,

𒁹𒁹, 𒌋𒌋𒌋𒁹𒁹, 𒌋𒌋𒁹𒁹𒁹.

This grouping helps to interpret the number as $(2 \times 60^2) + (32 \times 60) + 23$. We would write this number as 9143, which means $(9 \times 10^3) + (1 \times 10^2) + (4 \times 10) + 3$. Since our system is based on powers of 10, it is called a *decimal* system of numeration. The Babylonian system, based on powers of 60, is called a *sexagesimal* system of numeration.

One difficulty with this notation was apparent when we introduced grouping symbols in the numeral above. Without them, there is an inherent ambiguity, and one cannot be sure whether 𒌋𒌋𒁹𒁹𒁹 stands for 23 (or for $(20 \times 60) + 3$, or any other combination of products of powers of 60. Only the context could show which was meant, although archeologists have found that sometimes spacing was used to give some indication as to which "places" were intended.

It was by the careful study of ancient records, particularly those such as tables showing a pattern, that scholars were able to reconstruct this system. The reader should check the accompanying table of multiples of 7 (Table 3.2), recognizing that ambiguities are possible. For example, we can tell from the context that 𒁹 𒁹𒁹𒁹 must be interpreted as

$$(1 \times 60) + 3,$$

not as 4.

It was a situation such as this that led archaeologists studying Babylonian inscriptions to suspect that powers of 60 played an important role in their numeration system. In a particular tablet under inspection, multiples of 8 were found: 8, 16, . . . , 56. These were interpreted easily through 7×8. But the next entry, where 64 was expected, seemed out of place. Then someone interpreted 𒁹𒁹𒁹𒁹𒁹 as $(1 \times 60) + 4$ and the mystery was solved. The succeeding entries then verified this theory.

In records of the later periods a symbol was introduced to indicate the absence of a particular power of 60. Thus, $(2 \times 60^2) + (0 \times 60) + 21$ might be written 𒁹𒁹•𒌋𒌋𒁹 with the dot playing the role of our symbol 0. However,

Table 3.2 Multiples of Seven

⟨ 𒀹𒀹𒀹 𒀹	⟨⟨⟨ 𒀹𒀹𒀹 𒀹𒀹𒀹
⟨⟨ 𒀹	𒀹 𒀹𒀹𒀹
⟨⟨ 𒀹𒀹𒀹 𒀹𒀹𒀹 𒀹𒀹	𒀹 ⟨
⟨⟨⟨ 𒀹𒀹𒀹 𒀹𒀹	𒀹⟨ 𒀹𒀹𒀹 𒀹𒀹𒀹 𒀹
⟨⟨⟨ ⟨ 𒀹𒀹	𒀹⟨⟨ 𒀹𒀹𒀹 𒀹
⟨⟨⟨ ⟨ 𒀹𒀹𒀹 𒀹𒀹𒀹 𒀹𒀹𒀹	𒀹⟨⟨⟨𒀹

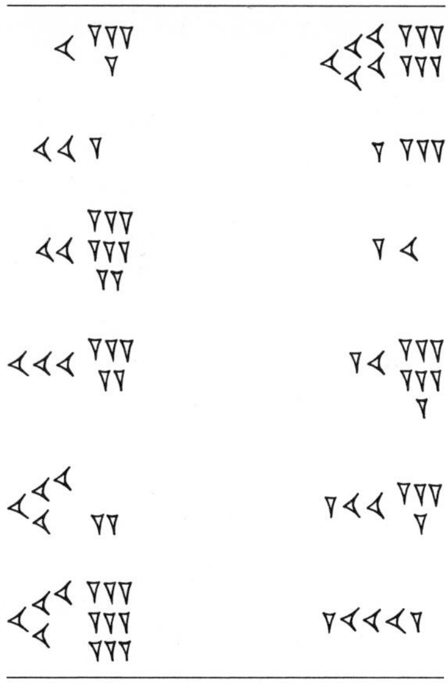

there was no indication that such a symbol was ever used at the right of
the numeral to indicate that smaller powers of 60 did not occur. There was
never a way to be sure that the above symbols represented $(2 \times 60^2) +
(0 \times 60) + 21$, which is 7,221, or $(2 \times 60^3) + (0 \times 60^2) + (21 \times 60) + 0$,
which is 217,260. In fact, many other interpretations may occur to the reader,
where certain contexts make them more reasonable.

We can now list the characteristics of the Babylonian system.

(a) It has only two symbols, 𒀹 and ⟨.
(b) Numbers are expressed in powers of 60.
(c) It is repetitive (for numbers 1 to 59).
(d) It is additive.
(e) A place value property is employed, although the lack of a symbol
    for zero and a method of separating powers of 60 often leads to
    ambiguity.

In spite of the inadequacies and ambiguities of the Babylonian system
of numeration, a great step forward had been made. To quote Neugebauer:
"The invention of this place value notation is undoubtedly one of the most
fertile inventions of humanity. It can be properly compared with the invention of the alphabet as contrasted to the use of thousands of picture-signs
intended to convey a direct representation of the concept in question."

It is worth noting that these Babylonian clay tablets with their cuneiform
symbols are on display in many museums. You should look for them when
the opportunity presents itself.

EXERCISE 3.3

1. Record in our decimal notation the least numbers represented by the following:

(a)  (c)  (e) 

(b)  (d)  (f) 

2. Make a table showing in Babylonian notation the multiples of 8 less than 100.
3. Write in cuneiform symbols the Babylonian representation of the numbers given in our decimal notation by:
   (a) 95 (b) 325 (c) 425 (d) 122
4. How would you set your watch when 26,516 seconds had elapsed since midnight? Now write the above number of seconds in the notation of a Babylonian student.
5. A family with three children used  quarts of milk in April. At 25 cents per quart, what was their month's bill?

3.4 THE ROMAN SYSTEM OF NUMERATION

The Roman numerals are still in use as dates on monuments and public buildings, in chapter headings, as volume numerals, in outlines, and other places where a formal numbering seems suitable.

Let us consider the properties that characterize this system. Different symbols, chiefly letters of the Roman alphabet, are used to represent powers of 10, with extra symbols for a number halfway between these powers.

Our Numeral	1		10		100		1000
		5		50		500	
Roman Numeral	I		X		C		M
		V		L		D	

The origin of these symbols is not certain. V may have begun as a representation of the five fingers, to be abbreviated finally to V. X might have evolved from two V's, put point to point. It seems reasonable to think that C and M are the initial letters of centum (hundred) and mille (thousand).

The principle of repetition was used until the next symbol was indicated; that is, no more than four of any one symbol would be used. When the symbols were written in order of equal or increasing value from right to left, the values were to be added; that is, CXXIII = 123. Besides this additive principle, that of subtraction was also introduced. A symbol for a smaller number placed to the left of that for a larger indicated subtraction. For example,

IV was written instead of IIII,
IX instead of VIIII,
XL instead of XXXX,
XC instead of LXXXX, and so on.

If this principle were used with more than two symbols in juxtaposition, the meaning would be ambiguous. Thus, IXL might be interpreted as either $50 - 9 = 41$ or $(50 - 10) - 1 = 39$. Hence definite rules for the use of this principle existed:

I could precede only V or X;
X could precede only L or C;
C could precede only D or M.

This subtractive principle, while shortening the writing, did make the operations of addition and substraction more complicated. The simple ideas of gathering together and replacing by a new symbol used in addition, and the reverse used in subtraction, presented difficulty. The only way to meet this was to change to the longer form.

Notice that the use of the subtractive principle makes necessary the proper arrangement of the symbols. Hence *sequence,* although not place value, is a characteristic of the Roman system. The following is an example.

$$\text{MCMLXV} = 1000 + (1000 - 100) + 50 + 10 + 5 = 1965$$

The Romans saved themselves from having to invent new symbols for very large numbers by using a principle of multiplication. A bar drawn over a numeral multiplied the number by 1000, two bars multiplied it by a thousand thousand, or 1,000,000. For example,

$\overline{\text{X}}$ is $10 \times 1000 = 10{,}000$;
$\overline{\text{V}}$ is $5 \times 1000 = 5000$;
$\overline{\overline{\text{V}}}$ is $(5 \times 1000) \times 1000 = 5{,}000{,}000$.

Thus, we have the following characteristics of the Roman system of numeration:

(a) It uses the seven symbols I, V, X, L, C, D, and M.
(b) It is repetitive.
(c) It is additive in general.
(d) It has a subtractive principle expressed by the order of the symbols.
(e) It has a multiplicative principle for expressing larger numbers.

EXERCISE 3.4

1. Write our numerals for:

(a) CMV
(b) DCLVIII
(c) MDXCV
(d) $\overline{\text{XIX}}$
(e) MMDCCCCXXXXV

2. Express in Roman numerals. If the subtractive principle applies, represent the number both with and without its use.

(a) 946	(d) 3724
(b) 1095	(e) 26,000
(c) 563	(f) 1944

3. Carry out the following additions and subtractions, using only Roman numerals. It is sometimes helpful to change from the subtractive form as in this example.

```
Add: XXIV =   XXIIII
      XIX =   XVIIII
            ____________
              XXXVIIIIIIII
            = XXXVVIII
            = XXXXIII
            = XLIII
```

(a) VIII + IV	(d) XXXIV − XIII
(b) XXIV + XLVII	(e) XXXIV − XXIX
(c) Add: CCXXVII XXXIV CLXVI	(f) CCLI − CLXII

3.5 THE HINDU-ARABIC SYSTEM OF NUMERATION

The great river valleys of Africa and Asia seem to have provided the settings for man's first experiments in civilization. Historically, the Nile, the Tigris and Euphrates, the Ganges, and the Yellow River all played an important part. We have learned something about how the Egyptians, who lived along the Nile, recorded their numbers in hieroglyphics. The Sumerian-Babylonian-Assyrian peoples, who successively occupied the valleys of the Tigris and Euphrates, contributed their clay records in cuneiform symbols. The Chinese had little communication with the others and their influence, however small or large it may have been, is not a matter of record. But the people of India, along the Ganges, have left an indelible mark upon our present system of numeration.

It is indeed difficult to give proper credit where it is due. Scholars must study records from many of the ancient peoples, try to estimate the time in history when they were written, and then conjecture as to the amount of influence one had upon another. There is no sure way to decide all these questions and what follows must be read in that light.

There seems little doubt that the *digits*

$$1, 2, 3, 4, 5, 6, 7, 8, 9$$

evolved from those used by the Hindus. They appear in inscriptions from the third century B.C. during the reign of the great Buddhist king, Asoka. These and other early examples from India contained no zero. Some time later, possibly between A.D. 200 and 600, the Hindus became acquainted with the work of Greek astronomers in which the symbol 0 was used for zero. They also knew about the Babylonian sexagesimal positional system which, although it had by this time a symbol for an absent digit in the interior of a number symbol, had none to show that one was lacking at the end. It seems plausible that the Hindus thus combined their own nine digits, the zero symbol of the Greeks, and the positional system of the Babylonians to form the notation for numbers that is known to us today. The symbols used were subject to a good deal of change in appearance due to differences in writing. The invention of the printing press and movable type served somewhat to standardize them. But now numerals on bank checks present an even different form.

But how did this Hindu system reach and become accepted by the Western world? Actually, it moved by a rather roundabout route.

In the seventh and eighth centuries the Arabs, strong in their Islamic faith, built the fabled "Arabian Nights" city of Baghdad near the site of ancient Babylon, absorbing the culture of the conquered peoples and carrying it with them as they moved toward the west. During this period, at about A.D. 825, an Arabic mathematician, al-Khowarizmi, wrote a little book describing the arithmetic of the Hindus. It seems certain that at this time the Asiatic Indians used the same method of writing positive integers and fractions that we have today. This book has been lost but a translation into Latin made by an English monk, Adelard of Bath (circa 1120), is still extant. It is by way of this small book, *Liber Algorismi de Numero Indorum,* that the Western world learned of the Hindu-Arabic numerals. The word "algorism" or "algorithm" (frequently used in arithmetic), meaning a process of calculation, can be traced to the name of the Arabic mathematician al-Khowarizmi.

The highly nomadic Arab people carried with them whatever knowledge they acquired as conquerors, moving toward the west along the north shore of Africa and finally across to the Spanish peninsula. Here they established the Moorish universities where much of the classical culture was preserved during the years prior to the Renaissance in Europe. The city of Cordoba with its magnificent library and university attracted thousands of students from all parts of Europe. Here al-Khowarizmi's book was studied and from here its influence spread. In 1202 Leonardo of Pisa, an Italian, wrote a book on arithmetic, *Liber Abaci,* where he uses the "figurae Indorum" which he had learned from a Moorish teacher.

Until recently these numerals and the system that uses them have been called *Arabic* since they came directly from the Arabs. But later and more careful study has traced them to their Hindu originators. So they are now called *Hindu-Arabic,* sharing the names of their inventors and preservers.

After studying the characteristics of other systems, it is not difficult to discover those of the Hindu-Arabic system:

(a) It uses 10 symbols (digits), 0, 1, 2, 3, 4, 5, 6, 7, 8, 9.
(b) Numbers are expressed in powers of 10.
(c) It is additive.
(d) It has place value.

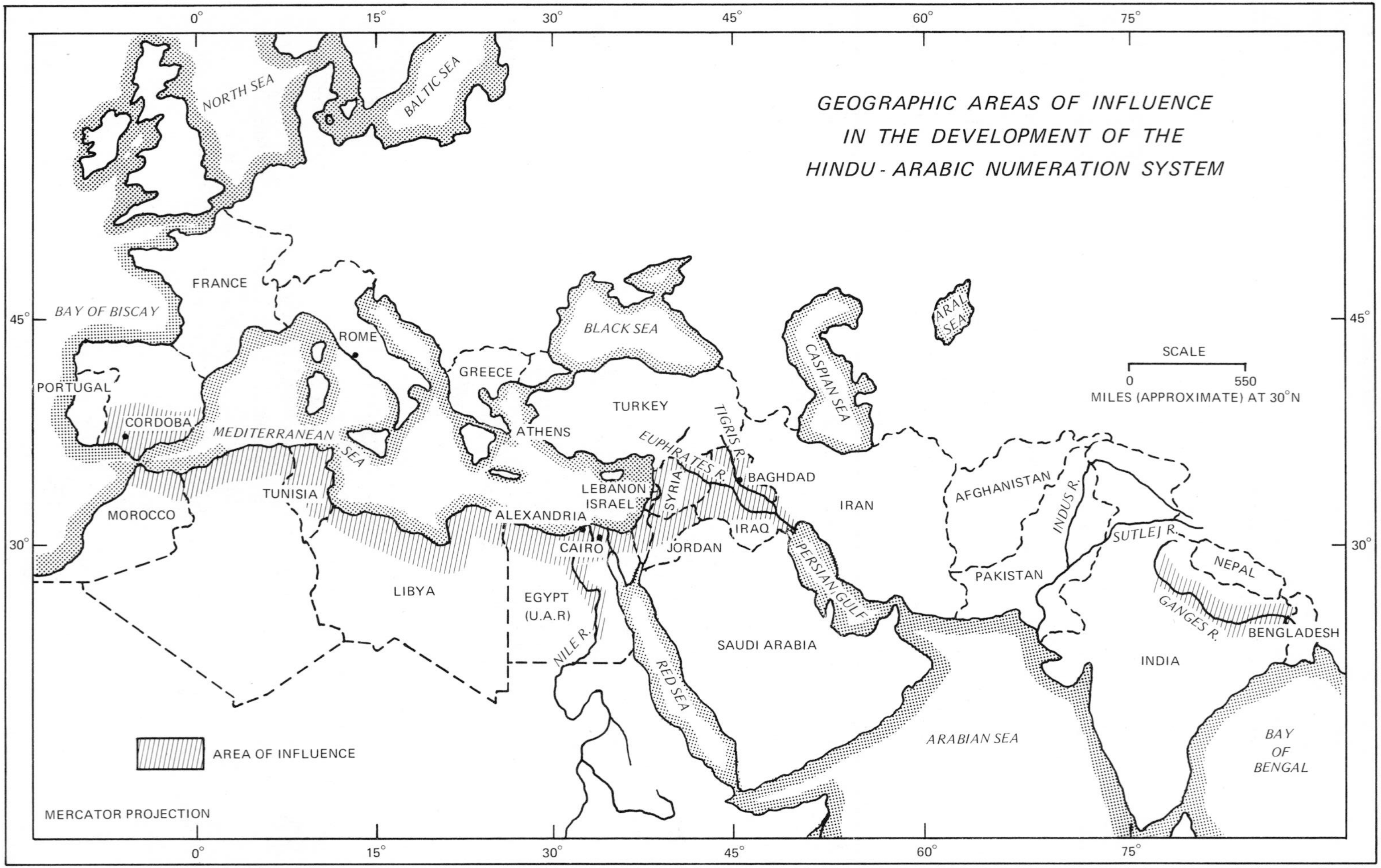
GEOGRAPHIC AREAS OF INFLUENCE
IN THE DEVELOPMENT OF THE
HINDU - ARABIC NUMERATION SYSTEM
SCALE
0
550
MILES (APPROXIMATE) AT 30°N
NORTH SEA
BALTIC SEA
BAY OF BISCAY
FRANCE
PORTUGAL
CORDOBA
ROME
MEDITERRANEAN SEA
GREECE
ATHENS
BLACK SEA
TURKEY
CASPIAN SEA
ARAL SEA
TIGRIS R.
EUPHRATES R.
BAGHDAD
SYRIA
LEBANON
ISRAEL
ALEXANDRIA
CAIRO
JORDAN
IRAQ
IRAN
AFGHANISTAN
INDUS R.
SUTLEJ R.
NEPAL
PAKISTAN
GANGES R.
BENGLADESH
INDIA
MOROCCO
TUNISIA
LIBYA
EGYPT
(U.A.R)
NILE R.
RED SEA
SAUDI ARABIA
PERSIAN GULF
ARABIAN SEA
BAY OF BENGAL
AREA OF INFLUENCE
MERCATOR PROJECTION
0°
15°
30°
45°
60°
75°

A more detailed study of the characteristics of this system will be made later in this chapter.

But Europe did not adopt this new system without a struggle. Rome had exerted a great deal of influence for over 2000 years—in commerce, in science, and in theology. The Roman numerals were easy to write. Most people needed to learn only I, V, X, L, and C. It was easier to see III than 3 and VIII than 8. Addition, the most used of all operations, was easier to perform with these symbols than with the new "heathen" numerals. The battle between the "abacists," who believed in the status quo, and the "algorists," who wished to adopt the new system, was a bitter one and lasted for several hundred years. It seems incredible that a fully developed positional system of notation has been in general use only about 400 years.

The next question might well be: "Is this the end?" The electronic computer, with its binary system using only two digits, may very well be leading man into a new system of counting. It remains to be seen what use and necessity will demand and what study and imagination will produce.

EXERCISE 3.5

1. Does 𓍢∩∩ represent the same number as ∩∩𓍢? Do 𒐕𒐕𒌋, 𒌋𒐕𒐕, and 𒐕𒌋𒐕 represent the same number? Do XXI and XIX? Explain your answers.
2. Write 375 and 573:
   (a) In Egyptian hieroglyphics
   (b) In Babylonian cuneiform symbols
   (c) In Roman numerals
3. Do the same as in Problem 2 for the number 307.
4. How many different symbols must one use to write numbers less than 1000 using the notation of Egypt, of Babylonia, and of Rome?

3.6 NUMBER AND NUMERAL

Earlier in this chapter you learned that if you had lived along the Nile River around 3000 B.C., you would have given the answer to 2 + 2 by making these strokes, IIII, either on papyrus or by chipping them out in stone. But, had you grown up as a Babylonian, you would have used your stylus to imprint in the soft clay the answer to 2 + 2 quite differently. It would probably have looked like this 𒐘. You, as a young Roman, would have given the answer as IV. Here, in our country, a person would write 4. Each of these shows a different way of representing the number four.

Symbols used to represent numbers, whether they are pictures on the wall of an Egyptian temple, indentations on a clay tablet from an ancient Babylonian city, or a New York stock report in the morning's paper, are *numerals*. Note that such symbols as

IV, 4, 𒐘, and IIII

are *not numbers;* they are numerals which *represent* a number. The notion of number was developed in Chapter 2 as a common property of equivalent sets. Thus, number is essentially an abstract idea. One *thinks a number* but *writes a numeral.* A number continues to exist, even though its numeral is erased.

A number may be represented by many numerals. Often, numerals are referred to as *names for* a number. We speak of a number as having many names. That is,

$$\left\{\text{IV}, 4, 2 + 2, \text{IIII}, \begin{smallmatrix}\nabla\nabla\nabla\\ \nabla\end{smallmatrix}\right\}$$

is a set of equivalent representations or names for the same number.

Also, word names may be given for the number property of this set. An English-speaking person names this number as *four.* A person from Spain would call it *cuatro,* a Frenchman would say *quatre,* while the Swahili word for four is *ine.*

3.7 EARLY RECORD KEEPING

Through the years, man has searched for more and more efficient ways to record numbers, as was pointed out earlier in this chapter. You studied the numeration systems used by the Egyptians, the Babylonians, the Romans, and the Hindus. You will recall that these systems have some characteristics in common. Each has a set of symbols and a set of rules governing the manner in which these symbols are to be used to record numbers. The symbols and the rules are the distinguishing features of the systems.

Of course, these numeration systems were not invented over a short period of time. They were the results of centuries of development. There was a time when man probably had little need for a systematic method of keeping a record of his few possessions. It was sufficient for him to drop a pebble in his pouch, cut a notch in a stick, or make a mark in clay for each animal that he owned. Establishing a one-to-one correspondence between his possessions and pebbles, notches, or marks was a simple, but adequate, method of record keeping. In case it was necessary to tell his neighbor how many animals he owned, he merely had to lift his pouch of pebbles, hold up his notched stick, or point to the marks in the clay, and say, "I have this many animals."

However, as life became more complicated for man, this simple method of record keeping did not suffice. As his few animals increased to large herds, his bag of pebbles became too heavy to carry around, his stick had to be too long to hold all of the notches, and the marks in the clay became too many. It is quite likely that man began to use a large pebble in place of a certain number of the smaller pebbles and a deeper cut notch instead of several of the smaller ones. The greatest difficulty inherent in this method was that he always needed to remember how many pebbles were represented by the large pebble or how many notches were represented by the large notch. Too, without an agreement with his neighbors as to how many notches the large notch would represent, man would probably encounter difficulty in communicating with his friends when bragging about how many animals he had killed in the hunt. So the people of a locality needed to make such an agreement.

The invention of symbols for numbers and names for the symbols is quite a step forward from holding up a bag of pebbles. You have seen that the number of symbols invented by various peoples of early cultures has varied from two symbols of the Babylonians to the increasing number of the Egyptians.

The study of primitive cultures brings to light the universality of finger counting. Because it was so practical and convenient it influenced man in his search for efficient schemes for recording number. The use of ten symbols (frequently called digits) in the decimal system, no doubt, was influenced by man having ten fingers. Twelve-fingered men probably would have invented a duodecimal system. With a limited number of symbols, the idea of place value developed quiet naturally.

From the experience of the Babylonians a place value system without a symbol for zero produced ambiguities that greatly weakened the system. Consequently, the zero symbol is accepted as an essential part of a place value numeration system.

3.8 THE FIST NUMERATION SYSTEM

Since the fingers of man served as early counters, do you think that a numeration system could be invented by using the fingers of only one hand? Counting fingers on one hand is illustrated in the following manner:

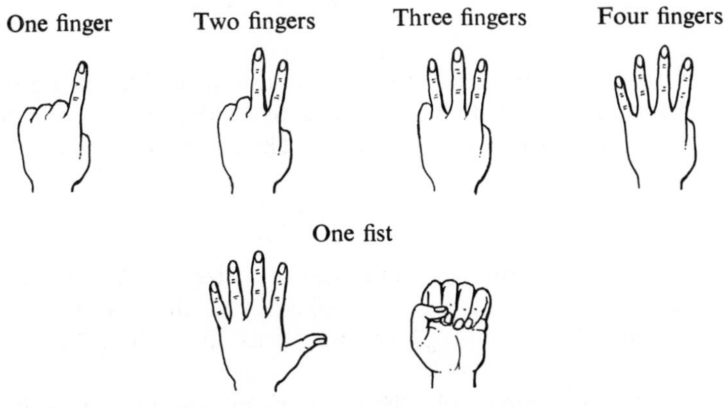

For convenience, the conventional symbols 0, 1, 2, 3, and 4 will be used. The "follows" relation of the symbols has been established from the order relation of the numbers they represent.

Look at the following tables, and fill in the omitted part in each.

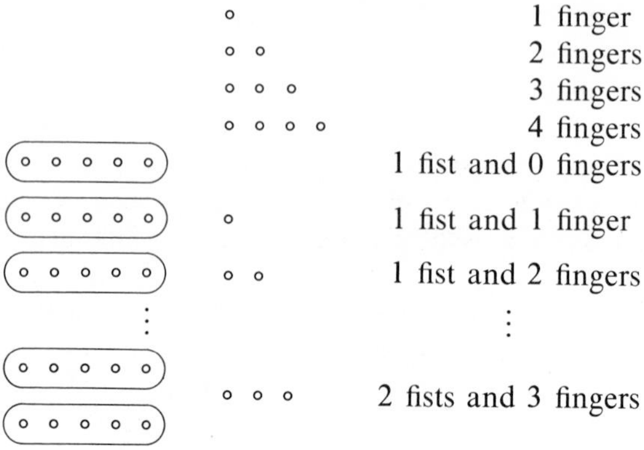

Fist	*Finger*
	1
	2
	3
	4
1	0
1	1
1	2
⋮	⋮
4	4

A numeral such as 11 without the use of the identifying words, fist and finger, or without its being related to a table such as the preceding one, could have various interpretations. The meaning is made clear by writing the numeral as

$$11_{\text{fist}}$$

where the subscript "fist" indicates the "size" of the set being used as the basic subset. This is usually spoken of as the *base* used in writing the numeral. To read "11_{fist}" we say, "1 fist and 1 finger" or "one, one, base fist."

In the preceding table, the last numeral was 44_{fist}. This illustrates 4 fists and 4 fingers of elements.

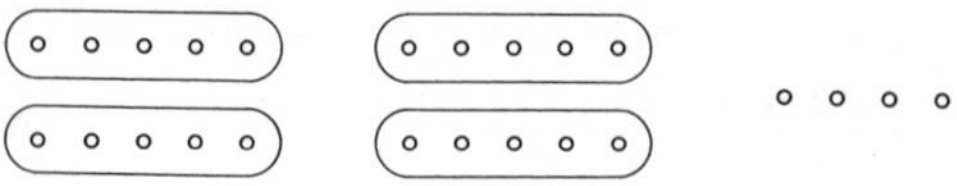

What numeral could we use if one more element is joined to the above? With the joining of this element, another fist is formed.

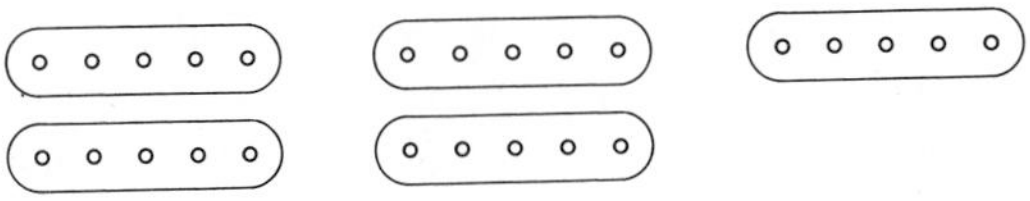

It seems natural that we use these to make a larger group. This is now a "fist of fists."

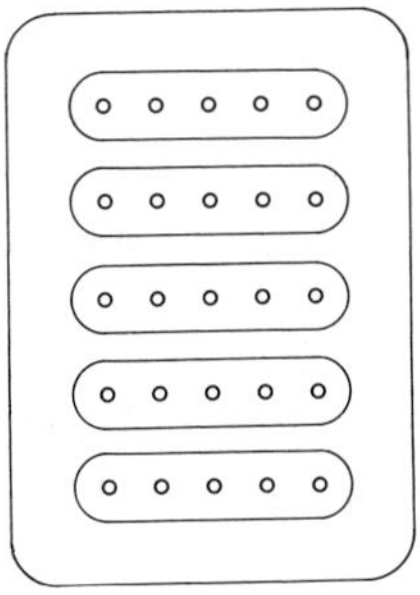

There are no fists and no fingers left after forming this group, and so the numeral is

1 fist of fists, 0 fist, 0 finger

or

$$100_{\text{fist}}.$$

The headings for our table must be extended as shown. Make a table like this and fill in the omitted parts.

Fist of Fists	*Fist*	*Finger*
1	0	0
1	0	1
1	0	2
⋮ 1	⋮ 0	⋮ 4
1	1	0
⋮ 1	⋮ 4	⋮ 4
⋮	⋮	⋮
4	4	4

We may use other names for the headings in the preceding table. The name "five" might well be used instead of "fist." Then, a "fist of fists" would be changed to a "five of fives" or just "five-fives." Instead of fingers, we would use "ones." One might use other names, such as pennies, nickels, and quarters.

Word names for the headings become cumbersome when it is necessary to extend the table. What name would you give to the place immediately to the left of a fist of fists? Since symbols are more convenient to use, we shall use 5 to represent a fist and 5 to represent five. Five-fives will then be represented by 5×5. The next place to the left will be $5 \times 5 \times 5$. With each succeeding position to the left, another factor 5 is used. This can become difficult to manage. So, we shall make use of still other symbols called *exponents.* When 5 is used as a factor three times, the product is written 5^3, and read, "five to the third *power* or five to the power three." (Compare the explanation in Section 3.2.) When 5 is used as a factor twice, the product is written 5^2. It is often read "five squared" rather than "five to the second power." How would you use an exponent to write 5?

We shall use 5^0 as another name for 1. Observing the descending nature of the exponents of the fives in the table as one looks from left to right, it seems quite natural that this symbol be used for 1. You may remember from your study of high school algebra the properties which exponents obeyed under multiplication and division. If so, you should be able to justify the statement that $5^0 = 1$. This does not constitute a proof. In mathematics we say that 5^0 is *defined* as being equal to 1. In fact, any whole number except 0 to the power 0 is 1.

The following table with alternative headings is to be used as an aid in reading the paragraphs below.

Alternative headings	5^2	5^1	5^0
	5×5	5	1
	Five-Fives	*Five*	*One*
	Fist of Fists	*Fist*	*Finger*
Examples: (a)		3	3
(b)		4	2
(c)	1	3	4
(d)	3	0	1

Instead of speaking of the numerals appearing in the table as "base fist" numerals, we shall now say they are numerals written in *base five.* The numeral for (a) in the table is written, 33_{five}; that for (b), 42_{five}; etc. In reading 33_{five}, we may say, "three groups of five and three ones," or we may say, "three, three, base five." In 33_{five}, the left 3 has 5 times the value of the right 3. Likewise, each 3 to the left in 3333_{five} has 5 times the value of the 3 immediately to its right.

Sometimes for the purposes of a discussion it is desirable to write a numeral in *expanded notation*—that is, the complete description of the number represented:

$$\begin{aligned} 33_{\text{five}} &= (3 \times 5^1) + (3 \times 5^0) \\ &= (3 \times 5) + (3 \times 1) \\ 301_{\text{five}} &= (3 \times 5^2) + (0 \times 5^1) + (1 \times 5^0) \\ &= (3 \times 25) + (0 \times 5) + (1 \times 1) \end{aligned}$$

Write 42_{five} and 134_{five} in expanded notation.

EXERCISE 3.8

1. Write the following in expanded notation using powers of five:
   (a) 31_{five} (b) 203_{five} (c) 1230_{five} (d)40134_{five}
2. Illustrate how this set of marks should be grouped to write the numeral in base five:

   o o o o o o o o o o o o o o o o
   o o o o o o o o o o o o o o o o o o

   Write the numeral.
3. The numeral that follows 24_{five} is 30_{five}. Write the numeral that follows each of these numerals.
   (a) 4_{five} (b) 314_{five} (c) 344_{five} (d) 4444_{five}
4. Suppose that you have three quarters, eight nickels, and six pennies in your pocket; use base five to write a numeral that will indicate your financial status.
5. Does 34_{five} represent the same number as 43_{five}? Explain.
6. Which of the following represent odd and which represent even numbers? (Can you justify your selections?)
   (a) 111_{five} (b) 11_{five} (c) 101_{five} (d) 1110_{five}
7. Count the people enrolled in your class, and write the result as a base five numeral.

3.9 THE DOZEN NUMERATION SYSTEM

Many articles are purchased in lots of a dozen, a gross, or a great gross. We speak of a dozen of eggs and a gross of pencils. This suggests grouping by dozens. For example, a basket of eggs may be arranged in groups of dozens by placing them in the standard egg cartons:

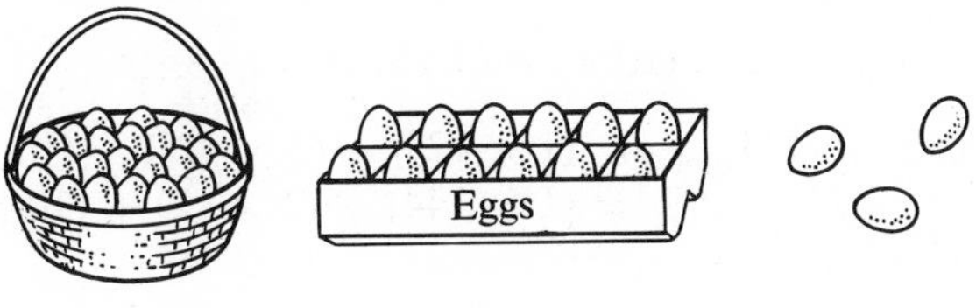

The diagram shows 1 dozen and three eggs. Since the size of the group we are using is a dozen (which may be given the name "twelve"), the numeral may be written 13_{dozen} or 13_{twelve}. This numeral is written in *base twelve.*

In the fist numeration system (base five) we used five symbols (digits). Hence, one would expect to use twelve digits in the dozen (base twelve) numeration system. If we so choose, we may use the standard digits 0, 1, 2, 3, 4, 5, 6, 7, 8, 9, but this set is two digits short of the number required. So we are forced to invent two digits. The shape of these digits may take any form. Suppose that we agree on T and E. Now, the set of digits which we shall use is

$$\{0, 1, 2, 3, 4, 5, 6, 7, 8, 9, T, E\}.$$

This sequence of the digits follows the order of the numbers where "T" stands for ten and "E" stands for eleven.

The positional value of the digits in a base dozen (twelve) numeral is shown by the following table:

12^3	12^2	12^1	12^0
$12 \times (12 \times 12)$	12×12	12	1
Twelve-(Twelve-Twelves)	*Twelve-Twelves*	*Twelve*	*One*
Dozen of (Dozen of Dozens)	*Dozen of Dozens*	*Dozen*	*Single*
	1	2	T

The numeral $12T_{twelve}$ in the table represents how many elements? Does the grouping of eggs in the set below tell us that we have 1 dozen of dozens, 2 dozens, and T singles?

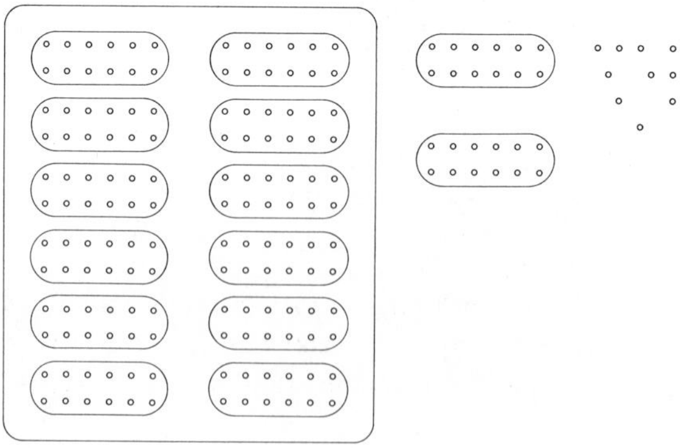

Using expanded notation, we have

$$\begin{aligned} 12T_{twelve} &= (1 \times 12^2) + (2 \times 12^1) + (10 \times 12^0) \\ &= (1 \times 144) + (2 \times 12) + (10 \times 1). \end{aligned}$$

Although we have been calling this sytem based on twelve the dozen numeration system, it is more often called the *duodecimal numeration system.* About the middle of the eighteenth century this system was suggested by a Frenchman. Many people today believe that our present system of notation should be replaced with the universal adoption of a system based on twelve. This movement has gained sufficient strength that a very active society exists today which advocates this change. It is called the Duodecimal Society of America.

EXERCISE 3.9

1. How many digits are used in a positional numeration system based on twelve? Why are 10 and 11 not used for the last two digits?
2. What is the next larger power of twelve following 12^3? Is there a greatest power of 12?
3. A stock room boy records the number of pencils received as 9 great gross, 15 gross, and 18. How should he record this amount in base twelve? (A "great gross" is twelve gross.)
4. Express the following with a base twelve numeral:
   (a) $(8 \times 12^3) + (T \times 12^2) + (0 \times 12^1) + (3 \times 12^0)$
   (b) $(\mathrm{E} \times 12^5) + (9 \times 12^3)$
5. Count the people enrolled in your class and write the result as a base twelve numeral. Is the numeral for the number of students in class the same as when expressed in base five? (Refer to Problem 7 of Exercise 3.8.) Is the number the same in each case?

3.10 THE COMPUTER NUMERATION SYSTEM

By now, you are no doubt convinced that a numeration system may be invented around almost any number base. Yet, would you dare suggest one having a base of two? That is just exactly what a German mathematician, Gottfried Leibniz, did. Now, some three hundred years later, this system has proven its real worth with the advent of the electronic computer, as the use of two symbols is particularly adaptable to the "off" and "on" of the switch of an electric circuit. We shall use the symbols 0 and 1, respectively, for the names of these two conditions. The numerals we write using these two symbols are called *binary numerals,* and the system is called the *binary numeration system.*

In this system using only two symbols to record the number of elements in a set, how many elements will form the basic subset? In preparation for writing the numeral for the number of members in each of the following collections, check to see whether the grouping is correct as shown.

(a) 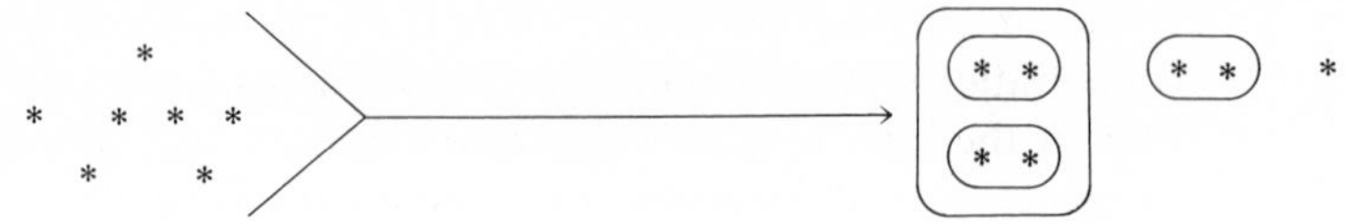

(b) 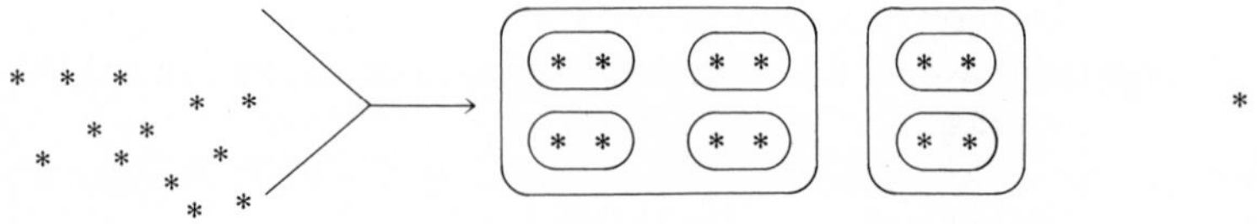

Show how the members of the following collection should be grouped into subsets preparatory to writing the numeral in base two.

(c) 

In writing the numerals for the number of elements in each of the collections (a), (b), (c), positional value is used as in the other systems. The significance of the value of a digit as it is related to its position in the numeral is shown by the following table.

	2^4	2^3	2^2	2^1	2^0
	$2 \times 2 \times 2 \times 2$	$2 \times 2 \times 2$	2×2	2	1
(a)			1	1	1
(b)		1	1	0	1
(c)	1	0	1	1	0

In the numeral for (a), 111_{two}, each 1 has how many times the value of the 1 which follows it? Each position represents a power of two. Using expanded notation, we have:

(a) $111_{\text{two}} = (1 \times 2^2) + (1 \times 2^1) + (1 \times 2^0)$

(b) $1101_{\text{two}} = (1 \times 2^3) + (1 \times 2^2) + (0 \times 2^1) + (1 \times 2^0)$

(c) $10110_{\text{two}} = (1 \times 2^4) + (0 \times 2^3) + (1 \times 2^2) + (1 \times 2^1) + (0 \times 2^0)$

EXERCISE 3.10

1. Count the number of people enrolled in your class, using base two. Compare this numeral with those when base twelve and base five were used.
2. If "off" is even and "on" is odd, starting with an "off" position, how many "flips of the switch" will always produce an even? How many will produce an odd? Using base two, write the numerals to record the counting of these flips. Can

you see a pattern in the numeral that helps in identifying the "even-ness" or the "odd-ness" of the number which they represent?

3. Write the following in expanded notation using powers of two:

   (a) 11_{two} (b) 101_{two} (c) 1000_{two} (d) 1100_{two}

4. Which of the numerals in the preceding problem represent an odd number?
5. Counting in base two, what numeral follows each of these?

   (a) 1_{two} (b) 11_{two} (c) 101_{two} (d) 1101_{two}

6. Give a base two name for the number that is one less than each of these.

   (a) 101_{two} (b) 10_{two} (c) 100_{two} (d) 1010_{two}

3.11 THE DECIMAL NUMERATION SYSTEM

The numeration system in most common use today is one having ten digits. The selection of ten digits was, no doubt, due to man's having ten fingers (digits). Any other number of fingers would probably have determined a different number of digits. The historical development of this, the *Hindu-Arabic system,* was traced earlier in this chapter.

Since ten basic symbols (digits) are used, it is called the *decimal numeration system,* from the Latin word *decem,* which means ten. Your study of numeration systems using a base of five, twelve, or two should help you to have a better understanding of this base ten system. The positional value principle used in writing the numerals is based on powers of ten. The ten symbols that we use are 0, 1, 2, 3, 4, 5, 6, 7, 8, 9. The names assigned to these symbols are zero, one, two, etc.

Before writing a decimal numeral for the number of elements in a set, we arrange the elements into as many subsets of ten elements as possible. If as many as ten of these subsets are formed, then they are in turn grouped into a subset consisting of ten tens. This process of making subsets out of subsets will be extended depending on the number of elements in the set. The formation of subsets of ten is shown here.

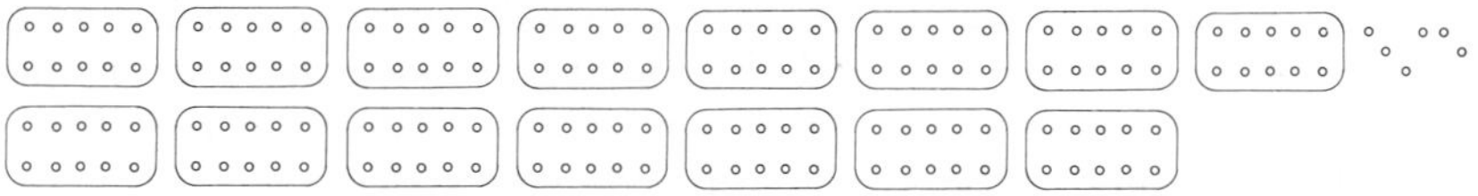

The formation of ten-tens is shown here.

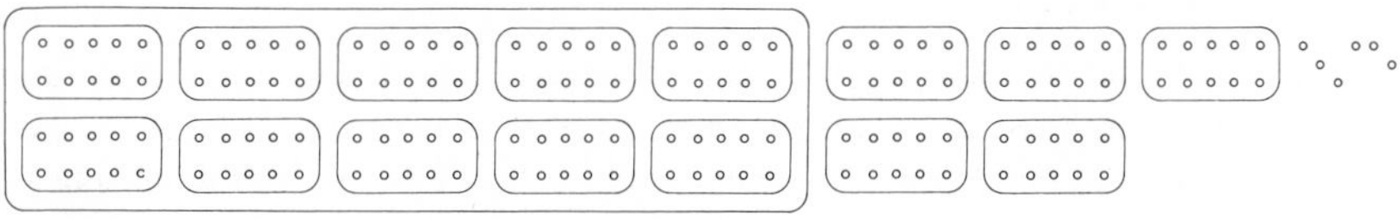

This plan of formation of the subsets is not new. It is similar to those used with the other numeration systems you have studied in this chapter. As with the other systems, the positional values are shown by the table. The numeral (a) given in the table is the name for the number of elements in the set shown below.

	Ten Thousand	*Thousand*	*Hundred*	*Ten*	*One*
	10,000	1000	100	10	1
	10^4	10^3	10^2	10^1	10^0
	$10 \times 10 \times 10 \times 10$	$10 \times 10 \times 10$	10×10	10	1
(a)			1	5	6

Take note of the word names that are given in the top row of the table. These are used when a word name for a decimal numeral is to be given. We generally write 156 without designating the base unless the situation is such that it becomes necessary for clarification. Using the expanded notation, 156 may be written

$$(1 \times 10^2) + (5 \times 10^1) + (6 \times 10^0)$$

or

$$(1 \times 100) + (5 \times 10) + (6 \times 1).$$

The last line leads to the name

one hundred, five tens, six ones,

which is read "one hundred fifty-six."

Some people believe that the Hindu-Arabic system is the most efficient of all numeration systems, and that it is the only one which will ever be needed. Now that you have studied other systems, do you agree with this? The breathtaking scientific and technological changes occurring in our world today make one wonder about the future adequacy of our own numeration system. Just as our system is far better than the Egyptian, the Babylonian, the Roman, and others that we might cite, someday you or someone else may invent a numeration system which will meet the demands of another era far more effectively than our decimal system.

EXERCISE 3.11

1. Count the number of people enrolled in your class using decimal numerals.
2. Write in base five, in base twelve, in base two, and in base ten the numerals for the number of people enrolled in class. Express each in expanded notation.
3. Express in expanded notation the following:
   (a) 578 (c) 671,248
   (b) 93,827 (d) 4060

4. What is the value of the 7 in each of these decimal numerals?
   (a) 807 (c) 378
   (b) 27,500 (d) 25,763
5. Illustrate the meaning of 123_{ten} by grouping the elements of a set into the necessary subsets.
6. What is the largest number that can be expressed in the decimal system with five digits?
7. Write all the decimal numerals possible using only the digits 2, 1, and 4 once and only once. Which numeral represents the greatest counting number? Which represents the least?

3.12 CONVERSION OF NUMERALS TO DIFFERENT BASES

If we count the number of students enrolled in a class, each time recording the number with a different base, each numeral will be different, but the number will be the same. We shall now consider methods of changing from a numeral in one base to a numeral in another base.

Here we shall indicate the base of a numeral only in those situations where there might be ambiguity if it were omitted. As is customary, the base of a decimal numeral is not indicated; that is, thirty-five is written as 35, not 35_{ten}. Since 1_{five}, 2_{five}, 3_{five}, and 4_{five} represent the same numbers as 1, 2, 3, and 4, respectively, the word for the base may be omitted in each case. Does this apply to other bases? If so, for what numbers in each?

Suppose that a number is represented by 102_{five}. Is it possible to find the decimal name for this number without resorting to the process of grouping by tens? This is accomplished by using expanded notation (Section 3.8):

$$\begin{aligned} 102_{\text{five}} &= (1 \times 5^2) + (0 \times 5^1) + (2 \times 5^0) \\ &= (1 \times 25) + (0 \times 5) + (2 \times 1) \\ &= 25 + 0 + 2 \\ &= 27 \end{aligned}$$

Similarly,

$$\begin{aligned} 11011_{\text{two}} &= (1 \times 2^4) + (1 \times 2^3) + (0 \times 2^2) + (1 \times 2^1) + (1 \times 2^0) \\ &= (1 \times 16) + (1 \times 8) + (0 \times 4) + (1 \times 2) + (1 \times 1) \\ &= 16 + 8 + 0 + 2 + 1 \\ &= 27 \end{aligned}$$

Also, we expect that there should be a way to change a numeral in base ten to one in any other base. For example, let us change 27 to a base two numeral. We shall use the following base two table:

2^5	2^4	2^3	2^2	2^1	2^0
32	16	8	4	2	1

Is it possible to arrange 27 elements into subsets having 32, 16, etc., elements in each? Upon investigation, we see that 27 elements cannot be arranged into a subset of 32, or 2^5, elements. However, it is possible to arrange them into a subset of 16 elements with 11 remaining:

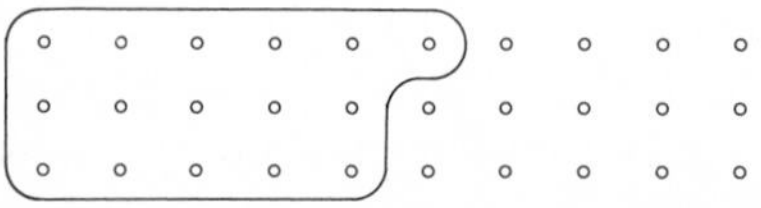

$$16\overline{)27} \quad 16 \quad 1 \quad \overline{11}$$

These 11 elements are next arranged into the next smaller subset, that is, 8 elements, with 3 remaining:

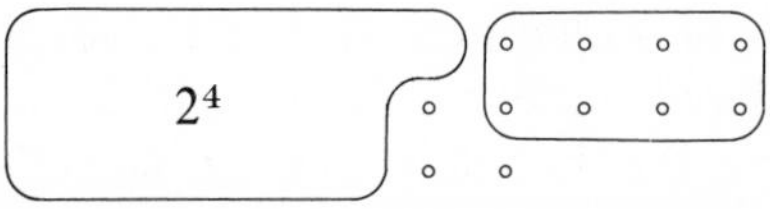

$$8\overline{)11} \quad 8 \quad 1 \quad \overline{3}$$

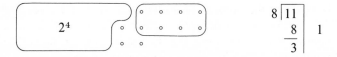

We see that there cannot be a subset of 4 elements and so we look to the next smaller subset. The 3 elements may be arranged into one subset of 2 elements with 1 remaining:

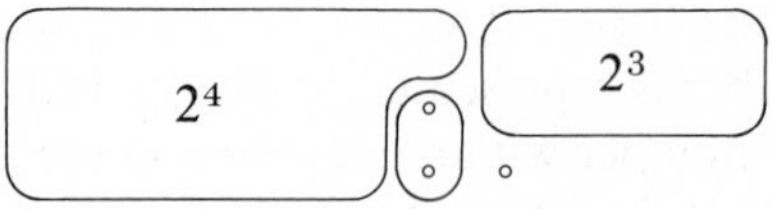

$$2\overline{)3} \quad 2 \quad 1 \quad \overline{1}$$

Following is a summary of the entire procedure:

FORMATION OF SUBSETS

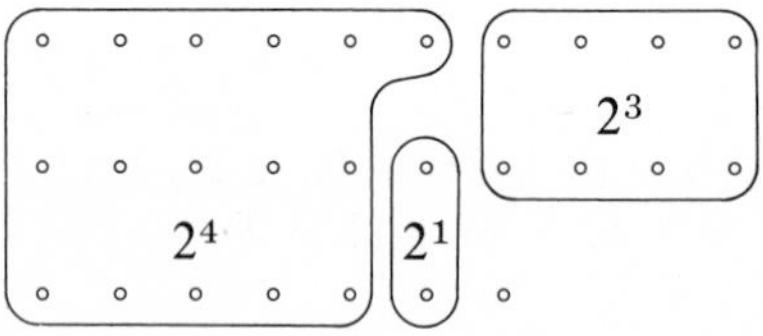

ALGORITHM

$16\overline{)27}$, 16 → 1
$8\overline{)11}$, 8 → 1
$4\overline{)3}$, 0 → 0
$2\overline{)3}$, 2 → 1
1

RECORD OF NUMERAL

2^4	2^3	2^2	2^1	2^0
1	1	0	1	1

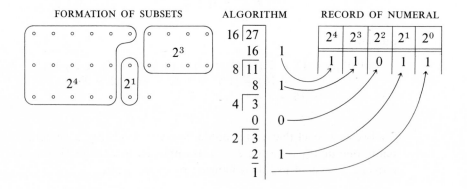

Since the base two numeral in the table and the original numeral in base ten are two different names for the same number, we may write

$$27 = 11011_{\text{two}}.$$

Let us change 27 to a base five numeral.

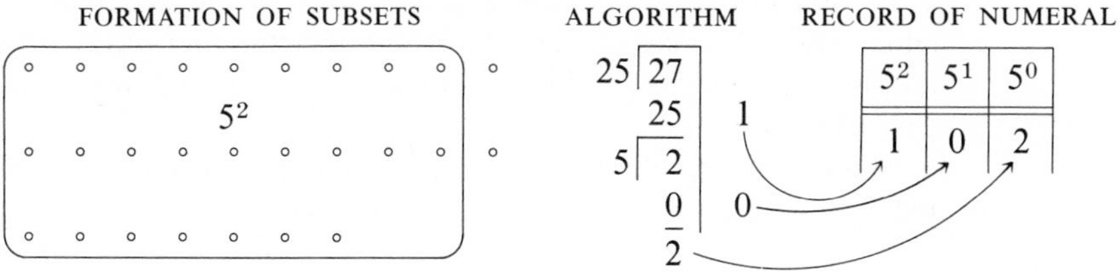

Thus, we may write

$$27 = 102_{\text{five}}.$$

The conversion of 273 to a base twelve numeral is now demonstrated.

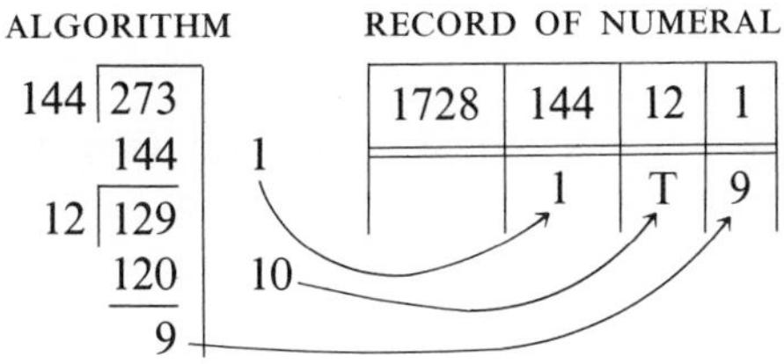

Hence, $273 = 1T9_{\text{twelve}}$.

EXERCISE 3.12

1. Express 125 in the following bases.
   (a) five
   (b) seven
   (c) twelve
   (d) two
   (e) three
   (f) nine
2. Express each of the following sets of numerals in base four.
   (a) 32, 33, 34, 35
   (b) 62, 63, 64, 65, 66
   (c) 258, 256, 254, 252
   (d) 1021, 1025, 1029
3. Complete the following with the correct numerals.
   (a) $1221_{\text{three}} = \underline{\qquad}_{\text{five}}$
   (b) $356_{\text{seven}} = \underline{\qquad}_{\text{four}}$
   (c) $ET80_{\text{twelve}} = \underline{\qquad}_{\text{nine}}$
   (d) $11011_{\text{two}} = \underline{\qquad}_{\text{eight}}$
4. What base a must have been used if $25_a = 21$?
5. The sum of 13 and 5 is 21. What base was used in writing these numerals?
6. Each of the following numerals represent thirteen if the correct base is indicated. Supply the base for each.
   (a) 31 (b) 23 (c) 16 (d) 14
7. Show the algorithm and the recording of the numeral in converting 65 to a base five numeral.

3.13 ARITHMETIC IN DIFFERENT BASES

Even though our ordinary arithmetic is done in the decimal numeration system, a study of some aspects of arithmetic with other bases is interesting, and it is also helpful in developing a deeper insight into our decimal positional numeration system.

We shall now investigate the addition of numbers when they are represented by numerals written in bases other than ten. Let us find

$$23_{\text{five}} + 14_{\text{five}}:$$

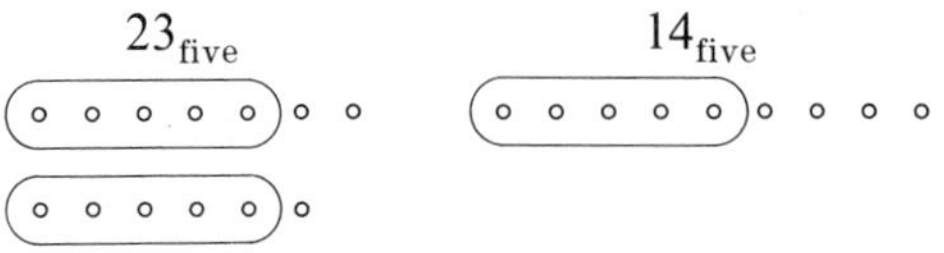

If these two sets are joined, how many elements are there in the new set? Observe that when these sets are joined, a new subset of five elements may be formed:

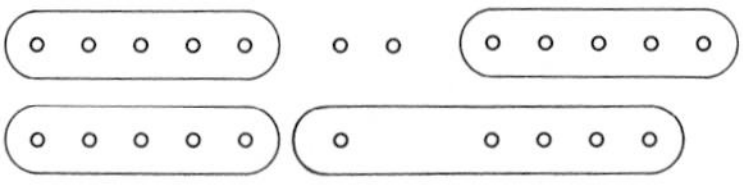

Using this organization of the elements in the new set, we write the numeral as

$$42_{\text{five}}.$$

Thus, we have

$$23_{\text{five}} + 14_{\text{five}} = 42_{\text{five}}.$$

In an addition problem, one should not have to construct the sets each time. We can use the same algorithm as for adding base ten numerals provided we keep in mind the base of the system of numeration we are using and use the "basic addition facts" for that system. Many of these do not have the same numerals as in base ten. For example, $4_{\text{five}} + 3_{\text{five}} = 12_{\text{five}}$, while $4 + 3 = 7$. In $4_{\text{five}} + 3_{\text{five}}$, try to think how much must be added to 4 to make 5. By renaming 3 as $1 + 2$, we have $4 + (1 + 2)$. Now we can think of this as one group of five and two, or 12_{five}.

We can use the addition grid for base five and add by columns as we do in base ten. To find the sum of $23_{\text{five}} + 14_{\text{five}}$, we can write:

+	0	1	2	3	4
0	0	1	2	3	4
1	1	2	3	4	10
2	2	3	4	10	11
3	3	4	10	11	12
4	4	10	11	12	13

fives	ones	
2	3	
1	4	
1	2	$(3 + 4 = 12_{five})$
3	0	$(20_{five} + 10_{five} = 30_{five})$
4	2	

Another way to write the problem is:

$$\begin{aligned} 23_{five} &= 2 \text{ fives } 3 \text{ ones} \\ 14_{five} &= \underline{1 \text{ five } \ 4 \text{ ones}} \\ & 3 \text{ fives } 1 \text{ five } 2 \text{ ones} = 4 \text{ fives } 2 \text{ ones} = 42_{five}. \end{aligned}$$

Do you have still other ways to suggest?

The answer for this problem may be checked by changing to base ten:

$$\begin{aligned} 23_{five} &= (2 \times 5) + (3 \times 1) = 13_{ten} \\ \underline{14_{five}} &= (1 \times 5) + (4 \times 1) = \underline{\ 9_{ten}} \\ 42_{five} &= (4 \times 5) + (2 \times 1) = 22_{ten} \end{aligned}$$

This has been only an introduction to what can be done in arithmetic with other bases. Multiplication, subtraction, and division are all fascinating when done with bases other than ten. Ability to do these operations in other bases greatly enriches one's understanding of what happens when working with base ten.

EXERCISE 3.13

1. Construct an addition grid for base two. How many addition facts must one learn? Use the grid to find the sum of:
   (a) $1001_{two} + 110_{two}$
   (b) $1011_{two} + 111_{two}$
   (c) $1101_{two} + 1101_{two}$
2. Check each answer in the preceding problem by changing the numerals to base ten.
3. Place one of the symbols, <, >, or =, in each blank to make a true sentence.
   (a) 234_{five} ——— 69_{ten}
   (b) 11111_{two} ——— 81_{ten}
   (c) 360_{ten} ——— $2E0_{twelve}$
   (d) 1101_{two} ——— 11_{twelve}
   (e) 471_{eight} ——— 621_{ten}
   (f) 2112_{three} ——— 2112_{ten}
   (g) 3411_{five} ——— 3411_{seven}
   (h) $217T_{twelve}$ ——— 454_{six}
   (i) 9278_{ten} ——— 5476_{eight}
   (j) 11011_{two} ——— 11011_{ten}

4. Invent symbols for a positional numeration system whose base is three, then write the numeral in this base for:

   (a) 3_{four} (d) 27_{ten}
   (b) 3_{ten} (e) 12_{five}
   (c) 9_{ten} (f) 12_{twelve}

5. Write your weight in:
   (a) base two numerals
   (b) base twelve numerals
   (c) base ten numerals
   (d) base five numerals

6. What are the advantages and the disadvantages in having a numeration system based on two rather than ten?

7. Copy and complete the following "counting" table:

Base : *Ten*	*Twelve*	*Eight*	*Seven*	*Five*	*Three*	*Two*
1						
2						
3						
⋮						
50						

   (a) Are there any numerals that are alike except for the base subscript that represent the same number?
   (b) Cite the two-digit numeral that represents the greatest number in each of the bases listed in the table.

8. Which of these represent odd numbers and which even numbers?

   (a) 10_{seven} (f) 111_{four}
   (b) 11_{seven} (g) 10_{two}
   (c) 111_{seven} (h) 11_{two}
   (d) 10_{four} (i) 111_{two}
   (e) 11_{four}

   Are you able to make any generalization concerning the numerals of odd and even numbers?

9. Give the sum of the following in the base of either addend:
   (a) $12_{five} + 21_{three}$
   (b) $352_{seven} + 11101_{two}$
   (c) $444_{five} + 111_{two}$
   (d) $T48_{twelve} + 167_{eight}$

10. Construct a multiplication grid for base five as was done for addition. Use these multiplication facts to find the product of each of the following pairs of numbers.

EXAMPLE $23_{\text{five}} \times 32_{\text{five}}$

$$
\begin{array}{rll}
 & 23_{\text{five}} & \\
 & \underline{32_{\text{five}}} & \\
11 & 2 \times 3 & = 11_{\text{five}} \\
40 & 2 \times (2 \times 5) & = 40_{\text{five}} \\
140 & (3 \times 5) \times 3 & = 140_{\text{five}} \\
\underline{1100} & (3 \times 5) \times (2 \times 5) & = 1100_{\text{five}} \\
1341_{\text{five}} & &
\end{array}
$$

(a) $21_{\text{five}} \times 2_{\text{five}}$

(b) $21_{\text{five}} \times 21_{\text{five}}$

(c) $12_{\text{five}} \times 3_{\text{five}}$

(d) $32_{\text{five}} \times 12_{\text{five}}$

(e) $24_{\text{five}} \times 43_{\text{five}}$

(f) $104_{\text{five}} \times 33_{\text{five}}$

4

Sets of Points

4.1 THE CONCEPT OF A SET OF POINTS

The concept of a set was introduced in Chapter 2. It would be helpful at this time for you to review in that chapter the discussion given concerning the meaning of set, the description of a set, and the operations on sets.

In Section 2.1 a set was described as being any well-defined collection of objects or ideas. The objects or ideas were identified as elements of the set. In developing the concept of a set, elements were often chosen from the physical world, our alphabet, or the abstract elements called numbers. Numbers as elements of sets will have an important role in our study of the systems of the natural numbers, of the whole numbers, of the integers, of the rational numbers, and finally of the real numbers.

In this chapter the sets that will be discussed are those whose elements consist of *points*. A point is an abstraction as is a number, and we can no more see a point than we can see a number. However, we use concrete representations to assist in understanding, and just as a numeral is used to represent a number, a dot will be used to represent a point. A numeral and a dot are both merely marks on this paper. Each is a concrete or physical representation of the abstract concepts of number and of point. Each is a representation of the thing but not the thing itself. A numeral and a dot may be erased, but the number and the point that each represents remain.

Since it is just as conceivable to have a set of points as a set of numbers, we shall consider some of the basic sets of points in the sections that follow. We shall discuss how they are represented and the intersection and union of these sets.

4.2 SOME BASIC SETS OF POINTS AND THEIR REPRESENTATIONS

A set of points may consist of no elements, a single element, or many elements. Obviously, the set of points consisting of no elements is the empty set. A set of points consisting of a single element is represented by using a small dot and naming it by a capital letter, as $\cdot A$. This may be read as "the set consisting of point A" or merely as "the point A." Also the sets represented in these diagrams are sets of points.

In Section 2.8, a number was assigned to each of the sets of equivalent sets. The elements in these sets could be counted (Section 2.9), and the cardinal number indicated the number of elements belonging to these sets. In other words, it identified the common number property of the set of equivalent sets. For example, it is not difficult to establish that there are 15 people in the room, 5 points in a set, or 5000 fruit trees in the orchard. We can say that the grains of sand on Waikiki Beach may be counted, even though it would be difficult to do so.

These are examples of discrete and finite sets. A set is *discrete* if its elements are individually distinct; a set is *finite* if it is possible to assign to it a cardinal number.

On the other hand, some sets have no cardinal number assigned to them. The "counting" here continues and has no end. We might say that there is always one more element in the set. Such sets are said to be *infinite sets.* An example of one kind of infinite and discrete sets is the set of numbers

$$\{1, 2, 3, 4, \ldots\}.$$

Examples of infinite and nondiscrete sets, the kind we shall study in this chapter, are sets of points:

space, plane, and *line.*

While a dot is used to represent a point, no form of representation is used for the infinite set of points called *space.* We think of space as the *universal set of points.*

The set of points called a *plane* is a subset of space. A plane is a particular set of points of space that one intuitively regards as having the property of "flatness" or "smoothness." Plane M is represented in either of these ways:

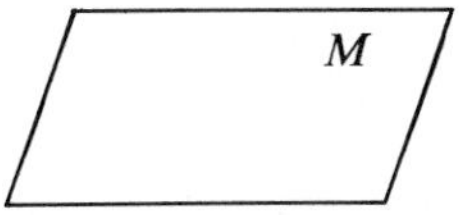

Even though a boundary is used in the pictures, this must not convey the idea that planes are bounded. A plane has no breaks and no bounds and the elements are not counted. It may also be represented in the following manner, which indicates more clearly that it is infinite in extent:

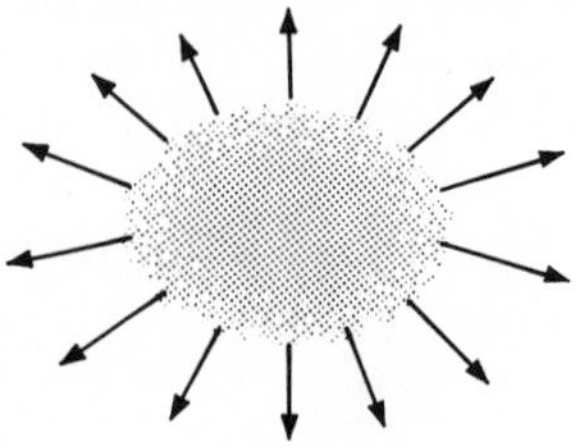

Representations of sets of points are frequently referred to as *geometrical figures.*

A *line* is a particular subset of a plane, which by agreement has the property of "straightness." Because of this agreement, the word "line" will imply a *straight line.* Since a line is a subset of a plane, and a plane is a subset of space, this leads us to make the statement that a line may also be regarded as a subset of space. A line is represented by this geometrical figure:

The arrow at each end implies that the set has no bounds, no ends. In other words, we think of a line as "going on forever." Two points of this set may be identified as shown, and the set may then be referred to as "the line *AB*" or "the line *BA*." To write this, we use the line symbol "↔" over the letters *AB* or *BA*, as

$$\overleftrightarrow{AB} \quad \text{or} \quad \overleftrightarrow{BA}.$$

These are read as "line *AB*" and "line *BA*." Since these are two different names for the same set of points, either name may be used when referring to this line. Only two points are used in giving a name for a line. If C identifies another point on $\overleftrightarrow{AB}$, we would not use $\overleftrightarrow{ABC}$ as a name. Instead, any of these names is acceptable: $\overleftrightarrow{AC}$, $\overleftrightarrow{BC}$, $\overleftrightarrow{AB}$, etc.

It is sometimes more convenient to identify a particular line by using a single lowercase letter on the mark representing the line. Then this line is called "the line *m*."

4.3 ABSTRACTING THE CONCEPT OF A SET OF POINTS FROM PHYSICAL MODELS

The concept of number was an abstraction made from equivalent discrete finite sets. We shall now develop intuitive notions from the physical world concerning the basic sets of points listed in Section 4.2. Relating mathematical concepts such as these to physical phenomena, with due regard for their limitations, frequently helps in understanding the mathematical abstractions when they are first being developed.

The notion of a *point* may be clarified by relating this concept to a position on a map where a dot is used to locate a town. Other examples from the physical world are the tip of a needle, a corner of the room where two walls and the floor converge, and a place on the map where a line of latitude crosses a line of longitude. These and many others relate the mathematical point with the physical model. One must realize, of course, that these "things" are not points, but are only representations selected from the world about us. In fact, there is nothing in the physical world that can be used to represent satisfactorily the abstract idea of point.

We now call on the physical world to assist us in developing the concept of a *line*. This set of points may be related to a string held taut, a straight wire, a crease in a sheet of paper, or the "join" of two walls in the room. Again, these are regarded as representations from which the notion of line may be abstracted. These models from the physical world present certain difficulties, as the physical model necessarily has end points. Now that we have become accustomed to using the physical models as only representations, their use should present no difficulties in our thinking.

Since a set of points called a *plane* has no boundaries, any physical models are inadequate. However, a cardboard, a floor or a wall of a room, a piece of flat plastic are all models that help in developing the notion of flatness or smoothness, a property of the plane.

We repeat that care must always be taken in using physical models for sets of points so that one does not assume that the sets have the physical properties of the models. It is quite easy to think of a plane as having a boundary from the bounded cardboard, and of a line as having a beginning and an end from the crease in a sheet of paper. Never think of the model as the thing itself, but only as a physical representation. Neither is a dot nor a long mark *the* point or *the* line. Each is a representation of a set of points.

At this time, discussion of measure related to the plane and the line is not introduced. Even though the model certainly has "thickness" along with length and width, the concept of measure is postponed until later.

EXERCISE 4.3

1. Make a drawing to represent:

   (a) A point (b) A line (c) A plane

2. Assign capital letters to each of the sets in Problem 1, and tell how each is read.
3. Consider this representation of a line:

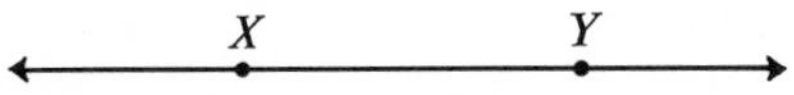

(a) Use the line symbol to denote its name.
(b) If it is possible to denote this same line with another name, give it.

4. Three points have been identified in this line.

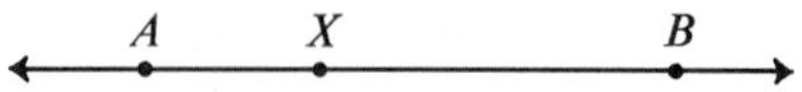

(a) Using the line symbol, give all the names for this line using the points *A*, *X*, *B*.
(b) With only these three points, how many names are possible?
(c) How many names are possible using:
(1) Two points? (2) Four points? (3) Five points?

5. Cite physical models which might be used as representations of:
(a) A point (b) A line (c) A plane

4.4 INTERSECTION OF SETS OF POINTS

Both intersection and union operations on sets were defined in Chapter 2. These operations now will be applied to particular pairs of point-sets, such as two lines, two planes, and a line and a plane. The first operation on sets of points to be considered is the *intersection* operation. The union operation on sets of points will be discussed in the next section.

The intersection operation on two lines may result in (a) the empty set, (b) a set consisting of a single element, or (c) an infinite set. These intersections are represented in Figures 4–1 to 4–4.

The intersection of the two lines represented in Figure 4–1 is the empty set. Lines m and n are subsets of plane M. Since $m \cap n = \varnothing$, these lines are said to be *parallel.*

In Figure 4–2, the intersection of the two lines p and q is also the empty set, $p \cap q = \varnothing$. There is no plane of which both p and q are subsets. In this case we say that p and q are *skew lines.*

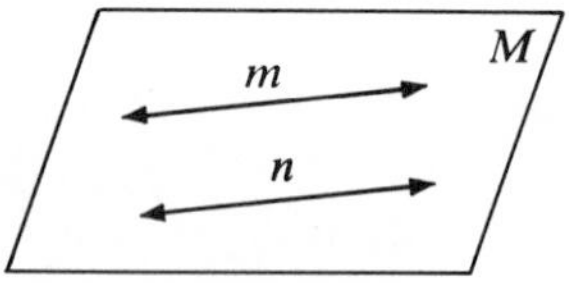

Figure 4–1

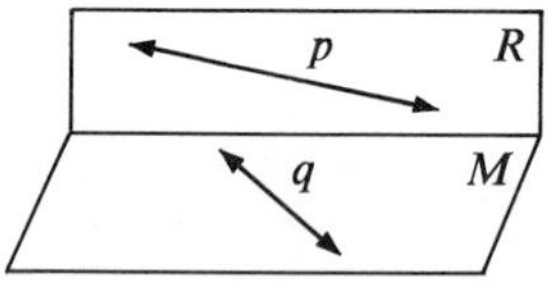

Figure 4–2

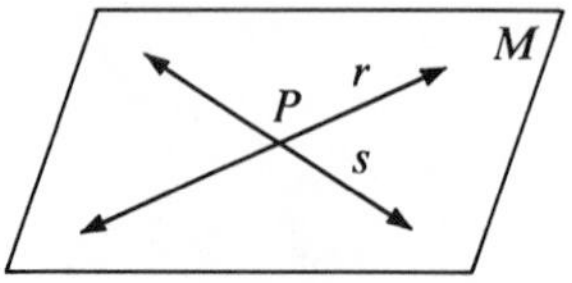

Figure 4–3

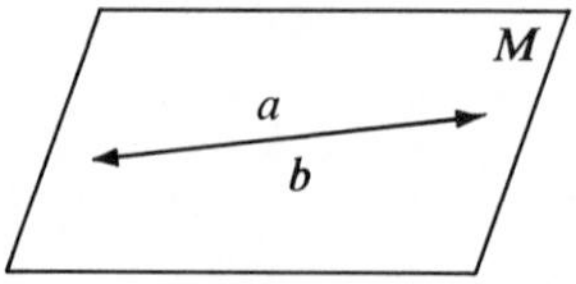

Figure 4–4

In Figure 4–3, we have lines *r* and *s*, subsets of plane *M*. The intersection of these lines is the single point *P*. Here, lines *r* and *s* are two different sets of points. Neither is a subset of the other. That is, $r \not\subset s$ and $s \not\subset r$. We speak of them as *distinct* lines.

In Figure 4–4, we have a representation of the intersection of two *nondistinct* lines, *a* and *b*. Lines that are nondistinct represent the same (identical) set of points. That is, $a \subset b$ and $b \subset a$. Hence, their intersection is that infinite set of points, the line represented in Figure 4–4.

In Figures 4–1 and 4–2, the two distinct lines do not intersect; hence their intersection is the empty set. In Figure 4–3, the two distinct lines do intersect; hence their intersection is a nonempty set. Observe that all pairs of lines either intersect or do not intersect. In each case we may describe their intersection as either nonempty or empty.

What can be said about the intersection of two planes? The intersection may be (a) the empty set or (b) an infinite set. Unlike the intersection of two lines, the intersection of two planes will never be a set having a single element. Why?

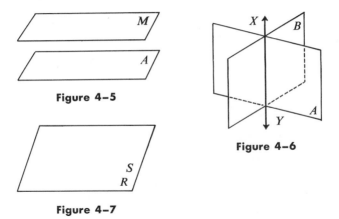

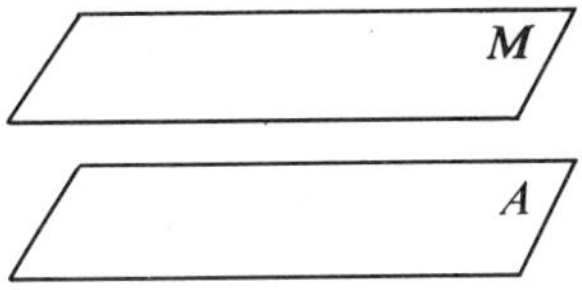

Figure 4–5

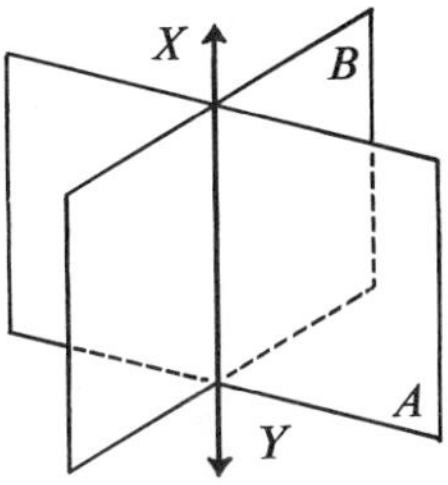

Figure 4–6

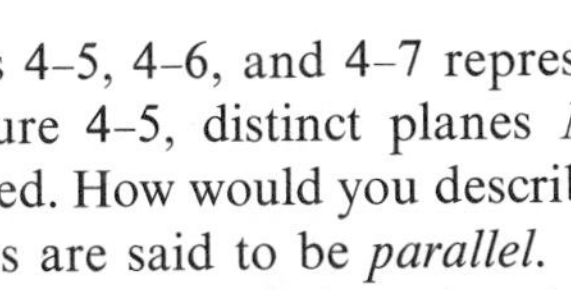

Figure 4–7

Figures 4–5, 4–6, and 4–7 represent the intersection of two planes.

In Figure 4–5, distinct planes *M* and *A*, each a subset of space, are represented. How would you describe their intersection? With $M \cap A = \varnothing$, the planes are said to be *parallel*.

Figures 4–6 and 4–7 show that the intersection of two planes is an *infinite set*. Figure 4–6 represents the intersection of two distinct planes *A* and *B*. Their intersection is the set of points identified as line *XY*. Figure 4–7 represents the intersection of the two nondistinct planes *R* and *S*. Their intersection is the plane that is the same as the nondistinct planes *R* and *S*.

We now consider the intersection of a line and a plane. The intersection of a line and a plane may be (a) the empty set, (b) a set consisting of a single element, or (c) an infinite set.

Figure 4–8

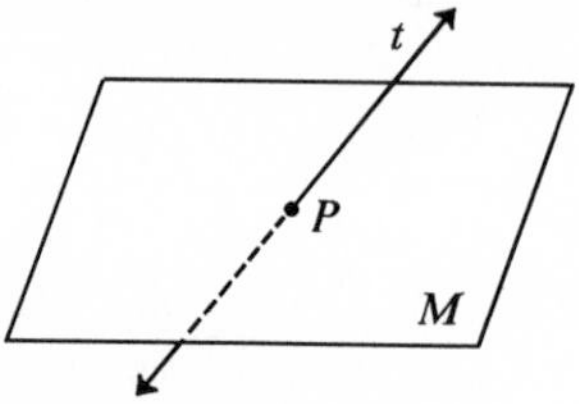

Figure 4–9

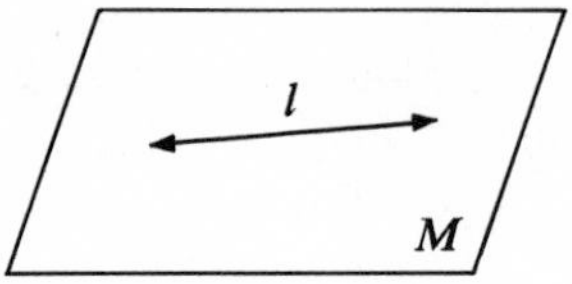

Figure 4–10

These three intersections are represented in Figures 4–8, 4–9, and 4–10.

Figure 4–8 represents the intersection of a line and a plane when the intersection is the empty set. When $t \cap M = \varnothing$, line t is said to be *parallel* to plane M.

Other intersections of a line and a plane are represented in Figures 4–9 and 4–10. The intersection of line t and plane M in Figure 4–9 is the single point P. Line l in Figure 4–10 is a subset of plane M, and hence the intersection is line l, the set of points in common. Since this is a line, this intersection is an infinite set.

EXERCISE 4.4

1. Describe and illustrate the different intersections of:

   (a) Two lines (b) Two planes (c) A line and a plane

2. Describe the intersection of each of the above pairs of sets when only distinct sets are considered.

3. Consider this figure:

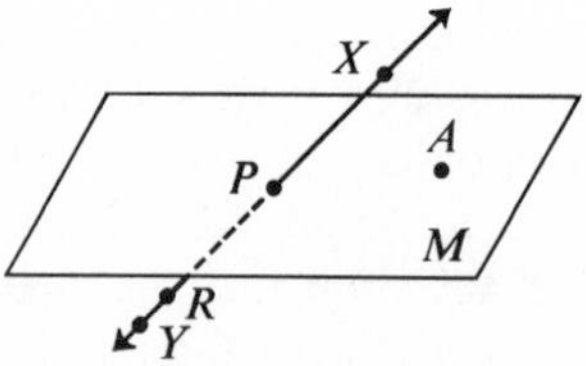

   (a) Is point A an element of plane M?
   Is it an element of $\overleftrightarrow{XY}$?

   (b) Is point R an element of plane M?
   Is it an element of $\overleftrightarrow{XY}$?

   (c) What point is an element of both plane M and $\overleftrightarrow{XY}$? Are there more?

4. Consider this representation of two planes:

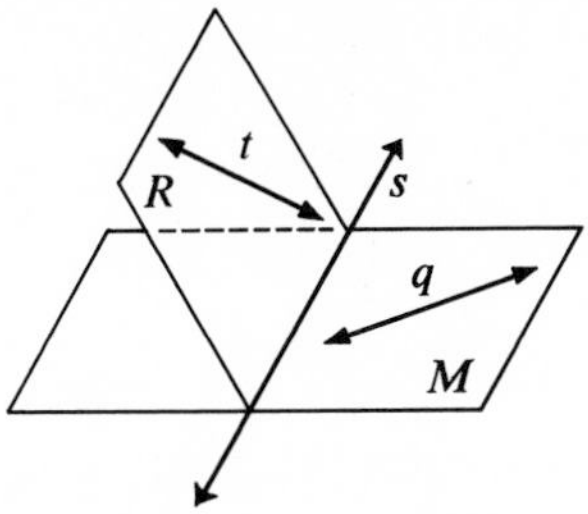

(a) Is line t a subset of plane R?
Is it a subset of plane M?
(b) Is line q a subset of plane R?
Is it a subset of plane M?
(c) What line is a subset of both plane R and plane M?
(d) Are there more lines that are contained in both planes R and M not represented in the drawing? If so, make a drawing to illustrate.
(e) Are there more planes that contain line s? Illustrate with a drawing.

5. Describe the intersection of:
(a) A plane and space (b) A line and space

6. Is it correct to say, "Not all planes intersect, but all planes have an intersection"? Explain.

7. Locate three points X, Y, and Z on your paper such that they are not elements of the same line.
(a) Make a drawing to represent $\overleftrightarrow{XY}$, $\overleftrightarrow{XZ}$, $\overleftrightarrow{YZ}$.
(b) What is $\overleftrightarrow{XY} \cap \overleftrightarrow{YZ}$? $\overleftrightarrow{XY} \cap \overleftrightarrow{XZ}$? $\overleftrightarrow{XZ} \cap \overleftrightarrow{YZ}$?

4.5 BETWEENNESS AND SEPARATION; UNION OF SETS OF POINTS

Suppose that you want to trace a path along m from A to B in Figure 4–11. Would you go through point P? Do you think of point P being *between* points A and B? Now follow path r in Figure 4–12 in going from point A

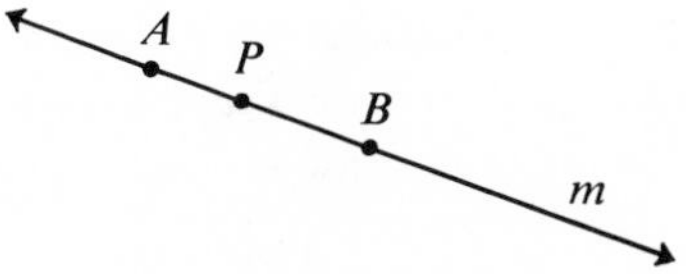

Figure 4–11

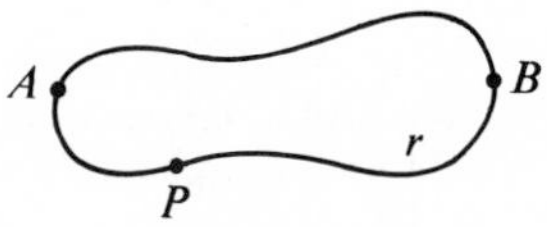

Figure 4–12

to point B. Did you go through point P? Is it possible to take a route along r that would not take you through P? Along which path, m or r, is it easier to determine whether point P is between the two points A and B? Obviously m, a (straight) line, is the simpler to use. When the points A, B, and P are elements of a (straight) line, one intuitively believes that one of the points must be between the other two. Therefore, we shall limit our discussion of betweenness of points to those contained in a line.

Example 1 Consider the drawing:

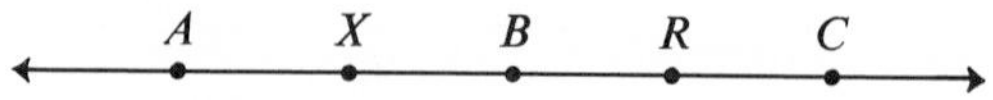

Which point is between points X and R? From the preceding agreement, point B is between points X and R, whereas points A and C are not.

Suppose that we consider line m with the point P between the points A and B:

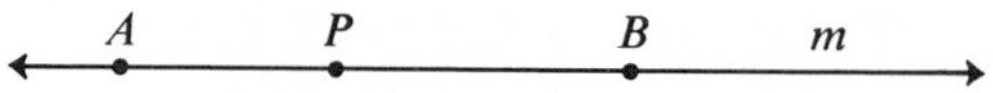

Point P *separates* line m into two *half-lines,* one containing point A and the other containing point B. It is convenient to refer to the half-line that contains point A as the set of points on the A-side of point P. Can you identify the B-side of point P?

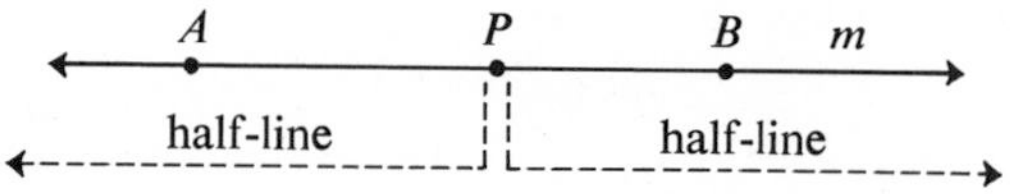

Neither of the half-lines contains point P. Point P separates the line into two half-lines and may be thought of as their boundary, but it is not an element of either. In other words, the intersection of the two half-lines is the empty set, since there is no element common to both half-lines.

Recall that the *union* of two sets was defined to be the set consisting of all elements belonging to either set or to both. This definition can be extended to three or more sets. We now observe that the union of the two half-lines and the point P is the line m.

The set of points consisting of the union of point P and a half-line is a *ray*. That is, a ray is a half-line together with the separation point, now called the *end point* of the ray. The ray represented in the drawing is a subset of line m, and the half-line on the B-side of P is a subset of the ray:

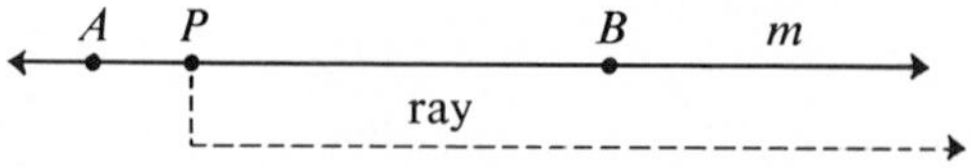

The ray with point P as the end point and point B an element of a half-line is called "ray *PB*" and is denoted by $\overrightarrow{PB}$. In identifying a ray, the letter associated with the end point is given first. That is, $\overrightarrow{BP}$ does not name the same set of points as does $\overrightarrow{PB}$, as shown in the following drawing.

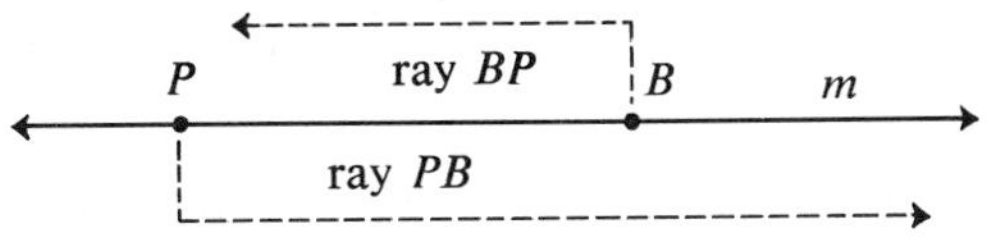

Example 2 Consider the drawing:

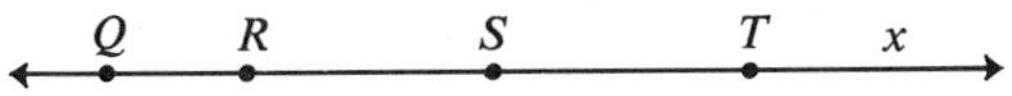

(a) List the rays which may be identified with these letters.
Some of these are $\overrightarrow{RQ}$, $\overrightarrow{QR}$, $\overrightarrow{RS}$, $\overrightarrow{ST}$. Name others.

(b) Is $\overrightarrow{RQ}$ the same as $\overrightarrow{QR}$?
No; the sets of points are different. The ray *RQ* consists of the point *R* and all points in the half-line on the *Q*-side of point *R*, whereas the ray *QR* consists of the point *Q* and all the points in the half-line on the *R*-side of point *Q*. Make a drawing to illustrate these two sets.

(c) Give another name for the same set of points as $\overrightarrow{RS}$.
Ray *RT* names the same set of points as $\overrightarrow{RS}$. Explain.

(d) What is $\overrightarrow{SR} \cup \overrightarrow{ST}$?
This consists of all the points that are either in $\overrightarrow{SR}$ or in $\overrightarrow{ST}$, or in both. This is line *x*.

(e) What is $\overrightarrow{SQ} \cap \overrightarrow{RT}$?
The points common to both rays are the points *S* and *R* and all the points between.

Let us take a more careful look at the last question in Example 2. Is there a particular name that we give to $\overrightarrow{SQ} \cap \overrightarrow{RT}$—that is, the points *R* and *S*

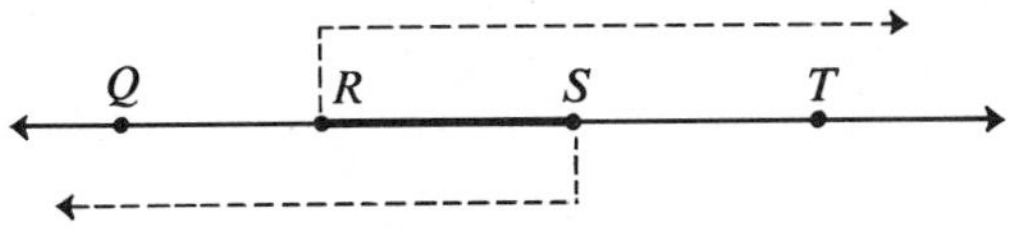

and all the points between? This set of points is represented in the above illustration by the heavier portion. This set of points is called *line segment RS*. The notation used to denote the *line segment RS* is $\overline{RS}$ or $\overline{SR}$. The points *R* and *S* are called the *end points* of the line segment *RS*.

Example 3 In this drawing, is point P an element of $\overline{AB}$?

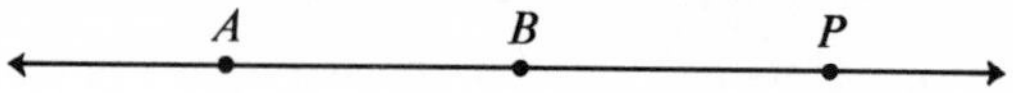

Since point P is not between points A and B, it is not an element of the line segment AB.

Example 4 This figure represents several sets of points, each a subset of m.

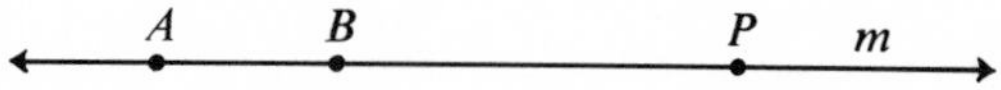

(a) Show how to refer to this line by using the notation agreed upon. Any one of the following would be correct to use: $\overleftrightarrow{AB}$, $\overleftrightarrow{BA}$, $\overleftrightarrow{AP}$, $\overleftrightarrow{PA}$, $\overleftrightarrow{BP}$, $\overleftrightarrow{PB}$.

(b) Using the points A, B, and P, name all the line segments and indicate that each is a subset of the line by using the proper notation.

$$\overline{AB} \subset \overleftrightarrow{AP}, \qquad \overline{BP} \subset \overleftrightarrow{AP}, \qquad \overline{AP} \subset \overleftrightarrow{AB}.$$

Observe that other names may be used for each of the line segments and for the line. That is, $\overline{BA} \subset \overleftrightarrow{BP}$ says the same thing about the sets of points as does $\overline{AB} \subset \overleftrightarrow{AP}$. Observe that $\overline{BA}$ and $\overline{AB}$ name the same set of points, while $\overleftrightarrow{AP}$ names the same set as $\overleftrightarrow{BP}$.

Since a point of a line separates the line into two half-lines, does it seem reasonable to expect that a line of a plane would separate the plane into two half-planes and that a plane would separate space into two half-spaces?

First consider a plane P and a line n that is a subset of plane P and two distinct points, A and B, that are in the plane but not in the line. If $\overline{AB} \cap n = \varnothing$, then the points A and B are on the same side of line n (see Figure 4–13). But if $\overline{AB} \cap n$ is nonempty, then the points A and B are on opposite sides of line n (see Figure 4–14). Can you find the A-side and the B-side of line n in Figure 4–14? The set of points on the A-side of line n is called a *half-plane,* while the set of points on the B-side of line n is the other half-plane. The line n belongs to neither half-plane. It separates the two half-planes and is the boundary of each. The union of the line and the two half-planes is the plane P.

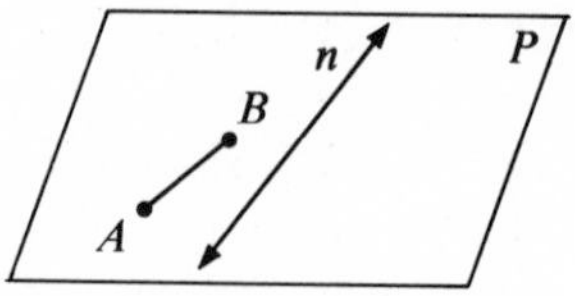

Figure 4–13

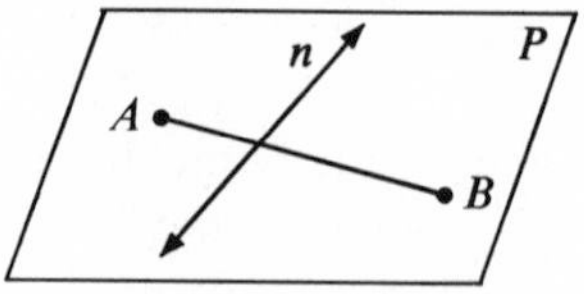

Figure 4–14

A physical model of the separation of a plane into two half-planes may be constructed. Let a sheet of paper represent the plane. Fold the sheet of paper forming a crease. This crease represents a line. Note the two half-planes that are represented.

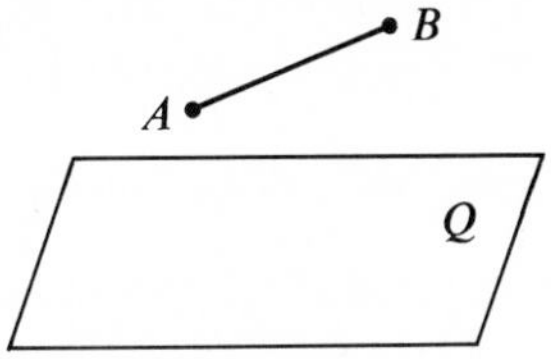

Figure 4–15

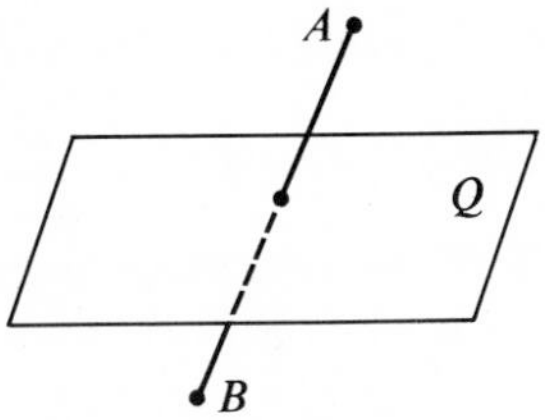

Figure 4–16

Now consider plane Q and let A and B be two distinct points not in plane Q. Suppose that the intersection of $\overline{AB}$ and plane Q is the empty set. Then the points A and B are on the same side of plane Q (see Figure 4–15). If the intersection of $\overline{AB}$ and plane Q is nonempty, then since A and B are distinct and not in Q, points A and B are on opposite sides of the plane (see Figure 4–16). The set of points on the A-side of plane Q is a *half-space,* and the set of points on the side opposite plane Q from A (or the set of points on the B-side of plane Q) is another half-space. The plane is not contained in either of the half-spaces. It separates space into two half-spaces and is the boundary of each. The union of the plane and the two half-spaces is the set of points that we call space.

EXERCISE 4.5

1. Given a line n with certain points identified by the letters A, B, R, Q, P:

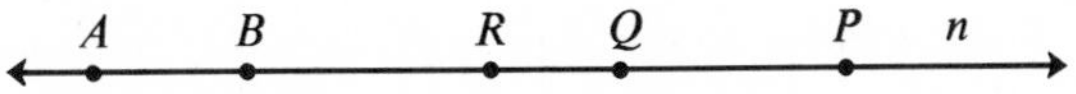

Indicate which of the following statements are true and which are false. Be prepared to justify your decision.

(a) $\overline{AB} = \overline{BA}$
(b) $\overrightarrow{AR} = \overrightarrow{RA}$
(c) $\overrightarrow{BQ} \subset n$
(d) $\overrightarrow{BR} \neq \overrightarrow{BP}$
(e) $\overline{AR} \subset \overrightarrow{BQ}$
(f) $\overrightarrow{QR} \subset \overrightarrow{RP}$
(g) $\overrightarrow{RA} \cap \overrightarrow{AB} = \overline{AR}$
(h) $\overrightarrow{RP} \cup \overrightarrow{QA} = \overline{RQ}$
(i) $\overline{QR} \cup \overline{RB} = \overline{QR}$
(j) $\overline{AB} \cap \overline{RP} = \overline{BP}$
(k) $\overline{BR} \subset \overleftrightarrow{RP}$
(l) $\overrightarrow{AP} = \overline{AP}$
(m) $\overrightarrow{AB} = \overrightarrow{AR} = \overrightarrow{AQ} = \overrightarrow{AP}$
(n) $\overrightarrow{BA} \cup \overrightarrow{RA} = \overrightarrow{RB}$

2. An intelligent ant travels back and forth from X to Y along a "straight" route that also contains B.

(a) How will the ant determine whether or not B is between X and Y?

(b) Suppose that B is between X and Y. If the ant leaves Y and goes to X, how many times must he pass through B to get back on the Y-side of B? The X-side of B?

(c) If B is between X and Y, what side of B will the ant be on if he starts at Y and passes through B three times? 4 times? 5 times? 6 times? 13 times? 36 times? 102 times? 513 times?

(d) Make a generalization concerning the number of times he must pass through B to be back on the Y-side of B. Do the same for him to be on the X-side of B. Assume in each case that the ant begins his trip at Y.

(e) Consider a "trip" to be made when the ant has gone from Y to X or when he has gone from X to Y. If the ant makes a series of trips beginning at Y, what is the number of trips which will always bring him back home? to X?

3. Use the accompanying figure with $\overleftrightarrow{XY}$ and points A, B, C, and D in plane M to answer the following questions:

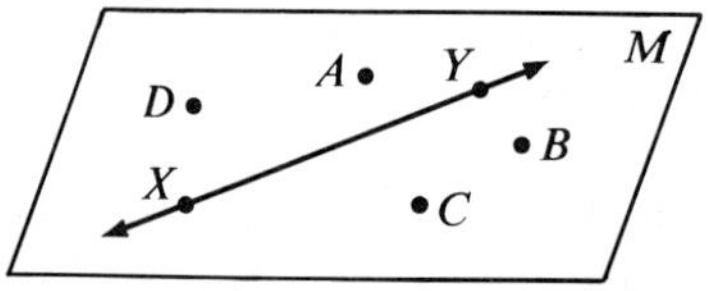

(a) What is $\overline{BC} \cap \overleftrightarrow{XY}$?

(b) Point B is on what side of $\overleftrightarrow{XY}$?

(c) Discuss the intersection of $\overline{AD}$ and $\overleftrightarrow{XY}$.

(d) On what side of $\overleftrightarrow{XY}$ is point D?

(e) Point A and point C are on opposite sides of $\overleftrightarrow{XY}$; that is, they are in different half-planes. Explain.

4. Given line m with four points named P, Q, R, and T:

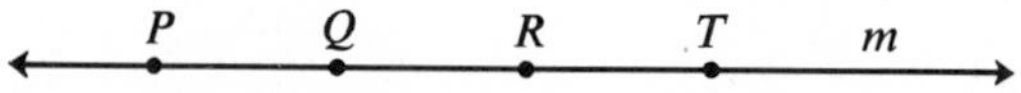

(a) Identify two rays which satisfy these descriptions:
   (1) Their union is line m.
   (2) Their intersection is $\overline{QR}$.
   (3) Their intersection is point R.
   (4) Their intersection is the empty set.
   (5) Their union contains the points P, Q, R, T, but not line m.

(b) Identify two segments in line m which satisfy these descriptions:
   (1) Their intersection is a point.
   (2) Their intersection is a line segment.

(3) Their union is a line segment.
(4) Their intersection is the empty set.

5. Draw a representation of a line containing points named A, B, and P such that P is between A and B.

(a) What is $\overline{AP} \cap \overline{AB}$?
(b) What is $\overline{PB} \cap \overline{AP}$?
(c) What is $\overrightarrow{PB} \cap \overline{PA}$?
(d) What is $\overline{PB} \cup \overline{AB}$?
(e) What is $\overline{PB} \cup \overrightarrow{AP}$?
(f) What is $\overrightarrow{PA} \cap \overrightarrow{AB}$?
(g) What do we call the set of points that is not contained in $\overrightarrow{PB}$?

6. Draw $\overline{AB}$ and $\overline{XY}$ such that $\overline{AB} \cap \overline{XY} = \varnothing$ and $\overleftrightarrow{AB} = \overleftrightarrow{XY}$.

4.6 ANGLES

Consider the drawing, which represents $\overleftrightarrow{ZX}$, $\overleftrightarrow{XY}$, and $\overleftrightarrow{YZ}$, each a subset of plane M:

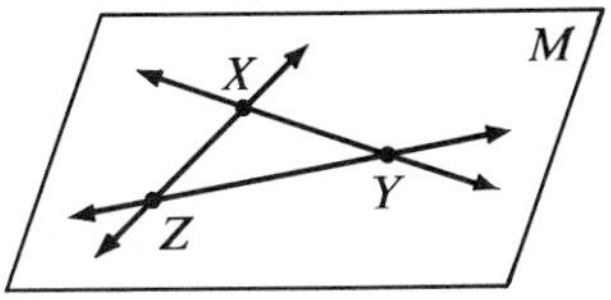

Since $\overrightarrow{ZX}$ is a subset of $\overleftrightarrow{ZX}$, it is also a subset of plane M. Likewise, $\overrightarrow{ZY}$ is a subset of plane M.

Now we shall consider a particular subset of plane M, namely, $\overrightarrow{ZX} \cup \overrightarrow{ZY}$. This subset is called the *angle XZY*. The symbol for angle is "$\angle$." Using this symbol, we write "angle XZY" as "$\angle XZY$." It is read "angle X, Z, Y."

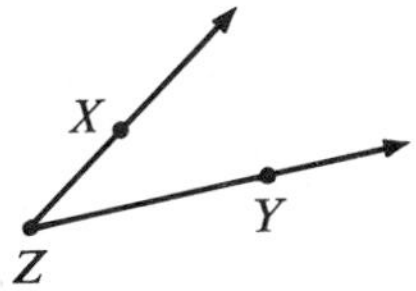

The common end point of the two rays is the *vertex* of the angle, and the two rays are the *sides* of the angle. When denoting an angle, the letter designating the vertex is always written between the other two letters. In the above example, Z is written between X and Y, as $\angle XZY$. Another name for this set of points is $\angle YZX$. Since $\angle XZY$ and $\angle YZX$ name the same set of points, we write

$$\angle XZY = \angle YZX.$$

Definition 4.6 An *angle* is the union of two rays (not subsets of the same line) with a common end point.

Notice that the rays are not to be contained in the same line, and so this definition does not admit that we have the so-called "straight angle" nor the "zero angle." We use the above definition because of the simplicity of this idea of an angle. If we should eliminate the restriction, "not subsets of the same line" from Definition 4.6, there would be an ambiguity in identifying as an angle the set of points illustrated here:

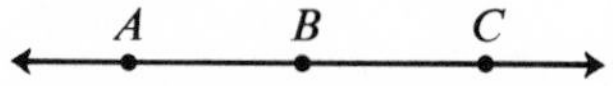

Does the figure represent $\overleftrightarrow{AC}$, $\angle ABC$, $\angle ACB$, or $\angle BAC$? From this, one might say that a line and an angle are the same set of points. In the initial development of the concepts of a ray, a line, and an angle we find it important that a distinction between these concepts be made, and so we prefer not to introduce any ambiguities at this point. This is the reason for the restriction "not subsets of the same line" in the definition.

An angle separates the plane containing it into two parts. The two parts are illustrated by the shaded portions, one with dots and the other with crosshatch:

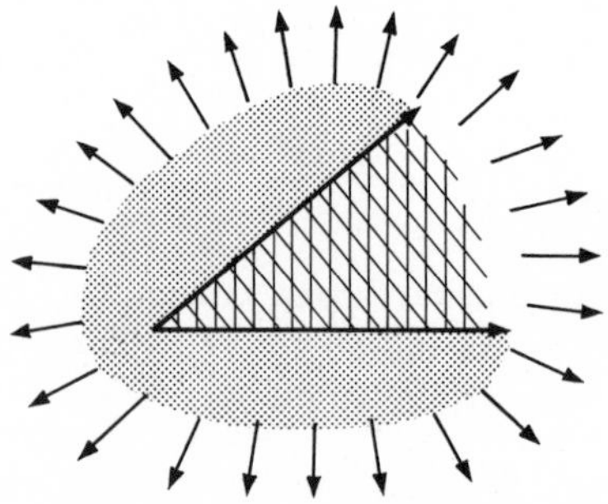

Since a plane has no boundary, these parts will have no boundary except the angle itself. The cross-hatched part is called the *interior* of the angle and the dotted part is called the *exterior* of the angle. The angle belongs to neither the interior nor the exterior. The angle is thought of as the boundary of these two parts just as the line was thought of as the boundary of the two half-planes.

Since a ray is a subset of a line, we shall consider $\overleftrightarrow{ZX}$ and $\overleftrightarrow{ZY}$ with subsets $\overrightarrow{ZX}$ and $\overrightarrow{ZY}$ in discussing the interior and the exterior of $\angle XZY$. Here we shall develop the concept of the interior and the exterior of an angle as the intersection of two half-planes.

The shaded portion in Figure 4–17 represents the set of points on the Y-side of $\overleftrightarrow{ZX}$, while the shaded portion in Figure 4–18 represents the set of points on the X-side of $\overleftrightarrow{ZY}$. The intersection of these two half-planes (the Y-side of $\overleftrightarrow{ZX}$ and the X-side of $\overleftrightarrow{ZY}$) is the *interior of* $\angle XZY$, represented

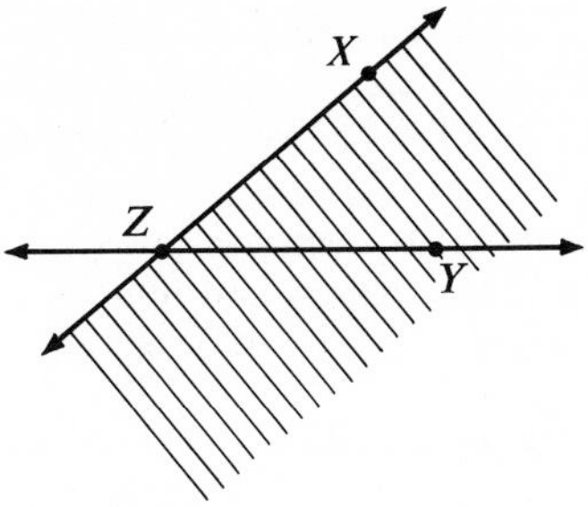

Figure 4–17

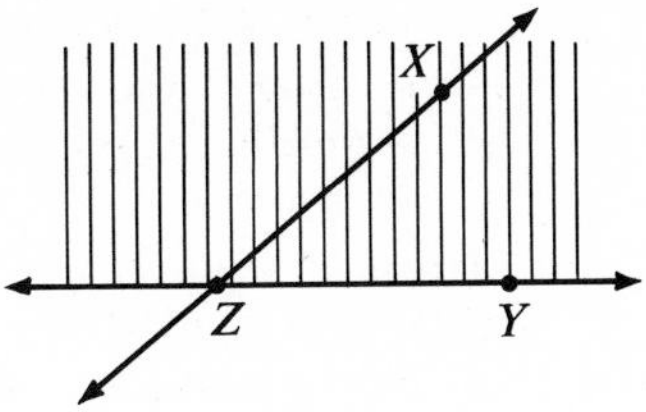

Figure 4–18

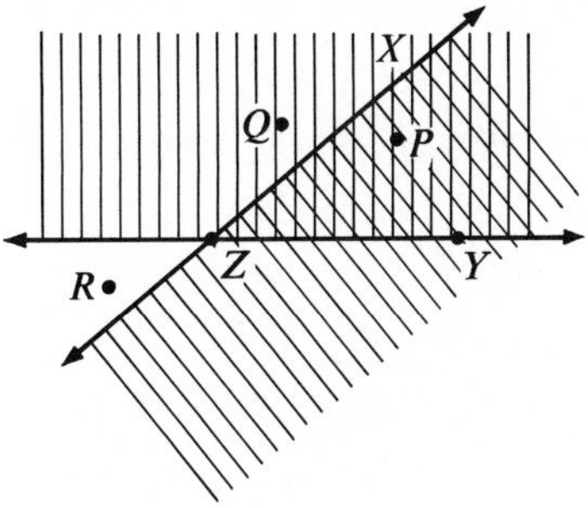

Figure 4–19

by the crosshatch in Figure 4–19. The set of all points other than those contained in the angle or in the interior is the *exterior of* $\angle XZY$. Point P is an element of the interior, while points Q and R are elements of the exterior of $\angle XZY$. Since points X, Z, and Y are elements of $\angle XZY$, they belong to neither the exterior nor the interior of the angle. The angle belongs to neither its interior nor its exterior.

In this study of an angle as a set of points, its measure was not introduced. This will be done later. At that time we shall be going from the nonmetric study of the angle to the metric.

EXERCISE 4.6

1. Name the angles found in this drawing.

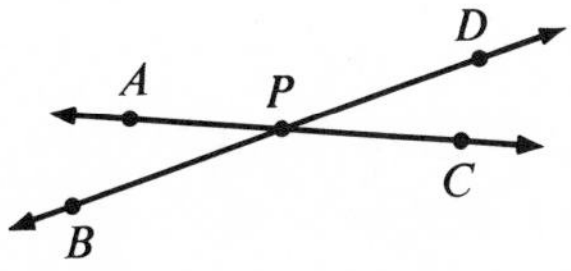

2. The union of the exterior and the interior of an angle is not the plane which contains the angle. Explain.
3. Locate three points P, Q, R not all elements of the same line.
   (a) Make drawings to represent angle PQR and angle QPR. Is it correct to write $\angle PQR = \angle QPR$? Why?

(b) Locate point B that is in the interior of $\angle PQR$. Is point B in the interior of $\angle QPR$ also? Discuss.
(c) If possible, locate a point that is in the interior of $\angle PQR$ and in the exterior of $\angle QPR$.

4. A pair of angles may have several different intersections. Select the following statements that describe the possible intersections and illustrate with a sketch.
(a) The intersection is the empty set.
(b) The intersection is exactly one point.
(c) The intersection is exactly two points.
(d) The intersection is exactly three points.
(e) The intersection is exactly four points.
(f) The intersection is a ray.
(g) The intersection is a line segment.

5. Use the accompanying figure in describing the following sets:
(a) $\overrightarrow{BR} \cap \overleftrightarrow{RC}$
(b) $\angle CBR \cap \overleftrightarrow{CR}$
(c) $\overrightarrow{BR} \cup \overrightarrow{BC}$
(d) $\overline{BC} \cup \angle RBC$
(e) $\overrightarrow{BR} \cap \overrightarrow{BC}$
(f) $\overline{BC} \cap \angle RBC$
(g) $\overleftrightarrow{RC} \cap \overrightarrow{BC}$
(h) $\overrightarrow{BC} \cap \overrightarrow{BA}$
(i) $\overrightarrow{BC} \cup \overrightarrow{BA}$
(j) $\overleftrightarrow{BR} \cap \angle RBC$
(k) $\angle RBC \cap \overleftrightarrow{RC}$
(l) $\overline{CB} \cup \overrightarrow{BA}$
(m) $\overline{RC} \cap \overrightarrow{BA}$

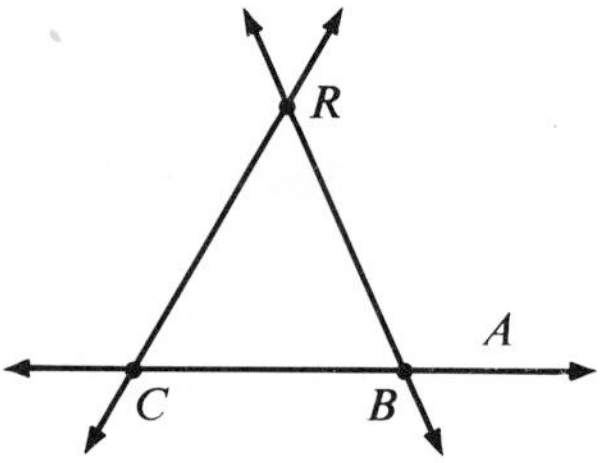

6. Refer to the figure in describing the following sets:
(a) $\overrightarrow{YX} \cap \overrightarrow{BC}$
(b) The intersection of $\angle XYZ$ and $\angle ABC$
(c) The intersection of the interior of $\angle ABC$ and the interior of $\angle XYZ$
(d) The intersection of the exterior of $\angle XYZ$ and $\overrightarrow{BY}$
(e) $\overline{BX} \cup \overline{XY} \cup \overline{YB}$

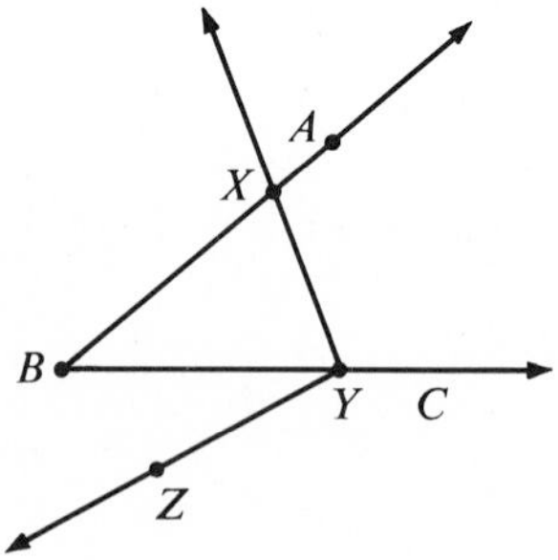

4.7 SIMPLE CLOSED CURVES

So far we have studied the following sets of points as subsets of space: lines, line segments, rays, angles, and planes. There are many other interesting sets of points to be studied, some of which are triangles, squares, rectangles, and circles.

In this course we choose to discuss what we mean by a *curve* in a very informal manner, and we shall limit the discussion to curves, each of which is a subset of a plane. First, let us agree on what we mean by a *simple curve.* For our purposes, we shall think of a simple curve in a plane as a set of points having two end points (distinct or nondistinct) which may be represented by a mark that can be traced from one end point to the other without intersecting the mark. That is, the curve is *simple* when it does not intersect itself.

Use your pencil to test which of these drawings represent simple curves and which do not.

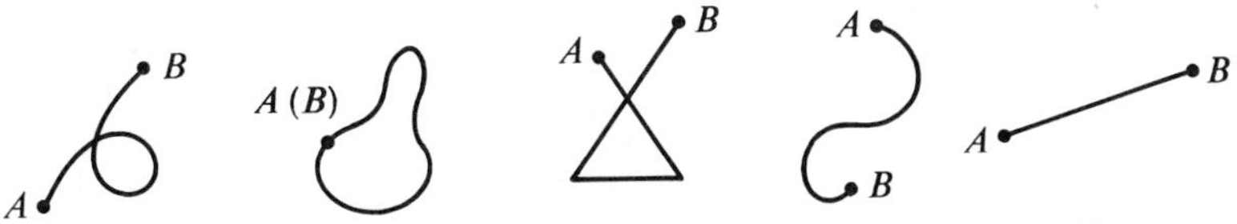

(1) (2) (3) (4) (5)

You will notice that the end points in each are distinct except in drawing (2). Here the end points are identical; that is, the curve starts and stops at the same point. When the end points are not distinct, the curve is said to be *closed,* and we call it a *closed curve.* The curve represented in (2) does not intersect itself; hence it is also a simple curve. So a simple curve whose end points are identical is called a *simple closed curve.*

Some examples of a simple closed curve are shown here:

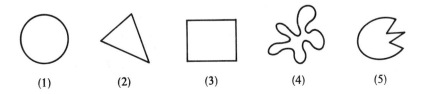

(1) (2) (3) (4) (5)

Is it possible to have a curve that is a simple curve but not closed? Closed but not a simple curve? Not closed and not a simple curve? Make drawings to illustrate which of these you think are possible. How would you name the curves represented below?

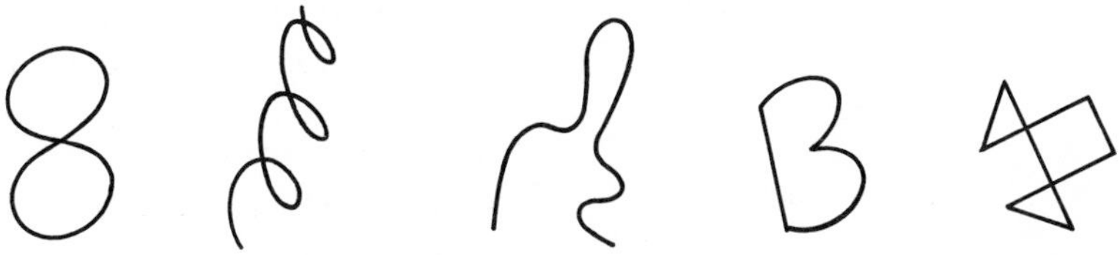

(1) (2) (3) (4) (5)

It is important to note that every simple closed curve in a plane separates the plane into a set of points called the *interior* and another set of points called the *exterior*. The union of the simple closed curve, its interior, and its exterior is the set of points that is the plane. The simple closed curve does not belong to either the exterior or the interior. It is the boundary of each of these sets. The union of the interior and the simple closed curve will be called the *region* of the curve. It is important to remember that the boundary (the simple closed curve) of the region is included in the region.

EXERCISE 4.7

1. Draw three representations of each of the following:
   (a) A simple curve, but not closed.
   (b) A closed curve, but not a simple curve.
   (c) A curve that is not a simple curve and not a closed curve.
   (d) A curve that is a simple curve and a closed curve (a simple closed curve).
2. Indicate by shading the set of points in both the exterior of curve a and the interior of curve b. How does the operation intersection of sets of points apply here?

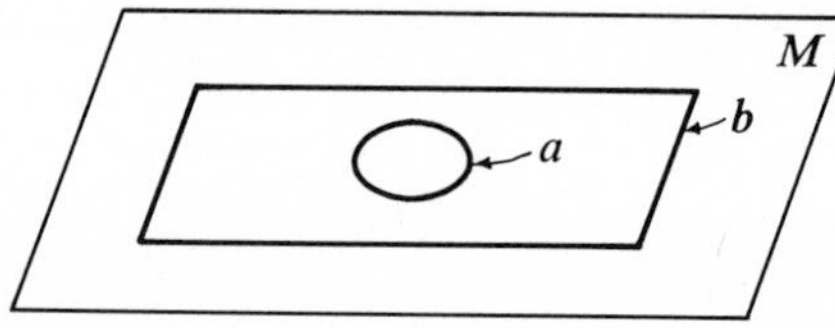

3. Let s represent a simple closed curve as in the drawing:

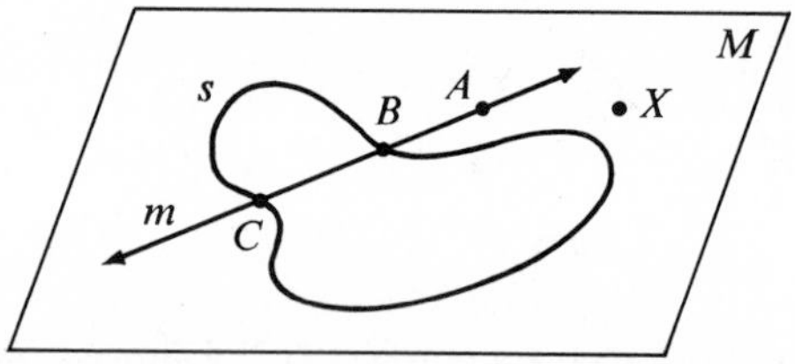

   (a) Shade the interior of s on the X-side of line m.
   (b) What is the intersection of line m with the region of s?
   (c) What is the intersection of line m and the simple closed curve s?
   (d) What is the intersection of $\overrightarrow{BA}$ and the interior of s?
4. Draw a representation of two simple closed curves such that the intersection of their regions is:
   (a) One region
   (b) Two different regions
   (c) Three different regions
   (d) The empty set
5. Draw two simple closed curves that intersect in two and only two distinct points. How many simple closed curves are represented in this drawing?

6. (a) How many simple closed curves are represented in the drawing below?
   (b) Locate a point that is in the curve ABC.
   (c) Locate a point that is in the interior of curve XYZ.
   (d) Locate a point that is in curve ABC *and* in the interior of curve XYZ.
   (e) Locate a point that is in curve ABC *and* in curve XYZ.
   (f) Describe the location of point A with respect to curve XYZ.
   (g) Describe the set of points that includes P but not A.

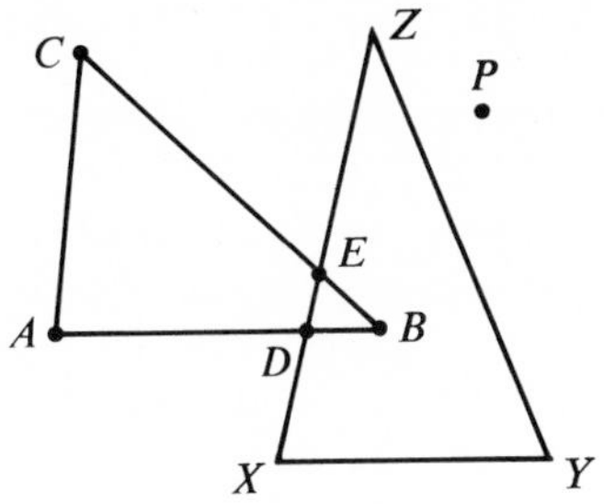

4.8 TRANSFORMATIONS OF THE PLANE

Consider the points A, B, and C in plane M. Suppose that it is possible for these points to move upon a given command. Let this command assign

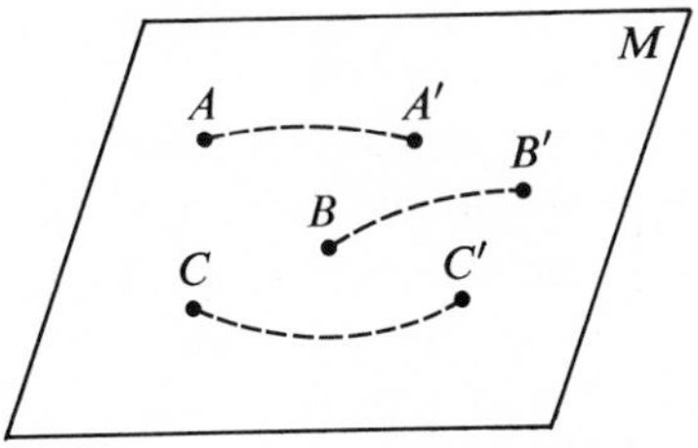

A, B, and C to new positions A', B', and C', respectively. This will be denoted by $A \to A'$, $B \to B'$, and $C \to C'$. The command actually does more. It assigns *every point* of the plane to *exactly one* position. This is a *transformation* of the plane onto itself. A discussion of some simple transformations of geometric figures in a plane follows.

A fascinating activity for children is to make ink images by paper folding. The usual procedure is to draw a picture in ink, then quickly fold the

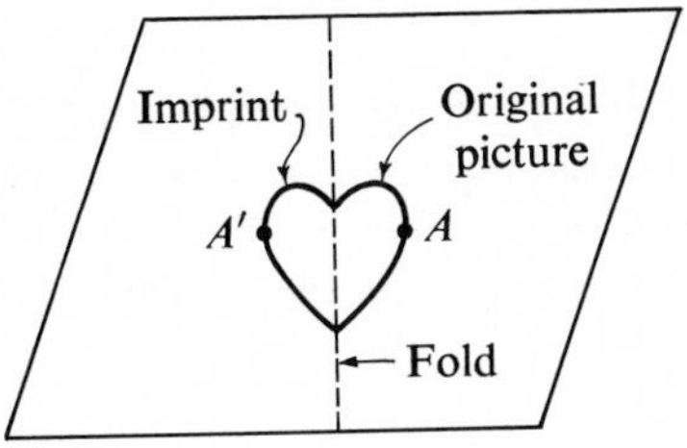

paper before the ink dries. The result is an imprint or image of the original picture. Observe that each "point" of the original picture has an image "point" in the imprint. For example, A has for its image A'.

Turning our attention to plane M, consider the points A and B. Imagine that these points are "inked" and a fold is made along f. Images of points A and B are determined. That is, the image of A is A' and B is B'. This "folding" gives a bonus. That is, *every point* of plane M will have exactly one image point. What is the image of a point in f, such as point D? Can you see that D is its own image?

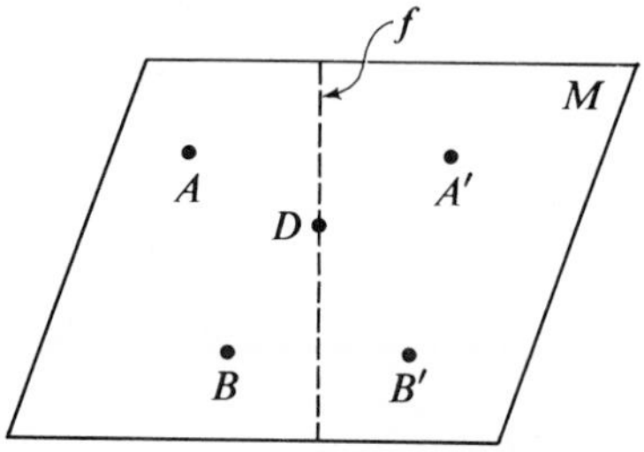

Making assignments of points to points as done in this discussion is called a *reflection in a line.* A reflection in a line is a mapping. The points and their images are symmetric with respect to f. Line f is called the *line of symmetry.*

Look again at the heart picture which was made by an imprint process. The imprint may be thought of as a mirror image of the original picture. The fold, f, is the line of symmetry of the heart. The image of A is A' when reflected in f. Observe that when A' is reflected in f, its image is A. The heart is symmetric with respect to the fold, f.

Consider the following figures:

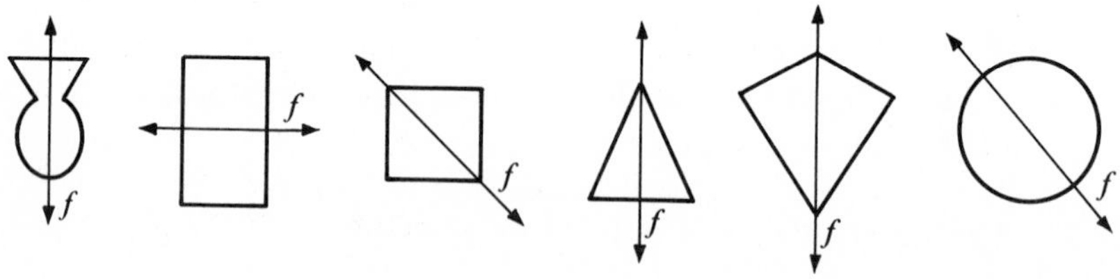

Each reflected in f produces a mirror image that is the same set of points. These figures are said to be *symmetric with respect to a line.* Do any of them have other lines of symmetry?

Let us consider some examples of reflections in a line which lead to other ideas.

Example 1 Given the line segment XY, a subset of plane M. The result of reflecting $\overline{XY}$ in f is $\overline{X'Y'}$. Observe that Y' is just as far from f as is Y. Is this true of X and X'? The point Y' is the image of Y, and X' is the image of X.

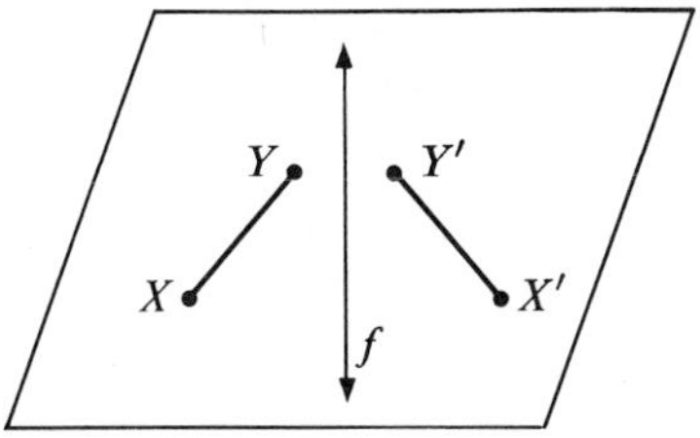

Where are the images of the other points of $\overline{XY}$? Imagine the plane being folded along f. Would $\overline{XY}$ fit exactly on $\overline{X'Y'}$? These line segments are said to be *congruent*. Using the symbol for "is congruent to," we write, $\overline{XY} \cong \overline{X'Y'}$.

Example 2 Given $\triangle PQR$, a subset of plane M. The reflection of the triangle in f is $\triangle P'Q'R'$. Explain why you think the given triangle and its image should be congruent. It may be written, $\triangle PQR \cong \triangle P'Q'R'$.

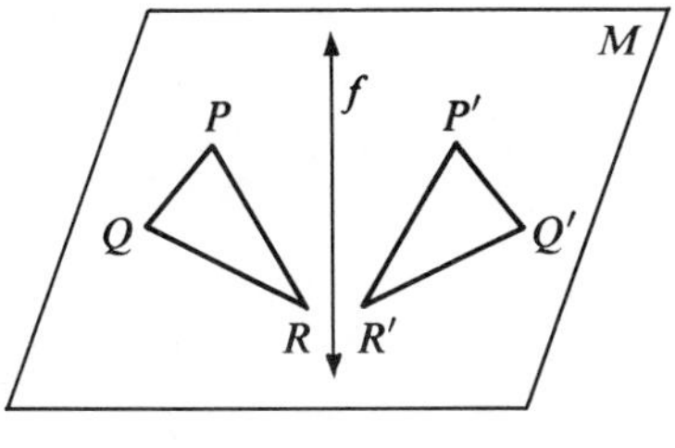

Example 3 Given $\angle ABC$, a subset of plane M. Is $\angle A'B'C'$ the image of $\angle ABC$ under a reflection in f? Does every point of $\overrightarrow{BA}$ have an image in $\overrightarrow{B'A'}$? Is it true that $\angle ABC \cong \angle A'B'C'$? Test this by "folding" the plane along f.

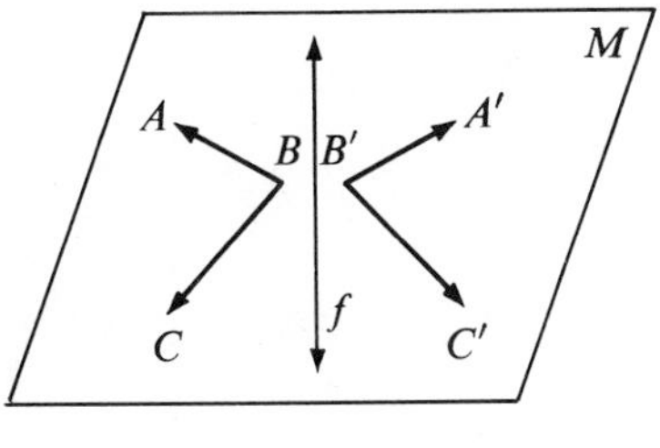

Example 4 Given line n, a subset of plane M. Consider reflecting line n in f. What is the image of P in this reflection? Where are the images of the other points on n located? It appears that the result of reflecting line n in f is line n.

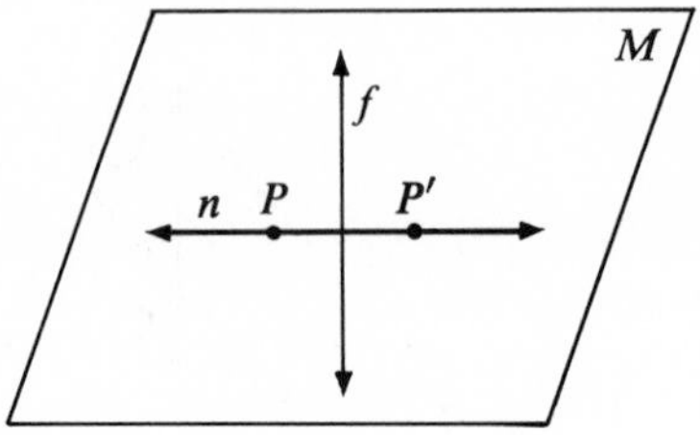

Now, let n assume the role of the line of symmetry instead of f. If f is reflected in n, the image of f is itself. Whenever there are two lines such that either is its own reflection in the other, the two lines are said to be *perpendicular* to each other. The symbol, $\perp$, stands for "is perpendicular to." We write, $n \perp f$ and $f \perp n$. It is read, "n is perpendicular to f" and "f is perpendicular to n." As a result of these perpendicular lines, $\angle XYZ$ is a *right angle.* In fact, there are four right angles represented in the drawing. Can you find them? One of the right angles is indicated by the symbol as shown. This

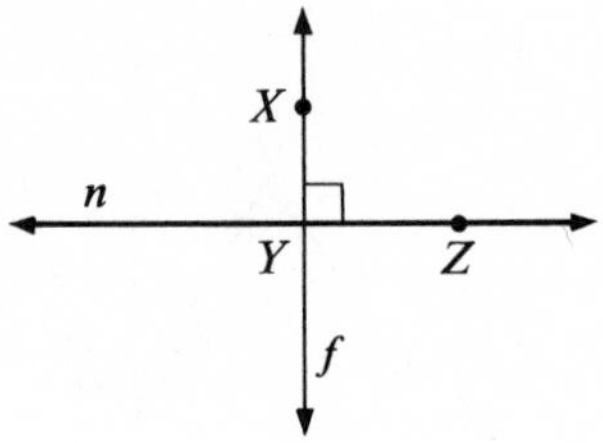

symbol may be used to call attention either to perpendicular lines or to a right angle.

Example 5 Given P, Q, and R, points on line n with Q between P and R. Reflect n in f. Describe the location of Q' in relation to the images of P and R.

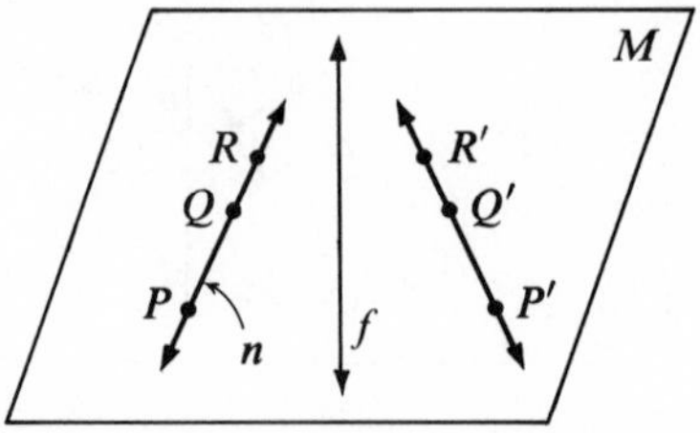

Notice that the points P, Q, and R are collinear, lie in the same line, and so do their images.

What observations is one likely to make from the preceding examples? (1) If a geometric figure is reflected in a line, its image is congruent to the original figure. (2) If collinear points are reflected in a line, their images are also collinear. (3) Betweenness is preserved.

Now, we shall consider another kind of symmetry in a plane. It is *symmetry in a point* or *point symmetry.*

To find the image of X under symmetry in P first draw a line through X and P. Then locate X' on $\overleftrightarrow{XP}$ so that it is just as far from P as X is from P. How would you locate the image of any other point in M, such as Y?

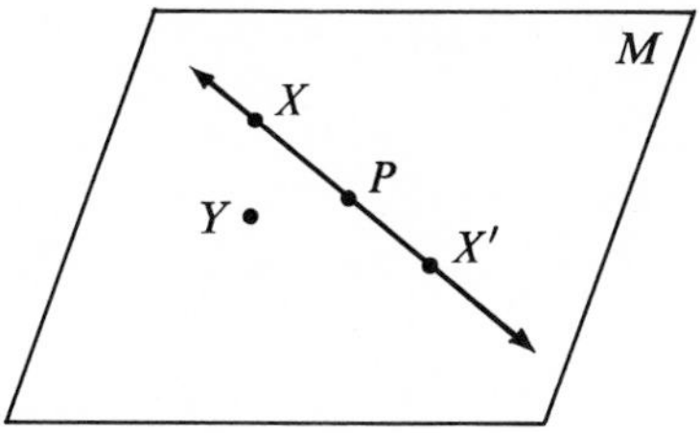

Will every point in plane M have exactly one image under symmetry in P? What is the image of P? Is P its own image? Symmetry in a point, just as reflection in a line, is a mapping. It is a mapping of all the points of a plane onto all the points of a plane.

Consider the following examples of symmetry in a point (Examples 6 through 9).

Example 6 Given $\overline{XY}$, a subset of plane M. Do you agree that the image of every point of $\overline{XY}$ under symmetry in P is $\overline{X'Y'}$? Observe the location of the images of the end points of $\overline{XY}$. Can you think of a way that these two line segments might be fitted together? Can you see that $\overline{XY} \cong \overline{X'Y'}$?

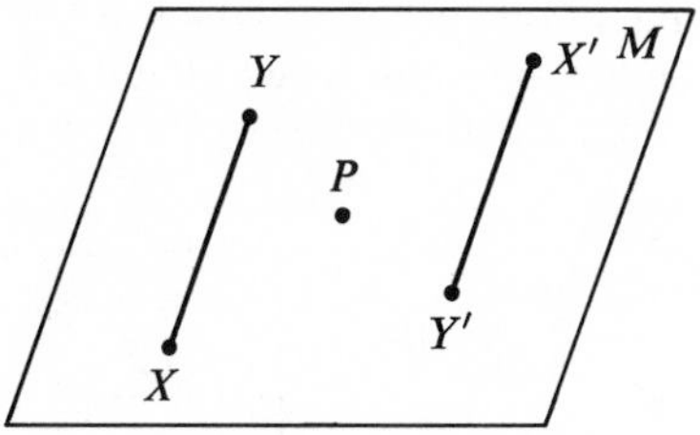

Example 7 Find the image of the L shaped figure under the symmetry in *P*. Do you agree that the ⅂ shaped figure is the correct image?

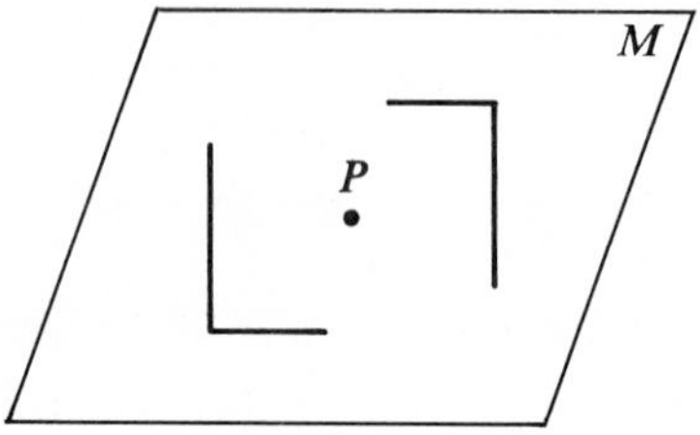

Example 8 Find the image of the letter *M* under the symmetry in *P*. What letter is the image of *M*? Discover another way that you could find the image. Consider rotating *M* about *P* a "half turn".

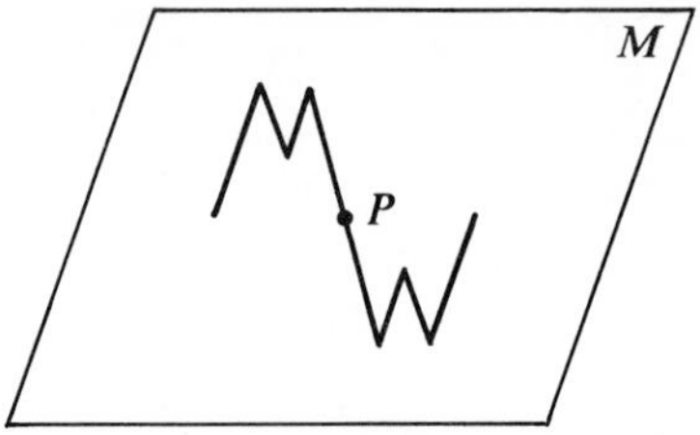

Example 9 Find the image of $\triangle XYZ$ under the symmetry in *P*. What is the image of *X*? of *Y*? of *Z*? Is $\triangle XYZ \cong \triangle X'Y'Z'$? Is it possible to make a "half turn" of $\triangle XYZ$ around *P* and land on $\triangle X'Y'Z'$?

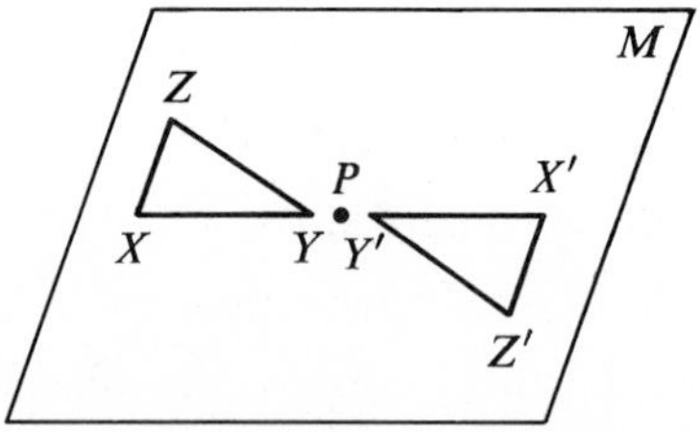

Would the observations made concerning reflections in a line also apply to symmetry in a point? Review them and find situations that illustrate each.

Under symmetry in point *P*, the image of each point of the square is the square itself. The square is said to be symmetric with respect to point *P*.

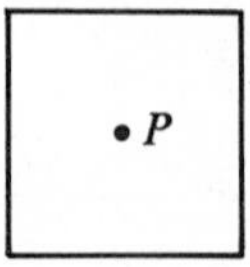

The geometric figures that follow are symmetric with respect to a point. Can you find the point of symmetry?

Another transformation of the plane was suggested by the preceding examples of symmetry in a point. It is *rotational symmetry*. Let us consider the following examples of this type of symmetry (Examples 10 to 13).

Example 10 Given the figure, a subset of plane M. Give this figure a half turn about point P in a counterclockwise direction. Do you see that the result is as if there had been no turning?

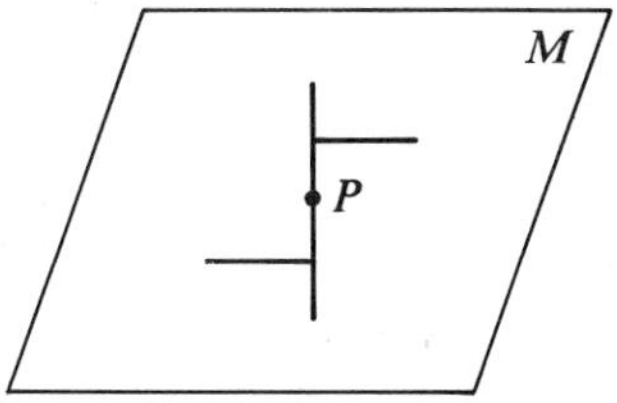

Example 11 Given the set of points, a subset of plane M. Make a one-third turn about point P in a counterclockwise direction. What does the figure look like now? Does it look the same as before you began the rotation?

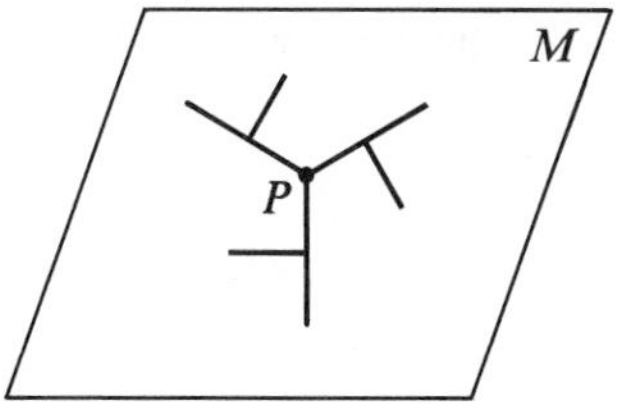

Example 12 Given the figure, a subset of plane M. What amount of turn about P will produce the same figure? Try a one-quarter turn about P. Would you say that this figure is its own image under this rotation? If so, it is said to have rotational symmetry.

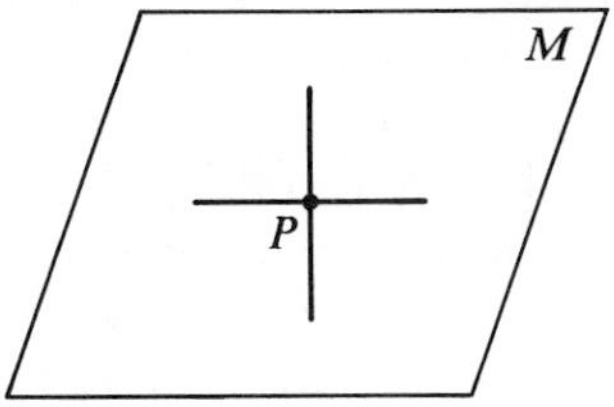

Example 13 Given the point X as shown in the illustration. Locate the image of X under a one-quarter rotation about P in a counterclockwise direction. Is it X'? What is the image of P? Would every point in the plane have an image under this rotation?

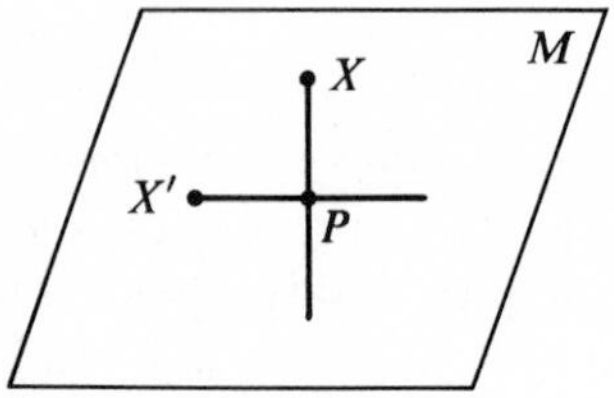

From the examples, we see that rotation in a point maps every point of the plane onto the plane. If there is a point and a rotation about this point that maps the set of points onto itself, we say that this set of points has *rotational symmetry*. However, the zero rotation and also the complete rotation are excluded.

Do any of the observations made concerning line and point symmetry hold for rotational symmetry?

Still another transformation of the plane is *translation*. This may be thought of as a "slide" in some particular direction. Consider the following examples.

Example 14 Let A and B be marbles on a large table. A strong wind blows in a northeasterly direction. At the end of a second the marbles will be at positions A' and B', denoted by $A \to A'$ and $B \to B'$. If other marbles were on the table, would they move to new positions in the manner that A and B moved?

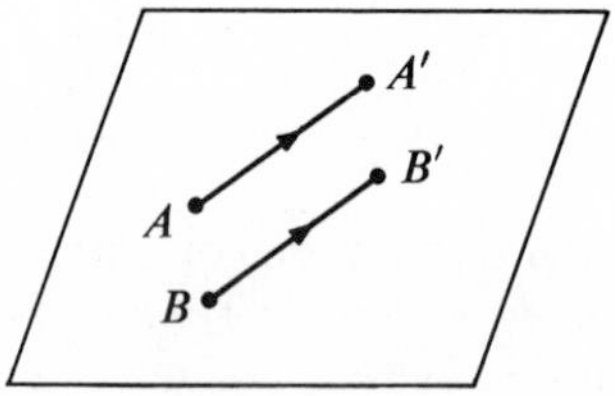

Example 15 Given $\triangle XYZ$, a subset of plane M. Slide the triangle a distance of 2 inches in the direction indicated by the arrows. The new location of $\triangle XYZ$ is $\triangle X'Y'Z'$. In this translation, $X \to X'$, $Y \to Y'$, and $Z \to Z'$. The image of each of the points X, Y, and Z is X', Y', and Z', respectively. Is any point its own image? Under this transformation, does every point in the plane have one and only one image?

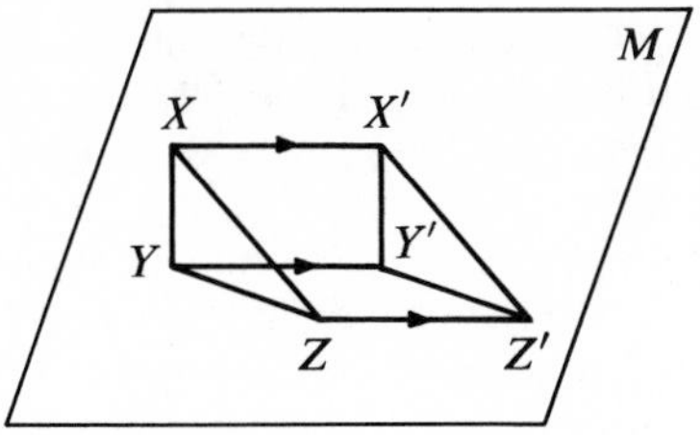

The preceding examples lead us to conclude that a *translation* maps every point of the plane onto the plane. A translation is defined when magnitude (distance) and direction are given.

EXERCISE 4.8

1. Locate a point that is its own image in
   (a) a reflection in line t.
   (b) a symmetry in point Q.
   (c) a rotational symmetry about point R.
   (d) a translation.
2. Select a letter of the alphabet that
   (a) is symmetric in a point.
   (b) is symmetric in a line.
   (c) has rotational symmetry.
3. Locate point P which maps $\overleftrightarrow{XY}$ onto itself under point symmetry. Is there only one point?
4. Locate line m which maps $\overleftrightarrow{XY}$ onto itself under line symmetry. What relationship exists between $\overleftrightarrow{XY}$ and m?
5. (a) Under what transformations in a plane is there a point that is its own image?
   (b) Describe the location of such a point when one exists.
6. Name the transformations that map a set of points onto a congruent set of points.
7. (a) Find the image of $\overrightarrow{XY}$ by reflecting it in f.
   (b) Compare $\angle YXZ$ and $\angle Y'XZ$. Ray XZ is said to *bisect* $\angle YXY'$.

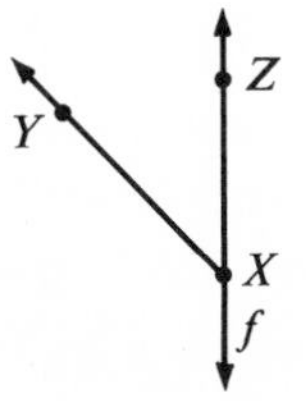

8. (a) Find the image of $\overline{PX}$ that is symmetric in P.
   (b) Compare $\overline{PX}$ and its image $\overline{PX'}$. Point P is said to *bisect* $\overline{XX'}$.

9. Identify which statements are true and which are false. If the statement is false, be prepared to explain why it is false.
   (a) Every point symmetry may be thought of as rotational symmetry of a one-half turn.
   (b) Every rotational symmetry may be thought of as point symmetry.
   (c) A figure has rotational symmetry if any amount of turn about a fixed point gives the same figure.
   (d) No point in the plane is its own image under a translation mapping.
10. Draw a picture of each of the following. Find a line (f) or a point (P) in which a symmetry will map the figure onto itself. If the figure has both line and point symmetry, locate f and P.
    (a) A piece of typing paper.
    (b) A circular rug.
    (c) The Christian star.
    (d) The star of David.
    (e) A square cracker.
    (f) A butterfly.
    (g) The cross section of a cell of a honey comb.
    (h) The cross section of an I-beam.
    (i) A kite.
11. By using two different transformations, reflection in a line and rotation about a point, one followed by the other, $\triangle XYZ$ may be taken into its image $\triangle X'Y'Z'$. All reflections must be made through a side and all rotations about a vertex of $\triangle XYZ$. Remember that the amount of rotation is never equal to 360 degrees. See if you can find more than one way.

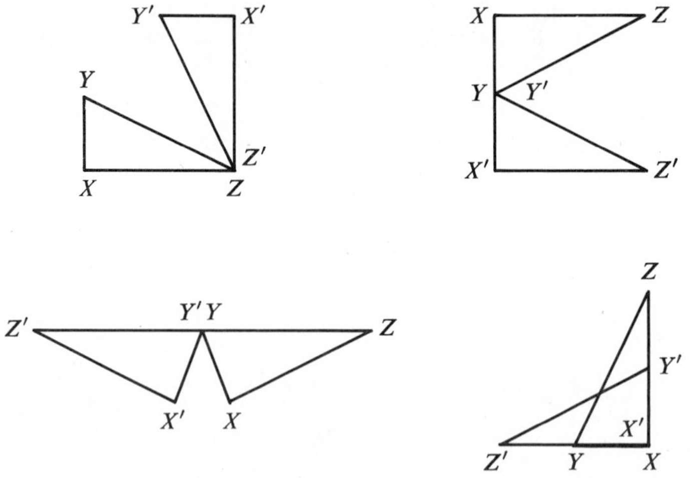

4.9 SPECIAL SIMPLE CLOSED CURVES

A simple closed curve formed by the union of three or more line segments is called a *polygon.* A simple closed curve formed by the union of three line segments is called a *triangle,* and one formed by the union of four line segments is a *quadrilateral.* A *pentagon* and a *hexagon* are simple closed curves formed by the union of five and six line segments, respectively. Probably you have already observed that the triangle, the quadrilateral, the pentagon, and the hexagon may be defined in terms of the polygon. That is, a triangle may be defined as a polygon having three line segments. The line segments are called *sides* and the intersection of each pair of line segments is a *vertex* of the polygon. A polygon is identified by the capital letters placed at its vertices (plural for vertex).

There are different types of triangles and quadrilaterals. Some of the more common types of triangles are as follows: *equilateral,* which has three congruent sides; *isosceles,* which has at least two congruent sides; *scalene,* which has no two sides congruent; and *right,* which has one right angle. Of course, a right triangle may also be either scalene or isosceles. The drawings that follow illustrate these types of triangles:

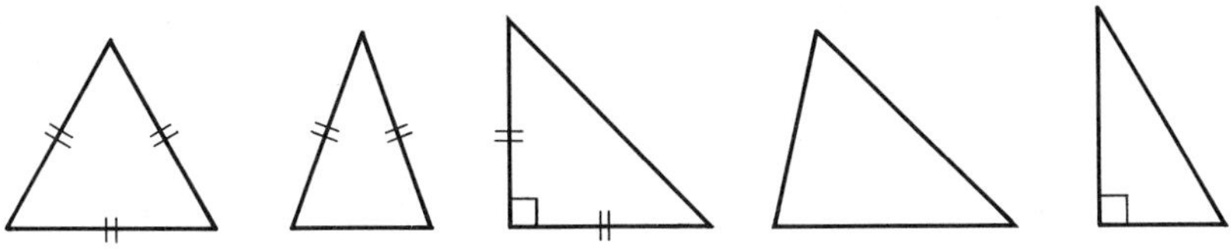

Equilateral | Isosceles | Right Isosceles | Scalene | Right Scalene

Observe the markings used to indicate the congruent sides, also that used to indicate the right angle.

Quadrilaterals that are more common than others are listed as: *parallelogram,* which has opposite sides parallel; *rectangle,* which has opposite sides parallel and four right angles; *square,* which has opposite sides parallel, four right angles, and four congruent sides. The drawings represent each of these quadrilaterals.

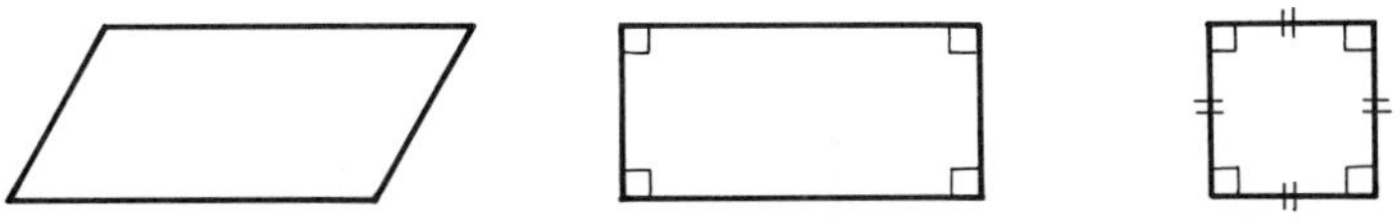

Parallelogram | Rectangle | Square

By studying these drawings, you should discover that the opposite sides of the parallelogram and the rectangle are congruent.

The following diagram establishes the hierarchy for defining each of these special simple closed curves.

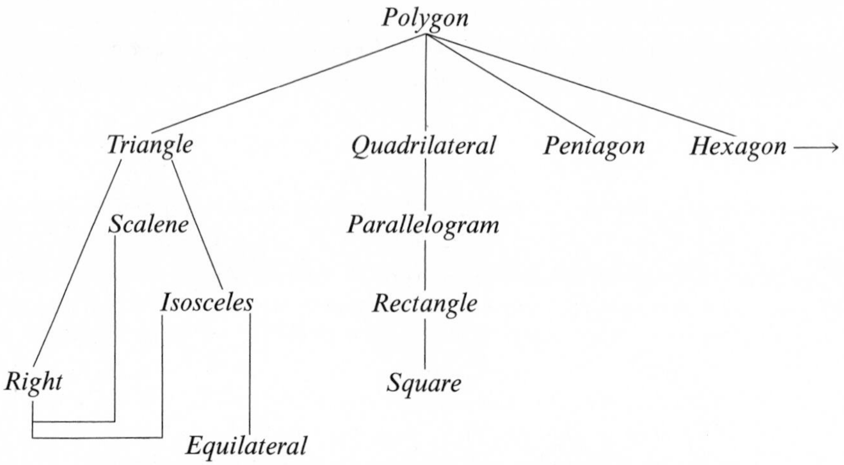

It is possible to formulate the most economical definitions from this. For example, a square may be defined as a rectangle having all sides congruent. By classifying the square as a rectangle, we accept all of the defining properties of the rectangle which in turn include those of the parallelogram, the quadrilateral, and the polygon.

Notice the arrow to the right of "hexagon" in the diagram. This means that there are polygons having more than six sides, such as an octagon, a decagon, a dodecagon, etc. Can you tell the number of sides each has by its name?

EXERCISE 4.9

1. With the information indicated below, is it possible to identify each of these simple closed curves?

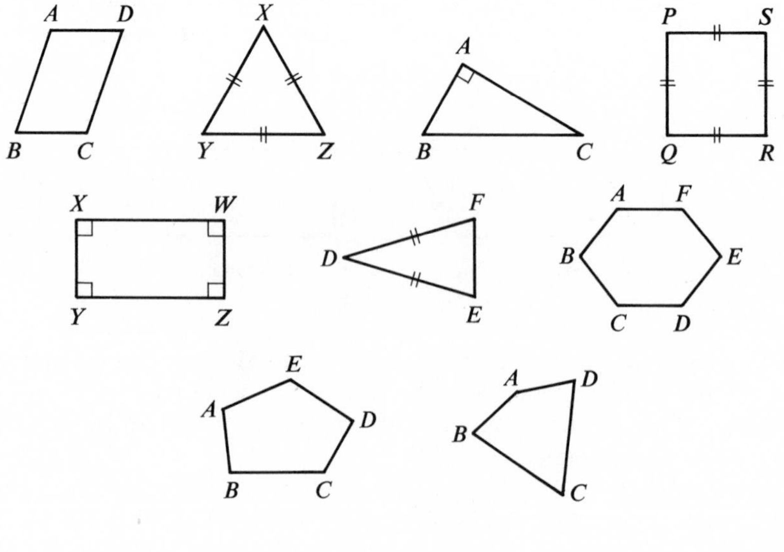

2. State the defining properties that must be included if a rectangle is classified as
   (a) a parallelogram.
   (b) a quadrilateral.
   (c) a polygon.
3. List those polygons shown in Problem 1
   (a) having four right angles.
   (b) having only one right angle.
   (c) having only four sides.
   (d) having only three sides.
   (e) having two congruent sides.
4. Identify the following statements as either true or false.
   (a) An isosceles triangle is equilateral.
   (b) An equilateral triangle is isosceles.
   (c) Some right triangles are isosceles.
   (d) All squares are rectangles.
   (e) Some rectangles are squares.
5. Locate the line of symmetry which may be used to show that the parts listed below each drawing are congruent.

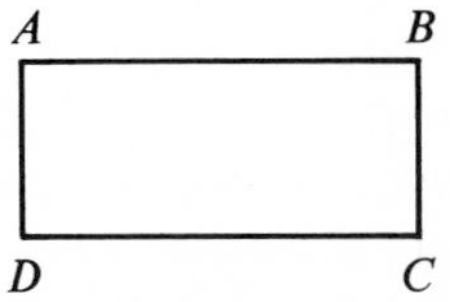

$\overline{AB} \cong \overline{DC}$

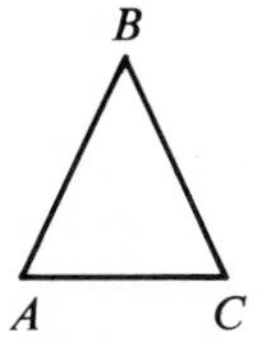

$\overline{AB} \cong \overline{BC}$

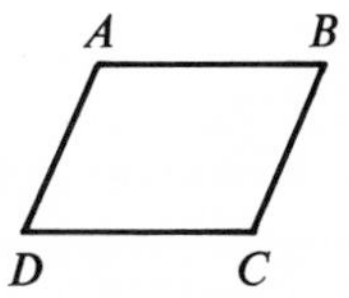

$\overline{AB} \cong \overline{BC}$

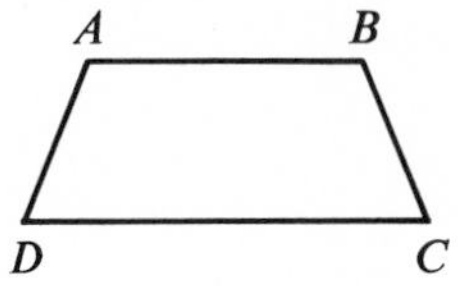

$\overline{AD} \cong \overline{BC}$

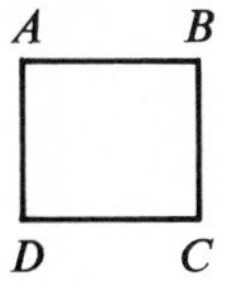

$\angle A \cong \angle C$

$\angle A \cong \angle C$

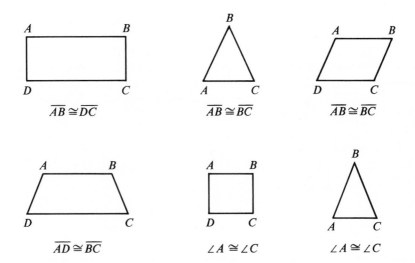

6. Locate the point of symmetry which may be used to show that the diagonal, *AB*, subdivides the polygon into two congruent triangles.

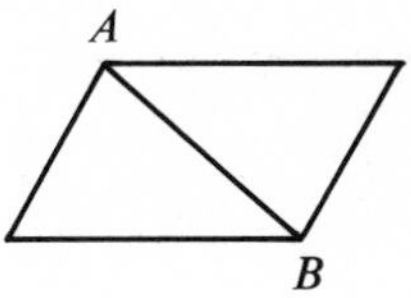

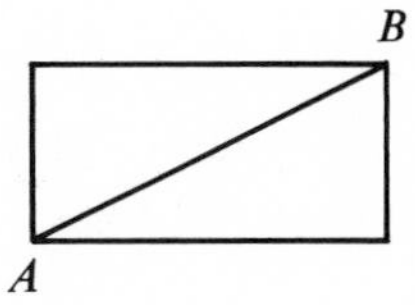

7. Locate the point and amount of rotation which may be used to show that the sides are congruent in each polygon.

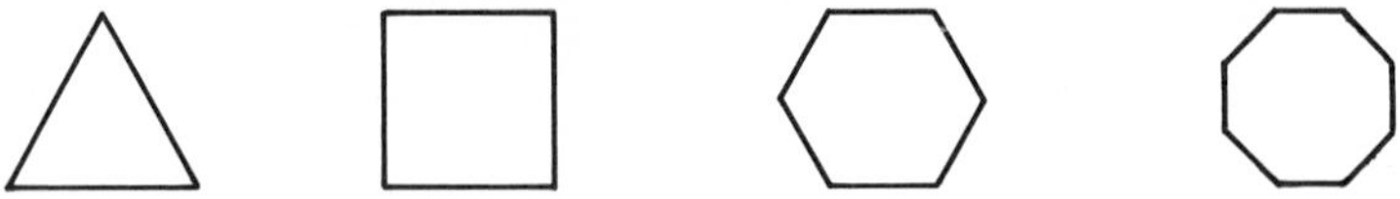

4.10 CONVEX AND NONCONVEX SETS OF POINTS

Each of the two figures below represents a region, a set of points consisting of the union of a curve and its interior. The line segment XY is a subset of region A, whereas the line segment PQ is not a subset of region B. Why?

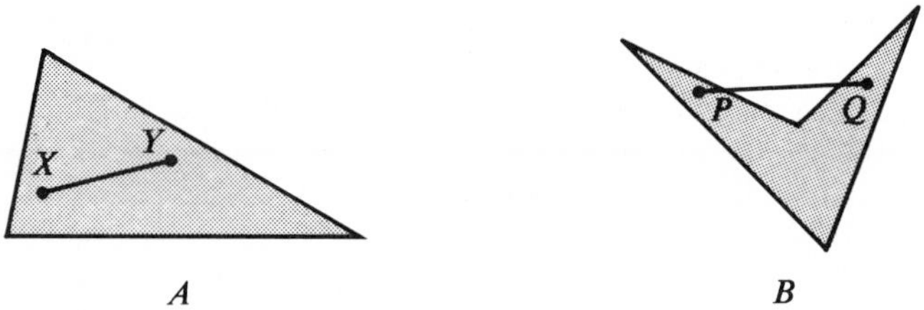

If any other pair of distinct points is selected in region A, will the line segment having these as end points lie entirely within the region? That is, will the line segment be a subset of the region? The answer is "yes"; however, this is certainly not possible in region B. With these distinguishing features, we call region A a *convex set* of points and region B a *nonconvex set*. This suggests the following definition:

Definition 4.10 A set of points is a *convex set* if for each pair of distinct points belonging to the set, the line segment having these as end points is a subset of the given set.

Consider the regions represented in Figure 4–20. Applying the preceding definition, do you see that each is a convex set?

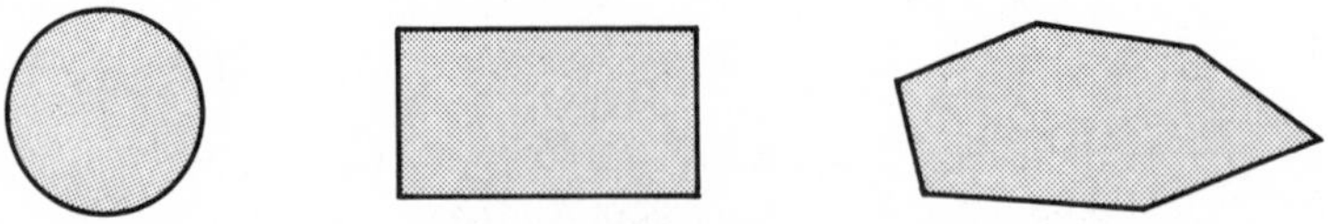

Figure 4–20

The regions represented in Figure 4–21 are examples of nonconvex sets. In each of these regions there is shown *at least* one line segment the end points of which belong to the region, but some points of which do not belong to the region. The line segment is not a subset of the region. Hence, the region is not convex.

From the preceding discussion, one might be led to conclude that regions

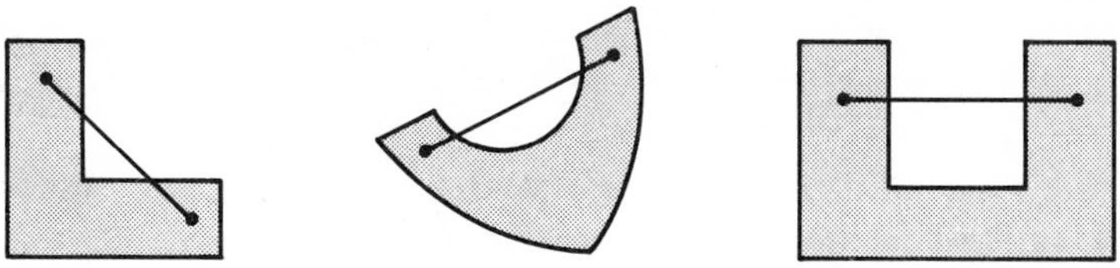

Figure 4–21

are the only sets of points that may be convex. This is not the case, however. Consider a line. Would you say that it is a convex set? The definition should direct you to an affirmative answer. Other examples of convex sets that are not regions are a ray, the interior of an angle, and a half-plane. Can you think of others?

Now, let us consider the set of points called a circle. Recall that a circle is a simple closed curve and does not include the interior. Does the circle qualify as a convex set? Obviously not, since one can find at least one line segment whose end points belong to the circle but which is not a subset of the circle.

We have discovered, then, that a circle is not a convex set, but a circular region is convex. Can you give examples of other nonconvex simple closed curves whose regions are convex? Would you say that each of the polygons studied in Section 4.9 is not convex, yet each region is a convex set?

EXERCISE 4.10

1. Which of the following are convex sets? If some are not convex, explain.
   (a) Circle
   (b) Angle
   (c) Rectangular region
   (d) Interior of an angle
   (e) Square
   (f) Ray
   (g) Half-plane
   (h) Region of a circle
   (i) Polygon
   (j) Line segment
2. Identify the statements that are true and those that are false. Justify your decision.
   (a) A triangle and its region are both convex sets.
   (b) If a region is convex, then the simple closed curve that is its boundary is a convex set.
   (c) The region of every simple closed curve is a convex set.
   (d) Every simple closed curve is a nonconvex set of points.
3. Demonstrate with a drawing that the regions suggested by the following descriptions are not convex sets.
   (a) A crescent shape
   (b) A bean shape
   (c) A star shape
   (d) An irregular shaped lake
4. Make a drawing of a simple curve not closed
   (a) that is a convex set.
   (b) that is a nonconvex set.
5. Is the intersection of two convex sets also convex? Explain.
6. Is the union of two convex sets also convex? Explain.

5

The System of Natural Numbers

5.1 THE SET OF NATURAL NUMBERS

The notion of number was developed in Chapter 2 as the common property of equivalent sets. This property is frequently referred to as the *cardinality* of the set. The cardinal number (Section 2.8) assigned to a set indicates how many elements belong to the set, and these were ordered (Section 2.9) as

$$0 < 1 < 2 < 3 \cdots .$$

In counting we customarily use the sequence

$$1, 2, 3, 4, \ldots .$$

We shall refer to these as the *counting numbers* or the *set of natural numbers*

$$N = \{1, 2, 3, 4, \ldots\}.$$

The order of the natural numbers was established above, that is,

$$1 < 2 < 3 < 4 \cdots ,$$

with each number one more than its predecessor except the first number, 1. Note also that the least element belonging to N is 1, and that there is no greatest element. This concept of order is sufficient until later when a formal definition will be given.

5.2 THE ADDITION OPERATION IN THE SET OF NATURAL NUMBERS

With the establishment of a set of elements (the set of natural numbers), our next undertaking is to define an operation on any two of the elements of this set N. Consider the following example:

Example 1 On Lesley's birthday, her grandmother gave her a dress, a coat, and a hat. Her aunt gave her a book and a pen. How many birthday presents did she receive? Lesley could put all of her presents together on her bed and count them. She would then be able to tell us that she received 5 presents. Suppose instead that Lesley has taken her book and pen to school, and the gifts from grandmother are in her closet. Suppose that she thinks of these as two separate sets,

$$G = \{\text{dress, coat, hat}\}$$

and

$$A = \{\text{book, pen}\}.$$

One would expect that Lesley has already determined the number of presents received from grandmother and the number from her aunt. By counting the elements in G, she finds that the number is 3. The number of elements in A is 2. Now, she says, "I received 3 presents from grandmother and 2 from my aunt, and I have 5 in all." At this point Lesley is not interested in *what* her presents were, but only in *how many* (the number) she received.

This example illustrates two operations: the *union operation on two nonempty, disjoint, discrete, finite sets* and a *different operation on two natural numbers.* The joining of G and A illustrates the union of these two sets:

$$G \cup A = \{\text{dress, coat, hat, book, pen}\}.$$

Here, we begin with the disjoint sets G and A, operate on them with $\cup$, and we get another set. But when we begin with the numbers 3 and 2, and assign to them the new number, 5, we call this *addition.* The symbol for addition is "+." We write

$$3 + 2 = 5$$

and read it, "3 *plus* 2 is equal to 5," or "the number represented by 3 + 2 is the same as the number represented by 5." Remember that the symbol "=" was used in Section 2.8 to connect names for the same number. We do not say, "3 and 2" for "3 + 2." The natural numbers 3 and 2 are called *addends,* and the natural number 5 is the *sum* of these two addends.

Notice that when we operated on the *numbers* assigned to sets G and A, we were no longer interested in the particular elements which formed these sets, as we were when we operated on the sets.

Let us summarize the preceding discussion, representing the elements of sets G and A by letters in order to make our work less cumbersome:

$$G = \{d, c, h\} \quad \text{and} \quad A = \{b, p\},$$

and so

$$G \cup A = \{d, c, h\} \cup \{b, p\} = \{d, c, h, b, p\},$$

where we have the *union* operation on *sets* G and A. On the other hand,

$$\text{N}(G) = \text{N}\{d, c, h\} = 3 \quad \text{and} \quad \text{N}(A) = \text{N}\{b, p\} = 2,$$

and so we have

$$\begin{array}{ccccc} N\{d, c, h\} & + & N\{b, p\} & = & N\{d, c, h, b, p\} \\ 3 & + & 2 & = & 5 \end{array}$$

where

$$N(G \cup A) = N\{d, c, h, b, p\} = 5.$$

Hence we have the *addition* of the *natural numbers* 3 and 2.

The sets we considered in Example 1 were disjoint sets; that is, their intersection was the empty set. Now let us investigate the addition of two numbers assigned to two nondisjoint sets.

Example 2 Given

$$P = \{b, c, d\} \quad \text{and} \quad Q = \{y, d\}; \qquad P \cap Q \neq \varnothing.$$

Now

$$\{b, c, d\} \cup \{y, d\} = \{b, c, d, y\}$$

seems to imply that

$$N\{b, c, d\} + N\{y, d\} = N\{b, c, d, y\}$$

or

$$3 + 2 = 4.$$

This is disturbing, since in Example 1 we found $3 + 2$ to be equal to 5. Next let us consider

$$P = \{b, c, d\} \quad \text{and} \quad R = \{c, d\}; \qquad P \cap R \neq \varnothing.$$

Now

$$\{b, c, d\} \cup \{c, d\} = \{b, c, d\}$$

seems to imply that

$$N\{b, c, d\} + N\{c, d\} = N\{b, c, d\}$$

or

$$3 + 2 = 3.$$

Again, this result contradicts that of Example 1.

Finding several sums for the addends 3 and 2 as in Examples 1 and 2 is contradictory to the way we wish to define addition. Instead, we shall demand that there be one and only one sum. This is the requirement for addition to be an operation. This sum will be unique if we require that these numbers be associated with sets that are disjoint. Compare Examples 1 and 2. This leads to the following definition:

Definition 5.2 The *addition operation on the natural numbers a and b* is the assignment of exactly one natural number c to the ordered pair (a, b), such that

$$N(A) + N(B) = N(A \cup B),$$

where A and B are nonempty, discrete, finite sets, $A \cap B = \varnothing$, and

$N(A) = a$, $N(B) = b$, and $N(A \cup B) = c$. That is,

$$(a, b) \xrightarrow{+} c$$

or

$$a + b = c.$$

It is important to note that the union operation here is performed on *two* discrete, finite sets, and the addition operation is performed on *two* natural numbers. Consequently, we say that these operations are *binary* operations. When we add 2 to 3, we think of the *ordered number pair* (3, 2) as matched with 5 under the (binary) operation addition. This may be written as

$$(3, 2) \xrightarrow{+} 5$$

and read, "The ordered pair (3, 2) is matched with 5 under the operation addition." This form for describing the addition operation is called the *ordered pair form.* The form

$$3 + 2 = 5$$

is called a *mathematical sentence.*

We may also say that "5" is *another name* for "3 + 2"; that is, "5" and "3 + 2" are two different names for the same number. Sometimes it will be convenient to *rename* 5 as 3 + 2 or 3 + 2 as 5.

Before continuing our discussion, let us consider further the meaning of *ordered pair.* The idea of ordered pair may be illustrated by the method used in locating a town on a map. On a particular road map, the town of Norman, Oklahoma, is located by giving the notation C–6. Since letters of the alphabet are printed along the two opposite borders of the map and numerals representing the natural numbers along the other two borders, it is clear what is meant by the symbol, C–6. This is an example of an ordered pair and in mathematics it would be written (C, 6). The C is called the *first component* and the 6 is called the *second component* of the ordered pair.

It is obvious that (C, 6) is a pair, but what is the significance of the word *ordered?* Here we have a map of Oklahoma with numerals replacing the letters of the alphabet. Now, when the pair of numbers (3, 6) is given to

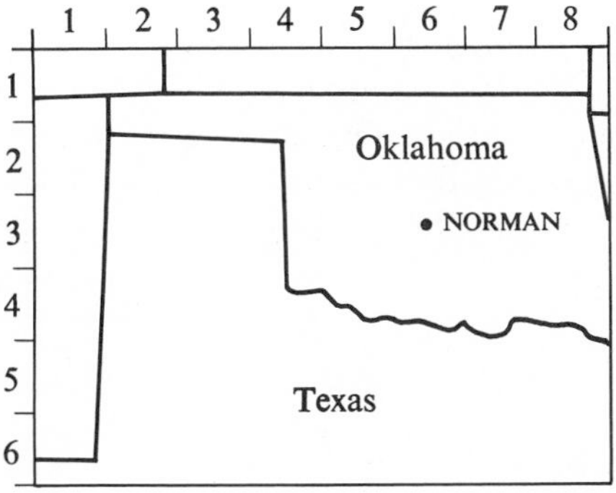

locate Norman on the map, one needs to know from which border each number is read. Suppose that we agree in this discussion that the first component be taken from the "vertical" border. Locate the position (3, 6) on the map. Now locate the position (6, 3) on the map. Are the two positions the same? Obviously not, since (3, 6) is the location of a town in Oklahoma, while (6, 3) is a location in Texas. Does the *order* of these numbers (which number is the *first* component and which is the *second* component) make any difference? Obviously the order of the components is important; hence, the expression *ordered pair* will be used from now on to describe (a, b).

EXERCISE 5.2

1. If $X = \{\text{chair, stove, box, cat}\}$ and $Y = \{\text{hat, coat}\}$, find:
   (a) $X \cup Y$ (d) $N(X) + N(Y)$
   (b) $N(X)$ (e) $N(X \cup Y)$
   (c) $N(Y)$ (f) Are sets X and Y disjoint?
2. Given A, the set of letters of the alphabet used in spelling the word *chair,* and B, the set consisting of the natural numbers between 26 and 35. Are sets A and B disjoint? Use the procedure of Example 1 to find $N(A) + N(B)$.
3. If $A = \{a, f, g, h, i\}$ and $B = \{f, a, c, e\}$, why does $N(A) + N(B)$ not equal $N(A \cup B)$?
4. Invent pairs of sets from which the following ordered pair forms for addition may be derived.
   (a) $(2, 4) \xrightarrow{+} 6$ (c) $(7, 2) \xrightarrow{+} 9$
   (b) $(3, 3) \xrightarrow{+} 6$ (d) $(2, 7) \xrightarrow{+} 9$
5. For some pairs of sets R and S, $N(R) + N(S) \neq N(R \cup S)$. Invent a pair of such sets R and S, and then show that $N(R) + N(S) \neq N(R \cup S)$.
6. Explain how addition may be defined as a mapping. Is it an onto or an into mapping? Is it reversible? What is the domain? the set of images? Make an arrow diagram to illustrate your discussion.

5.3 THE BASIC ADDITION FACTS FOR THE NATURAL NUMBERS

We shall now display the *basic facts of addition* in an *addition grid,* which gives the sums of pairs of elements in the subset $\{1, 2, 3, \ldots, 9\}$ of N. Refer to the addition grid at the top of page 117. Notice that the result of the binary operation addition on the elements of an ordered pair, say (7, 4), is found at the intersection of the row† beginning with 7 along the left vertical line and the column headed 4 along the top horizontal line. The 7-row and the 4-column intersect at the number 11. Hence,

$$(7, 4) \xrightarrow{+} 11$$

or

$$7 + 4 = 11.$$

†We shall agree that "rows" are horizontal and "columns" are vertical.

ADDITION GRID

	Second Component of Ordered Number Pair									
First Component of Ordered Number Pair	+	1	2	3	4	5	6	7	8	9
	1	2	3	4	5	6	7	8	9	10
	2	3	4	5	6	7	8	9	10	11
	3	4	5	6	7	8	9	10	11	12
	4	5	6	7	8	9	10	11	12	13
	5	6	7	8	9	10	11	12	13	14
	6	7	8	9	10	11	12	13	14	15
	7	8	9	10	11	12	13	14	15	16
	8	9	10	11	12	13	14	15	16	17
	9	10	11	12	13	14	15	16	17	18

Observe that the set N does not include the number 0, and yet some of the entries (the sums) in the grid have a 0—for example, 10. Recall from Sections 3.3 and 3.5 the necessity of having a 0 digit in a place value numeration system. One must conclude that 0 used in the numeral 10 is one of the ten digits used to record numbers in the decimal place value notation, and not an element of N. The number 0 was introduced in Section 2.8 and operations with it will be discussed in a later chapter.

Notice the *symmetry* of the addition grid. It is possible to match like numerals by folding the grid along the diagonal from the upper left-hand corner to the lower right-hand corner. In which portions of the grid do you find the sum $8 + 5$ and the sum $5 + 8$? Where are the sums $5 + 5$, $1 + 1$, $8 + 8$, $3 + 3$ found in the grid? After finding $6 + 6$ in the grid, where would you look to find $6 + 7$?

EXERCISE 5.3

1. Use the addition grid to find the number that is matched with the following ordered pairs of numbers:

(a) (3, 8)	(d) (2, 6)	(g) (2, 8)	(j) (4, 5)
(b) (9, 2)	(e) (6, 7)	(h) (8, 2)	(k) (6, 6)
(c) (1, 7)	(f) (7, 6)	(i) (4, 4)	(l) (7, 7)

2. Describe where the numbers that match with the ordered pair (b, b), $b \in N$ and $b < 10$, are located in the addition grid.
3. Explain how the number matched with the ordered pair $(b, b + 1)$, $b \in N$ and $b < 9$, is located in the addition grid by using the result of Problem 2.

4. Give all the ordered pairs of natural numbers (a, b), $a < 10$, and $b < 10$ that are matched with the following numbers under the addition operation.

   (a) 11 (b) 7 (c) 16 (d) 18 (e) 14

5. Which number (or numbers) found in the addition grid of natural numbers has the greatest number of ordered pairs matched with it? Which has the least number of ordered pairs?

6. If $(a, b) \xrightarrow{+} n$, then what must (b, a) be matched with? The addition grid is an example of what type of symmetry? Explain.

5.4 THE PROPERTIES FOR ADDITION IN THE SET OF NATURAL NUMBERS

The set of numbers which we call the natural numbers has been described, and addition on any pair of these numbers has been defined as an operation. We shall now undertake to establish some properties that we expect these natural numbers to obey under the operation addition.

It seems appropriate to demand that the sum of any two natural numbers be exactly one natural number as required by the definition of the addition operation. However, if "exactly one" (uniqueness) is deleted, the demand is less strict. Then, all that is expected is that the sum of each pair of natural numbers be found in N. This leaves the door open for each pair to have many sums assigned to it. The requirement that for each $a \in N$ and $b \in N$, then $(a + b) \in N$ is called *closure for addition.*

Since both *closure* and *uniqueness* are needed to meet the requirements for an *operation,* we shall not identify closure as a distinct property.

Would you expect that any set of numbers should obey the closure property for addition? Let us investigate a subset of N, the set

$$A = \{1, 3, 5, 7, \ldots\}.$$

Do the elements of set A obey this property when any two of its elements are added? When 3 is added to 1, the sum, 4, is not an element of set A. Since

$$(1, 3) \xrightarrow{+} 4$$

requires a natural number, 4, which does not exist in set A, this set does not obey the closure property for addition. All we need in order to prove that a general statement is false (in this case, "A has closure for addition") is to show one example that contradicts it. This is called *proof by counterexample.*

Now let us consider another subset of N, the set

$$B = \{2, 4, 6, 8, \ldots\}.$$

Does set B have closure for addition? By trying several examples, $4 + 6$, $8 + 12$, $2 + 10$, we find that the sum for each of these is an element of set B. After selecting several pairs and finding no counterexample among them, we are given reasonable assurance that set B has closure for addition.

Proof that this is the case requires that we show that *every* pair of numbers in set B obeys this property. Needless to say, it would be an impossible task to verify this for every pair of numbers in this set. Regardless of how extensive the verification is, it is not proof.

We call this justification of a certain conclusion by showing its truth in many cases *inductive reasoning*. It is an extremely important part of scientific investigation. Most theorems in mathematics and laws in chemistry or physics are first discovered as educated guesses or hunches. There would be little creative science if it were not for imaginative, forward-looking inductive reasoning.

After a certain conclusion seems reasonable in the light of many verifications, then comes the search for a "proof." Whatever form this proof may take, it is concerned with what is called *deductive reasoning*. In the familiar proofs of theorems in geometry, one assumes certain knowledge and from this certain conclusions are drawn. If basic assumptions and rules for drawing conclusions are stated in advance, then the stage is set for a formal deductive proof to follow.

As stated before, two forms of reasoning are employed in making a generalization or drawing a conclusion: inductive, or discovery, reasoning and deductive reasoning that leads to a proof. Another form of proof which is useful in certain cases is proof by counterexample, as used to show that set A *did not* have closure for addition.

Later in this chapter (Section 5.8) when properties for the operations on the natural numbers have been established, we shall be able to set up a deductive proof for the conclusion that the set $B = \{2, 4, 6, 8, \ldots\}$ has closure under addition—a conclusion arrived at so far by inductive reasoning.

Use the addition grid in Section 5.3 to find the sum for

$$7 + 4$$

and the sum for

$$4 + 7.$$

Did you find the sum to be the same in both instances? Is the same number for x found for the two ordered pair forms, $(7, 4) \xrightarrow{+} x$ and $(4, 7) \xrightarrow{+} x$? A study of this grid reveals that in finding the sum of any pairs formed from numbers in the grid, the order of the number pair makes no difference. The property illustrated is called the *commutative property for addition*. It is extended to all natural numbers through the use of sets as shown in the following development.

In Problem 1(c) of Exercise 2.5, you proved for sets X and Y that $X \cup Y = Y \cup X$. If sets X and Y are disjoint sets, Definition 5.2 may be applied. So, since

$$X \cup Y = Y \cup X, \qquad X \cap Y = \emptyset,$$

we have

$$\mathrm{N}(X \cup Y) = \mathrm{N}(Y \cup X)$$

and

$$\mathrm{N}(X) + \mathrm{N}(Y) = \mathrm{N}(Y) + \mathrm{N}(X),$$

by Definition 5.2. If we let $N(X) = a$ and $N(Y) = b$, then we have

$$a + b = b + a.$$

Formally, we have:

The Commutative Property for Addition

For any two numbers $a \in N$ and $b \in N$,

$$a + b = b + a.$$

This may also be written in the ordered pair form: If $a, b, c \in N$ and $(a, b) \xrightarrow{+} c$, then $(b, a) \xrightarrow{+} c$.

To know that all pairs of natural numbers obey the commutative property for addition is significant for children, for they realize that as a result of this property the number of "basic addition facts" to be learned is reduced.

Up to this time, we have spoken only of the addition of *two* (a pair of) numbers. Since addition is a binary operation, this was our only choice. This presents a problem when it is necessary to add three numbers, as in $7 + 4 + 6$. The property which we shall introduce next will assist us in this situation. This is called the *associative property for addition.*

Referring again to Chapter 2, we find that in Problem 3(a) of Exercise 2.5 we proved that $(R \cup T) \cup S = R \cup (T \cup S)$. Consider R, S, and T as disjoint sets. Using the definition for addition, we can show that†

$$[N(R) + N(T)] + N(S) = N(R) + [N(T) + N(S)].$$

Let $N(R) = a$, $N(T) = b$, and $N(S) = c$. Then we have

$$(a + b) + c = a + (b + c).$$

Formally, we have:

The Associative Property for Addition

For any three numbers $a \in N$, $b \in N$, and $c \in N$,

$$(a + b) + c = a + (b + c).$$

The associative property gives directions for adding three numbers, such as 7, 4, and 6. This property tells us that 4 can be added to 7, then 6 added to this sum; or 6 can be added to 4, then that sum added to 7. Do you think that one of these ways is easier than the other? Whenever a sum is represented as $7 + 4 + 6$ without any parentheses, we shall agree that parentheses should be around the first two numerals, that is,

$$7 + 4 + 6 = (7 + 4) + 6.$$

†The symbol used "[ ]", called *brackets,* and that used below, "( )", called *parentheses,* indicate that the operation on the elements enclosed by them is to be performed first; the symbols may be thought of as punctuation marks.

In case we wish to add 4 and 6 instead of 7 and 4, we may apply the associative property for addition as

$$(7 + 4) + 6 = 7 + (4 + 6).$$

For convenience we shall say

$$7 + (4 + 6) = (7 + 4) + 6$$

is also an example of the associative property. In either case two numbers are being added at a time; that is, the requirement that addition be a binary operation is fulfilled. This makes it possible to find the sum of three numbers. Another use that we may make of the associative property for addition is illustrated in the example. (Review the use of expanded notation in Section 3.12.)

Example 1 Stanley needs to add 9 and 7. He cannot remember the sum and finds that he has misplaced his addition grid. He then decides to use the associative property for addition to help him in finding this sum. He knows that another name for 7 is 1 + 6, and so he writes:

$9 + 7 = 9 + (1 + 6)$	*Renaming* 7 *from the* + *grid*
$= (9 + 1) + 6$	*Associative property for addition*
$= 10 + 6$	*Renaming* (9 + 1) *from the* + *grid*
$= 16$	*Renaming* (10 + 6) *by decimal notation*

Hence,

$$9 + 7 = 16.$$

We have now set up two properties which we shall demand that the set of natural numbers obey under the binary operation addition—namely, *commutativity* and *associativity*. It should be noted that in Section 2.6 Venn diagrams were used to illustrate the above properties for union of sets.

EXERCISE 5.4

1. Identify the property used in each of the following:
   (a) $6 + 4 = 4 + 6$
   (b) $(2 + 9) + 8 = 2 + (9 + 8)$
   (c) $(3 + 1) + 7 = 7 + (3 + 1)$
   (d) $5 + (3 + 2) = 5 + (2 + 3)$
   (e) $[(4 + 5) + 6] + 3 = [4 + (5 + 6)] + 3$
2. The addition grid displayed in Section 5.3 shows 81 "basic addition facts." By using the commutative property for addition, we can reduce this number to how many?
3. An addition problem is often made simpler by using the properties in Section 5.4. Also they often assist us in finding sums "mentally." Sometimes it is convenient to "rename," or write another name for, a number. Justify each step in the following computations by giving the property or by stating that another name has been given, as the case may be. Study the example. Recall that the parentheses in (6 + 3) + 4 prescribe that 3 is to be added to 6 first, then 4 added to that sum.

EXAMPLE	$(6 + 3) + 4$	
	$4 + (6 + 3)$	*Commutative property for addition*
	$(4 + 6) + 3$	*Associative property for addition*
	$10 \quad + 3$	*Renaming* $(4 + 6)$ *from the* $+$ *grid*
	13	*Renaming* $(10 + 3)$ *by decimal notation*

(a) $6 + 9$
$(5 + 1) + 9$
$5 + (1 + 9)$
$5 + \quad 10$
$10 + 5$
15

(c) $7 + 6$
$(3 + 4) + 6$
$3 + (4 + 6)$
$3 + \quad 10$
13

(e) $4 + (8 + 6)$
$(8 + 6) + 4$
$8 + (6 + 4)$
$8 + \quad 10$
18

(b) $5 + 6$
$5 + (5 + 1)$
$(5 + 5) + 1$
$10 \quad + 1$
11

(d) $(8 + 7) + 2$
$2 + (8 + 7)$
$(2 + 8) + 7$
$10 \quad + 7$
17

4. Given $P = \{1, 3, 5, 7, \ldots\}$. Does closure exist for finding the average of each pair of numbers? Is this an operation? Explain.

5.5 THE CARTESIAN PRODUCT OF SETS

Up to this time we have discussed only two operations on sets, namely, union and intersection. The union operation on disjoint sets was used in developing the concept of the addition operation on numbers. We now wish to develop the multiplication operation on numbers. To do this we shall use a third operation on sets, which is called the *Cartesian product* operation. This operation differs from the union and intersection operations in the following way. "P union Q" and "P intersection Q" each form a unique set of *elements* from P and Q, while the Cartesian product operation on P and Q forms a unique set of *ordered pairs* whose *components* are elements of P and Q.

Given any two nonempty sets (not necessarily disjoint)

$$P = \{s, t\} \quad \text{and} \quad Q = \{n, p, r\}.$$

The *Cartesian product* of P and Q is the set

$$P \text{ X } Q = \{(s, n), (s, p), (s, r), (t, n), (t, p), (t, r)\},$$

where the symbol "X" is used to denote the Cartesian product. This is read, "The Cartesian product of P and Q is the set whose elements are the ordered pairs (s, n), (s, p), ..." "P X Q" may also be read, "P cross Q."

The set of ordered pairs in this product consists of all the pairs which can be formed from P and Q by using the elements of P as first component and the elements of Q as second component. A convenient method of finding all of the ordered pairs in P X Q is to fill in a grid as follows:

	n	p	r
s			
t			

	n	p	r
s	(s, n)	(s, p)	(s, r)
t	(t, n)	(t, p)	(t, r)

Will $P \text{ X } Q$ give the same set as $Q \text{ X } P$? Write out a grid and see.

Definition 5.5 The *Cartesian product* of the sets P and Q, $P \text{ X } Q$, is the set consisting of all the ordered pairs whose first component is an element of P and whose second component is an element of Q.

Since to a pair of sets there is assigned exactly one set, then the Cartesian product is a binary operation.

EXERCISE 5.5

1. Use $P = \{s, t\}$ and $Q = \{n, p, r\}$ in the following:
   (a) Set up a grid showing the ordered pairs of $Q \text{ X } P$.
   (b) Explain the difference between $P \text{ X } Q$ and $Q \text{ X } P$.
   (c) Is the Cartesian product operation a binary operation? Explain.
   (d) For any two sets A and B, would there ever be common elements in the Cartesian product of $A \text{ X } B$ and $B \text{ X } A$?
2. Given $A = \{a, b, c\}$ and $N = \{1, 2, 3\}$:
   (a) Find $A \text{ X } N$. (b) Find $N \text{ X } A$.
   (c) After $A \text{ X } N$ is found, is there an easy way to find $N \text{ X } A$? Explain.
3. Given $R = \{n, s, t\}$ and $T = \{a\}$:
   (a) Find the Cartesian product of R and T.
   (b) How many elements are there: In R? In T? In $R \text{ X } T$?
4. Give the number of elements in the Cartesian product of each of the following:
   (a) $P = \{a\}$ and $Q = \{x\}$
   (b) $P = \{a, b\}$ and $Q = \{x, y, z\}$
   (c) $P = \{a, b, c\}$ and $Q = \{x, y, z, d\}$
   (d) Is it possible to give the number of elements in the Cartesian product without forming the ordered pairs?
5. Given the set $A = \{3, 4, 5, 6\}$:
   (a) How many elements are there in $A \text{ X } A$?
   (b) Are there any elements in $A \text{ X } A$ of the form (b, b) where b is any element of A?
   (c) List the elements in $A \text{ X } A$.

5.6 THE MULTIPLICATION OPERATION IN THE SET OF NATURAL NUMBERS

We shall now use the Cartesian product of two nonempty, discrete, finite sets (not necessarily disjoint) in the development of the multiplication operation on natural numbers. If

$$P = \{s, t\} \quad \text{and} \quad Q = \{n, p, r\},$$

then

$$\{s, t\} \text{ X } \{n, p, r\} = \{(s, n), (s, p), (s, r), (t, n), (t, p), (t, r)\}.$$

When we consider the *number* of elements of each of these sets, we no longer use the Cartesian product operation, but another and more familiar operation that we call *multiplication:*

$$\text{N}\{s, t\} \times \text{N}\{n, p, r\} = \text{N}\{(s, n), (s, p), (s, r), (t, n), (t, p), (t, r)\}.$$

Using the more familiar symbols for numbers, the above becomes the mathematical sentence

$$2 \times 3 = 6.$$

The ordered pair form is written

$$(2, 3) \xrightarrow{\times} 6$$

and is read, "The ordered pair (2, 3) is matched with 6 under the operation multiplication." The components of the ordered pair (2, 3) are called *factors,* and the matching number 6 is called the *product.*

Notice that the symbol "×" is used to designate the multiplication operation on numbers, while the symbol "X" is used to designate the Cartesian product operation on sets. The preceding discussion is used as a basis for the following definition of multiplication:

Definition 5.6 The *multiplication operation on the natural numbers a and b* is the assignment of exactly one natural number c to the ordered pair (a, b), such that

$$\text{N}(A) \times \text{N}(B) = \text{N}(A \text{ X } B),$$

where A and B are nonempty, discrete, finite sets, and $\text{N}(A) = a$, $\text{N}(B) = b$, and $\text{N}(A \text{ X } B) = c$. That is,

$$(a, b) \xrightarrow{\times} c$$

or

$$a \times b = c.$$

From this definition we see that the multiplication operation is a *binary* operation on the components of an ordered pair of natural numbers which is matched with a *unique* natural number. Let us explore this interpretation of multiplication in the following example.

Example 1 The Norman Little League baseball team has 4 catchers and 3 pitchers. How many different batteries may be formed by the coach? Suppose that we form the Cartesian product:

		Pitchers		
		a	*b*	*c*
Catchers	*r*	•	•	•
	s	•	•	•
	t	•	•	•
	v	•	•	•

Instead of showing the batteries as ordered pairs (which is not significant here), we record each with a dot in the grid. Our interest in this problem is in determining the *number* of different batteries that are possible. From the definition,

$$\mathrm{N}(C) \times \mathrm{N}(P) = \mathrm{N}(C \text{ X } P).$$

But $\mathrm{N}(C) = 4$, $\mathrm{N}(P) = 3$, and $\mathrm{N}(C \text{ X } P) = 12$. Since $\mathrm{N}(C \text{ X } P)$ is the number of batteries represented by the dots in the grid, the number of different batteries is determined by multiplying 4 by 3,

$$4 \times 3 = 12.$$

EXERCISE 5.6

1. Given $A = \{a, b, c\}$ and $B = \{5, 7, 9, 3\}$:
   (a) Find $A \text{ X } B$. (b) Find $B \text{ X } A$.
   (c) Is $A \text{ X } B$ the same as $B \text{ X } A$? Explain.
   (d) Is $\mathrm{N}(A \text{ X } B)$ the same as $\mathrm{N}(B \text{ X } A)$? Explain.
2. Construct two nonempty sets A and B and show that $\mathrm{N}(A) \times \mathrm{N}(B) = \mathrm{N}(A \text{ X } B)$.
3. Given $R = \{2, 4\}$, $V = \{6, 3, 2\}$, and $P = \{a, b, c\}$:
   (a) Is the following a true sentence for the given sets R, V, and P?

   $$\mathrm{N}(R \text{ X } V) \times \mathrm{N}(P) = \mathrm{N}(R) \times \mathrm{N}(V \text{ X } P)$$

   Use the given sets to verify your answer. Be alert to the different operation symbols, "X" and "×," used in this problem.
   (b) Use the given sets to show whether the following is true:

   $$[\mathrm{N}(R) \times \mathrm{N}(V)] \times \mathrm{N}(P) = \mathrm{N}(R) \times [\mathrm{N}(V) \times \mathrm{N}(P)]$$

4. An ice cream parlor serves vanilla, chocolate, and strawberry ice cream in yellow, white, blue, and pink bowls. How many different ways may the ice cream be served? Record the different ways in a grid as done in the example in Section 5.6. Write the mathematical sentence, and then express it in the ordered pair form.
5. Use Definition 5.6 to determine how many different couples are possible with 15 boys and 12 girls attending the Junior Class dance.
6. Is multiplication an onto or an into mapping? Is it reversible? Identify its domain and its set of images. Make an arrow diagram to illustrate your discussion.

5.7 THE BASIC MULTIPLICATION FACTS FOR THE NATURAL NUMBERS

From the Cartesian product operation on sets we develop the *basic facts of multiplication* for the subset $\{1, 2, 3, 4, 5, 6, 7, 8, 9\}$ of N. These products are recorded in the following multiplication grid:

MULTIPLICATION GRID

	Second Component of Ordered Number Pair									
First Component of Ordered Number Pair	×	1	2	3	4	5	6	7	8	9
	1	1	2	3	4	5	6	7	8	9
	2	2	4	6	8	10	12	14	16	18
	3	3	6	9	12	15	18	21	24	27
	4	4	8	12	16	20	24	28	32	36
	5	5	10	15	20	25	30	35	40	45
	6	6	12	18	24	30	36	42	48	54
	7	7	14	21	28	35	42	49	56	63
	8	8	16	24	32	40	48	56	64	72
	9	9	18	27	36	45	54	63	72	81

To find c in $(5, 8) \xrightarrow{\times} c$, we use the same procedure as that given for the addition grid. Here the pair of numbers 5 and 8 are matched with 40. This may be written in either form:

$$(5, 8) \xrightarrow{\times} 40$$

or

$$5 \times 8 = 40.$$

Recall that when we say "5 times 8 *equals* 40" or write "$5 \times 8 = 40$," we mean that 5×8 is another name for 40; that is, 5×8 and 40 are two different names for the same number.

Like the addition grid, the multiplication grid is *symmetric* with respect to the diagonal from the upper left-hand corner to the lower right-hand corner. Make your own test of this by selecting several pairs of factors from the grid. The products represented along the diagonal are the result of using what kind of factors? When two factors are the same, we often refer to their product as a *perfect square* or, informally, as a "square number."

EXERCISE 5.7

1. Use the grid to find the number that is matched with each of the following ordered pairs of numbers under the operation multiplication:

(a) (3, 1) (e) (5, 7) (i) (4, 4)
(b) (8, 1) (f) (7, 5) (j) (4, 5)
(c) (5, 1) (g) (4, 8) (k) (7, 7)
(d) (1, 1) (h) (8, 4) (l) (6, 7)

2. Locate in the multiplication grid the numbers that match the ordered pair

$$(b, b),\ b \in N, \text{ and } b < 10.$$

3. Explain how the number matched with

$$(b, b + 1), \quad b < 9,$$

is located in the multiplication grid by using the result of Problem 2.

4. Give *two* ordered pairs found in the multiplication grid that are matched with each of the following numbers:

(a) 36 (d) 21 (g) 56
(b) 35 (e) 28 (h) 5
(c) 72 (f) 7 (i) 63

5. Find a number in the grid that is matched with only one ordered pair under multiplication. Are there any others? If there are, give them. Describe their location in the grid.
6. If $(a, b) \xrightarrow{\times} n$, then what must (b, a) be matched with? The multiplication grid is an example of what type of symmetry? Explain.

5.8 THE PROPERTIES FOR MULTIPLICATION IN THE SET OF NATURAL NUMBERS

Now that another operation on natural numbers has been defined, the multiplication operation, we shall establish the properties that we expect this set of numbers to obey under this operation.

From the definition for multiplication of two natural numbers, we have $N(A) \times N(B) = N(A \text{ X } B)$, A and B nonempty, discrete, and finite. The Cartesian product, $A \text{ X } B$, exists and is unique, and $N(A \text{ X } B)$, the number assigned to the finite set, exists and is unique. Hence, the definition for the multiplication operation in N ensures the existence and uniqueness of a product in N. Observe that "existence" implies closure; that is, for any two numbers, $a \in N$ and $b \in N$, then $(a \times b) \in N$.

From your investigation of the multiplication grid in Section 5.7, you should have observed that (5, 8) and (8, 5) have the same matching number, 40. This is as it should be. Let us explore this further by referring to the Cartesian product of two nonempty and nonequal sets, A and B. In Problem 1 of Exercise 5.6 you showed that $A \text{ X } B \neq B \text{ X } A$, but that $N(A \text{ X } B) = N(B \text{ X } A)$. By Definition 5.6, $N(A) \times N(B) = N(A \text{ X } B)$ and $N(B) \times N(A) = N(B \text{ X } A)$. Therefore,

$$N(A) \times N(B) = N(B) \times N(A).$$

If $N(A) = 5$ and $N(B) = 8$, this means that $5 \times 8 = 8 \times 5$. In fact, we shall expect all natural numbers to obey the following:

The Commutative Property for Multiplication

For any two numbers $a \in N$ and $b \in N$,

$$a \times b = b \times a.$$

This may also be written in the ordered pair form: If $a, b, c \in N$ and $(a, b) \xrightarrow{\times} c$, then $(b, a) \xrightarrow{\times} c$.

The use of this property, as in addition, reduces the number of "basic multiplication facts" that children need to learn.

Since multiplication is a binary operation, and we may be required to find a product of three factors, it seems reasonable to expect (as in addition) an associative property. That is, for the three natural numbers 4, 3, and 6

$$(4 \times 3) \times 6 = 4 \times (3 \times 6),$$

which shows two different ways to find the product of $4 \times 3 \times 6$. Again we shall make an agreement concerning the placement of parentheses when none are shown. That is,

$$4 \times 3 \times 6 = (4 \times 3) \times 6.$$

This property for the natural numbers may be developed from the Cartesian product of three nonempty sets. While it is not true that the Cartesian products $(A \text{ X } B) \text{ X } C$ and $A \text{ X } (B \text{ X } C)$ are the same, it may be verified that

$$\text{N}[(A \text{ X } B) \text{ X } C] = \text{N}[A \text{ X } (B \text{ X } C)].$$

Then from the definition of multiplication we have:

$$\text{N}(A \text{ X } B) \times \text{N}(C) = \text{N}(A) \times \text{N}(B \text{ X } C),$$
$$[\text{N}(A) \times \text{N}(B)] \times \text{N}(C) = \text{N}(A) \times [\text{N}(B) \times \text{N}(C)].$$

Letting $\text{N}(A) = a$, $\text{N}(B) = b$, and $\text{N}(C) = c$, we have

$$(a \times b) \times c = a \times (b \times c).$$

Formally, we have:

The Associative Property for Multiplication

For any three numbers $a \in N$, $b \in N$, and $c \in N$,

$$(a \times b) \times c = a \times (b \times c).$$

For example:†

$$(5 \times 3) \times 2 = 5 \times (3 \times 2)$$
$$15 \times 2 = 5 \times 6 \qquad \textit{Renaming } (5 \times 3) \textit{ and } (3 \times 2)$$
$$30 = 30 \qquad \textit{Renaming } (15 \times 2) \textit{ and } (5 \times 6)$$

†Here the reader is to assume the usual algorithm for multiplication.

Hence, the associative property for multiplication holds for these three numbers and we shall expect all the natural numbers to obey this property.

Another property that the elements of N obey connects the two operations addition and multiplication. Applications of this property are illustrated in the following examples.

Example 1 Find the perimeter of a rectangle having dimensions of 6 and 4.

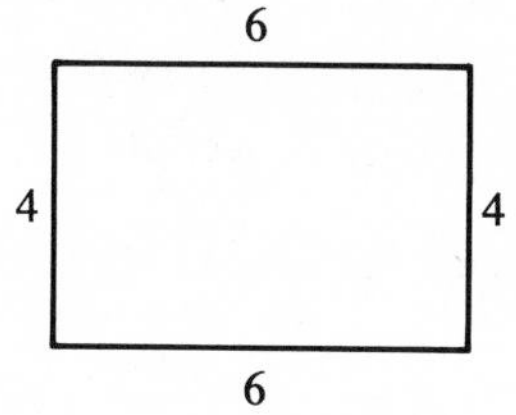

The perimeter may be found by adding the measures of two sides, $(4 + 6)$, and then multiplying this by 2, that is,

$$(4 + 6) \times 2.$$

Another method is to multiply each measure by 2, and then add these results, that is,

$$(4 \times 2) + (6 \times 2).$$

Since each method will give the same perimeter, we may show this by writing

$$(4 + 6) \times 2 = (4 \times 2) + (6 \times 2).$$

Example 2 Find an easy way to write a simpler name for

$$(8 \times 22) + (8 \times 78).$$

$(8 \times 22) + (8 \times 78)$	$= 176 + 624$	*Renaming* (8×22) *and* (8×78)
	$= 800$	*Renaming* $(176 + 624)$
But, $8 \times (22 + 78)$	$= 8 \times 100$	*Renaming* $(22 + 78)$
	$= 800$	*Renaming* (8×100)

Since the latter is the easier, you would, no doubt, choose it in preference to the former method. But, since each is equal to 800, we may write

$$(8 \times 22) + (8 \times 78) = 8 \times (22 + 78)$$

and use either of the methods for finding the simpler name. We may write the preceding sentence also as

$$8 \times (22 + 78) = (8 \times 22) + (8 \times 78).$$

Examples 1 and 2 give instances of the use of the *distributive property for multiplication over addition*. Here again we shall expect all natural numbers to obey this property. Its usefulness will be seen in many problem situations; however, it is probably most helpful in the elementary school mathematics program in developing the algorithm for multiplication of numbers represented by two or more digits. This will be discussed in Chapter 7.

Since this property connects the operations of addition and multiplication, we expect to be able to develop it from a combination of the union and the Cartesian product operations on sets. It is possible to show that

$$(A \cup B) \times C = (A \times C) \cup (B \times C),$$

A, B, and $C \neq \varnothing$ and finite, $A \cap B = \varnothing$. Using the definitions for both addition and multiplication, we have:

$$\mathrm{N}(A \cup B) \times \mathrm{N}(C) = \mathrm{N}(A \times C) + \mathrm{N}(B \times C),$$
$$[\mathrm{N}(A) + \mathrm{N}(B)] \times \mathrm{N}(C) = [\mathrm{N}(A) \times \mathrm{N}(C)] + [\mathrm{N}(B) \times \mathrm{N}(C)].$$

Let $\mathrm{N}(A) = a$, $\mathrm{N}(B) = b$, $\mathrm{N}(C) = c$, where $a, b, c \in N$, and then

$$(a + b) \times c = (a \times c) + (b \times c).$$

Formally, we have:

The Distributive Property for Multiplication over Addition

For any three numbers $a \in N$, $b \in N$, and $c \in N$,

$$(a + b) \times c = (a \times c) + (b \times c).$$

If the commutative property for multiplication is applied, then the above form becomes

$$c \times (a + b) = (c \times a) + (c \times b).$$

For practical purposes, either form will be called the distributive property for multiplication over addition. Examples 1 and 2 are illustrations of both forms.

When there is no chance of ambiguity, we shall agree that the symbol for multiplication, $\times$, may be omitted and $a(b + c)$ may be written for $a \times (b + c)$. Also, $7(2 + 9)$ is a shorter form for $7 \times (2 + 9)$. In general, pq means $p \times q$, but, of course, 53 does not mean 5×3.

With this property we are now equipped to formulate a deductive proof for the statement already conjectured (Section 5.4) through inductive reasoning, that is, that the set

$$B = \{2, 4, 6, 8, \ldots, 2k, \ldots\}, k \in N,$$

has closure for addition.

Proof: Let *any* two elements of set B be $2s$ and $2t$, $s, t \in N$. Then

$$2s + 2t = 2(s + t)$$

by the distributive property for multiplication over addition. But $(s + t)$ is an element of N, since N has closure for addition. Let $s + t = r$. Then $2r$ is also an element of B, and therefore the sum of *any* two elements of B is an element of this set. This proves that set B has closure for addition.

There is one number in the set of natural numbers, N, that has a unique quality under the operation of multiplication. In Problem 3 of Exercise 5.5, it was discovered that the number of elements in $R \text{ X } T$, where T contained only a single element, was the same as the number of elements in R. Obviously, this is true however many elements may be included in set R. This brings us to the conclusion that

$$\text{if } a \in N, \quad \text{then } a \times 1 = a$$

or, in the ordered pair form,

$$\text{if } a \in N, \quad \text{then } (a, 1) \xrightarrow{\times} a.$$

The commutative property for multiplication permits us to write also

$$1 \times a = a$$

and

$$(1, a) \xrightarrow{\times} a.$$

This tells us that when any natural number is multiplied by 1 or when 1 is multiplied by any natural number, the product is that natural number. For instance, $8 \times 1 = 8$, $1 \times 8 = 8$, $(7 + 2) \times 1 = (7 + 2)$. This unique quality of the number 1 is described formally as follows:

The Identity Element for Multiplication

There exists an element, 1, belonging to N, such that for every $a \in N$,

$$a \times 1 = 1 \times a = a.$$

The number 1 is called the *identity element for multiplication* or the *multiplicative identity element.*

EXERCISE 5.8

1. Identify the property which makes each of the following sentences true:
   (a) $5 \times 6 = 6 \times 5$
   (b) $8 \times 1 = 8$
   (c) $(6 + 3) + 2 = 6 + (3 + 2)$
   (d) $(6 \times 3) \times 2 = 6 \times (3 \times 2)$
   (e) $5 \times (3 \times 2) = 5 \times (2 \times 3)$
   (f) $6 \times (4 \times 7) = (4 \times 7) \times 6$
   (g) $(5 + 3) \times 2 = (5 \times 2) + (3 \times 2)$
   (h) $(7 \times 1) + 5 = 7 + 5$
   (i) $4 + 3 = 3 + 4$
   (j) $(2 + 6) \times 1 = 2 + 6$
   (k) $3 \times (6 + 14) = (3 \times 6) + (3 \times 14)$

(l) $12 + (5 \times 3) = (5 \times 3) + 12$
(m) $(7 + 1)4 = (7 \times 4) + (1 \times 4)$
(n) $1(23 \times 15) = (23 \times 15)$
(o) $[(7 \times 9) + 5] + 2 = [5 + (7 \times 9)] + 2$
(p) $[(2 \times 3) \times 5] \times 6 = [2 \times (3 \times 5)] \times 6$

2. Identify the properties that would make the computation of the following easier. Complete the computation, justifying each step.†
   (a) $2 \times (5 \times 32)$
   (b) $(62 \times 4) \times 25$
   (c) 14×3 (HINT: Rename 14 as $10 + 4$, and then use the distributive property.)
   (d) 32×7
   (e) $(25 + 59) + 75$
   (f) $(50 \times 57) \times 2$
   (g) $55 + (92 + 45)$
3. Given $A = \{a, b\}$, $B = \{c, d, e\}$, $C = \{f\}$, show that

$$(A \cup B) \text{ X } C = (A \text{ X } C) \cup (B \text{ X } C),$$

   and how this may be used as a basis for the distributive property for multiplication over addition.
4. Construct the required sets and by using the definition for either addition or multiplication, show that the following sentences are true:
   (a) $3 + 1 = 4$ (c) $2 + 4 = 6$ (e) $4 + 2 = 6$
   (b) $4 \times 1 = 4$ (d) $2 \times 3 = 6$ (f) $3 \times 2 = 6$
5. Show how the distributive property may be used to find
   (a) 6×7, if 6×6 is known.
   (b) 8×9, if 8×8 is known.
   (c) 7×8, if 7×7 is known.
6. Refer to the addition grid for all pairs of numbers selected from $\{1, 2, 3, \ldots 9\}$ in Section 5.3.
   (a) What is the least number of sums that must be computed if the commutative property for addition is applied in constructing the grid?
   (b) What sum in the grid has the fewest matching number pairs? Which has the most?
   (c) If the commutative property for addition is disregarded, how many sums will need to be found to completely fill the grid?
7. Refer to the multiplication grid for all pairs of numbers selected from $\{1, 2, 3, \ldots, 9\}$ in Section 5.7.
   (a) What is the least number of products that must be computed if the commutative property for multiplication is applied in constructing the grid?
   (b) By applying both the commutative property for multiplication and the multiplicative identity, how many products must be computed in constructing the grid?
   (c) If both the commutative property and the identity element for multiplication are disregarded, how many products must be found to fill in the grid?
   (d) What product in the grid has the fewest matching number pairs? Which has the most?
   (e) Why do you think the products along the diagonal (from upper left to lower right) are called "square numbers"?

†Here the reader is to assume the usual algorithms for addition and multiplication.

8. Test the following sets for closure under both operations *addition* and *multiplication:*
   (a) $\{1, 2, 3\}$ (d) $\{1, 4, 9, 16, 25, \ldots\}$
   (b) $\{1, 3, 5, 7, \ldots\}$ (e) $\{10, 5, 15\}$
   (c) $\{2, 4, 6, 8, \ldots\}$ (f) $\{1, 2, 4, 8, 16, \ldots\}$
   List a few more elements that are implied by "..." in the sets given in parts (b), (c), (d), and (f).
9. List properties to justify these statements:
   (a) $8 + (8 \times 4) = 8(1 + 4)$
   (b) $(2 + 5) + 6(2 + 5) = (1 + 6) \times (2 + 5)$
10. Use inductive reasoning, deductive reasoning, or proof by counterexample to affirm or deny the truth of:
    (a) $B = \{2, 4, 6, 8, \ldots\}$ has closure for multiplication.
    (b) $C = \{3, 5, 7, \ldots\}$ has closure for addition.
11. Is addition distributive over multiplication? Give an example.

5.9 NUMBER SYSTEMS

In Chapter 2, a *binary operational system* was defined. A *number system* may consist of one or more operational systems. That is, it consists of a set of elements (numbers), one or more binary operations uniquely defined on these elements, and the properties these elements obey under the defined operations. The relation of equality is considered to be a part of the system. This relation, along with the properties it obeys, will be discussed in a later chapter.

Much of the discussion in this textbook will be concerned with the development of different number systems, each built from the preceding one, until the final *structure* of the *real number system* is evident. It is a gradual growth, and the reader should be alert to the similarities and differences of the succeeding systems.

5.10 THE SYSTEM OF NATURAL NUMBERS

With the preceding section in mind, we shall first define the *system of natural numbers.* We have a set of elements

$$N = \{1, 2, 3, 4, \ldots\},$$

the relation of equality, and two binary operations, addition and multiplication, defined. These operations on any two of the elements of N obey the following properties:

1. The operations of addition and multiplication are commutative.
2. The operations of addition and multiplication are associative.
3. Multiplication is distributive over addition.
4. There exists an identity element for multiplication.

The following is a diagram of the *system of natural numbers:*

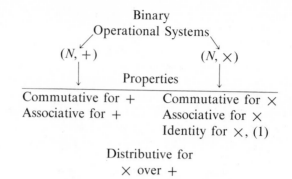

Binary
Operational Systems

$(N, +)$ ↙ ↘ $(N, \times)$

↓ Properties ↓

Commutative for +	Commutative for ×
Associative for +	Associative for ×
	Identity for ×, (1)

Distributive for
× over +

Is this system adequate for all of our needs? Does it seem strange that there is an identity element for multiplication, but none for addition? The answer to this question will be found in the next chapter.

In Chapter 3 you studied about numeration systems and in this chapter a number system has been introduced. What is the difference between a *numeration system* and a *number system?*

6

The System of Whole Numbers

6.1 THE SET OF WHOLE NUMBERS

Since the system of natural numbers has an identity element for the multiplication operation (Section 5.8) but none for the addition operation, we regard this as a deficiency. In order to eliminate this deficiency, we make an extension of the set of numbers

$$N = \{1, 2, 3, \ldots\}$$

so as to provide a number that will be an *identity element for the addition operation.* For the present we shall call this number z. The union of the set consisting of only this new number and the set of natural numbers is a unique set that we shall call the *set of whole numbers;* that is,

$$\{z\} \cup \{1, 2, 3, \ldots\} = \{z, 1, 2, 3, \ldots\}.$$

So, the set of whole numbers is identified as

$$W = \{z, 1, 2, 3, \ldots\}.$$

If this new number z is to be an *additive identity element,* then when it is added to any number in W, the sum will be that number; that is,

$$\text{For each } a \in W, (a, z) \xrightarrow{+} a \text{ and } (z, a) \xrightarrow{+} a,$$
$$\text{or} \quad a + z = z + a = a.$$

The set of natural numbers is a subset of the set of whole numbers:

$$N \subset W.$$

This being the case, the set N still has all the definitions and properties previously discussed. We shall expect the set W to have these properties and possibly to acquire some more because of the new number z. Here we shall investigate the definitions for addition and multiplication and their properties, now using the elements of set W. Since z is the only new element, it is the only one that needs to come under our consideration.

6.2 THE ADDITION OPERATION IN THE SET OF WHOLE NUMBERS

The definition of addition in the set of natural numbers was developed from the union of disjoint, nonempty, discrete, finite sets. We shall now explore this definition when the empty set is allowed. Let A represent any nonempty set and $\varnothing$ represent the empty set. Consider the union of A with $\varnothing$. Refer to Problem 5(a) in Exercise 2.5.

$$A \cup \varnothing = \varnothing \cup A = A$$
$$\mathrm{N}(A) + \mathrm{N}(\varnothing) = \mathrm{N}(\varnothing) + \mathrm{N}(A) = \mathrm{N}(A)$$

If a is the natural number assigned to set A and 0 (*zero*) is the number assigned to $\varnothing$ (Section 2.8), then we have

$$a + 0 = 0 + a = a.$$

But we are searching for a number z such that $a + z = z + a = a$. We therefore conclude that our new number z is none other than the number 0 which we assigned to the empty set. So, the set of whole numbers is

$$W = \{0, 1, 2, 3, 4, \ldots\}.$$

Suppose that we consider the union of the empty set with itself. We have:

$$\varnothing \cup \varnothing = \varnothing,$$
$$\mathrm{N}(\varnothing) + \mathrm{N}(\varnothing) = \mathrm{N}(\varnothing),$$
$$0 + 0 = 0.$$

Formally, we have:

The Identity Element for Addition

There exists an element, 0, belonging to W, such that for each $a \in W$,

$$a + 0 = 0 + a = a.$$

The number 0 is called the *identity element for addition* or the *additive identity element.*

Addition was defined as a binary operation in the set of natural numbers. We find that with the inclusion of 0 there is no evidence of the lack of closure and uniqueness. Hence, we are prepared to accept addition in the set of whole numbers as a binary operation.

The definition for the operation, addition of whole numbers, follows.

Definition 6.2 The *addition operation on the whole numbers a and b* is the assignment of exactly one whole number c to the ordered pair (a, b) such that

$$\mathrm{N}(A) + \mathrm{N}(B) = \mathrm{N}(A \cup B),$$

where A and B are discrete, finite sets, $A \cap B = \varnothing$, and $\mathrm{N}(A) = a$, $\mathrm{N}(B) = b$, and $\mathrm{N}(A \cup B) = c$. That is,

$$(a, b) \xrightarrow{+} c$$

or

$$a + b = c.$$

Compare this definition with the one which defines addition in the set of natural numbers (Definition 5.2). What justification can you give that Definition 5.2 applies only to nonempty sets, while Definition 6.2 does not have this restriction?

Example 1 Let X represent the set of birds around a feeder. This set consists of a jay, a cardinal, and a mockingbird. Let Y represent the set of birds consisting of a sparrow and a thrush which are in an oak tree. How many birds are there around the feeder when those in the tree join the ones at the feeder? Using Definition 6.2 or Definition 5.2, we have $3 + 2 = 5$. One must keep in mind that the sets of birds are not added, they are joined. The numbers of the birds in the two sets are added in determining the number of birds in the union of the two sets. This is the provision of the definition.

Example 2 Given $P = \{\mathcal{E}, \%, \#, \$, *\}$ and $Q = \{\ \}$. Since $\mathrm{N}(P) = 5$ and $\mathrm{N}(Q) = 0$, we have by Definition 6.2 $\mathrm{N}(P \cup Q) = 5 + 0 = 5$. That is, $(5, 0) \xrightarrow{+} 5$.

6.3 THE BASIC ADDITION FACTS FOR THE WHOLE NUMBERS

The addition grid for the whole numbers is an extension of the one constructed for the natural numbers (Section 5.3). It includes the sums of all the ordered number pairs in that grid and the sums with the additive identity as one or both of the addends. This grid is a record of the sums of all the ordered number pairs of this subset of W:

$$D = \{0, 1, 2, 3, 4, 5, 6, 7, 8, 9\}.$$

The grid may be filled in with these sums by selecting sets to which these ordered number pairs are assigned and then applying the definition for addition for whole numbers.

ADDITION GRID

+	0	1	2	3	4	5	6	7	8	9
0	0	1	2	3	4	5	6	7	8	9
1	1									
2	2									
3	3									
4	4									
5	5									
6	6									
7	7									
8	8									
9	9									

By filling in the remainder of the grid, you will see that the natural number grid for addition has been bordered on top and down the left side with the sums when at least one addend is the additive identity element. By a careful look at the grid, you should observe that the whole numbers conform to the same properties for addition that the natural numbers obey. These will be discussed more formally in the next section.

6.4 THE PROPERTIES FOR ADDITION IN THE SET OF WHOLE NUMBERS

Recalling the properties of the natural numbers for addition, we shall expect that the new number 0 (the *additive identity*) will obey these properties also. By admitting the empty set, Definition 6.2 tells us that there exists a unique whole number for each pair of whole numbers under the operation addition. That is, whenever any two whole numbers are added, one and only one whole number is obtained.

Since the elements of N, a subset of the whole numbers W, obey the commutative property for addition, we shall expect the elements of W to do likewise. Consequently, we must check to see if it is possible for 0 to fit into this requirement. The development in Section 6.2 shows that with $\text{N}(A) = a$ and $\text{N}(\varnothing) = 0$, $a, 0 \in W$, we have $a + 0 = 0 + a$. This justifies our claim that *the set of whole numbers has the commutative property for addition.*

Recall that the elements of N, a subset of W, obey the associative property for addition. Again, we verify that the elements of W obey this property:

$$\text{For each } a, b, c \in W, \qquad (a + b) + c = a + (b + c).$$

Since this property holds (Section 5.4) as long as a, b, and c belong to N, a subset of W, we must check it for the new element, 0. What occurs, for example, if b is 0? Then we have:

$$\text{For each } a, c \in W, \qquad (a + 0) + c = a + (0 + c).$$

Since $a + 0 = a$ and $0 + c = c$, we have

$$a + c = a + c,$$

and the property holds in this instance. The number 0 may be selected for any a, b, or c in $(a + b) + c = a + (b + c)$. Show that the associative property for addition holds for each of these selections. Now we can say that *the set of whole numbers has the associative property for addition.*

EXERCISE 6.4

1. Fill in the addition grid shown in Section 6.3.
   (a) How many "sums" are needed including those already in the grid?
   (b) What addition property makes the grid symmetric with respect to the "upper left to the lower right" diagonal?
   (c) Using the commutative property for addition and the additive identity element, what is the minimum number of basic addition facts that need to be learned?
2. Justify each of the following sentences.

   EXAMPLE $(6 + 0) + 4 = 6 + 4$ Additive identity element

   (a) $(3 + 8) + 0 = 0 + (3 + 8)$
   (b) $(7 + 6) + 0 = 7 + (6 + 0)$
   (c) $[(4 + 8) + (2 + 3)] + 0 = (4 + 8) + (2 + 3)$
   (d) $5 + 0 = 0 + 5$
   (e) $(0 + 9) + (6 + 4) = 9 + (6 + 4)$
3. Find the whole number for b which will make each of the following true. Is this always possible?

   (a) $(3, b) \xrightarrow{+} 5$ (c) $(0, b) \xrightarrow{+} 7$ (e) $(4, b) \xrightarrow{+} 3$
   (b) $(b, 3) \xrightarrow{+} 5$ (d) $(b, 0) \xrightarrow{+} 0$ (f) $(4, b) \xrightarrow{+} 0$

6.5 THE PROCESS OF SUBTRACTION IN THE SET OF WHOLE NUMBERS

Each of the statements in Problem 3 of Exercise 6.4 may be changed from the ordered pair form to a mathematical sentence (recall Section 5.2). That is, $(3, b) \xrightarrow{+} 5$ may be written as the mathematical sentence

$$3 + b = 5.$$

Finding the number b that will make the sentence *true* may be thought of as asking: "What number added to 3 will give the sum 5?" The process of finding the missing addend b is called *subtraction*. The number that will make this sentence true is 2. Such a number is said to be a *solution* of the sentence. Other numbers make *false* sentences, such as $3 + 1 = 5$, $3 + 3 = 5$.

We might have expected that there should be a solution for each of the sentences in Problem 3. However, Problems 3(e) and 3(f) have no solutions in the set of whole numbers. That is, the system of whole numbers does not have closure under subtraction and so subtraction is not called an operation. However, we may discuss subtraction in the set of whole numbers with certain restrictions which we shall discover.

Subtraction may also be developed in a manner similar to that used in developing addition. We use an operation on sets as an approach. Suppose that we have a set A consisting of 3 elements and would like to have a set C consisting of 5 elements. It is necessary to find a set X (such that $A \cap X = \varnothing$) that can be joined with the set of 3 elements to form the set of 5 elements. This requires the union operation:

$$A \cup X = C,$$
$$\mathrm{N}(A) + \mathrm{N}(X) = \mathrm{N}(C),$$
$$3 + b = 5.$$

This mathematical sentence has a missing addend, indicated here by b. We can describe this by the form

$$5 - 3 = b,$$

where "$5 - 3$" is read "five *minus* three" and is called the *difference*. We see that the two sentences are related, and we write:

$$\text{If } 3 + b = 5, \quad \text{then } 5 - 3 = b.$$

We can also say that the matching number for (5, 3) under subtraction, $(5, 3) \xrightarrow{-} b$, is the same as the missing addend in $(3, b) \xrightarrow{+} 5$. Thus, subtracting 3 is said to *undo* adding 3.

This missing addend may be found by using the addition grid developed in Section 6.3. Start with the row of the first component, 3. Proceed horizontally in the 3-row to find the entry 5. Now move vertically to find the numeral, 2, which heads this column. This represents the missing addend. Thus we have found the number that can be added to 3 to give 5. It is 2.

Next we consider the question: "Is there a whole number n that can be added to 6 to give the sum 4?" This may be written as the mathematical sentence:

$$6 + n = 4.$$

If there is such a number, then $4 - 6$ is equal to this whole number. Using the addition grid in the manner described in the preceding paragraph, we see that there is no whole number that can be added to 6 to give 4. That is, there is no whole number that can be matched with (4, 6) under subtraction. Notice, in this case, that the second component of the ordered number pair (4, 6) is greater than the first, while in the preceding example, the second component of the ordered number pair (5, 3) is less than the first component. Which of these ordered pairs has a matching number under subtraction? We have now discovered the restriction that must be placed on the numbers in the ordered number pair (a, b) for subtraction in the set of whole numbers; b must be less than or equal to a.

Definition 6.5 For $a, b, n \in W$ such that $b + n = a$,

the *difference* $a - b$ is equal to n.

That is,

$$\text{if } b + n = a, \text{ then } a - b = n.$$

When the set of whole numbers has been extended to the set of integers in a later chapter, the restriction on a and b will be lifted and subtraction as an *operation* will be defined.

EXERCISE 6.5

1. (a) Apply Definition 6.5 to express each of the following as an addition sentence.
   (1) $15 - 7 = n$ (4) $13 - 0 = q$ (7) $14 - 2 = z$
   (2) $10 - 6 = n$ (5) $(17, 3) \xrightarrow{-} y$ (8) $(9, 10) \xrightarrow{-} x$
   (3) $(16, 8) \xrightarrow{+} p$ (6) $11 - 5 = x$ (9) $8 - 12 = p$
   (b) Use the addition grid to find the missing addend of each of the addition sentences. Does the + grid yield the missing addend for each sentence?
2. (a) Complete each statement with a related subtraction form by applying Definition 6.5.

   EXAMPLE If $7 + n = 12$, then $\underline{12 - n = 7.}$

   (1) If $5 + n = 7$, then ______. (4) If $4 + 1 = a$, then ______.
   (2) If $6 + x = 6$, then ______. (5) If $b + 0 = 3$, then ______.
   (3) If $b + 4 = 9$, then ______. (6) If $3 + x = 2$, then ______.
   (b) Explain why Definition 6.5 eliminates the necessity for learning "subtraction facts."

6.6 THE MULTIPLICATION OPERATION IN THE SET OF WHOLE NUMBERS

In arriving at a definition for multiplication of natural numbers from the Cartesian product of two sets, the empty set was excluded. Since 0 is the number assigned to this set, we shall develop the multiplication of whole numbers from the Cartesian product of any two sets, at least one being the empty set.

Consider the Cartesian product of set A and $\varnothing$, $A = \{a, b, c\}$.

$$\{a, b, c\} \text{ X } \{\ \ \} = \{\ \ \}$$
$$N\{a, b, c\} \times N\{\ \ \} = N\{\ \ \}$$

Since the number assigned to $\{a, b, c\}$ is 3 and the number assigned to $\varnothing$ is 0, we have

$$3 \times 0 = 0$$

or

$$(3, 0) \xrightarrow{\times} 0.$$

Let set A be any set and $N(A) = a$; then we are aware of:

The Role of Zero in Multiplication

For each $a \in W$, $a \times 0 = 0 \times a = 0$.

This says that if at least one factor is 0, then the product is 0. Do you see that if the product is 0, then at least one factor must be 0?

If set A is any set and $N(A) = a$, does it seem reasonable to say that for $a \in W$, $a \times 0 = 0$? Does this include the case where $a = 0$—that is, $0 \times 0 = 0$?

The definition for the operation for multiplication of natural numbers was established in Chapter 5. The preceding discussion established multiplication using the new number 0 as a factor. We are now prepared for the following definition of the multiplication operation for the set of whole numbers.

Definition 6.6 The *multiplication operation on the whole numbers a and b* is the assignment of exactly one whole number c to the ordered pair (a, b) such that

$$N(A) \times N(B) = N(A \text{ X } B),$$

where A and B are discrete, finite sets, and $N(A) = a$, $N(B) = b$, and $N(A \text{ X } B) = c$. That is,

$$(a, b) \xrightarrow{\times} c,$$

or

$$a \times b = c.$$

Is it possible for either A or B or both sets to be empty? Is the operation, as defined, binary? Observe that the Cartesian product operation is performed on two sets, while multiplication is performed on two numbers.

6.7 THE BASIC MULTIPLICATION FACTS FOR THE WHOLE NUMBERS

Since $N \subset W$, the grid for the basic multiplication facts may be constructed by joining a row and a column of the "0-facts" to the already established multiplication grid for the natural numbers.

MULTIPLICATION GRID

×	0	1	2	3	4	5	6	7	8	9
0	0	0	0	0	0	0	0	0	0	0
1	0									
2	0									
3	0									
4	0									
5	0									
6	0									
7	0									
8	0									
9	0									

6.8 THE PROPERTIES FOR MULTIPLICATION IN THE SET OF WHOLE NUMBERS

Following the discussion of the properties for addition in Section 6.4, we wish the whole numbers to have all of the properties of the natural numbers. Our concern is whether the number 0 will comply.

All of the natural numbers of the set of whole numbers obey the commutative property for multiplication. Can we justifiably demand that 0 also obey this property? Following the discussion given in Section 6.6, it can be shown that for any nonempty set A,

$$A \text{ X } \varnothing = \varnothing \text{ X } A.$$

Applying Definition 6.6, we have

$$a \times 0 = 0 \times a.$$

This assures us that *the set of whole numbers has the commutative property for multiplication.*

Since the natural numbers obey the associative property for multiplication, we shall expect the number 0 to conform to this also. With a, b, and c natural numbers, we know that

$$(a \times b) \times c = a \times (b \times c).$$

It is left to you to demonstrate that if any of the numbers a, b, or c is 0, then this property still holds. We can now say that *the set of whole numbers has the associative property for multiplication.*

Since the distributive property already holds for natural numbers, we shall consider only instances where zero is involved. For each $a, b \in W$ is

$$a(b + 0) = (a \times b) + (a \times 0) \text{ true?}$$

Renaming $(b + 0)$ as b (0 is the additive identity), and renaming $(a \times 0)$ as 0, we have $ab = ab$. The reader may verify any other possibilities. Thus *the set of whole numbers has the distributive property for multiplication over addition.*

Does the multiplicative identity element for the natural numbers carry over to the set of numbers that includes 0? Find

$$\varnothing \text{ X } \{p\}.$$

Apply the definition for multiplication of whole numbers. This results in

$$0 \times 1 = 0.$$

This convinces us that there exists in this set of numbers, W, the *multiplicative identity element* 1, such that for any $a \in W$,

$$a \times 1 = a.$$

EXERCISE 6.8

1. Make a multiplication grid.
   (a) Is this grid symmetric with respect to a diagonal? Identify the diagonal. What property makes the grid symmetric?
   (b) Using this property and considering the roles that 0 and 1 play in multiplication, find the least number of basic multiplication facts that need to be learned.
2. Given $A = \{r, s, t, u, v\}$:
   (a) Find $A \text{ X } \{\ \ \}$.
   (b) What numbers are associated with sets

   $$A, \quad \{\ \ \}, \quad \text{and} \quad A \text{ X } \{\ \ \}?$$

   (c) Use the definition for multiplication to write a statement about these numbers.
   (d) Demonstrate that

   $$A \text{ X } \{\ \ \} = \{\ \ \} \text{ X } A.$$

   (e) Construct a nonempty set B. Substitute it for $\{\ \ \}$ in part (d), and demonstrate whether the statement is now true.
3. Construct three sets and show that the associative property for multiplication holds for the numbers assigned to these sets.
4. Make a multiplication grid that includes only those products that you think are necessary to know.
   (a) What properties were used to eliminate some of the "basic facts"?
   (b) Are the products recorded along the diagonal (from the upper left to the lower right) of the grid necessary? The product of what number pairs are recorded here?

5. Find the product of 7×7 in the grid, and then use the distributive property for multiplication over addition and the multiplicative identity element to find 7×8.
6. Justify each of the following sentences by naming the property.
   (a) $(9 \times 0) \times 4 = 9 \times (0 \times 4)$
   (b) $(6 \times 3) \times 1 = 1 \times (6 \times 3)$
   (c) $(7 \times 8) + (3 \times 8) = (7 + 3) \times 8$
   (d) $(0 \times 4) \times 3 = 3 \times (0 \times 4)$
   (e) $[(9 \times 6) + (2 \times 6)] \times 1 = (9 \times 6) + (2 \times 6)$
7. With $1 \times 0 = 0$ given, show that $2 \times 0 = 0$ by renaming 2 and applying the distributive property for multiplication over addition. Justify each step in the development by naming the property that applies.
8. Find the whole number for b that will make each of the following true. Is this always possible?
   (a) $(2, b) \xrightarrow{\times} 6$
   (b) $(b, 2) \xrightarrow{\times} 6$
   (c) $(0, b) \xrightarrow{\times} 8$
   (d) $(b, 0) \xrightarrow{\times} 0$
   (e) $(b, b) \xrightarrow{\times} 9$
   (f) $(b, b) \xrightarrow{\times} 5$
   (g) $(3, b) \xrightarrow{\times} 3$
   (h) $(5, b) \xrightarrow{\times} 2$

6.9 THE PROCESS OF DIVISION IN THE SET OF WHOLE NUMBERS

Each of the statements in Problem 8 of Exercise 6.8 may be changed from the ordered pair form to a mathematical sentence. That is,

$$(2, n) \xrightarrow{\times} 6$$

may be written as the mathematical sentence

$$2 \times n = 6.$$

It is possible to use the multiplication grid to find the missing factor n in the same manner in which the addition grid was used to find the missing addend. The process of finding the missing factor n is called *division.* Now if

$$2 \times n = 6,$$

we may write a related sentence, using the division symbol, "$\div$,"

$$6 \div 2 = n$$

or

$$(6, 2) \xrightarrow{\div} n.$$

These are read, "6 divided by 2 is equal to n" and "the ordered pair (6, 2) is matched with n under division"; 6 is called the *dividend,* 2 the *divisor,* and n the *quotient.* In this case, the quotient is 3. From this discussion we see that dividing by 2 is *undoing* multiplying by 2. The relatedness of the multiplication and division sentences is expressed in the following manner:

$$\text{If } 2 \times n = 6, \text{ then } 6 \div 2 = n.$$

On the other hand,

$$(5, n) \xrightarrow{\times} 2$$

or

$$5 \times n = 2$$

has no solution in the set of whole numbers. Use the multiplication grid for whole numbers to check this. Hence (5, 2) is not a possible pair to use for division as long as we are restricted to the set of whole numbers. Thus, division cannot be defined as an operation. Do you see why? Division may be defined for *certain* pairs of numbers in W. The restrictions on these pairs of numbers are evident from Definition 6.9 and the discussion that follows.

Definition 6.9 For $a, b, n \in W$. $b \neq 0$, such that $b \times n = a$,

$$\text{the } quotient\ a \div b \text{ is equal to } n,$$

that is,

$$\text{if } b \times n = a, \text{ then } a \div b = n.$$

Why is $b \neq 0$ included in the definition of division? The following examples should make this clear. By the definition of division,

$$8 \div 0 = n$$

means that we must find a number n which when multiplied by 0 gives the product 8. In other words, is it possible to find a number n which will make the following sentence true?

$$0 \times n = 8$$

By Section 6.6, if a factor is 0, the product cannot be a nonzero number. Hence, $8 \div 0 = n$ has no solution.

Let us now consider $0 \div 0 = n$. This means that we must find a number n that will make the following sentence true:

$$0 \times n = 0.$$

Since any number n will make the sentence true, the solution is not unique.

We now see that difficulties arise when division by 0 is attempted; that is, there is no number for $8 \div 0$ and every number for $0 \div 0$. Consequently, we choose to exclude 0 as a divisor, or we may say that division by zero has no meaning.

What is the solution of $0 \div 5 = n$? This requires that we find a number n which when multiplied by 5 will give 0:

$$5 \times n = 0.$$

Here the solution, 0, exists and is unique.

We summarize the three cases as follows:

$a \div 0 = \bar{n}$,	$a \neq 0$	*No solution*	*Meaningless*
$0 \div 0 = n$		*Infinite number of solutions*	
$0 \div a = n$,	$a \neq 0$	*One solution,* 0	

As in the case of subtraction, division is defined for only certain pairs of whole numbers, and Problems 8(c), 8(f), and 8(h) of Exercise 6.8 have no solutions in the set of whole numbers. Thus, we see that the set of whole numbers does not have closure under division, and hence division is not considered to be an operation in W. After the set W has been extended in a later chapter, it may be possible to find a solution for the sentence in Problem 8(h), and division may then be considered to be an operation for the extended set. Problem 8(f) will require further study. In the meantime we may investigate sentences involving whole numbers that might be expected to occur in the multiplication grid.

EXERCISE 6.9

1. (a) Apply Definition 6.9 to express each of the following as a multiplication sentence.

   (1) $28 \div 7 = n$ (4) $0 \div 7 = n$ (7) $(2, 12) \overset{\div}{\Rightarrow} p$

   (2) $35 \div 5 = n$ (5) $81 \div 9 = q$ (8) $15 \div 4 = n$

   (3) $(72, 8) \overset{\div}{\Rightarrow} p$ (6) $(2, 1) \overset{\div}{\Rightarrow} x$ (9) $15 \div 0 = q$

   (b) Use the multiplication grid to find the missing factor of each of the multiplication sentences. Is it possible to find the missing factor from the $\times$ grid for each?

2. (a) Complete each statement with a related division form by applying Definition 6.9.

   EXAMPLE If $4 \times n = 20$, then <u>$20 \div 4 = n$.</u>

   (1) If $7 \times n = 21$, then______. (4) $3 \times 4 = a$

   (2) If $8 \times n = 32$, then______. (5) $15 \times n = 16$

   (3) If $b \times 6 = 42$, then______. (6) $9 \times n = 5$

   (b) Explain why Definition 6.9 eliminates the necessity for learning "division facts."

6.10 ORDER OF WHOLE NUMBERS AND THE NUMBER LINE

The concept of order of the natural numbers was developed in Chapter 2 from ordered nonequivalent sets along with the numbers assigned to them. That is, since $\{p, q, r\}$ contained fewer members than $\{x, y, z, w, n\}$, we said that

$$N\{p, q, r\} < N\{x, y, z, w, \text{n}\}$$

or

$$3 < 5.$$

Likewise, $\varnothing$ has fewer members than $\{p\}$, and $N(\varnothing) < N\{p\}$ or $0 < 1$. So we have the order of the whole numbers (see also Section 2.8) as:

$$0 < 1 < 2 < 3 < 4 < 5 < 6 \cdots.$$

We shall now match these numbers in this order with "equally spaced" points of a line:

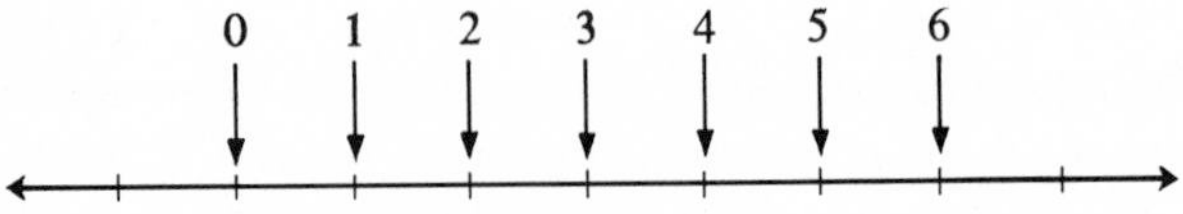

Such a matching produces the *number line*. Every element in set W corresponds to a point in the line. It is important that you realize that there are many more points—in fact, an infinite number of them—each of which will later have a matching number. Can you imagine an infinite number of points between any two whole numbers, and the kind of numbers that might be matched with them? What about the points represented to the left of 0? Shall we have numbers for them? The drawing here represents a portion of the number line showing the set of whole numbers.

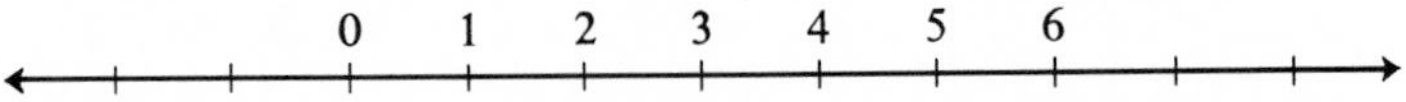

With the numbers "lined up" in this manner on the number line, we observe that a number is greater than those to the left of it and less than those to the right of it. Zero is greater than what number? The least whole number is 0, but there is no greatest number in this set. The number line is quite satisfactory in giving us a *picture* of the order of the whole numbers, but we now seek a *definition* which we may use to determine the order of any two numbers of W. Since the set of natural numbers is a subset of the set of whole numbers, whatever statement about order we use for W must apply to N. The following definition for order of the whole numbers is consistent with the idea of order already established for the natural numbers.

Definition 6.10(a) For each $a, b, c \in W$ and $c \neq 0$,

$$\text{if } a + c = b, \text{ then } a < b.$$

Example 1 Place $<$ between 23 and 14 so as to make a true sentence. We seek a value for c such that either $14 + c = 23$ or $23 + c = 14$. It is clear that $c = 9$ makes the first sentence true, while there is no number in set W that will make the latter sentence true. Then, by definition, $14 < 23$.

Example 2 Is there a whole number c that can be added to $(8 + 7)$ to make 15? If 0 is added to $(8 + 7)$, the sum is 15, which might lead us to declare that $(8 + 7)$ is less than 15. But Definition 6.10(a) excludes $c = 0$. Hence,

$(8 + 7) \nless 15$. Does it seem appropriate to say, "If $(8 + 7) \nless 15$, then $(8 + 7) = 15$ or $(8 + 7) > 15$"?

It is important to realize how the order symbols, $<$ and $>$ (Section 2.9), are related. If $12 < 24$, then this may be written $24 > 12$. That is:

Definition 6.10(b) For each $a, b \in W$,

$$\text{if } a < b, \quad \text{then } b > a.$$

The statement $4 < 7 < 8$ means $4 < 7$ *and* $7 < 8$. It is also possible to write this as $8 > 7 > 4$ or $8 > 7$ *and* $7 > 4$.

Definition 6.10(c) For each $a, b, c \in W$,

$$\text{if } a < c < b, \quad \text{then } a < c \quad \textit{and} \quad c < b;$$
$$\text{if } b > c > a, \quad \text{then } b > c \quad \textit{and} \quad c > a.$$

The statement "$(2 + 3)$ is either less than *or* equal to (2×4)" is written

$$(2 + 3) \leq (2 \times 4).$$

It may be written,

$$(2 + 3) < (2 \times 4) \quad \textit{or} \quad (2 + 3) = (2 \times 4).$$

This means that it is either one or the other, but not both. Then, would you say that $(2 + 3) \leq (2 \times 4)$ is a true statement? To answer this question, first check to see if $(2 + 3)$ is less than (2×4). Finding that this is true, but that $(2 + 3)$ is not equal to (2×4), we see that the statement $(2 + 3) \leq (2 \times 4)$ holds since the symbol $\leq$ is to be interpreted as "either . . . or." What about $(3 + 5) \leq (2 \times 4)$? This is also true since $=$ holds and $<$ does not.

Formally, we have:

Definition 6.10(d) For each $a, b \in W$,

$$\text{if } a \leq b, \quad \text{then } a < b \quad \textit{or} \quad a = b;$$
$$\text{if } a \geq b, \quad \text{then } a > b \quad \textit{or} \quad a = b.$$

6.11 THE =, <, AND > RELATIONS

The relation "equals" (=), is called an *equivalence relation.* Are other relations such as "is less than" (<) and "is greater than" (>) also equivalence relations?

What is meant by an equivalence relation? Such a relation must obey the following properties, where R stands for the relation and a, b, and c belong to set S.

Equivalence Relation Properties

Reflexive property:	For each $a \in S$, $a \mathrm{R} a$.
Symmetric property:	For each $a, b \in S$, if $a \mathrm{R} b$, then $b \mathrm{R} a$.
Transitive property:	For each $a, b, c \in S$, if $a \mathrm{R} b$ and $b \mathrm{R} c$, then $a \mathrm{R} c$.

We shall now see whether the equals relation qualifies as an equivalence relation in the whole number system. Set S becomes set W while a, b, and c are numbers of W. The relation R becomes the equals relation, =. The properties now say the following:

Reflexive property:	For each $a \in W$, $a = a$.
Symmetric property:	For each $a, b \in W$, if $a = b$, then $b = a$.
Transitive property:	For each $a, b, c \in W$, if $a = b$ and $b = c$, then $a = c$.

The reflexive property now states that every whole number is equal to itself. If a first whole number is equal to a second, the symmetric property states that the second number is equal to the first. If a first whole number is equal to a second and that second number is equal to a third, by the transitive property the first number is equal to the third. Since there appears to be no instance in which these properties are not fulfilled, we accept the *equals relation* as an *equivalence relation.* Note: If one instance can be cited that does not hold, this is sufficient evidence to *disprove* any generalization. The one instance is called a *counterexample.* This method of *proof by counter-example* was discussed in the preceding chapter.

Is the $<$ relation an equivalence relation in set W? Using replacements for a, b, and c the whole numbers 2, 4, and 7, respectively, and for R the $<$ relation, we shall test to see whether each of the properties holds for these numbers.

Reflexive:	$2 < 2$
Symmetric:	if $2 < 4$, then $4 < 2$
Transitive:	if $2 < 4$ and $4 < 7$, then $2 < 7$

Are each of these sentences true? Does the relation, $<$, for 2, 4, 7 hold for the reflexive, symmetric, and transitive properties? Is this a counterexample?

Is the relation $>$ an equivalence relation? Devise a procedure for testing to see whether this relation holds for each of the properties.

An arrow diagram is convenient in testing for equivalence relations. To indicate $a \mathrm{R} b$, draw an arrow from a to b as,

Then, $b \mathrm{R} a$ is indicated by

The three properties are represented in the following manner:

1. The reflexive property, $a \mathrm{R} a$, has a "loop" on a.

2. The symmetric property, if $a \mathrm{R} b$ then $b \mathrm{R} a$, is a "two-way street."

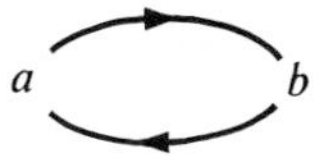

3. The transitive property, if $a \mathrm{R} b$ and $b \mathrm{R} c$, then $a \mathrm{R} c$, is a "short-cut" from a to c.

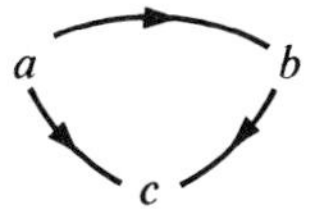

Arrow diagrams have been constructed for the *equals* and the *is less than* relations where a, b, and c are elements of W.

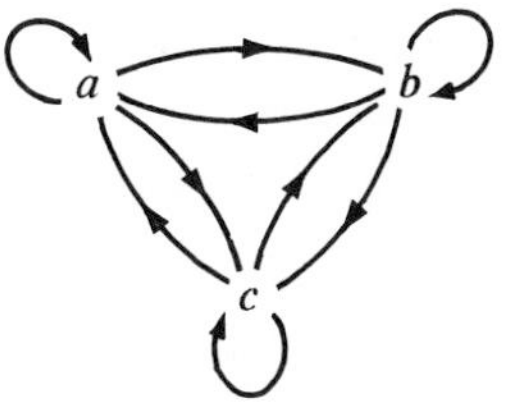

Equals Relation

Figure 6–1

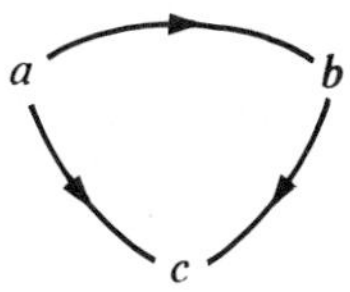

Is Less than Relation

Figure 6–2

In Figure 6–1 the reflexive property holds since each element has a "loop"; the symmetric property holds since each pair of elements has a "two-way street"; the transitive property holds since each "route" has a "shortcut." Hence, the equals relation is an equivalence relation as illustrated. Consider Figure 6–2. The arrows indicate that the "route" from a to b then b to c has a "short-cut" from a to c. This shows that the transitive property holds. But, since there is no "loop" on each element and no "two-way streets," we conclude that only the transitive property holds for the relation, $<$. Hence, this is not an equivalence relation.

Suppose that the equals ($=$) relation does not exist between two numbers a and b; then the "not equals" ($\neq$) relation must exist. That is, two numbers are either equal or not equal. In case the two numbers are not equal ($\neq$), then one is either less than ($<$) the other or greater than ($>$) the other.

For example, consider the three pairs of numbers represented by these numerals,

$$(2 + 3) \text{ and } 5;$$
$$(2 \times 3) \text{ and } (3 + 5);$$
$$(1 + 2) \text{ and } (1 + 0);$$

and answer the following questions.

Is $2 + 3 = 5$?	Is $2 \times 3 = 3 + 5$?	Is $1 + 2 = 1 + 0$?
Is $2 + 3 < 5$?	Is $2 \times 3 < 3 + 5$?	Is $1 + 2 < 1 + 0$?
Is $2 + 3 > 5$?	Is $2 \times 3 > 3 + 5$?	Is $1 + 2 > 1 + 0$?

Notice that in each case only one relation exists between each pair of numbers. It is never said that two numbers are $=$, $<$, and $>$. Any two numbers are related in *one and only one of these ways.* Formally, we have:

The Trichotomy Property

For each $a, b \in W$,

$$a = b \quad \text{or} \quad a < b \quad \text{or} \quad a > b.$$

The trichotomy property applies to the set of natural numbers also and should be added to the list of the properties of the system of natural numbers given in Section 5.10.

By the trichotomy property, if a is not greater than b, written $a \not> b$, then either $a = b$ or $a < b$. In other words,

$$\text{if } a \not> b, \quad \text{then } a \leq b.$$

Similarly,

$$\text{if } a \not< b, \quad \text{then } a \geq b.$$

How should the symbol "$\not\geq$," which means "not $\geq$," be interpreted? In writing $a \not\geq b$, we are eliminating the two relations "$>$" and "$=$" between a and b. Then, by the trichotomy property, the only remaining relation between a and b is $a < b$. Hence,

$$\text{if } a \not\geq b, \quad \text{then } a < b.$$

Similarly,

$$\text{if } a \not\leq b, \quad \text{then } a > b.$$

EXERCISE 6.11

1. Fill in the blanks with one of the symbols $=$, $<$, or $>$ to make a true statement of each. Be able to justify the use of each.
   (a) 8______32
   (b) $(3 + 8)$______$(10 + 1)$
   (c) $7 \times (4 + 6)$______34
   (d) (9×8)______(25×0)
   (e) $(6 + 0)$______(2×3)

(f) $(8 + 2) \times 1$______$2 \times (5 + 8)$
(g) $5 \times (6 + 2)$______$(5 \times 6) + (5 \times 2)$
(h) $(4 + 9) + 0$______$4 + 9$
(i) $2 \times 3 \times 4$______5×5
(j) $(9 \times 4) \times 0 \times (7 \times 5)$______$(18 \times 0) \times 8$

2. Decide whether each statement is true or false.
   (a) $17 \neq 25$
   (b) $6 \times 4 \not< 6 \times 3$
   (c) $18 \not> (3 \times 5) + 2$
   (d) $5 + 7 = 3 \times 4$
   (e) $3 \times 0 \leq 10 \times 0$
   (f) $17 \not< 25$
   (g) $(6 \times 4) > (6 \times 3)$
   (h) $18 > [(3 \times 5) + 2]$
   (i) $(5 + 7) \not> (3 \times 4)$
   (j) $(3 \times 0) = (10 \times 0)$
   (k) $17 < 25$
   (l) $(6 \times 4) \neq (6 \times 3)$
   (m) $18 \not< [(3 \times 5) + 2]$
   (n) $(5 + 7) \leq (3 \times 4)$
   (o) $(3 \times 0) \not> (10 \times 0)$
3. Given that 6 is less than 8:
   (a) How are $6 + 2$ and $8 + 2$ related?
   (b) How are $6 + b$ and $8 + b$ related when b is a whole number? Will the relation hold for all whole numbers?
   (c) How are 6×2 and 8×2 related?
   (d) How are $6 \times b$ and $8 \times b$ related when b is a whole number? Will the relation hold for all whole numbers? Discuss.
4. Given that $(2 + 4)$ is equal to 6:
   (a) How are $(2 + 4) + 3$ and $6 + 3$ related?
   (b) Is it true for each $b \in W$, $(2 + 4) + b = 6 + b$?
   (c) How are $(2 + 4) \times 3$ and 6×3 related?
   (d) Is $(2 + 4) \times b = 6 \times b$ for each $b \in W$?
5. Prove by counterexample that $>$ is not an equivalence relation.
6. Show that $>$ is not an equivalence relation by constructing an arrow diagram similar to Figures 6–1 and 6–2.

6.12 THE SYSTEM OF WHOLE NUMBERS

The system of whole numbers consists of two operational systems, $(W, +)$ and $(W, \times)$. That is, it includes a set of elements,

$$W = \{0, 1, 2, 3, \ldots\},$$

and the binary operations, addition $(+)$ and multiplication, $(\times)$. The equivalence relation, $=$, is a part of the system. The addition and multiplication operations obey the following:

Commutative Properties

For each $a, b \in W$, $a + b = b + a$.
For each $a, b \in W$, $a \times b = b \times a$.

Associative Properties

For each $a, b, c \in W, (a + b) + c = a + (b + c)$.
For each $a, b, c \in W, (a \times b) \times c = a \times (b \times c)$.

Distributive Property for Multiplication over Addition

For each $a, b, c \in W, (a + b) \times c = (a \times c) + (b \times c)$;

or $$c \times (a + b) = (c \times a) + (c \times b).$$

Identity Elements

For each $a \in W$, there exists a unique element $0 \in W$ such that $a + 0 = 0 + a = a$.
For each $a \in W$, there exists a unique element $1 \in W$ such that $a \times 1 = 1 \times a = a$.

Role of Zero in Multiplication

For each $a \in W, a \times 0 = 0 \times a = 0$.

Trichotomy Property

For each $a, b \in W, a = b$ or $a < b$ or $a > b$.

The following diagram shows that the *system of whole numbers* consists of two binary operational systems and properties that these operations obey:

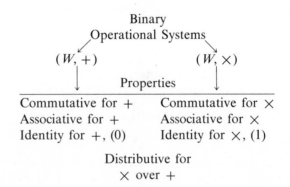

Binary
Operational Systems
↙ ↘
$(W, +)$ $(W, \times)$
↓ Properties ↓

Commutative for $+$	Commutative for $\times$
Associative for $+$	Associative for $\times$
Identity for $+$, (0)	Identity for $\times$, (1)

Distributive for
$\times$ over $+$

EXERCISE 6.12

1. In what way does the set of whole numbers differ from the set of natural numbers?
2. Does the system of natural numbers have an identity element for addition? Does it have an identity element for multiplication? Identify the elements if they exist.

3. Does the system of whole numbers have an identity element for addition? Does it have an identity element for multiplication? If either exists, identify it.
4. Explain the difference between the system of natural numbers and the system of whole numbers. (HINT: Make a table showing the similarities and the differences; see Sections 5.10 and 6.12.)

6.13 SOLUTION SETS OF SENTENCES

Earlier we have considered some simple mathematical sentences and whether certain of these sentences were true or false. Verify that the following sentences are true:

$$5 + 2 = 7 \qquad (3 \times 2) > (2 + 3)$$
$$(4 + 0) \ngtr 4 \qquad 9 \neq (3 \times 4)$$

Verify that the following sentences are false:

$$9 + 8 = 6 \qquad (6 \times 0) \nless 6$$
$$(2 \times 5) \times 1 \neq (2 \times 5) \qquad 15 > (7 \times 4)$$

Is it possible to tell whether the sentences

$$3 + b = 7 \quad \text{and} \quad 3 \times 2 > \square$$

are true or false? If the letter b and the *frame,* $\square$, stand for any number in W, these sentences may be made either true or false by the selection of one of the numbers from W.† Before the selections are made, however, sentences such as $3 + b = 7$ and $3 \times 2 > \square$ are neither true nor false and are called *open sentences.* The set of numbers that make an open sentence true is called the *solution set* of that sentence. Also, we may say that the elements of the solution set *satisfy* the open sentence, and if the solution set consists of a single element, that element may be called the *solution* of the sentence.

Before trying to find the solution set of a sentence, we must know from what set the numbers are to be selected. In this discussion it is W.

Example 1 From the set of whole numbers, find the solution set of

$$8 + \square = 11.$$

Suppose that the first whole number selected is 0. On putting 0 in place of the frame, we have $8 + 0 = 11$. Since this sentence is false, we know that 0 is not an element of the solution set. By trying other numbers we find that only when 3 is selected do we have a true sentence. That is, $8 + 3 = 11$ is true. After such investigations, one soon surmises that the solution set consists of only one number, 3. Using the roster notation, we describe this set:

$$\{3\}$$

†The use of the frame $\square$, in which numerals may be written, is helpful in teaching children to find solution sets of sentences.

The set builder form to describe the solution set is

$$\{x \in W : 8 + x = 11\} \quad \text{or, more simply,} \quad \{x \in W : x = 3\}.$$

The solution set may be represented on the number line:

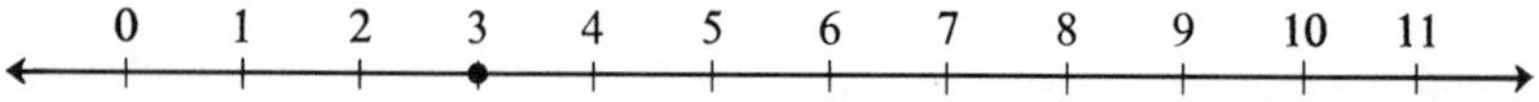

The representation of the solution set on the number line is called the *graph* of the solution set.

Example 2 From the set of whole numbers, find the solution set of

$$6 > \square.$$

If 4 is selected for $\square$, the sentence becomes $6 > 4$, which is true. So 4 is included in the solution set. Other selections from W for $\square$ which make this sentence true are 1, 0, 3, 2, 5. We describe the solution set of this sentence as

$$\{0, 1, 2, 3, 4, 5\}.$$

What is the greatest whole number in the solution set? What is the least? The solution set may also be described by using the set builder notation:

$$\{x \in W : 6 > x\}.$$

This is the graph of the solution set on the number line:

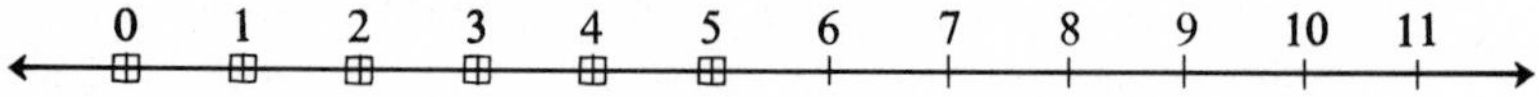

Example 3 Find the solution set from W for

$$3 < b < 8.$$

Definition 6.10(c) allows us to write $3 < b < 8$ as $3 < b$ *and* $b < 8$. It is possible to write $3 < b$ as $b > 3$ by Definition 6.10(b). The solution set of $b > 3$ is

$$\{4, 5, 6, 7, 8, 9, 10, \ldots\},$$

while the solution set of $b < 8$ is

$$\{7, 6, 5, 4, 3, 2, 1, 0\}.$$

The solution set of $3 < b < 8$ consists of the elements that are common to the preceding sets. That is, the solution set of $b > 3$ *and* $b < 8$ is

$$\{4, 5, 6, 7, 8, 9, \ldots\} \cap \{7, 6, 5, 4, 3, 2, 1, 0\} = \{4, 5, 6, 7\}.$$

The solution set may be described by using either one of the set builder forms:

$$\{b \in W: 3 < b < 8\} \quad \text{or} \quad \{b \in W: b > 3\} \cap \{b \in W: b < 8\}.$$

On the number line, we shall illustrate the solution set of $b > 3$ with triangles and that of $b < 8$ with circles. This will help in determining those elements that belong to both sets.

Example 4 Represent on the number line the solution set of

$$(x + 5) \leq 8, \qquad x \in W.$$

By Definition 6.10(d) this may be written

$$(x + 5) < 8 \quad \text{or} \quad x + 5 = 8.$$

Find the solution set of each sentence independently. Represent the solution set of $(x + 5) < 8$ on the number line with squares and that of $x + 5 = 8$ with dots. The solution sets are $\{0, 1, 2\}$ and $\{3\}$, respectively:

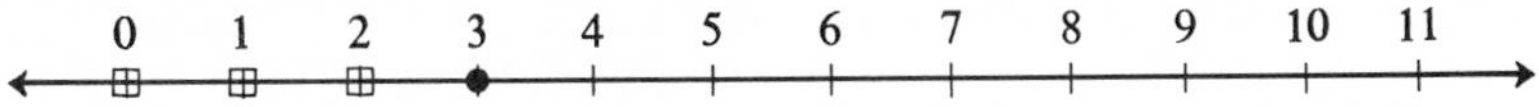

The solution set of $(x + 5) \leq 8$ consists of the elements in the union of these two sets. That is, the solution set is

$$\{0, 1, 2\} \cup \{3\} = \{0, 1, 2, 3\}.$$

We may express the solution set in any one of these ways also:

$$\{x \in W: (x + 5) \leq 8\},$$
$$\{x \in W: (x + 5) < 8 \quad \text{or} \quad x + 5 = 8\},$$
$$\{x \in W: (x + 5) < 8\} \cup \{x \in W: x + 5 = 8\}.$$

If we had used the set of natural numbers from which to make our selections in Example 4 rather than the set of whole numbers, that is,

$$(x + 5) \leq 8, \quad x \in N,$$

what change would there have been in the solution set?

EXERCISE 6.13

1. Change the following verbal sentences into mathematical open sentences. Either a letter or a frame may be used. Consider each to be in the set of whole numbers unless otherwise stated.

   EXAMPLE Two plus some number is six. The mathematical sentence is written $2 + b = 6$, or it might be written $2 + \square = 6$.

   (a) What number plus eight is equal to fifteen?
   (b) Three times some number plus one is greater than ten.

(c) Six times some number plus three is less than or equal to fifteen.
(d) A number is greater than zero and less than seven.

2. Express each of the following mathematical open sentences as verbal sentences.

EXAMPLE $(8 + x) \geq 5$. This says, "Eight plus some number is greater than or equal to five."

(a) $7 + b = 15$
(b) $(7 + b) \leq 15$
(c) $b + 3 \neq 9$
(d) $5 < b < 8$

3. Find the solution set of each of the following sentences. Give the solution set in the roster notation. Then show its graph on the number line.

(a) $a < 6, a \in W$
(b) $a < 6, a \in N$
(c) $a \leq 6, a \in W$
(d) $1 < b < 5, b \in W$
(e) $1 < b \leq 5, b \in W$
(f) $1 \leq b \leq 5, b \in W$
(g) $5 > q$ and $q > 8, q \in W$
(h) $y \geq 7$ or $y < 7, y \in W$
(i) $y \geq 7$ and $y < 7, y \in W$
(j) $k \geq 7$ and $k \leq 7, k \in W$
(k) $(y + 6) < 12, y \in N$
(l) $(z + 2) < 7$ or $(z + 1) > 3, z \in W$
(m) $\{r \in W : (r + 2) < 7\} \cap \{r \in W : (r + 1) > 3\}$
(n) $z > 5$ and $z < 9, z \in W$
(o) $\{b \in W : b > 5\} \cup \{b \in W : b < 9\}$
(p) $b + 1 = 1 + b, b \in W$
(q) $b + 1 = b, b \in W$
(r) $(b + 1) > b, b \in W$
(s) $b + 5 \neq 7, b \in W$

4. The following questions refer to parts of Problem 3 above.
(a) Discuss the solution sets of the sentences in (a), (b), and (c) with respect to subsets, equivalent sets, and the number of elements in each set.
(b) Discuss the solution sets of the sentences in (d), (e), and (f) in the same manner as above.
(c) What number is not in the solution set of (s)?
(d) What numbers are elements of the solution set of (p)? Of (r)? Of (q)?
(e) Are there any elements in the solution set of (g)?

7

Arithmetic of the Whole Numbers

7.1 FACTORS AND MULTIPLES

The definition of the binary operation of multiplication makes it possible for every pair of whole numbers to be matched with *one and only one* whole number. However, several number pairs may be matched with the same number. Is this contradictory to our definition of multiplication? For example, $(4, 3) \xrightarrow{\times} 12$, $(2, 6) \xrightarrow{\times} 12$, and $(12, 1) \xrightarrow{\times} 12$. Are there others? Each member of a pair has been identified as a *factor* of the matched number (Section 5.6). All of the factors of 12 may be collected to form a set, which is denoted by

$$F_{12} = \{1, 2, 3, 4, 6, 12\}.$$

We also speak of the matched number, 12, as a *multiple* of 4, 6, 3, or any one of its factors. We now state formally:

Definition 7.1 For each $a, b, c \in W$, if $a \times b = c$, then a and b are *factors* of c, and c is a *multiple* of a and of b.

If a number is a multiple of 3, then it must have 3 as one of its factors; that is, 6 is a multiple of 3 since $3 \times 2 = 6$. Other multiples of 3 are 3, 9, 18. Even 0 is a multiple of 3, since $3 \times 0 = 0$. The first six multiples of 3 may be found by multiplying 3 by the first six whole numbers:

$$\begin{array}{lll} 3 \times 0 = 0 & 3 \times 1 = 3 & 3 \times 2 = 6 \\ 3 \times 3 = 9 & 3 \times 4 = 12 & 3 \times 5 = 15 \end{array}$$

What is the least multiple of 3? What is the greatest? The set of multiples of 3 may be shown as

$$M_3 = \{(3 \times 0), (3 \times 1), (3 \times 2), (3 \times 3), (3 \times 4), (3 \times 5), \ldots\}$$

or

$$M_3 = \{\ 0,\quad 3,\quad 6,\quad 9,\quad 12,\quad 15,\quad \ldots\}.$$

What does the symbol "..." imply here?

Since 1 is the multiplicative identity element, that is, for each b,

$$b \times 1 = b,$$

each whole number b is a multiple of itself and of 1 by Definition 7.1. For example, 3 is a multiple of 3 and also of 1, as these are factors of 3. Is 0 a multiple of 0 and of 1? Of what other numbers is 0 a multiple?

In $4 \times 3 = 12$, the factors, 4 and 3, may be called the *divisors* of the product, 12. We say that 4 *divides* 12 and 3 *divides* 12 or that 12 *is divisible by* each of its factors, 4 and 3. In general, each factor is called a divisor of the product if the product is a nonzero whole number. If the product is zero, then only the nonzero factor is called a divisor of the product. For example, in the case of $4 \times 0 = 0$, it is true that both 4 and 0 are factors of 0, but only the factor 4 (the nonzero factor) is called a divisor of 0.

EXERCISE 7.1

1. Answer each of the following questions and be able to justify your answer.
   (a) Is 0 a multiple of 7? Of 0?
   (b) Is 0 a factor of 7? Of 0?
   (c) Is 0 a divisor of 7? Of 0?
   (d) Is there a number which is a multiple of only itself? If so, name it.
2. Find the set of all the factors of 20. How many are there? What is the least factor? What is the greatest factor?
3. Find the set of all the factors of 13. How many are there? What is the least factor? What is the greatest factor?
4. (a) Make a list of the following:
      (1) the first 8 multiples of 2
      (2) the first 11 multiples of 2
   (b) Denoting the multiples of 2 by M_2, use the roster notation form to describe this set.
   (c) List the least element of M_2. Is there a greatest element?
5. (a) Use the roster notation form to describe the set M_1, the multiples of 1.
   (b) What is the least element? What is the greatest element?
6. (a) List the first 5 elements in the set M_{10}, the multiples of 10.
   (b) Denote the set of factors of 10 by F_{10} and the roster notation form.
   (c) How many factors of 10 are there? How many multiples of 10 are there?
   (d) What is the least factor? What is the least multiple?
   (e) What is the greatest factor? What is the greatest multiple?
7. Let F_{24} be the set of all the factors of 24, F_{30} be the set of all the factors of 30, and F_{35} be the set of all the factors of 35.
   (a) Find F_{24}, F_{30}, and F_{35}.

(b) Find $F_{24} \cap F_{30}$ and $F_{30} \cap F_{35}$.
(c) What is the greatest element in each intersection set? What is the least element?
(d) How may the elements of the intersection of two sets be described?

8. Denote the multiples of 2 by M_2, the multiples of 3 by M_3, and the multiples of 6 by M_6.
   (a) Use the roster notation form to describe M_2, M_3, and M_6.
   (b) Find $M_2 \cap M_3$. How is it related to M_6?
   (c) What is the least common element greater than 0 found in $M_2 \cap M_3$? What is the least element greater than 0 found in M_6?
   (d) Is there a greatest element in $M_2 \cap M_3$? Explain. (HINT: Is M_6 another name for $M_2 \cap M_3$?)
9. (a) Find the set of all the factors of 28, denoting the set by F_{28}.
   (b) Find the sum of all the factors less than 28. Is the sum 28? A special name is given to numbers such as 28. They are called *perfect numbers.*
   (c) Test to see if 6 is a perfect number.
   (d) Is there another perfect number less than 50?

7.2 PRIME NUMBERS AND COMPOSITE NUMBERS

Make a list representing the whole numbers greater than 0 and less than 25:

1	2	3	4	5	6	7	8
9	10	11	12	13	14	15	16
17	18	19	20	21	22	23	24

Encircle 1, and then go through this list marking out the multiples of 2 that are greater than 2, the multiples of 3 greater than 3, and the multiples of 5 greater than 5. Will any member of the list have more than one mark through it? Your list should now look like this:

①	2	3	4	5	~~6~~	7	~~8~~
~~9~~	~~10~~	11	~~12~~	13	~~14~~	~~15~~	~~16~~
17	~~18~~	19	~~20~~	~~21~~	~~22~~	23	~~24~~

If you should mark out the multiples of 4, 6, 7, each greater than 4, 6, 7, would there be any new ones marked out? Can you explain why? The numbers "sifted" out, except 1, are called *composite numbers,* while the numbers that remain unmarked are called *prime numbers.* Notice that the composite numbers have factors that are prime numbers.

This process of "sifting" out certain numbers is called the "Sieve of Eratosthenes." Eratosthenes was a Greek scholar who lived before 200 B.C. While his name is most often remembered because of the sieve, he had many and varied interests. He was tutor to the son of Ptolemy III of Egypt and served as librarian for the great University of Alexandria. For a thousand years from the time of Euclid this university was the center of Greek learning.

Using the preceding sieve, let P_{24} represent the set of prime numbers less than 25 and C_{24} represent the set of composite numbers less than 25.

$$P_{24} = \{2, 3, 5, 7, 11, 13, 17, 19, 23\}$$
$$C_{24} = \{4, 6, 8, 9, 10, 12, 14, 15, 16, 18, 20, 21, 22, 24\}$$

Observe that each prime number has two, and only two, distinct factors: itself and 1. The number 1 also might be considered to have only two factors, itself and 1. However, these are not distinct. This is a justification for excluding 1 from the set of primes. Can you give a good argument for excluding 1 from the set of composite numbers?

We now consider the entire set, W, of whole numbers greater than 0 instead of the subset that is greater than 0 and less than 25, and we shall write the set of prime numbers, P_W, and the set of composite numbers, C_W, as

$$P_W = \{2, 3, 5, 7, 11, 13, 17, 19, 23, \ldots\},$$
$$C_W = \{4, 6, 8, 9, 10, 12, 14, 15, 16, 18, 20, 21, 22, 24, \ldots\}.$$

Were we correct in using the symbol ". . ." in P_W and in C_W? Is there an infinite number of elements in each of these sets? Euclid, the Greek mathematician of geometry fame (third century B.C.), proved that there are infinitely many primes. A specific example, which illustrates his method of proof, follows.

Either there are infinitely many primes or there is a greatest one. Let us suppose that 11 is the last prime. We construct a number Q to be the product of all the prime numbers less than or equal to 11:

$$Q = 2 \times 3 \times 5 \times 7 \times 11$$

Let us consider the number $Q + 1$, which is certainly greater than 11. It is either prime or composite. If $Q + 1$ is prime, we have proved that there is a prime greater than 11. If $Q + 1$ is composite, it must have a prime number as a factor. But when $Q + 1$ is divided by

$$2, 3, 5, 7, \text{ or } 11,$$

the remainder is 1. Hence, the prime factor must be a prime number greater than 11. In either case there exists a prime number greater than 11. So 11 is not the last prime, and our assumption has led to a contradiction. You may wish to use this argument as a model and prove that for any prime, p, there is always a greater one.

A formal definition for prime numbers and composite numbers follows.

Definition 7.2 For each whole number $a > 1$, a is a *prime number* if its factors are *only* a and 1; all the whole numbers $a > 1$ that are not prime are *composite numbers.*

EXERCISE 7.2

1. Make a list of the prime numbers less than 100. Use the Sieve of Eratosthenes.
   (a) How many prime numbers are less than 50? How many are less than 100?
   (b) From this list select all the pairs of prime numbers which have a difference of 2. How many are there? Such pairs are called *twin primes.*
2. List all the factors of 30 that are prime numbers. List all the factors of 30 that are composite numbers.
3. List all the multiples of 2 less than 100. Would 0 be included? These and all other whole numbers that are multiples of 2 are called *even whole numbers;* the remaining whole numbers are called *odd whole numbers.* Every whole number is either even or odd.
   (a) How many even whole numbers less than 100 are prime? How many are composite?
   (b) Select several whole numbers. Multiply each by 2. The result in each case is either an odd or an even whole number. Identify each result as to whether it is an odd or an even whole number.
   (c) Select several whole numbers. Multiply each by 2 then add 1. Identify each result as to whether it is an odd or an even number.
   (d) Write a statement that will define an even whole number, and one that will define an odd whole number. (HINT: Consider any whole number a and use the results of parts (b) and (c) above.)
   (e) Make a general statement concerning the sum of:
      (1) two even numbers
      (2) two odd numbers
      (3) an even and an odd number
   (f) Give a deductive proof for each of the statements in part (e) above. (See discussion of deductive proof in Section 5.4 and Problem 10 of Exercise 5.8.)
4. If there are any prime numbers *between* the following pairs of whole numbers, list them:

   2 and 4; 6 and 12; 12 and 24; 8 and 16; 15 and 30.

   (a) Did you find at least one prime between each of these pairs?
   (b) The elements of each of these pairs are related in what manner? Find others similarly related by selecting a whole number and demonstrating how to find the number that should be paired with it.
   (c) These are instances of an interesting statement known as *Bertrand's postulate.* It states that between any whole number $a > 1$ and its double there is at least one prime number. After testing many pairs of such numbers, are you convinced that this postulate is true? Does this constitute a proof? What type of reasoning has been used?
5. Consider the even whole numbers 4, 6, 8, 12, 20.
   (a) Show that each is the sum of *two* prime numbers.
   (b) Is this true for all even numbers? If there are any restrictions, give them.
   (c) These are instances of another interesting statement known as *Goldbach's conjecture.* It states that every even whole number greater than 2 may be expressed as the sum of two prime numbers. Do the examples used bear this out? By testing this statement over and over, and finding each time that it holds, do you consider this as proof of the conjecture?

7.3 PRIME FACTORIZATION

Our experience with factors leads us to admit that every whole number has *at least* two factors. Each prime number has *only* two factors, itself and 1. A composite number may have many (at least more than two) factors. However, each composite number has one and only one *set of prime factors.* This comes from an important statement:

The Fundamental Theorem of Arithmetic

Every whole number greater than 1 can be expressed uniquely, except for order, as a prime or as a product of primes.

For instance, suppose that we ask two boys, Al and Ben, to express 30 as the product of prime factors. Al might write $30 = 2 \times 3 \times 5$, while Ben might write $30 = 3 \times 2 \times 5$. The prime factors which Al selected are the same as those selected by Ben. These are the only possible selections that they could have made. Notice that the boys arranged the primes differently.

When a number is expressed as the product of prime factors, and only prime factors, this product is called the *complete* or *prime factorization* of the number, and we say that the number has been completely factored into its primes. For example, in

$$78 = 2 \times 3 \times 13 \quad \text{and} \quad 60 = 2 \times 2 \times 3 \times 5,$$

both 78 and 60 have been completely factored. The product $2 \times 3 \times 13$ is the complete factorization of 78. What is the complete factorization of 60? The complete factorization of 60 expressed with exponents is

$$2^2 \times 3 \times 5.$$

A method which may be used in completely factoring a composite number into its prime factors is illustrated by the following factorization of 60. Select any pair of factors of 60, each being greater than 1, say 6 and 10. Select a pair of factors greater than 1 for each of the composite factors 6 and 10. The only factors for 6 greater than 1 are 2 and 3, and for 10 are 2 and 5. The process may be arranged in a "factor tree":

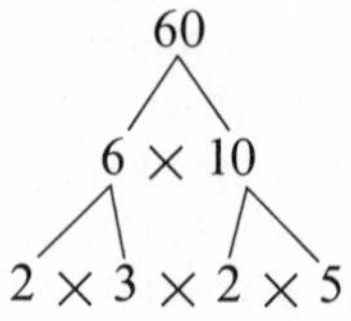

Since the factors in the last line are prime factors, we have completely factored 60.

Use 2 and 30 as the first pair of factors of 60, and complete a "factor tree." Are the resulting prime factors the same as those shown above?

Using the notation of Section 7.1, we describe the set consisting of *all the factors* of 60 as

$$F_{60} = \{1, 2, 3, 4, 5, 6, 10, 12, 15, 20, 30, 60\}.$$

Observe that this list includes prime factors as well as composite factors and 1. Is it possible to use the prime factors of 60 to help in finding this set? Of course, the elements of F_{60} may well be found by using "trial and error" based on one's knowledge of the multiplication facts. Instead, let us proceed to use the prime factors found in the complete factorization of 60:

$$60 = 2 \times 2 \times 3 \times 5 \quad \text{or} \quad 60 = 2^2 \times 3 \times 5$$

The factors consist of all the different primes and all the different products that may be made from these primes. Since 1 is a factor of each whole number, it is included in the set. Here is the set of all the factors of 60 derived from the prime factors:

$$F_{60} = \{1, 2, 3, 5, (2 \times 2), (2 \times 3), (2 \times 5), (3 \times 5), (2 \times 2 \times 3), (2 \times 2 \times 5), (2 \times 3 \times 5), (2 \times 2 \times 3 \times 5)\}$$

To begin the process of completely factoring a number, it is helpful to be able to select the first factor by inspection. This is possible in some special cases as will be pointed out.

In Exercise 7.2, the multiples of 2 were identified as *even* numbers. Hence, 2 is a factor of all even numbers. Is it possible to detect even numbers by their numerals? When written in decimal notation, it is observed that the ones digit is always either 0, 2, 4, 6, or 8. Then, we know by inspection that 956 has 2 as a factor, while 423 does not. Is there some rationale for this? Write 956 in expanded notation.

$$956 = (9 \times 10^2) + (5 \times 10) + (6 \times 1)$$

Since each of these addends has a factor 2, an application of the distributive property satisfies us that 956 has a factor 2. Since the first two addends, expressed as multiples of powers of 10 greater than 0, will always be even, we need only to be concerned with the ones digit. Using a similar argument, can you justify calling 5956 an even number? In other words, does it have 2 as a factor? If it does, then 5956 is divisible by 2.

A similar investigation may be used to establish that 423 is not divisible by 2. Writing 423 in expanded notation, we have

$$423 = (4 \times 10^2) + (2 \times 10) + (3 \times 1).$$

We already know that the first two addends are even. Is 3×1 even? Is it possible to apply the distributive property with 2 as a factor over this sum? Does 423 represent an even number? Notice that the ones digit is the determining factor as to the "even-ness" or "odd-ness" of a number, and whether the number is divisible by 2.

Since 423 has 3 as its ones digit, is this an indication that it is divisible by 3? The expanded notation above does not indicate this. However, we can rename the addends and rearrange them in such a way that we can answer this question.

$$\begin{aligned}
423 &= (4 \times 10^2) + (2 \times 10) + (3 \times 1) \\
&= (4 \times 100) + (2 \times 10) + (3 \times 1) \\
&= [4 \times (99 + 1)] + [2 \times (9 + 1)] + (3 \times 1) \\
&= [(4 \times 99) + (4 \times 1)] + [(2 \times 9) + (2 \times 1)] + (3 \times 1) \\
&= [(4 \times 99) + 4] + [(2 \times 9) + 2] + 3 \\
&= (4 \times 99) + (2 \times 9) + (4 + 2 + 3)
\end{aligned}$$

Notice that 100 and 10 have been renamed as (99 + 1) and (9 + 1), respectively. Why was this necessary? Are the first two addends in the last step divisible by 3? Would they be divisible by 3 for any number? Is the last addend, (4 + 2 + 3), divisible by 3? Since the sum is 9, the last addend is indeed divisible by 3. The first two addends for any three-digit number written in this form will be divisible by 3. Then for the number to be divisible by three, the last addend, consisting of the sum of the digits of the number, must also be divisible by 3. Use the same argument to show that 423 is also divisible by 9.

Are 121, 2121, and 36 each divisible by 3 and by 9?

Other well-known *divisibility tests* may be developed from arguments similar to those just given. A summary of four of these tests is as follows:

1. A number written in decimal notation is divisible by 2 if its ones digit is either 0, 2, 4, 6, or 8.
2. A number written in decimal notation is divisible by 3 (by 9) if the sum of its digits is divisible by 3 (by 9).
3. A number written in decimal notation is divisible by 5 if its ones digit is either 0 or 5.
4. A number written in decimal notation is divisible by 10 if its ones digit is 0.

Notice that each of these tests calls for the number to be written in decimal notation. Is this necessary? We shall consider the following examples.

Example 1 Is 12_{five} divisible by 2? In other words, does it represent an even number? Expressed in expanded notation,

$$12_{\text{five}} = (1 \times 5) + (2 \times 1).$$

The last addend is divisible by 2. However, since the first addend, (1 × 5), is not, we must admit that 12_{five} is not divisible by 2.

Example 2 Use the test for divisibility by 3 on 24_{eight}. Since the sum of the digits is 6 (a number divisible by 3), one might be tempted to say that 24_{eight} is divisible by 3. Written in expanded notation, this becomes

$$\begin{aligned}
24_{\text{eight}} &= (2 \times 8) + (4 \times 1) \\
&= 16 + 4 \\
&= 20.
\end{aligned}$$

Are the addends 16 and 4 each divisible by 3? Does 20 have a factor 3?

One must conclude from these examples that the four divisibility tests apply only when the number is written in base ten. It should be evident that these tests are most useful when it is necessary to find the prime factorization of a number.

EXERCISE 7.3

1. Write the prime factorization of the following numbers. From the prime factorization, write the set consisting of all the factors of each number:

   (a) 42 (c) 64 (e) 231 (g) 200 (i) 323

   (b) 35 (d) 105 (f) 120 (h) 301 (j) 1000

2. What factors of 40 do not appear in the prime factorization of 40?
3. Which of these represent an even number? Are the remaining ones odd?

   (a) $5 \times 2 \times 6$ (f) 202_{three} (k) 101_{two}

   (b) $7 \times 6 \times 3$ (g) $25 + 33$ (l) 111_{two}

   (c) $24 + 95$ (h) 111_{seven} (m) 222_{three}

   (d) $72 + 38$ (i) 111_{six} (n) 122_{three}

   (e) 31_{five} (j) 332_{six} (o) 22_{three}

4. Write the prime factorization of 334_{six}. Give the set of all the factors in base six.
5. Use the four divisibility tests to determine whether each of the following numbers in decimal notation has a factor 2, 3, 5, 9, or 10.

   (a) 129 (d) 135 (g) 3576

   (b) 117 (e) 807 (h) 1112

   (c) 6140 (f) 107

6. Give a test for divisibility by 6. Use this test to determine if 915,702 has a factor 6. Does 915,702 have 18 as a factor also? Explain.
7. Use an argument similar to that given in developing the divisibility test for 2 and for 3 to show that 215 is divisible by 5.
8. Make a list of the numbers less than 50 that have exactly two factors.
9. (a) If $b = 2^5$, how many factors does b have? List them.

   (b) If $b = p^4$, where p is prime, how many factors does b have? List them.

   (c) If $b = 5^m$, where $m \in W$ and is greater than 0, how many factors does b have?

   (d) If $b = pq$, where p and q are distinct primes, how many factors does b have? (NOTE: Recall that $pq = p \times q$.)

10. Are the following statements true for all primes? Discuss any that need certain restrictions to make them true. Is it possible to prove beyond a doubt that the ones you select are true by using several examples? Explain. Justify your selection of the false sentences with a counterexample for each.

    (a) The sum of two primes is always an even number.

    (b) The sum of a prime and a composite number is an odd number.

    (c) The sum of two composite numbers is an even number.

(d) The product of a prime number and 2 is an even number.
(e) The sum of a prime and 3 times that prime is an odd number.
(f) All natural numbers are either prime or composite.

7.4 GREATEST COMMON FACTOR

To find the set of the *common* factors of 36 and 30, we shall begin by completely factoring each:

$$36 = 2^2 \times 3^2,$$
$$30 = 2 \times 3 \times 5.$$

These complete factorizations are used to find the set of all the factors of 36 and of 30. We have:

$$\begin{aligned} F_{36} &= \{1, 2, 2^2, 3, 3^2, (2 \times 3), (2 \times 3^2), (2^2 \times 3), (2^2 \times 3^2)\} \\ &= \{1, 2, 4, 3, 9, 6, 18, 12, 36\}; \\ F_{30} &= \{1, 2, 3, 5, (2 \times 3), (2 \times 5), (3 \times 5), (2 \times 3 \times 5)\} \\ &= \{1, 2, 3, 5, 6, 10, 15, 30\}. \end{aligned}$$

The intersection of these two sets is the set consisting of the *common factors* of 36 and 30:

$$F_{36} \cap F_{30} = \{1, 2, 3, 6\}.$$

The number 1 is the least element in this set, while 6 is the greatest. Since 1 (the least element) is a common factor for any whole numbers that we might consider, it has no particular significance. However, the greatest element depends upon the numbers chosen. Hence, we call attention to the *greatest common factor,* and say that 6 is the greatest common factor of 36 and 30. This may be abbreviated as

$$\text{gcf}\{36, 30\} = 6.$$

An interesting observation is that the greatest common factor, 6, is a multiple of all the other common factors, 1, 2, 3.

It is usually more convenient to find the greatest common factor of two or more numbers directly from the prime factorization of each instead of using the set of all factors of each.

Example 1 Find the greatest common factor of 36 and 30. The complete factorizations are:

$36 = 2^2 \times 3^2$ which may be written $36 = (2 \times 3) \times 2 \times 3.$
$30 = 2 \times 3 \times 5$ which may be written $30 = (2 \times 3) \times 5.$

As can be seen, (2×3) is the greatest factor common to both 36 and 30.

Example 2 Find the greatest common factor of 90, 144, and 54.

$$90 = 2 \times 3^2 \times 5 \quad \text{which may be written} \quad 90 = (2 \times 3^2) \times 5$$
$$144 = 2^4 \times 3^2 \quad \text{which may be written} \quad 144 = (2 \times 3^2) \times 2^3$$
$$54 = 2 \times 3^3 \quad \text{which may be written} \quad 54 = (2 \times 3^2) \times 3$$

The greatest common factor of 90, 144, and 54 is (2×3^2). Give a simpler name for this.

Example 3 What is the greatest common factor of 35 and 24?

$$35 = 5 \times 7$$
$$24 = 2^3 \times 3$$

Obviously, 1 is the only common factor. Hence, it is the greatest common factor of 35 and 24.

Formally, we have:

Definition 7.4 The *greatest common factor* of two or more numbers greater than 0 is the product of the highest powers of the different prime factors common to all the numbers.

Since there are no prime factors common to 35 and 24 in Example 3, the numbers are said to be *relatively prime* even though neither is a prime number. When numbers are relatively prime, their greatest common factor is 1.

Since a factor of a number is also the divisor of that number, the greatest common factor of two or more numbers may also be called their *greatest common divisor,* gcd. For example, gcf{5, 15} may be designated as gcd{5, 15}.

EXERCISE 7.4

1. (a) Write the sets of all the factors of 8, 12, and 16.
   (b) Write the set of common factors for the above numbers.
   (c) What is the greatest common factor of 8, 12, and 16?
2. Express each number in completely factored form and then write the greatest common factor, or greatest common divisor, for each of the following:
   (a) 80, 48, 72 (c) 630, 60, 600, 720 (e) 250, 108, 375
   (b) 156, 585, 78 (d) 65, 84
3. What is the greatest common factor of 8 and 8? Of 1 and 8?
4. What is the greatest common factor of 1 and m where $m \in W$? Consider the special case when $m = 0$.

7.5 LEAST COMMON MULTIPLE

What do we mean by a multiple of a number? This was defined in Definition 7.1. Let us consider the set of multiples of 4, and the set of multiples of 6:

$$M_4 = \{0, 4, 8, 12, 16, 20, 24, 28, 32, 36, 40, 44, \ldots\},$$
$$M_6 = \{0, 6, 12, 18, 24, 30, 36, 42, 48, 54, 60, 66, \ldots\}.$$

Then the common multiples of 4 and 6 are the elements of

$$M_4 \cap M_6 = \{0, 12, 24, 36, \ldots\}.$$

Some other common multiples which would be included in this set would be 48, 60, 72. Since both M_4 and M_6 are infinite sets, it seems likely that their intersection also has infinitely many terms. Hence they have *no greatest common multiple.* The *least* number in this set of common multiples is zero. Since it will be an element in each intersection set of multiples, it has an insignificant role in the discussion of a least common multiple. The *least common multiple,* greater than 0, of 4 and 6 is 12. We write this as

$$\text{lcm}\{4, 6\} = 12.$$

The least common multiple will always refer to multiples greater than 0.

The search for the least common multiple of two or more numbers from the set of multiples of each, as was done above, often proves to be quite inconvenient. We shall now explore the use of prime factorization in determining the least common multiple.

Example 1 Find the least common multiple of 4 and 6.

$$4 = 2^2$$
$$6 = 2 \times 3$$

A multiple of 4 must contain the factors of 4, while a multiple of 6 must contain the factors of 6. So, in selecting a multiple of 4 and 6, we must require that it contain the factor 2 used twice as well as the factor 3. Does the choice of only these factors ensure that we have a multiple of both 4 and 6? Consider the following:

$$\underbrace{2 \times \mathrlap{\overbrace{2 \times 3}^{6}}2}_{4}\phantom{{}\times 3}$$

Notice that one of the factors 2 serves a double role. It is a factor of 4 and also of 6. Hence, the least common multiple of 4 and 6 is

$$2 \times 2 \times 3 \quad \text{or} \quad 12.$$

Example 2 Find the least common multiple of 16, 24, and 50.

$$16 = 2^4$$
$$24 = 2^3 \times 3$$
$$50 = 2 \times 5^2$$

The highest powers of the prime factors of 16, 24, and 50 must be included.

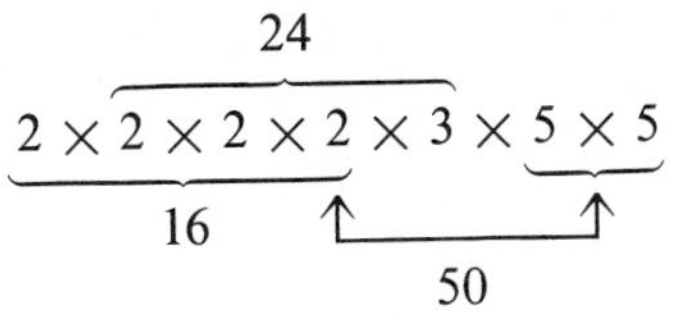

Observe that all necessary factors, and only those factors, have been used. With no extra factors, the least common multiple of 16, 24, and 50 is

$$2^4 \times 3^1 \times 5^2 = 1200.$$

The method of these examples is much more efficient than the tedious procedure of using the intersection of the sets of multiples.

Definition 7.5 The *least common multiple* of two or more numbers greater than 0 is the product of the highest power of each of the different prime factors of each of the numbers.

EXERCISE 7.5

1. Given the numbers 2, 3, 5, 6, and 10:
   (a) Find the multiples less than 50 of each number.
   (b) Use the above multiples to find the least common multiple of 2, 3, 6. Do the same for 2, 5, 10, and for 2, 6, 10.
   (c) What is the least common multiple of 2 and 3? of 2 and 5? Is there a simple way to find the least common multiple of prime numbers?
   (d) Is 20 a common multiple of 2 and 10? Is it the least common multiple?
2. Use the prime factorization method to find the least common multiple for the elements of each of these sets:

(a) $\{2, 3\}$	(f) $\{6, 18\}$	(k) $\{213, 102\}$
(b) $\{2, 5, 7\}$	(g) $\{32, 64\}$	(l) $\{9, 36, 15\}$
(c) $\{6, 10\}$	(h) $\{35, 32\}$	(m) $\{15, 60\}$
(d) $\{21, 1\}$	(i) $\{45, 15, 10\}$	(n) $\{4, 12, 48\}$
(e) $\{15, 12\}$	(j) $\{8, 10, 12\}$	(o) $\{105, 210\}$

3. From Problem 2 do you agree with any of the statements given below? Where you disagree, give a counterexample. Be able to explain why you agree with the others.
   (a) The least common multiple of prime numbers is the product of the numbers.
   (b) The least common multiple of composite numbers is the product of the numbers.
   (c) The least common multiple of any set of natural numbers, one of which is a multiple of the others, is that multiple.
   (d) The least common multiple of b and $b, b > 0$, is $b \times b$.
   (e) The least common multiple of 1 and $b, b > 0$, is b.
   (f) The least common multiple of any two distinct primes may be one of those primes.
4. Is there a greatest common multiple of two or more prime numbers? Of composite numbers? Explain.

5. Given the numbers 10 and 18:
   (a) What is the least common multiple?
   (b) What is the greatest common factor?
   (c) Compare the product of the least common multiple and the greatest common factor with the product of the two given numbers.
6. From Problem 5 does it seem reasonable to expect that for any two whole numbers, m and n, greater than 0, their product is equal to the product of their greatest common factor, $\text{gcf}\{m, n\}$, and their least common multiple, $\text{lcm}\{m, n\}$? Give a convincing argument for:

$$m \times n = \text{gcf}\{m, n\} \times \text{lcm}\{m, n\}$$

7. Given the pair of numbers 18 and 20:
   (a) Find the greatest common factor, gcf{18, 20}.
   (b) Find the least common multiple, lcm{18, 20}.
   (c) Find the product of 18 and 20.
   (d) By knowing the product, 18×20, and the gcf{18, 20}, is it possible to use this information to find the lcm{18, 20}?

7.6 THE ALGORITHMS FOR THE OPERATIONS ON WHOLE NUMBERS

As was mentioned in Section 3.5, the word *algorithm,* or *algorism,* has its origin in the name of an Arab mathematician, al-Khowarizmi, the man who described the arithmetic of the Hindus in a book that was later translated into Latin. As is often the case in translating from one language to another, the form of the name al-Khowarizmi was changed and became Algorithmi. From this translated form we have taken the word algorithm, or algorism, and we use it to mean a systematic procedure for computation.

The algorithms we use today for both addition and multiplication are based on the place value numeration system and the properties for the whole numbers. Because of our familiarity with these algorithms, we are likely to think that these are the only ones possible. However, one should be reminded that through the ages man has changed them in his search for more efficient ways of finding sums and products. And, as time goes on, it is possible that he will continue in his search for improvements.

The Addition Algorithm The basic addition facts for all pairs of whole numbers less than 10 are learned by children (see Section 6.3). These facts are based on the union of disjoint, discrete sets. However, as the number of elements in these sets is increased, a device is needed to find the number of elements in the union set. For instance, how may one determine the number of elements in the union of two disjoint sets having 24 elements and 35 elements? Must we depend on counting them? One immediately explains that the numerals 24 and 35 should be written as shown and then proceed to add the "ones," then the "tens":

$$\begin{array}{r} 24 \\ \underline{35} \\ 59 \end{array}$$

Many generations have learned and used this procedure, this *algorithm.* Too often this algorithm has been taught in such a way that it has little or no meaning for the student. Consequently, it is regarded as some mystical procedure that, when followed, will produce the answer. The student needs to understand why this addition algorithm works. The justification of it is based on our decimal system of numeration along with the associative and commutative properties for addition and the distributive property in the set of whole numbers. The justification follows:

$24 + 35 = [(2 \times 10) + (4 \times 1)] + [(3 \times 10) + (5 \times 1)]$	*Decimal expanded notation*
$= [[(2 \times 10) + (4 \times 1)] + (3 \times 10)] + (5 \times 1)$	*Associative property for addition*
$= [(3 \times 10) + [(2 \times 10) + (4 \times 1)]] + (5 \times 1)$	*Commutative property for addition*
$= [(3 \times 10) + (2 \times 10)] + [(4 \times 1) + (5 \times 1)]$	*Associative property for addition*
$= [(3 + 2) \times 10] + [(4 + 5) \times 1]$	*Distributive property for multiplication over addition*

At this point the "legality" of our well-known algorithm of adding the "ones" and the "tens" has been established. Continuing with the justification, we have:

$[(3 + 2) \times 10] + [(4 + 5) \times 1] = (5 \times 10) + (9 \times 1)$	*Renaming* (3 + 2) *and* (4 + 5) *from the basic addition facts*
$= 59$	*Renaming by applying place value notation*

The preceding development suggests the following algorithm:

$$\begin{array}{rl} 24 & \\ \underline{35} & \\ 9 & \ldots (4 + 5) \text{ ones} \\ \underline{50} & \ldots (2 + 3) \text{ tens} \\ 59 & \end{array} \qquad \text{or} \qquad \begin{array}{rl} 24 & \\ \underline{35} & \\ 50 & \ldots (2 + 3) \text{ tens} \\ \underline{9} & \ldots (4 + 5) \text{ ones} \\ 59 & \end{array}$$

The first form demonstrates the addition of the "ones" first, then the "tens"; while in the second, the "tens" are added before the "ones." Obviously, both procedures yield the same sum; hence, one is just as acceptable as the other.

In the standard or "adult" algorithm

$$\begin{array}{r} 24 \\ \underline{35} \\ 59 \end{array}$$

the partial sums 9 and 50 are not recorded. Instead, the final sum 59 is arrived at through knowledge of decimal place value.

We shall now justify the algorithm for the addition of two numbers in which the sum of the ones digit is greater than 9. For example:

$$\begin{array}{r} \textcircled{1} \\ 56 \\ \underline{38} \\ 94 \end{array}$$

This algorithm involves what is commonly called "carrying."

$56 + 38 = [(5 \times 10) + (6 \times 1)] + [(3 \times 10) + (8 \times 1)]$	*Renaming by decimal expanded notation*
$= [(5 \times 10) + (3 \times 10)] + [(6 \times 1) + (8 \times 1)]$	*Associative and commutative properties for addition*
$= [(5 + 3) \times 10] + [(6 + 8) \times 1]$	*Distributive property*
$= (8 \times 10) + (14 \times 1)$	*Renaming* (5 + 3) *and* (6 + 8)
$= (8 \times 10) + 14$	*Multiplicative identity*
$= (8 \times 10) + [(1 \times 10) + (4 \times 1)]$	*Renaming* 14 *by decimal expanded notation*
$= [(8 \times 10) + (1 \times 10)] + (4 \times 1)$	*Associative property for addition*
$= [(8 + 1) \times 10] + (4 \times 1)$	*Distributive property*
$= (9 \times 10) + (4 \times 1)$	*Renaming* (8 + 1)
$= 94$	*Renaming by decimal place value notation*

Hence, 56 + 38 = 94, and the algorithm is justified.

A careful study of the preceding suggests the following algorithm:

$$\begin{array}{rl} 56 & \\ \underline{38} & \\ 80 & \ldots.(5 + 3) \text{ tens} \\ \underline{14} & \ldots.(8 + 6) \text{ ones} \\ 94 & \end{array} \qquad \text{or} \qquad \begin{array}{rl} 56 & \\ \underline{38} & \\ 14 & \ldots.(8 + 6) \text{ ones} \\ \underline{80} & \ldots.(5 + 3) \text{ tens} \\ 94 & \end{array}$$

Here again the order of adding the tens and the ones appears insignificant.

However, if the algorithm is to be changed to the standard form, it is more convenient to add the ones first. In the standard form, the sum of the ones is renamed as tens and ones and only the ones are recorded. The following stages may be used in arriving at the standard algorithm:

STAGE I		STAGE II	STAGE III
		(10)	
56		56	56
38		38	38
4	> 14 renamed	4	94
10		90	
80		94	
94			

In the second stage, the ten has been recorded above the problem. Compare this with the recording of ten in the first stage; also with the illustration given at the beginning of the discussion of "carrying."

Applying the algorithm we shall find the sum of 236 and 195. Here we may write:

236		236
195		195
11 .... (6 + 5) ones		300 .... (2 + 1) hundreds
120 .... (3 + 9) tens	or	120 .... (3 + 9) tens
300 .... (2 + 1) hundreds		11 .... (6 + 5) ones
431		431

When the sum of the "ones," the "tens," or the "hundreds" is greater than 9, it is possible to shorten the form by adding first the "ones," then the "tens," and so on, renaming each of the sums as it is found and omitting the recording of them separately. This constitutes the procedure used in the standard algorithm:

$$\begin{array}{r} 236 \\ \underline{195} \\ 431 \end{array}$$

The Subtraction Algorithm As we saw in Section 6.5, subtraction was not an operation in the set of whole numbers. Yet, since subtraction is possible for *certain* pairs of whole numbers, it seems appropriate to discuss some subtraction algorithms that may be used with these pairs.

At this point it will be helpful to review Section 6.5 and be thoroughly familiar with Definition 6.5 and the problems in Exercise 6.5. In particular, recall that if

$$b + x = a, \quad \text{with } a, b, x \in W,$$

then

$$a - b = x.$$

As long as the addends, b and x, are less than 10, it is possible to use the addition grid as was done in Section 6.5. What is to be done when b or x or both are equal to or are greater than 10? Suppose, for example, that we wish to find the number x such that

$$23 + x = 38$$

or

$$(38, 23) \Rightarrow x.$$

The addition algorithm was justified on the basis of certain properties for the addition of whole numbers and the decimal place value numeration system. Since there are no comparable properties for subtraction, a different approach must be considered to justify the subtraction algorithm.

First, let us consider, in the light of our previous knowledge of addition and subtraction, whether the following sentence (a) is true:

(a) $$(6 + 2) - (4 + 1) = (6 - 4) + (2 - 1).$$

Determining what number is represented on each side of the = symbol, we have:

$$\begin{aligned}(6 + 2) - (4 + 1) &= 8 - 5 \\ &= 3\end{aligned}$$

and

$$\begin{aligned}(6 - 4) + (2 - 1) &= 2 + 1 \\ &= 3 .\end{aligned}$$

Hence, (a) is a true sentence. Evaluate the following sentences as indicated for (a), and determine if they are true.

(b) $(9 + 5) - (2 + 3) = (9 - 2) + (5 - 3)$
(c) $(3 + 8) - (1 + 3) = (3 - 1) + (8 - 3)$
(d) $(5 + 7) - (2 + 4) = (5 - 2) + (7 - 4)$

Observe the pattern of these mathematical sentences; it is this:

$$(\triangle + \square) - (\bigcirc + \square) = (\triangle - \bigcirc) + (\square - \square).$$

Construct other true mathematical sentences that follow this pattern. From these, we make the following generalization:

A Generalization for the Subtraction Algorithm

If $a, b, c, d \in W$, with $c \leq a$ and $d \leq b$, then

$$(a + b) - (c + d) = (a - c) + (b - d).$$

We shall now use this generalization to find the solution set of

$$38 - 23 = x.$$

First, we write:

$$\begin{aligned} 38 - 23 &= (30 + 8) - (20 + 3) && \textit{Renaming 38 and 23} \\ &= (30 - 20) + (8 - 3) && \textit{Generalization for the subtraction algorithm} \end{aligned}$$

Since (30 − 20) cannot be found from the addition grid, we need another property.

To find (30 − 20), we express it as follows:

$$30 - 20 = (3 \times 10) - (2 \times 10) \qquad \textit{Renaming by expanded notation}$$

This suggests the use of a form of the distributive property that distributes *multiplication over subtraction:*

$$\begin{aligned} (3 \times 10) - (2 \times 10) &= (3 - 2) \times 10 \\ &= 1 \times 10 \\ &= 10. \end{aligned}$$

Hence, we have

$$30 - 20 = 10.$$

Continuing the development, we have:

$$\begin{aligned} 38 - 23 &= (30 - 20) + (8 - 3) \\ &= 10 + 5 \\ &= 15. \end{aligned}$$

This development shows that it is "legal" to subtract the "ones" and then the "tens." Verify that 38 − 23 = 15 by showing that 23 + 15 = 38.

The preceding development suggests the following algorithm:

$$\begin{array}{rl} 38 & \\ \underline{23} & \\ 5 & \ldots\ldots (8-3) \text{ ones} \\ \underline{10} & \ldots\ldots (3-2) \text{ tens} \\ 15 & \end{array}$$

By eliminating some of the recording, this is shortened to the standard algorithm:

$$\begin{array}{r} 38 \\ \underline{23} \\ 15 \end{array}$$

From the generalization, we are able to find $(a + b) - (c + d)$ only when $c \leq a$ and $d \leq b$. What procedure must be used to find

$$56 - 19$$

where $9 > 6$ (that is, $d > b$)? We use the following procedure. Using both renaming and the generalization for the subtraction algorithm, we have:

$$\begin{array}{l}
56 = (50 + 6) = (40 + 10) + 6 = 40 + (10 + 6) = (40 + 16) \\
\underline{19} = \underline{(10 + 9)} = \underline{(10 \quad + \quad 9)} = \underline{(10 \quad + \quad 9)} \;\; = \underline{(10 + \;\; 9)} \\
\hphantom{56 = (50 + 6) = (40 + 10) + 6 = 40 + (10 + 6) = } (30 + \;\; 7)
\end{array}$$

Hence,

$$56 - 19 = 37.$$

Verify this by showing that $19 + 37 = 56$.

Do you see that the commonly called "borrowing" method has its origin in the preceding procedure? Relate the two methods.

There are many other interesting subtraction algorithms. Some are given in the following examples:

Example 1 Find another name for $50 - 18$.

Rename 50 as a sum of two addends as,

$$\begin{array}{rl}
50 = & 49 + 1 \\
18 = & \underline{18 \qquad\;} \\
& 31 + 1 = 32
\end{array}$$

Was there an advantage in renaming 50 as $49 + 1$?

Example 2 If a farmer has 200 sheep and he sells 137, how many does he have left?

Using the same procedure as in Example 1,

$$\begin{array}{rl}
200 = & 199 + 1 \\
137 = & \underline{137 \qquad\;} \\
& \;\;62 + 1 = 63
\end{array}$$

Example 3 John had \$56. After buying his schoolbooks, which cost \$19, how much money does he have?

$$\begin{array}{rl}
56 = 50 + 6 = & 49 + 7 \\
19 = \qquad\qquad\; & \underline{19 \qquad\;} \\
& 30 + 7 = 37
\end{array}$$

Observe that the decimal place value was not used in renaming 50, 200, and 56. Can you create other algorithms for subtraction?

The Multiplication Algorithm Since only products of whole numbers less than 10 may be found in the multiplication grid, a procedure is needed for finding the product of numbers equal to or greater than 10. The standard multiplication algorithm will be developed with accompanying justifications in the following examples.

Example 4 Find the product of 32 and 6.

$32 \times 6 = (30 + 2) \times 6$	*Rename* 32 *by decimal notation*
$= (30 \times 6) + (2 \times 6)$	*Distributive property*

The product (2×6) may be found in the multiplication grid; however, this grid does not contain the product (30×6). This product is established in the following manner:

$30 \times 6 = (10 \times 3) \times 6$	*Rename* 30 *by expanded notation*
$= 10 \times (3 \times 6)$	*Associative property for multiplication*
$= 10 \times 18$	*Rename* (3×6)
$= 180$	*Rename by decimal notation*

Hence,

$$\begin{aligned} 32 \times 6 &= (30 \times 6) + (2 \times 6) \\ &= \quad 180 \quad + \quad 12 \quad . \end{aligned}$$

The preceding discussion justifies the following algorithm:

```
 32
  6
---
180 . . . . (30 × 6)
 12 . . . . (2  × 6)
---
192
```

The 12 and 180 are called *partial products*. The following shortened version is considered to be the standard algorithm.

```
 32
  6
---
192
```

Do you see how this shorter algorithm came about?

Example 5 Give a justification for $15 \times 13 = 195$.

15×13	
$= 15 \times (10 + 3)$	*Rename* 13 *by decimal notation*
$= (15 \times 10) + (15 \times 3)$	*Distributive property*
$= [(10 + 5) \times 10] + [(10 + 5) \times 3]$	*Rename* 15 *by decimal notation*
$= [(10 \times 10) + (5 \times 10)] + [(10 \times 3) + (5 \times 3)]$	*Distributive property*

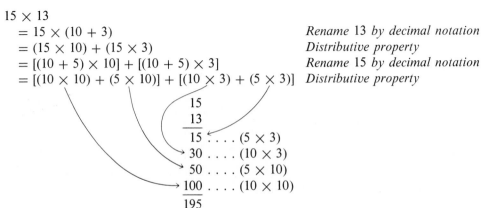

```
 15
 13
---
 15 . . . . (5 × 3)
 30 . . . . (10 × 3)
 50 . . . . (5 × 10)
100 . . . . (10 × 10)
---
195
```

The standard algorithm evolves from the above. Explanations are given.

```
 15
 13
----
      . . . . (5 × 3) = 1 ten 5 ones     . . . . 4 tens 5 ones
      . . . . (10 × 3) = 3 tens                        |
 45  <-------------------------------------------------
      . . . . (5 × 10) = 5 tens          . . . . 15 tens
      . . . . (10 × 10) = 10 tens                  |
150  <---------------------------------------------
----
195
```

A geometric model is effective in developing the standard algorithm. It relates the algorithm to the Cartesian product of two sets. You should review Sections 5.5 and 5.6. Here the ordered pairs of the Cartesian product of sets P and Q were arranged in an *array*. That is, they were arranged in rows and columns with each row having the same number of elements. The rows formed a set of equivalent sets, and so did the columns.

As we become interested in the number of ordered pairs, instead of the ordered pairs themselves, an appropriate representation of the array is as follows:

```
* * *
* * *
```

This array consists of two *rows* and three *columns*. It is by agreement that the rows are horizontal and the columns are vertical. This is a 2 × 3 array which contains 6 elements. Do you see that this is a model for 2 × 3 = 6? The first factor gives the number of rows in the array, while the second gives the number of columns. Designating the factors in this way is an arbitrary choice; however through usage, this has become accepted.

The discussion that follows shows the use of an array in developing an algorithm for 15 × 13. For this, a rectangular region is used for the array instead of one like that shown for 2 × 3. Is it possible to think of this region as representing an array having 15 rows and 13 columns?

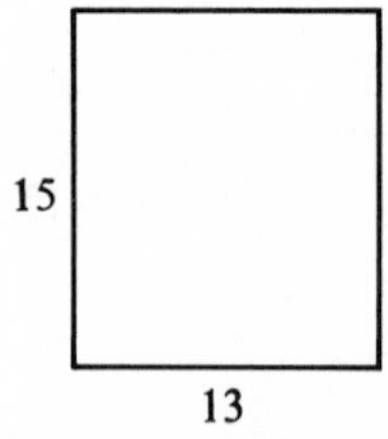

The renaming of 15 as (10 + 5) and 13 as (10 + 3) in Example 5 is represented by subdividing the region with the horizontal and vertical lines seen in the following diagram. The region (array) is thus subdivided into smaller regions (arrays) identified as (10 × 10), (5 × 10), (3 × 10), and

(5 × 3). The product of 15 and 13, (10 + 5) × (10 + 3), is the sum of these four products. Observe how the subdivisions of the region are related to the algorithms.

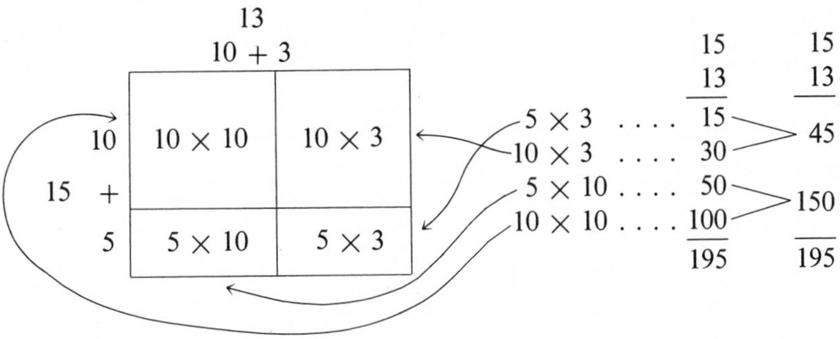

Compare the algorithms in Example 5 with these.

The Division Algorithm In the set of whole numbers, division has been referred to as a *process* instead of an *operation*. See Section 6.9. Relating division and multiplication as was done in this section, we find that some pairs of whole numbers do not have a matching number in W under division. For example, $7 \div 5 = n$ has no solution in W, since there is no whole number which can be multiplied by 5 to give 7.

Since problems involving division of whole numbers appear in the study of elementary arithmetic, it seems appropriate that a procedure for division in W be developed. The standard division algorithm will be developed in the examples and discussion that follow.

Example 6 Suppose that 36 chairs are to be arranged in rows. If there are to be 4 rows, how many chairs must there be in each row if all the chairs are to be used in the arrangement? This question can be expressed by the mathematical sentences:

$$36 \div 4 = n \quad \text{or} \quad (36, 4) \overset{\div}{\Rightarrow} n$$

From Definition 6.9 we know that we must find n such that

$$4 \times n = 36.$$

Since from the multiplication grid we have

$$4 \times 9 = 36,$$

we know that

$$36 \div 4 = 9.$$

Thus, the chairs will be arranged like this, with 9 chairs in each row:

* * * * * * * * *
* * * * * * * * *
* * * * * * * * *
* * * * * * * * *

All of the 36 chairs have been used in making the arrangement. There are no chairs remaining. A sentence describing this is

$$36 = (4 \times 9) + 0.$$

The 36 is called the *dividend,* 4 the *divisor,* 9 the *quotient,* and 0 the *remainder.*

Example 7 Now suppose that the number of chairs to be arranged in 4 rows is 38. If possible, we would like to have the same number of chairs in each row. That is, we would like the arrangement to look like an array as in Example 6.

We begin by making an estimate of the number of columns needed for the 38 chairs to be arranged in the 4 rows. Suppose our first estimate is 7. This means that we expect a 4 × 7 array to be made out of the 38 chairs. As the chairs are arranged in this array, it is found that 10 are not in the arrangement.

A sentence describing this model is

(i) $$38 = (4 \times 7) + 10.$$

Since this is a division problem, $38 \div 4 = n$, the 7 in the preceding sentence is the quotient and 10 is the remainder. The question is: Is 7 the best quotient?

We find that the array may be enlarged by using the 10 remaining chairs. They are then arranged in a 4 × 2 array. Two chairs remain.

A sentence describing this model is

$$38 = [4 \times (7 + 2)] + 2,$$

or

(ii) $$38 = (4 \times 9) + 2.$$

The 38 chairs have been arranged in 4 rows having 9 chairs in each row with 2 chairs left over.

In sentence (i), the remainder 10 is greater than the divisor 4; while in (ii), the remainder 2 is less than the divisor. When the latter occurs, the quotient is as great as possible. Hence, 9 in (ii) is the best possible quotient.

Examples 6 and 7 give evidence that a statement may be made about division for every pair of whole numbers (a, b). These statements are summarized as follows:

For $36 \div 4 = n$, we have

$$36 = (4 \times 9) + 0.$$

With the remainder 0, this becomes

$$36 = (4 \times 9).$$

For $38 \div 4 = n$, we have

$$38 = (4 \times 9) + 2.$$

We generalize these statements by stating the following which is known as Archimedes' Division Axiom:

A Generalization for the Division Algorithm

If $a, b \in W$, with $a \geq b$ and $b \neq 0$, then q and r can be found such that

$$a = (b \times q) + r,$$

and q is the greatest whole number such that $(b \times q) \leq a$ and $r < b$; q is called the *quotient*, and r is called the *remainder*.

Notice that if $(b \times q) = a$, then q assumes the role of the missing factor (Section 6.9) and $r = 0$; if $(b \times q) < a$, then $r \neq 0$. Why is it necessary that $b \neq 0$?

Now suppose that larger numbers are involved, and the multiplication grid does not supply the answer. Suppose that a custodian has been asked to arrange enough chairs to seat 290 children in 18 rows. Is it possible for him to find the number of chairs there should be in each row without going through the process of placing the chairs? From the preceding discussion and generalization, we have the sentence

(a) $$290 = (18 \times q) + r,$$

where q is to be the greatest number of chairs we can have in each of the 18 rows, and r is to be the number of chairs remaining after the arrangement has been completed. The condition which will ensure $r < 18$ is that the whole number q satisfy the sentence

(b) $$(18 \times q) \leq 290 < [18 \times (q + 1)].$$

Suppose that we try 15 for q in (b). Is the following sentence true?

(1) $$(18 \times 15) \leq 290 < [18 \times (15 + 1)]$$

Since $290 \nless [18 \times (15 + 1)]$, this sentence is not satisfied by 15 for q. Which of the following sentences is true?

(2) $(18 \times 16) \leq 290 < [18 \times (16 + 1)]$

(3) $(18 \times 17) \leq 290 < [18 \times (17 + 1)]$

Since (2) is true and (1) and (3) are not, we know that 16 is the required whole number for q. Now sentence (a) becomes

$$290 = (18 \times 16) + r.$$

The remainder r is found by subtracting (18×16) from 290; this gives 2. Thus, sentence (a) becomes

$$290 = (18 \times 16) + 2.$$

We may say, "When 290 *is divided by* 18, that is, $290 \div 18$, the quotient is 16 and the remainder is 2." Thus, the chairs may be arranged in 18 rows of 16 chairs each, with 2 chairs left over.

When a and b are large numbers, difficulty may be encountered in using the above procedure efficiently. Instead of attempting to find the greatest q at the outset, we may find q_1, q_2, q_3, etc., in successive stages such that

$$q_1 + q_2 + q_3 + \cdots + q_n = q.$$

We shall refer to q_1, q_2, etc., as *partial quotients*. After each partial quotient has been selected, we must examine the corresponding remainder to see if it is greater than or less than the divisor b. If the remainder is greater than b, then another partial quotient must be found from this remainder. This process continues until the generalization for the division algorithm is satisfied with $r < b$.

We shall use the same problem to demonstrate the stages in finding the partial quotients. That is, to find $290 \div 18$, we proceed as follows:

$$290 = [18 \times (q_1 + q_2 + q_3 + \cdots + q_n)] + r$$

Applying the distributive property for multiplication over addition, we have

$$290 = [(18 \times q_1) + (18 \times q_2) + \cdots + (18 \times q_n)] + r.$$

Let $q_1 = 10$; then

$$290 = (18 \times 10) + [(18 \times q_2) + \cdots + (18 \times q_n) + r]$$
$$= (18 \times 10) + [110], \quad \text{with the remainder } 110 > 18.$$

Since this remainder is greater than 18, there must be at least one other partial quotient in it. Let us choose $q_2 = 5$. Then

$$110 = (18 \times 5) + [(18 \times q_3) + \cdots + r]$$
$$= (18 \times 5) + [20], \quad \text{with the remainder } 20 > 18.$$

Hence,

$$290 = [(18 \times 10) + (18 \times 5)] + 20.$$

The remainder, 20, is reduced by finding a partial quotient, q_3, in 20. Let $q_3 = 1$. Then

$$20 = (18 \times 1) + 2, \quad \text{with the remainder } 2 < 18.$$

Hence,

$$290 = [(18 \times 10) + (18 \times 5) + (18 \times 1)] + 2.$$

At last we have a remainder, 2, that is less than the divisor 18. This is the requirement for r set up in the generalization for the division algorithm. By the distributive property, the above sentence becomes

$$290 = [18 \times (10 + 5 + 1)] + 2.$$

Thus, $q = 10 + 5 + 1$, and the sentence may be written

$$290 = (18 \times 16) + 2.$$

Thus, for $290 \div 18$, the quotient is 16 with a remainder 2.

The above procedure may be arranged so that we may conveniently keep a record of the steps taken. Are you able to locate the steps of that procedure in the following algorithm?

$$\begin{array}{r|l}
18\,\overline{)290} & \\
\underline{180} & q_1 = 10, \quad (18 \times 10) = 180; \quad 290 - 180 = 110, \text{ and } 110 > 18 \\
110 & \\
\underline{90} & q_2 = 5, \quad (18 \times 5) = 90; \quad 110 - 90 = 20, \text{ and } 20 > 18 \\
20 & \\
\underline{18} & q_3 = \underline{1,} \quad (18 \times 1) = 18; \quad 20 - 18 = 2, \text{ and } 2 < 18 \\
2 & q \;\, = 16
\end{array}$$

Such a procedure permits making estimates when selecting the partial quotients, keeping in mind that $(b \times q)$ be less than or equal to a. In case the estimate is much less than is necessary, this causes no difficulty in the above algorithm. However, the better the estimates, the shorter will be the algorithm.

Our interest then is to find a way to make an intelligent estimate for each partial quotient, so that the algorithm will be as short as possible. For ease in computation, we select multiples of powers of 10 for each partial quotient. If the selection is made at random, it is likely to be much too small or too large. Hence, we need to establish some safeguards to be used in making the selections. These are explained in the following discussion.

A search for the best partial quotients will be undertaken as we use the algorithm for $5683 \div 24$. The sum of these partial quotients, $q_1 + q_2 + q_3 + \cdots + q_n$, will give a number which when multiplied by 24 should give 5683. If it does not, then there will be a remainder.

We now begin a search for a number for q_1. Since 24×10^2 is equal to 2400 and 24×10^3 is 24,000, we know that q_1 must lie between 10^2 and 10^3. Seeing that 2400 is closer to but less than 5683, we try 2×10^2. But, $24 \times (2 \times 10^2)$ is less than 5683. So, we try 3×10^2 and find that $24 \times (3 \times 10^2)$ is greater than 5683. Hence, we select 2×10^2 as the largest acceptable multiple of 10^2 for q_1. This, then, is the best partial quotient that could be found. The test for q_1 is

$$24 \times (2 \times 10^2) \leq 5683 < 24 \times (3 \times 10^2).$$

The algorithm thus far is

$$\begin{array}{rr|l} & 24\,\overline{)5683} & \\ 24 \times 200 = & 4800 & q_1 = 200. \\ \cline{2-2} & 883 & \end{array}$$

Now we seek a number for q_2 which when multiplied by 24 gives 883. Since we have exhausted all multiples of 10^2, we look for some multiple of 10^1. We find that $24 \times (3 \times 10^1)$ is less than 883 and $24 \times (4 \times 10^1)$ is greater than the remainder 883. This means that 3×10^1 must be selected for q_2. The test for q_2 is

$$24 \times (3 \times 10^1) \leq 883 < 24 \times (4 \times 10^1).$$

Continuing with the algorithm,

$$\begin{array}{rr|l} & 24\,\overline{)883} & \\ 24 \times 30 = & 720 & q_2 = 30. \\ \cline{2-2} & 163 & \end{array}$$

A number when multiplied by 24 that gives 163 is the third partial quotient, q_3, which we seek. Since multiples of 10^1 have been exhausted, we try multiples of 10^0. Recall that 10^0 has been defined as 1. We find that $24 \times (6 \times 10^0)$ is less than 163 and $24 \times (7 \times 10^0)$ is greater than the remainder 163. Which number must we select? Can you tell by looking at this test for q_3?

$$24 \times (6 \times 10^0) \leq 163 < 24 \times (7 \times 10^0)$$

The algorithm continues with

$$\begin{array}{rr|l} & 24\,\overline{)163} & \\ 24 \times 6 = & 144 & q_3 = 6. \\ \cline{2-2} & 19 & \end{array}$$

Since the multiples of 10^0 (or 1) have been exhausted, and the remainder has become less than the divisor 24, our algorithm for whole numbers ends.

The following shows the complete algorithm with the tests for selecting each q.

$$\begin{array}{rr|rl} & 24\,\overline{)5683} & & \\ 24 \times 200 = & 4800 & 200, & 24 \times (2 \times 10^2) \leq 5683 < (3 \times 10^2) \\ \cline{2-2} & 883 & & \\ 24 \times 30 = & 720 & 30, & 24 \times (3 \times 10^1) \leq 883 < (4 \times 10^1) \\ \cline{2-2} & 163 & & \\ 24 \times 6 = & 144 & 6, & 24 \times (6 \times 10^0) \leq 163 < (7 \times 10^0) \\ \cline{2-3} & 19 & 236 & \end{array}$$

The remainder $19 < 24$, and the quotient $236 = 200 + 30 + 6$. Hence,

$$5683 = (24 \times 236) + 19.$$

The test for the largest acceptable partial quotient is

$$b \times (m \times 10^n) \leq a < b \times ((m + 1) \times 10^n)$$

where b is the divisor, a is the dividend, $m \in N$, and $n \in W$.

EXERCISE 7.6

1. Use different forms of the algorithm to find each sum:
   (a) 212 + 34 (d) 25 + 38
   (b) 631 + 235 (e) 278 + 365
   (c) 2345 + 1521 (f) 8345 + 6284
2. Justify the steps used in finding each of the given results by giving the appropriate generalizations and properties:
   (a) $56 \times 7 = 392$ (d) $249 + 364 = 613$
   (b) $35 + 28 = 63$ (e) $83 \times 24 = 1992$
   (c) $79 - 23 = 56$ (f) $48 - 10 = 38$
3. Demonstrate how the generalization for the subtraction algorithm can be applied to three-digit numerals, such as 985 − 123.
4. Write another mathematical sentence related to each of the following.
   (a) $9 - 4 = 5$ (e) $23 \times n = 46$
   (b) $6 \div 2 = 3$ (f) $253 - n = 54$
   (c) $8 + 5 = n$ (g) $14 - n = 14$
   (d) $6 + n = 15$ (h) $18 \times 82 = n$
5. Apply the generalization for the subtraction algorithm to find a simpler name for each of the following.
   (a) 64 − 23 (d) 92 − 38
   (b) 83 − 51 (e) 265 − 124
   (c) 43 − 27 (f) 643 − 58
6. Develop a distributive property that distributes division over addition and demonstrate how this property may be used to find the quotient of $36 \div 3$.
7. After giving the justification for Problem 2(e), relate this to the steps taken in the multiplication algorithm in finding 83×24.
8. (a) Use the *test for the largest acceptable partial quotient* to find all partial quotients for $7001 \div 16$.
   (b) Divide 7001 by 16 making use of the information found in (a).
   (c) What is the final remainder? Is it less than 16? What significance does this have?
   (d) Use the *generalization for the division algorithm* (Archimedes' Division Axiom) to write a sentence giving the result of (b).
   (e) Show how (b) could be used as a basis for establishing the *standard* algorithm which many students use today.

9. (a) In seeking the largest acceptable partial quotients for 6720 ÷ 16, which of these statements are true and which false?

q_1: $16 \times (3 \times 10^2) \leq 6720 < 16 \times (4 \times 10^2)$
q_1: $16 \times (4 \times 10^2) \leq 6720 < 16 \times (5 \times 10^2)$
q_2: $16 \times (1 \times 10^1) \leq 320 < 16 \times (2 \times 10^1)$
q_2: $16 \times (2 \times 10^1) \leq 320 < 16 \times (3 \times 10^1)$

(b) Name the partial quotients you would select from the above cases. Explain why you select these.
(c) How many partial quotients are there, and why only this many?
(d) Divide 6720 by 16 giving the complete algorithm which includes each partial quotient and its test.
(e) Write a sentence which shows the relatedness of the dividend, the divisor, the quotient, and the remainder. Explain how the quotient was determined. What is significant about this remainder?

10. Find the quotient and the remainder using the division algorithm as in Problem 9. Write the mathematical sentence for each result.

(a) 325 ÷ 15 (d) 1728 ÷ 12
(b) 804 ÷ 13 (e) 2584 ÷ 152
(c) 1608 ÷ 28 (f) 307,465 ÷ 203

7.7 FLOW CHARTS OF ALGORITHMS

Have you ever used a simple computer, a desk calculator, to add two numbers? Instructions must be given the machine by punching certain keys. The procedure you would use in adding 24 and 52 consists of the following:

1. Punch key 2 then key 4.
2. Punch the + key.
3. Punch key 5 then key 2.
4. Punch the + key.
5. Punch the T (Total) key.

The print-out shows the recorded data and the total.

24
52

76

Here, we used a set of instructions arranged in a particular sequence. It was used to "talk to" or program our desk calculator.

Each computer, whether simple or complex, requires a program before it can solve a problem. Thus a problem must be analyzed in order that a step-by-step procedure be developed. This procedure is an algorithm used to solve the problem. It is the program given to the computer.

When a chart is designed showing the steps of the algorithm in their proper sequence, we have a *flow chart*. Certain conventions will be used in constructing a flow chart. Each of the steps will be placed in a "box," the interior of a simple closed curve. The steps consist of the data to be used, what is to be done with the data, possible decisions, and the outcome. The shape of the box will be a signal as to the type of step enclosed. The following conventions will be used:

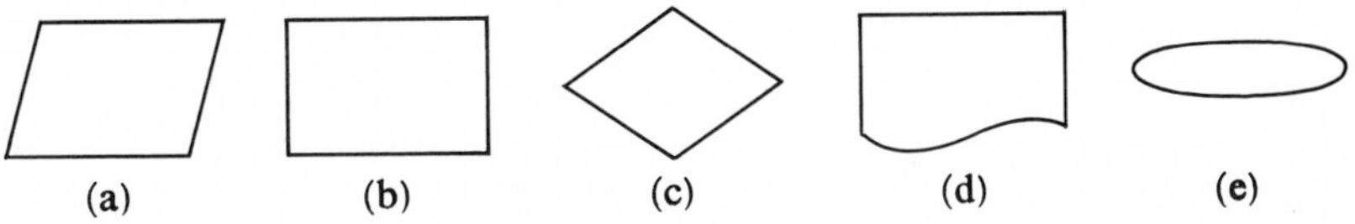

(a) The *input box* encloses data, such as essential facts and equipment.
(b) The *operation box* encloses instructions or operations to be performed on data.
(c) The *decision box* encloses a question requiring a decision.
(d) The *output box* encloses the outcome, a result of the operation(s) on the data.
(e) The *start/stop box* encloses the start and stop signals.

The start box indicates where the flow chart begins, and the arrows establish the sequence of the steps leading to the stop box.

The use of these conventions is demonstrated in the following examples:

Example 1 Construct a flow chart for a desk calculator for finding the sum of 24 and 52.

Using the instructions given to the calculator at the beginning of this section for finding the sum of 24 and 52, we have,

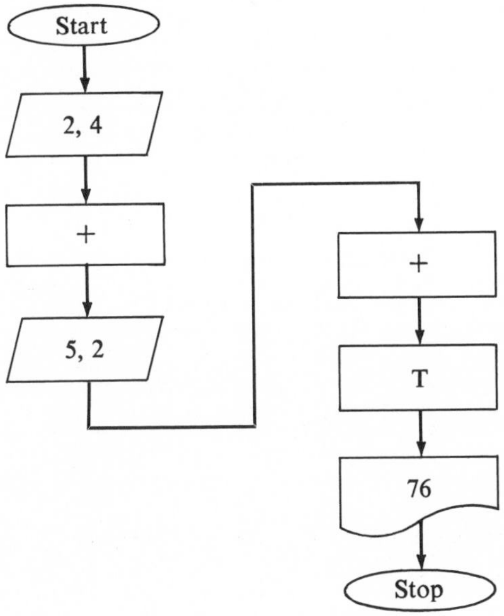

Example 2 Construct a flow chart of the algorithm for finding the average of 4 and 10.

In analyzing the problem, we find that the first step is to add 4 and 10. Then, this sum divided by 2 gives the average. Now that the problem has been analyzed and an algorithm has been developed, the following flow chart is constructed:

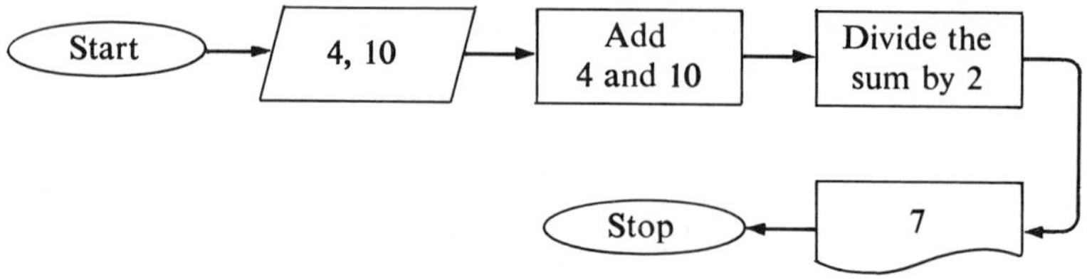

The analysis of a problem is certainly not a new approach to problem solving; however, the construction of a diagram or flow chart of the algorithm has been of prime interest since the advent of electronic computers. Even though the flow chart makes its greatest contribution here, it serves also as a tool in every-day problem solving situations. The housewife might find a flow chart useful as she goes through the steps in making a molded gelatin. The dentist might use one to explain to a young patient the steps in brushing teeth. These are demonstrated in the examples that follow.

Example 3 Construct a flow chart for making a molded gelatin. The recipe that includes the ingredients and instructions for combining them may be compared to the algorithm. The flow chart follows:

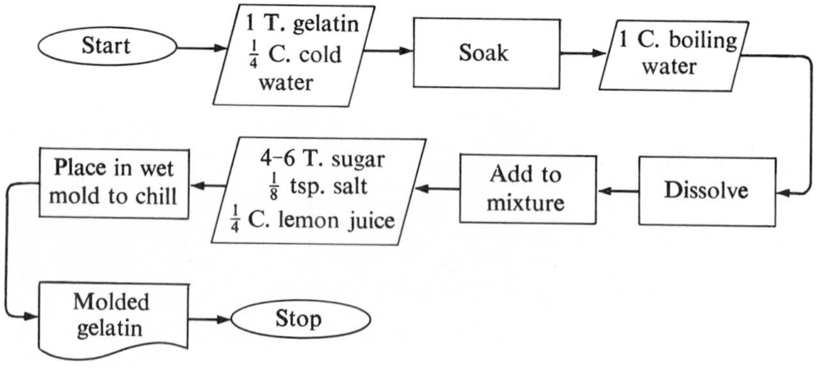

Example 4 Construct a flow chart for brushing teeth, which is recommended by John's dentist.

John's dentist suggested the following plan for brushing teeth.

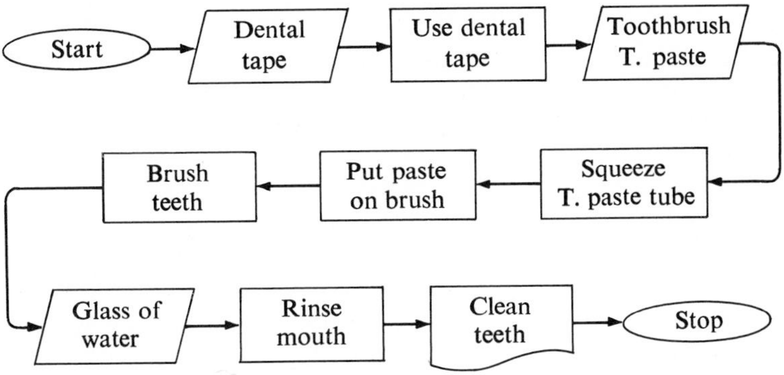

Not all flow charts are of the type demonstrated in the preceding examples, where the boxes are lined up. Some may be *branching,* as will be shown in the following examples. These include the decision box, which encloses a question requiring either a "yes" or a "no" answer.

Example 5 Construct a flow chart for finding whether sets A and B are identical or whether A is a proper subset of B.

Observe how the answer to the decision question affects the output.

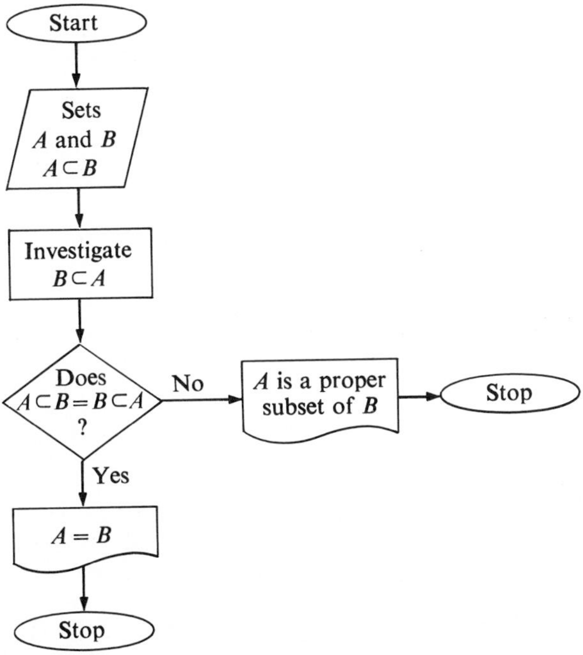

Some flow charts contain several branchings and loops. Observe the following example.

Example 6 Construct a flow chart of the algorithm for the addition of a 3-digit number to a 3-digit number.

The flow chart makes use of the symbols

$$\overset{2}{\square}\overset{1}{\square}\overset{0}{\square} + \overset{2}{\triangle}\overset{1}{\triangle}\overset{0}{\triangle},$$

where □ and △ are frames that hold the digits and 2, 1, and 0 designate the position of the digits in each number. Position of a digit will be referred to in the flow chart as x. If $x = 0$, the digits in the ones position are being considered; to increase this value of x by 1, the digits considered are in the tens position. With each addition of 1, digits are designated progressively to the left. Another symbol, ▽, will be used in which the "carry" numeral is placed.

Before beginning the flow chart, let us consider the addition of 256 and 972. Using □ for the digits of the first addend, 256; △ for the digits of the second addend, 972; ▽ for the "carry"; and ⬡ for the digits of the sum, we have,

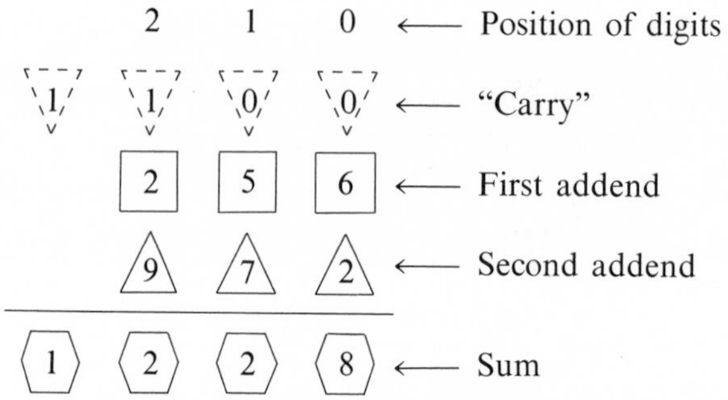

In studying the flow chart at the right, you should relate each of the instructions to the steps of the preceding algorithm.

EXERCISE 7.7

1. Identify each of the following as input, output, decision, or operation boxes.

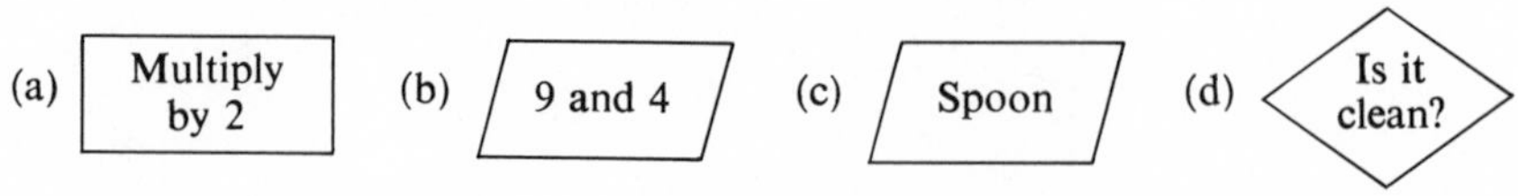

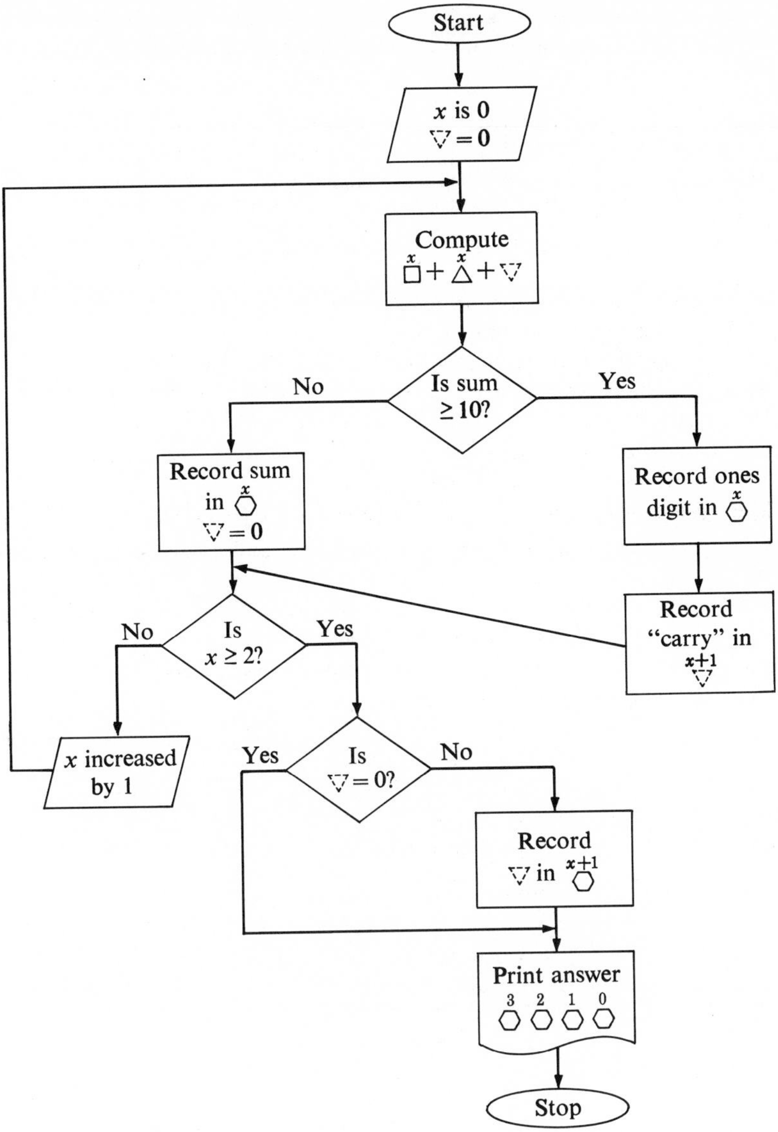
Start
x is 0
▽ = 0
Compute
□ + △ + ▽
No
Is sum
≥ 10?
Yes
Record sum
in ⬡
▽ = 0
Record ones
digit in ⬡
Record
"carry" in
x+1
▽
No
Is
x ≥ 2?
Yes
x increased
by 1
Yes
Is
▽ = 0?
No
Record
▽ in ⬡
Print answer
3 2 1 0
⬡ ⬡ ⬡ ⬡
Stop

(e) 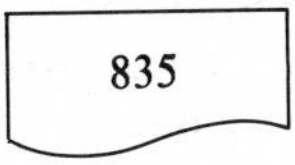
835

(f) 
Mow lawn

(g) 
Is he
taller than
6′?

(h) 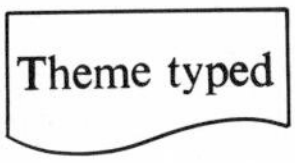
Theme typed

2. Fill in the diagram so that the given instructions will have the proper sequence. Hamburger meat; Make into patties; Grill; Cooked patties; Build fire; Put patties on grill

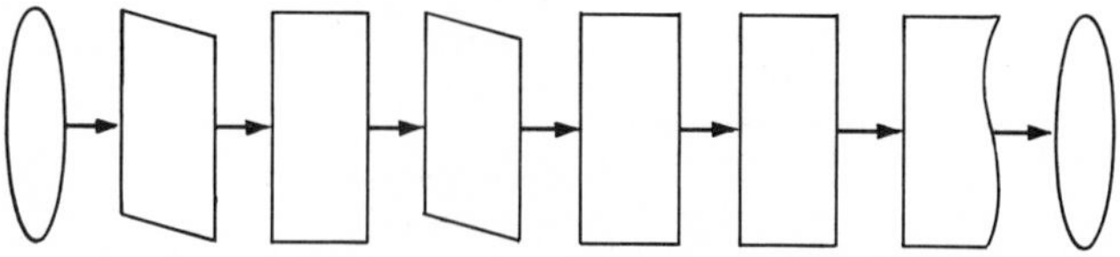

3. Arrange each set of instructions in a sequence to make a sensible flow chart. Place each instruction in the proper box.
   (a) Turn handle; Sharpener; Pencil; Sharp pencil; Pencil in sharpener
   (b) Start car; Car; Back car; Keys; Put in reverse; Out of garage
   (c) Ball; Referee's signal; Punt; Field goal; Goal posts
4. (a) Use the flow chart in Example 2 as a guide in making one for finding the average of any two whole numbers, x and y.
   (b) Make a flow chart for finding the average of 96, 81, and 90.
5. (a) Use the flow chart in Example 7 to guide you in finding the sum of 68 and 54.
   (b) What minimum change is necessary to convert the flow chart in Example 7 to one suitable for adding two 4-digit numbers?
6. State a question that should be put in the decision box in order to have the two outputs as indicated in the flow chart.

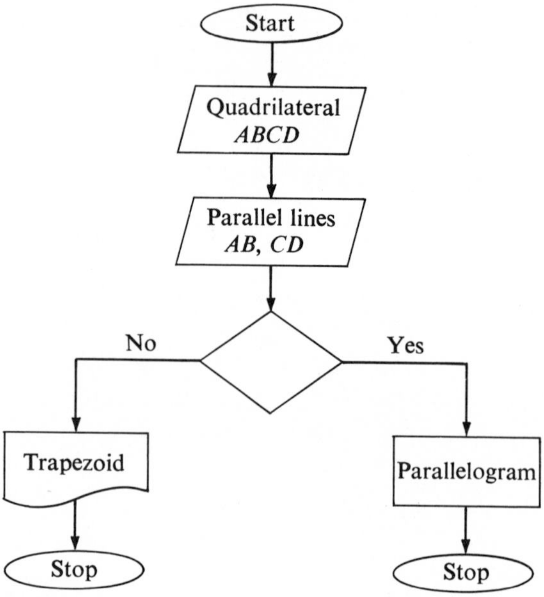

7. Convert the flow chart in Example 4 to a branching and looping one, by inserting decision boxes and loop-routes. Include appropriate questions in the decision boxes.

8

The System of Integers

8.1 THE SET OF INTEGERS

Up to now we have studied the development of two number systems—natural numbers and whole numbers. The set of natural numbers was extended to form the set of whole numbers when we demanded that there be an additive identity element, zero. Now we shall make a further demand and thus justify another extension.

In Chapter 6, we found that it is not possible to match every pair of whole numbers with a whole number under subtraction. For example, there is no whole number which when added to 6 gives 4. Thus, we found that in the set of whole numbers, for (a, b) to be matched with the unique number c, the second component, b, must be less than or equal to the first component, a. In the case of (4, 6), the component 6 is not less than the component 4.

Does it seem reasonable to demand that subtraction be an operation, that is, that it apply to all elements of *some* set? Such a set would have closure under subtraction, and the restriction placed on a and b in Section 6.5 would be removed.

Suppose that there exists some unique number which when added to 6 gives 4, that is, that there is some x such that

$$6 + x = 4.$$

What is this number x? Let us rename 6 as $4 + 2$; then we have

$$(4 + 2) + x = 4.$$

By the associative property for addition, we have

$$4 + (2 + x) = 4.$$

This sentence will be true by the additive identity property if $2 + x = 0$.

Obviously, there is no whole number which when added to 2 gives 0. So, to meet this need, we shall invent a number that satisfies this requirement. This invented number will be represented by the symbol $^-2$ and will have the property that

$$2 + {}^-2 = 0, \qquad \text{or} \qquad (2, {}^-2) \xrightarrow{+} 0.$$

We say that $^-2$ is the *additive inverse* of 2. We read $2 + {}^-2 = 0$ as, "2 plus the additive inverse of 2 equals 0."

Returning to the original question, and assuming the associative property, we have:

$$\begin{aligned} 4 + (2 + {}^-2) &= 4, \\ (4 + 2) + {}^-2 &= 4, \\ 6 + {}^-2 &= 4. \end{aligned}$$

Using the idea of subtraction introduced in Section 6.5, we can now write

$$4 - 6 = {}^-2.$$

In order to make subtraction always possible, we shall need an additive inverse of each whole number. We shall now invent these also. For instance, $^-6$ will be the additive inverse of the whole number 6, and $^-15$ the additive inverse of the whole number 15, such that

$$(6, {}^-6) \xrightarrow{+} 0 \qquad \text{and} \qquad (15, {}^-15) \xrightarrow{+} 0.$$

What new numbers will be paired with 8, 12, and 3 under addition to match with 0?

What new number is to be paired with 0 under addition to match with 0? We would expect this number to be $^-0$ such that

$$0 + {}^-0 = 0.$$

Yet, from the definition of addition of whole numbers, we have

$$0 + 0 = 0.$$

Our conclusion is that 0 is its own additive inverse; it is the only whole number which has this distinction.

We make the following generalizations concerning additive inverses:

(i) For each $n \in N$, there exists a unique number, the additive inverse of n, such that the sum of n and its additive inverse is 0.
(ii) The number 0 is the additive inverse of 0.

Now we have a set consisting of these new numbers—namely,

$$M = \{{}^-1, {}^-2, {}^-3, {}^-4, \ldots\}.$$

When this set of numbers is joined to the set of whole numbers, we have another set, called the *integer set:*

$$I = \{\ldots, {}^-4, {}^-3, {}^-2, {}^-1, 0, 1, 2, 3, 4, \ldots\}.$$

Thus, we have:

Definition 8.1(a) The *set of integers, I,* is the union of

(i) the set of natural numbers, N,
(ii) the set consisting of the additive inverse of each $n \in N$,
(iii) the set consisting of 0.

It is customary to refer to

$$M = \{^{-}1, {}^{-}2, {}^{-}3, \ldots\},$$

the set of additive inverses of elements of N, as the set of *negative integers.* We shall read $^{-}3$ as "negative 3." The elements of the remaining subset of I will be called the *nonnegative integers.* There are times when it is convenient to call the natural numbers the *positive integers.* These subsets of I are illustrated in the following diagram:

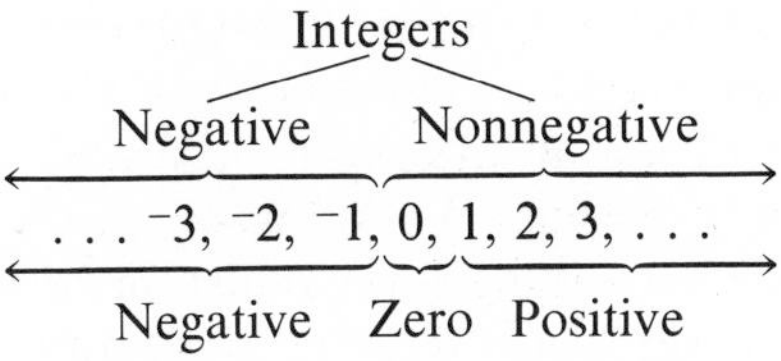

From an earlier discussion, $^{-}2$ was said to be the additive inverse of 2, since $2 + {}^{-}2 = 0$. If the commutative property for addition holds, then $^{-}2 + 2 = 0$. From this we see that 2 is the additive inverse of $^{-}2$. Hence, 2 and $^{-}2$ are additive inverses of each other. The arrows in the following diagram show the assignment of a number and its additive inverse.

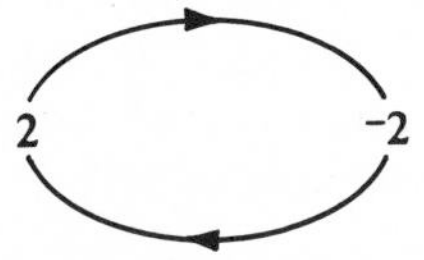

The frequent writing of the expression "additive inverse of" is simplified by using the symbol "$-$." The following examples show the substitution of the symbol for the expression.

Example 1 Rewrite each of the sentences by changing the expression, the additive inverse of, to the symbol, $-$.

(a) The *additive inverse of* 10 is $^{-}10$.

$$-10 \text{ is } {}^{-}10$$

(b) The *additive inverse of* $^{-}10$ is 10.

$$-\,{}^{-}10 \text{ is } 10$$

(c) The *additive inverse of* the *additive inverse of* $^{-}10$ is $^{-}10$.

$$-(-\,{}^{-}10) \text{ is } {}^{-}10$$

Example 2 Given the expression $-(-10)$.

(a) How is it read? We say, "the additive inverse of the additive inverse of 10."

(b) What is a simpler way to write this? Using the symbol for the additive inverse of, we begin with

$$-(-10),$$

then convert to

$$-(^{-}10) = 10.$$

Hence, a simpler way to write the given expression is to write 10.

What is the meaning of $-x$, where $x \in I$? Since the symbol, $-$, stands for the additive inverse of, $-x$ is read, "the additive inverse of x." It is never read negative x for the following reason. Suppose x represents 3, then $-x$ will be negative 3, ($^{-}3$). However, if x represents $^{-}3$, then $-x$ will be 3. So, $-x$ does not always represent a negative number, and we should not designate it as such. We think of $-x$ as open, neither negative nor nonnegative. It is waiting for a replacement of a number. Only after the replacement is made are we able to identify it as either negative or nonnegative.

Since the additive inverse of 5 is $^{-}5$, denoted by $-5 = {}^{-}5$, some prefer calling -5 negative 5 instead of the additive inverse of 5. However, the preceding discussion gives the danger in identifying $-x$ as negative x.

The generalization made earlier concerning additive inverses may now be extended to include *all integers.* We conclude that every integer is the additive inverse of a unique integer, if their sum is 0.

Definition 8.1(b) For each $n \in I$, n and $-n$ are additive inverses of each other, if and only if $n + -n = 0$.

EXERCISE 8.1

1. What is the additive inverse of each of the following integers?

(a) 7	(f) $-[-(^{-}5)]$
(b) $-(-^{-}3)$	(g) $(6 + 3)$
(c) 3	(h) $^{-}(6 + 3)$
(d) $^{-}1$	(i) 0
(e) $-(^{-}5)$	(j) $(8 + {}^{-}8)$

2. From the following list select the ordered pairs that are additive inverses.

(a) $(4, {}^{-}4)$	(g) $(-^{-}9, {}^{-}9)$
(b) $(^{-}2, 2)$	(h) $(-^{-}5, 5)$
(c) $((6 + 3), {}^{-}(6 + 3))$	(i) $(-^{-}(2 + 3), (2 + 3))$
(d) $(6, {}^{-}7)$	(j) $(^{-}15, 15)$
(e) $(3, -^{-}3)$	(k) $(0, 0)$
(f) $(-^{-}(1 + 1), {}^{-}(1 + 1))$	(l) $((5 + 3), {}^{-}(2 \times 4))$

3. Give another name for:

   (a) $1 + {}^{-}1$
   (b) $18 + {}^{-}18$
   (c) ${}^{-}10 + 10$
   (d) $({}^{-}7 + 7) + 8$
   (e) ${}^{-}(5 + 2) + (5 + 2)$
   (f) $-{}^{-}6 + (3 + 3)$
   (g) $(8 + {}^{-}8) + ({}^{-}6 + 6)$

4. How many ordered pairs of elements of W can be matched with 0 under addition? How many pairs from I?
5. Define the set of integers using the union operation on two sets of numbers.
6. What integer will make these sentences true?

   (a) ${}^{-}6 + \square = 0$
   (b) $-{}^{-}6 + \square = 0$
   (c) $(8 + 2) + {}^{-}(8 + 2) = \square$
   (d) ${}^{-}5 + -\square = 0$
   (e) $-{}^{-}5 + -\square = 0$
   (f) $7 + {}^{-}(6 + 1) = \square$
   (g) $\square + {}^{-}4 = 0$
   (h) $\square + -{}^{-}4 = 0$
   (i) $\square + (4 + 3) = 0$
   (j) $-\square + 3 = 0$
   (k) $-\square + -{}^{-}3 = 0$
   (l) $-\square + (2 + 6) = 0$

8.2 THE ADDITION OPERATION IN THE SET OF INTEGERS

Now that we have a new set of numbers, I, that includes the natural numbers, their respective additive inverses, and zero, we are ready to develop methods of finding sums of these integers. From the way in which the new elements (negative integers) were invented, we carry over to the set I the *operation* addition and the set of properties that the elements obey under this operation.

We shall consider several examples. In these we shall use our knowledge of the addition of whole numbers, the sum of an integer and its additive inverse, and the properties that we carry over from the whole number system.

Example 1

$9 + {}^{-}6 = \square$	
$(3 + 6) + {}^{-}6 = \square$	*Renaming the whole number* 9
$3 + (6 + {}^{-}6) = \square$	*Associative property for addition*
$3 \quad + \quad 0 \quad = \square$	*Additive inverse*
$3 = \square$	*Additive identity*

Hence,

$$9 + {}^{-}6 = 3.$$

Observe that the positive integer 9 was renamed so that the second addend was the additive inverse of ${}^{-}6$. The associative property for addition made possible the addition of the additive inverses. The sum of $9 + {}^{-}6$ followed.

The following problems are given in order that you may have some experience in finding sums as was done in Example 1 above.

(1) $5 + {}^{-}2 = \square$
(2) $7 + {}^{-}5 = \square$
(3) ${}^{-}8 + 10 = \square$
(4) ${}^{-}14 + 20 = \square$
(5) ${}^{-}15 + 25 = \square$
(6) $28 + {}^{-}21 = \square$

This development suggests the following definition:

Definition 8.2(a) If a and b are nonnegative integers with $b < a$, then

$$a + -b = x,$$

where x is a positive integer such that $b + x = a$.

In the system of whole numbers

$$\text{if } b + x = a, \text{ then } a - b = x.$$

Since the whole numbers may now be defined as the nonnegative integers, this gives direction in finding the sum x.

Next we turn our attention to the sum of two negative integers.

Example 2 $^-6 + {}^-9 = \square$

The additive inverse of $^-6$ is 6 and the additive inverse of $^-9$ is 9. If it can be shown that

$$(^-6 + {}^-9) + (6 + 9) = 0,$$

then $(^-6 + {}^-9)$ is the additive inverse of $(6 + 9)$. Since the sum of $6 + 9$ is 15, its additive inverse is $^-15$. Hence we wish to show that

$$^-6 + {}^-9 = {}^-15.$$

We shall now find the sum of $(^-6 + {}^-9)$ and $(6 + 9)$.

$(^-6 + {}^-9) + (6 + 9)$	$= (^-6 + {}^-9) + (9 + 6)$	*Commutative property for addition*
	$= {}^-6 + [^-9 + (9 + 6)]$	*Associative property for addition*
	$= {}^-6 + [(^-9 + 9) + 6]$	*Associative property for addition*
	$= {}^-6 + [0 + 6]$	*Additive inverse*
	$= {}^-6 + 6$	*Additive identity*
	$= 0$	*Additive inverse*

Thus, $(^-6 + {}^-9) + (6 + 9) = 0$, and hence $^-6 + {}^-9 = {}^-(6 + 9) = {}^-15$.

Apply the procedure used in Example 2 to find the sums expressed in the following problems.

(7) $^-1 + {}^-5 = \square$ (10) $^-27 + {}^-12 = \square$
(8) $^-6 + {}^-13 = \square$ (11) $^-27 + {}^-23 = \square$
(9) $^-32 + {}^-28 = \square$ (12) $^-8 + {}^-10 = \square$

From these problems and Example 2, you are probably ready to say that the sum of two negative integers is found by first finding the sum of their additive inverses, and then finding the additive inverse of this sum. Generalizing the result in Example 2, we have

Definition 8.2(b) If a and b are nonnegative integers, then

$$-a + -b = -(a + b).$$

We are now prepared to find the sum of a negative and a positive integer when the additive inverse of the negative integer is greater than the positive integer. Example 3 below serves as an illustration. Note that Definition 8.2(b) is used to rename $^-9$ as $^-6 + {}^-3$, where one addend is the additive inverse of 6. (Explore the possibility of renaming 6 by using Definition 8.2(a).)

Example 3

$$6 + {}^-9 = \square$$

$$6 + ({}^-6 + {}^-3) = \square \quad \textit{Using Definition 8.2(b) to rename } {}^-9$$

$$(6 + {}^-6) + {}^-3 = \square \quad \textit{Associative property for addition}$$

$$0 \quad + {}^-3 = \square \quad \textit{Additive inverse}$$

$$^-3 = \square \quad \textit{Additive identity}$$

Hence,

$$6 + {}^-9 = {}^-3.$$

Proceed as in Example 3 to find the sums expressed in the following problems.

(13) $^-4 + 3 = \square$

(14) $^-16 + 10 = \square$

(15) $7 + {}^-9 = \square$

(16) $^-5 + 3 = \square$

(17) $6 + {}^-8 = \square$

(18) $^-11 + 10 = \square$

(19) $^-25 + 20 = \square$

(20) $^-27 + 13 = \square$

Generalizing the result in Example 3 leads to the following:

Definition 8.2(c) If a and b are nonnegative integers with $b > a$, then

$$a + -b = -c.$$

where c is a positive integer and $a + c = b$.

Definitions 8.2(a), 8.2(b), 8.2(c), and Definition 6.2 for whole numbers (now called nonnegative integers), and Definition 8.1(b) complete the definitions for addition in the set of integers. They will be used in establishing the order in the integer set in the next section.

EXERCISE 8.2

Find the sum of each of the following by using the definition which applies and any properties necessary.

1. $9 + {}^-2$
2. $17 + {}^-5$
3. $^-14 + 25$
4. $^-21 + 27$
5. $^-7 + 6$
6. $^-10 + 3$
7. $9 + {}^-16$
8. $36 + {}^-40$
9. $^-8 + {}^-2$
10. $^-5 + {}^-1$

11. $^-12 + ^-27$
12. $^-6 + ^-13$
13. $^-15 + 28$
14. $^-32 + ^-38$
15. $(8 + ^-8) + ^-15$
16. $69 + 73$
17. $49 + ^-28$
18. $29 + ^-29$
19. $(23 + 13) + ^-13$
20. $^-65 + ^-21$
21. $^-11 + 10$
22. $^-14 + 0$
23. $(^-15 + ^-13) + 12$
24. $38 + (^-14 + ^-5)$
25. $(9 + 7) + ^-15$
26. $^-19 + (6 + ^-1)$
27. $^-8 + (^-3 + ^-7)$

8.3 ORDER OF INTEGERS AND THE NUMBER LINE

Since order has been established for the whole numbers (now called the nonnegative integers), our present concern is to fit the negative integers into an order scheme for the set of integers. The question to be resolved is, "Which is the lesser of any two integers?" We know that 3 is less than 8, but, for instance, which is less, 2 or $^-6$?

We are so familiar with the order of the nonnegative integers that we can easily name the greater or the less of any pair of these numbers. However, let us recall the statement of Definition 6.10(a). From this, 3 is less than 8 since 5 must be added to 3 to make the sum equal to 8. This may be written as

$$3 < 8 \text{ because } 3 + 5 = 8.$$

We now extend the definition of order to include the negative integers. Compare the following definition with Definition 6.10(a).

Definition 8.3(a) For each $a, b \in I$ and $c \in N$, if $a + c = b$, then

$$a < b.$$

Note that the number c that is added to the number a must be a nonnegative integer greater than 0; that is, it must be a natural number.

Example 1 Which is less, 2 or $^-6$? We ask the question, "Is there a number in N such that when it is added to 2, the sum is $^-6$?" The answer is "no." The next question is, "Is there a number in N that when added to $^-6$ will give the sum 2?" The answer here is "yes." Using Definition 8.3(a), we have:

If there exists $\square$ in N such that $^-6 + \square = 2$, then $^-6 < 2$.
But $^-6 + 8 = 2$; thus $^-6 < 2$.

Example 2 Is $^-8$ less than $^-3$?

If there exists $\square$ in N such that $^-8 + \square = ^-3$, then $^-8 < ^-3$.
But $^-8 + 5 = ^-3$; thus $^-8 < ^-3$.

Example 3 Is 0 less than $^-4$?

Since there is no number in N which when added to 0 gives the sum $^-4$, we cannot say that 0 is less than $^-4$.

Is $^-4$ less than 0?

If there exists $\square$ in N such that $^-4 + \square = 0$; then $^-4 < 0$.
But $^-4 + 4 = 0$; thus $^-4 < 0$.

Example 4 Construct a flow chart for finding whether $a < b$, $a, b \in I$.

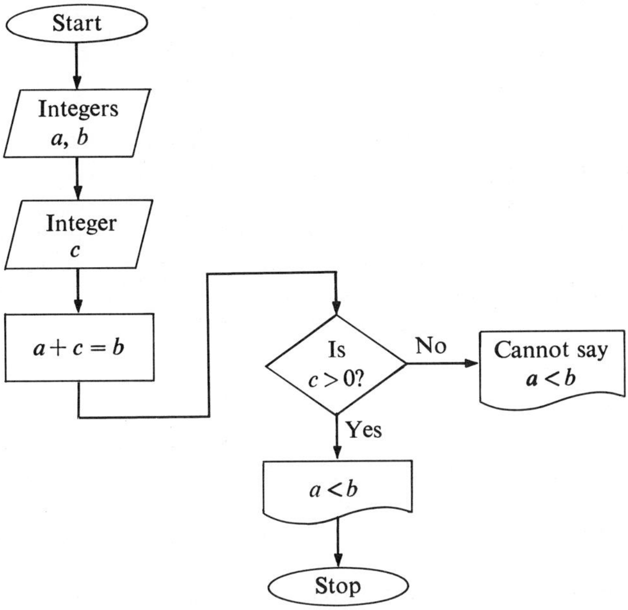

If the answer to the decision question: Is $c > 0$? is "no," does the flow chart give any direction as to what procedure must be followed from there? Since this diagram gives no such direction, the insert which follows provides the necessary instructions.

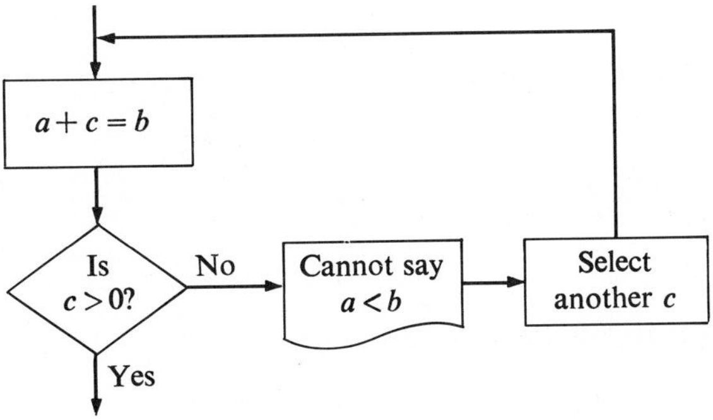

A "no" answer indicates that we cannot say that $a < b$. The chart now directs us to select another integer c, return to the box that says add c to a to give a sum b, then proceed to the decision box. Here the loop has been completed. If the answer to the question is still "no," the loop is repeated. When the answer becomes "yes," the looping comes to an end.

Using Definition 8.3(a), we can establish the order of the elements of the set of integers as:

$$\cdots\ {}^-4 < {}^-3 < {}^-2 < {}^-1 < 0 < 1 < 2 < 3 < 4 \cdots$$

Observe that there is no least and no greatest element. Each succeeding number in this sequence is 1 greater than the preceding number; that is, each successor is 1 greater than its predecessor. Thus:

$$\ldots, {}^-3 = {}^-4 + 1, {}^-2 = {}^-3 + 1, {}^-1 = {}^-2 + 1, 0 = {}^-1 + 1, 1 = 0 + 1, \ldots$$

We now look again at the number line introduced in the study of the whole number system (Section 6.10). With the order of the integers established, a one-to-one correspondence may be set up between the elements of the set of integers and the elements of a set of points on a line. The matching of the elements of I will follow the prescribed sequence, with the integers being matched with "equally spaced" points on the line. A representation of this matching is shown below:

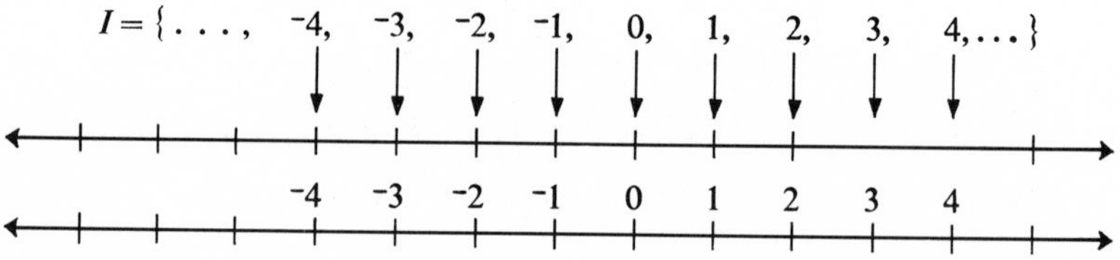

Thus, the number line of whole numbers has been extended to accommodate the negative integers. Observe that ${}^-1$ is opposite 1 on the number line.

Other useful definitions given in Chapter 6 concerning the order of whole numbers are now restated so as to apply to the integers:

Definition 8.3(b) For each $a, b \in I$, if $a < b$, then

$$b > a.$$

Definition 8.3(c) For each $a, b, c \in I$, if $a < c < b$, then

$$a < c \text{ and } c < b.$$

Definition 8.3(d) For each $a, b \in I$, if $a \leq b$, then

$$a < b \text{ or } a = b.$$

Also, if $a \geq b$, then

$$a > b \text{ or } a = b.$$

With the order of the integers and their positions on the number line established, it is customary to refer to the negative integers as those "less than" 0, and to the nonnegative integers as those "greater than or equal to" 0. Thus:

For each $a \in I$, if a belongs to the set of negative integers, then $a < 0$.
For each $a \in I$, if a belongs to the set of nonnegative integers, then $a \geq 0$.

Another important order idea is the trichotomy property, which was established for the whole numbers in Chapter 6. We now extend it to include the negative integers:

Trichotomy Property

For each $a, b \in I$, $\quad a = b \quad$ or $\quad a < b \quad$ or $\quad a > b$.

Find the integers that make this sentence true:

$$x > {}^{-}3.$$

You should find that the least integer is $^{-}2$, and there is no greatest. Check to see if this is the solution set of $x > {}^{-}3$.

$$A = \{{}^{-}2, {}^{-}1, 0, 1, 2, 3, \ldots\}$$

Next find the integers that make this sentence true:

$$-x < 3.$$

Suppose that $^{-}2$ is selected. Replacing x with $^{-}2$ we have

$$-{}^{-}2 < 3,$$

which becomes

$$2 < 3.$$

So, $^{-}2$ belongs to the solution set. In a similar manner, find other integers of the solution set. Do you find the solution set of $-x < 3$ to be

$$B = \{{}^{-}2, {}^{-}1, 0, 1, 2, 3, \ldots\}?$$

If so, $A = B$, which seems to justify writing,

$$\text{If } x > {}^{-}3, \text{ then } -x < 3.$$

Investigate to see whether we should be justified in writing these also.

If $x < {}^{-}3$, then $-x > 3$.
If $x > 3$, then $-x < {}^{-}3$.
If $x < 3$, then $-x > {}^{-}3$.

The following generalization is based on the preceding discussion:

Generalization 8.3

For each x and $a \in I$,

(i) if $x > a$, then $-x < -a$;
(ii) if $x < a$, then $-x > -a$.

The integer matched with a certain point on the number line is called the *coordinate* of that point. In the illustration below, the coordinate of the point labeled P is 3, that of the point labeled Q is $^{-}3$, and that of the point labeled R is 0:

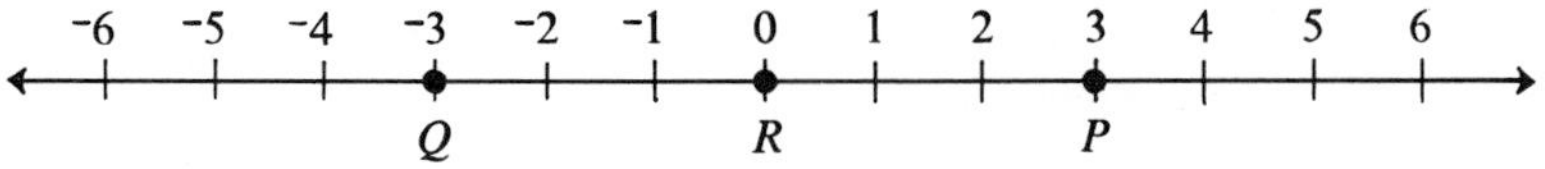

The number of intervals from point R to point P is called the *distance* from R to P. Thus, distance is a *number*. In the set of integers, distance is a nonnegative integer. The distance from point R to point P is 3; this is the same as the coordinate of P. The distance from point R to point Q is also 3; however, the coordinate of point Q is $^{-}3$. We can designate this distance to point Q in terms of its coordinate by writing $|^{-}3|$ to mean 3. The expression

$$|^{-}3| = 3$$

is read, "The *absolute value* of $^{-}3$ is 3." Of course, $|3|$ is also 3. The absolute value symbol "| |" may be used with any integer to represent a nonnegative integer. That is, $|^{-}5| = 5$, $|0| = 0$, and $|4| = 4$.

Definition 8.3(e) For each $a \in I$, the *absolute value of a*, denoted by $|a|$, is a nonnegative integer such that

$$\begin{aligned} &\text{if } a > 0, \text{ then } |a| = a; \\ &\text{if } a = 0, \text{ then } |a| = 0; \\ &\text{if } a < 0, \text{ then } |a| = -a. \end{aligned}$$

Example 5 Use Definition 8.3(e) to determine the absolute value of 8, 0, and $^{-}4$.

$$\begin{aligned} &\text{Since } 8 > 0, \text{ we have } |8| = 8. \\ &\text{Since } 0 = 0, \text{ we have } |0| = 0. \\ &\text{Since } {}^{-}4 < 0, \text{ we have } |{}^{-}4| = -({}^{-}4) = 4. \end{aligned}$$

Example 6 Use Definition 8.3(e) to determine $|^{-}2 + 7|$, $|2 + {}^{-}7|$, and $|^{-}3 + 3|$.

$|^{-}2 + 7| = |5| = 5$, since for $a > 0$, $|a| = a$.
$|2 + {}^{-}7| = |^{-}5| = -(^{-}5) = 5$, since for $a < 0$, $|a| = -a$.
$|^{-}3 + 3| = |0| = 0$, since for $a = 0$, $|a| = 0$.

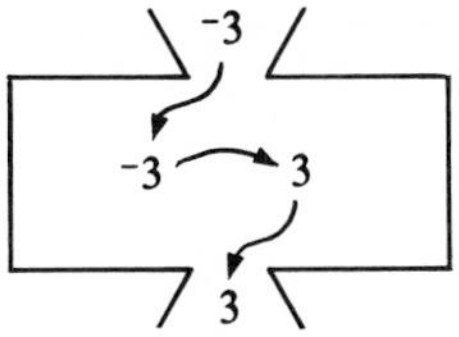

Absolute Value
Machine

Another interpretation of absolute value is presented by means of an "absolute value machine." It works in this manner. The integer $^{-}3$ goes into the *input*. Then, the machine "manufactures" 3, the additive inverse of $^{-}3$. The machine takes a "look" at the two numbers. It selects the greater number, and sends it out the *output* shoot. Hence, the number that comes out of the output is the absolute value of the number that went into the input. What happens when a positive integer is sent through the machine? What happens when 0 is sent through?

We have been using the equals symbol, =, for the set of integers in the same sense as we did for the set of whole numbers (Section 6.11), and we note that the relation which it represents obeys the three properties—reflexive, symmetric, and transitive.

Reflexive property: For each $a \in I$, $a = a$.
Symmetric property: For each $a, b \in I$, if $a = b$, then $b = a$.
Transitive property: For each $a, b, c \in I$, if $a = b$ and $b = c$, then $a = c$.

Hence, we can say that the *equality of integers* is an *equivalence relation.*

Are the "greater than" and the "less than" relations equivalence relations? Do they obey any of the properties of an equivalence relation? If so, which ones?

EXERCISE 8.3

1. Use Definition 8.3(a) and Examples 1, 2, and 3 to discuss the order of the following pairs of numbers.
   (a) 5 and 14
   (b) $^{-}9$ and $^{-}12$
   (c) 15 and $^{-}7$
   (d) 0 and $^{-}8$
   (e) $^{-}26$ and 0
   (f) $^{-}5$ and -2
   (g) $^{-}13$ and 13
   (h) $(8 + {}^{-}10)$ and $-({}^{-}6 + {}^{-}3)$
   (i) $({}^{-}2 + {}^{-}5)$ and $({}^{-}7 + 4)$

2. Use one of the symbols <, >, or = in the blank to make each of the following a true sentence.
   (a) 18______25
   (b) $^{-}18$______$^{-}25$
   (c) $^{-}18$______25
   (d) 18______$^{-}25$
   (e) $^{-}21$______35
   (f) 28______$^{-}54$
   (g) $(32 + {}^{-}12)$______$^{-}92$
   (h) $-({}^{-}9 + {}^{-}6)$______26
   (i) $-(6 + 7)$______$({}^{-}11 + {}^{-}2)$
   (j) $-({}^{-}15 + 8)$______$(14 + {}^{-}19)$
   (k) $-(6 + {}^{-}17)$______$(5 + 6)$
   (l) $({}^{-}2 + {}^{-}6)$______$-(10 + {}^{-}2)$

3. Use Definitions 8.3(b), 8.3(c), and 8.3(d) to write the following in a different form.

   EXAMPLE. $^-5 < 7$ may be written $7 > {}^-5$ by Definition 8.3(b).

   (a) $2 < 3$ and $3 < 4$ (HINT: See Definition 8.3(c).)
   (b) $-(3 + 5) \leq {}^-7$
   (c) $^-2 < b < 3$
   (d) $^-34 < {}^-29$
   (e) $V < 15$ or $V = 15$
4. Give the value of each of the following.
   (a) $|^-12|$
   (b) $|4 + 6|$
   (c) $|^-7 + 5|$
   (d) $|5 + {}^-3|$
   (e) $-|^-8 + {}^-5|$
   (f) $|16 + {}^-16|$
   (g) $--|13 + {}^-14|$
   (h) $|^-25 + 10|$
   (i) $|25 + {}^-10|$
   (j) $|a + b|$, a, b nonnegative integers
   (k) $|a + -b|$, a, b nonnegative integers, $b > a$
   (l) $|a + -b|$, a, b nonnegative integers, $b < a$
5. Given $A = \{a \in I : a > {}^-5\}$ and $B = \{b \in I : b < 5\}$:
   (a) List the elements in set A.
   (b) List the elements in set B.
   (c) List the elements in $A \cap B$.
   (d) List the elements in $A \cup B$.
   (e) Was it possible to list all of the elements in the preceding parts?
6. Given $P = \{p \in I : |p| < 3\}$ and $Q = \{q \in I : q < 2\}$:
   (a) List the elements in set P. What is the greatest element? What is the least element?
   (b) List the elements in set Q. What is the greatest element? What is the least element?
   (c) Use the roster notation to describe $P \cap Q$ and $P \cup Q$.
7. Given $R = \{r \in I : |r| \leq 3\}$ and $S = \{s \in I : |s| \geq 3\}$:
   (a) List the elements of set R. What is the greatest element? What is the least element?
   (b) List the elements of set S. What is the greatest element? What is the least element?
   (c) Describe $R \cap S$.
   (d) Describe $R \cup S$.
8. Discuss the integers that belong to set $A = \{a \in I : |a| < {}^-2\}$.

8.4 THE SUBTRACTION OPERATION IN THE SET OF INTEGERS

In Section 8.1 we invented the negative integers so that subtraction would always be possible. Definition 6.5 may thus be rewritten for the set of integers as follows:

Definition 8.4 For each $a, b, x \in I$, if $b + x = a$, then

$$a - b = x,$$

or

$$(a, b) \xrightarrow{-} x.$$

Subtraction is always possible in the set I; that is, the set I is closed under subtraction. Thus, subtraction is an operation for the set I.

By Definition 8.4, subtraction is accomplished in terms of addition. Now with addition defined for the set of integers, I, in Section 8.2, subtraction is done as shown in the following examples.

Example 1 Find the solution of $5 - {}^-2 = \square$.

$5 - {}^-2 = \square$ if ${}^-2 + \square = 5$.	*By Definition* 8.4
Since ${}^-2 + 7 = 5$, $5 - {}^-2 = 7$.	7 *is the solution of this sentence by Definition* 8.2(a).

Hence, $5 - {}^-2 = 7$. The solution set is $\{7\}$.

Example 2 ${}^-8 - {}^-7$ names what integer?

${}^-8 - {}^-7 = \square$ if ${}^-7 + \square = {}^-8$.	*By Definition* 8.4
Since ${}^-7 + {}^-1 = {}^-8$, ${}^-8 - {}^-7 = {}^-1$.	${}^-1$ *is the solution of this sentence by Definition* 8.2(b).

Hence, ${}^-8 - {}^-7$ is another name for ${}^-1$.

Example 3 What integer is matched with (5, 7) under the operation of subtraction?

$5 - 7 = \square$ if $7 + \square = 5$.	*By Definition* 8.4
Since $7 + {}^-2 = 5$, $5 - 7 = {}^-2$.	${}^-2$ *is the solution of this sentence by Definition* 8.2(c).

Hence, $5 - 7 = {}^-2$. The required integer is ${}^-2$.

EXERCISE 8.4

1. What integer will make each of the following a true sentence?
   (a) ${}^-5 - 3 = \square$ if $3 + {}^-8 = {}^-5$.
   (b) $5 - {}^-3 = \square$ if ${}^-3 + 8 = 5$.
   (c) ${}^-3 - {}^-7 = \square$ if ${}^-7 + 4 = {}^-3$.
   (d) $12 - {}^-8 = 20$ if ${}^-8 + \square = 12$.
   (e) $9 - 14 = {}^-5$ if $14 + \square = 9$.
   (f) $0 - {}^-4 = \square$ if ${}^-4 + \square = 0$.
   (g) ${}^-15 - {}^-23 = \square$ if ${}^-23 + \square = {}^-15$.
   (h) $6 - \square = {}^-4$ if $\square + {}^-4 = 6$.
   (i) $\square - 10 = 7$ if $10 + 7 = \square$.
   (j) $\square - {}^-22 = 18$ if ${}^-22 + 18 = \square$.
2. Use Definition 8.4 and justifications as in Examples 1, 2, and 3 to find the integer named by each of the following.

(a) $15 - 5$	(d) $3 - 5$	(g) $7 - {}^-2$
(b) $8 - 3$	(e) $10 - 25$	(h) ${}^-5 - 17$
(c) $9 - 15$	(f) $4 - 9$	(i) ${}^-3 - {}^-12$

(j) $8 - {}^-8$ (m) $0 - {}^-11$ (p) ${}^-1 - 18$

(k) $15 - 15$ (n) ${}^-11 - 0$ (q) $0 - 0$

(l) ${}^-23 - {}^-23$ (o) $18 - {}^-1$ (r) ${}^-29 - {}^-1$

3. The following problems have been designed to help you to make a generalization concerning a relationship between subtraction and addition of integers. Find the integer for each of the following.

   (a) $5 - 3$ and $5 + {}^-3$ (d) $2 - {}^-6$ and $2 + 6$

   (b) ${}^-6 - 10$ and ${}^-6 + {}^-10$ (e) $7 - 9$ and $7 + {}^-9$

   (c) $8 - {}^-2$ and $8 + 2$ (f) ${}^-3 - {}^-5$ and ${}^-3 + 5$

   What generalization can you make from the above?

4. Use the generalization developed in Problem 3 to find the integer for each of the following.

   (a) $5 - {}^-3$ (d) $8 - {}^-10$ (g) $-{}^-8 - {}^-({}^-1 + 9)$

   (b) ${}^-4 - {}^-9$ (e) ${}^-9 - 5$ (h) ${}^-(4 + 5) - (4 + 5)$

   (c) ${}^-3 - {}^-4$ (f) $5 - {}^-(4 + 7)$ (i) $({}^-3 + {}^-2) - {}^-(3 + 2)$

Generalization 8.4

For each $a, b, x \in I$,

$$\text{if } a - b = x, \text{ then } a + -b = x.$$

8.5 THE PROPERTIES FOR SUBTRACTION IN THE SET OF INTEGERS

Do the integers obey the commutative and associative properties under the subtraction operation? We investigate the commutative property under subtraction for a selected pair of integers $(8, {}^-2)$. Is it true that $8 - {}^-2 = {}^-2 - 8$? Since $8 - {}^-2$ and ${}^-2 - 8$ name different integers (10 and ${}^-10$), we write

$$8 - {}^-2 \neq {}^-2 - 8.$$

This counterexample is sufficient evidence that the subtraction operation is *not commutative.*

Similarly, we investigate the associative property for a particular set of integers. Is it true that $(9 - 4) - {}^-2 = 9 - (4 - {}^-2)$? We answer this question in the negative, since $(9 - 4) - {}^-2 = 7$ and $9 - (4 - {}^-2) = 3$. This counterexample is all that is needed to convince us that the subtraction operation in the set of integers is *not associative.*

8.6 THE MULTIPLICATION OPERATION IN THE SET OF INTEGERS

The multiplication operation for the whole numbers was defined in Definition 6.6. Since multiplication has been defined for the whole numbers (the nonnegative integers), a subset of the integers, we need only to concern ourselves with defining this operation for the negative integers. We shall assume that the negative integers obey the same properties for multiplication as did the whole numbers. We shall apply these properties in the examples that follow, as we evolve methods of finding products involving negative integers.

Example 1 What integer is matched with $(6, {}^-9)$ under the operation multiplication?

First let us consider the product of 6 and $(9 + {}^-9)$ where 9 is the additive inverse of ${}^-9$.

$6 \times (9 + {}^-9) = (6 \times 9) + (6 \times {}^-9)$	*Distributive property for multiplication over addition*
$6 \times 0 = (6 \times 9) + (6 \times {}^-9)$	*Additive inverse,* $9 + {}^-9 = 0$
$0 = (6 \times 9) + (6 \times {}^-9)$	*Renaming* (6×0)
$0 = 54 + (6 \times {}^-9)$	*Renaming* (6×9)

Since $54 + (6 \times {}^-9) = 0$, $(6 \times {}^-9)$ is the additive inverse of 54. But, the additive inverse of 54 is ${}^-54$. Hence,

$$(6 \times {}^-9) = {}^-54.$$

Find the products of the following, using the procedure given in Example 1.

(1) $8 \times {}^-3 = \square$
(2) ${}^-8 \times 3 = \square$
(3) $4 \times {}^-1 = \square$
(4) ${}^-4 \times 1 = \square$
(5) $-{}^-4 \times {}^-3 = \square$
(6) $4 \times --{}^-3 = \square$

These problems and Example 1 lead us to make the following definition.

Definition 8.6(a) If a and b, are nonnegative integers, then

$$a \times -b = -x,$$

where $a \times b = x$ and x is nonnegative.

We shall study the next example to help in forming a definition for the multiplication of two negative integers.

Example 2 What is the integer which names ${}^-3 \times {}^-4$?

We use a method similar to that of Example 1. Let us find the product of $(3 + {}^-3)$ and ${}^-4$:

$(3 + {}^-3) \times {}^-4 = (3 \times {}^-4) + ({}^-3 \times {}^-4)$	*Distributive property for multiplication over addition*
$0 \times {}^-4 = (3 \times {}^-4) + ({}^-3 \times {}^-4)$	*Additive inverse,* $3 + {}^-3 = 0$
$0 = {}^-12 + ({}^-3 \times {}^-4)$	*Renaming* $(0 \times {}^-4)$ *and* $(3 \times {}^-4)$

Since $^{-}12 + (^{-}3 \times {}^{-}4) = 0$, $(^{-}3 \times {}^{-}4)$ is the additive inverse of $^{-}12$. The additive inverse of $^{-}12$ is 12. Hence,

$$^{-}3 \times {}^{-}4 = 12.$$

Use the procedure given in Example 2 to find the product of the following.

(7) $^{-}7 \times {}^{-}8 = \square$
(8) $7 \times 8 = \square$
(9) $5 \times 3 = \square$
(10) $^{-}5 \times {}^{-}3 = \square$
(11) $^{-}6 \times {}^{-}3 = \square$
(12) $^{-}9 \times {}^{-}2 = \square$

These suggest the following definition for the multiplication of two negative integers.

Definition 8.6(b) If a and b are nonnegative integers, then

$$-a \times -b = x,$$

where $a \times b = x$ and x is nonnegative.

EXERCISE 8.6

1. Use the appropriate definition and properties to find the product of each of the following.
   (a) 15×3
   (b) $^{-}15 \times 3$
   (c) $15 \times {}^{-}3$
   (d) $^{-}15 \times {}^{-}3$
   (e) $^{-}6 \times (8 - 10)$
   (f) $(^{-}6 \times 4) \times 5$
   (g) $(^{-}2 \times {}^{-}4) \times {}^{-}7$
   (h) $(^{-}2 \times 4) \times {}^{-}7$
   (i) $^{-}1 \times {}^{-}1$
   (j) $^{-}1 \times {}^{-}1 \times {}^{-}1$
   (k) $^{-}1 \times {}^{-}1 \times {}^{-}1 \times {}^{-}1$
   (l) $^{-}2 \times (^{-}3 - 7)$
2. Prove that the following generalizations are true.
   (a) For each b, a nonnegative integer,
   $$b \times -1 = -b.$$
   (HINT: Use the procedure in Example 1.)
   (b) For each b, a nonnegative integer,
   $$-b \times -1 = b.$$
   (c) For each a, b, nonnegative integers,
   $$-(a + b) = -a + -b.$$
   (HINT: Use part (b).)
3. Give a simpler name for each of the following.
   (a) $^{-}6 \times {}^{-}3$
   (b) $^{-}5 \times (4 - 8)$
   (c) $(^{-}3 + 7) \times {}^{-}1$
   (d) $^{-}2 \times {}^{-}9 \times {}^{-}3$
   (e) $^{-}(7 + {}^{-}3)$
   (f) $-(^{-}8 + 2)$
   (g) $^{-}2 \times (^{-}3 - 7)$
   (h) $^{-}3 \times 3 \times {}^{-}3$
   (i) $(5 + {}^{-}5) \times 6$
   (j) $(6 - {}^{-}6) \times (6 - 7)$

8.7 UNIQUENESS AND CANCELLATION PROPERTIES

Since $6 + {}^-8$ and ${}^-1 + {}^-1$ are names for the same integer, we may write

$$6 + {}^-8 = {}^-1 + {}^-1.$$

Are the following sentences true? Verify each by finding the indicated sums.

$$\left.\begin{aligned}(6 + {}^-8) + 5 &= ({}^-1 + {}^-1) + 5\\ (6 + {}^-8) + {}^-3 &= ({}^-1 + {}^-1) + {}^-3\\ (6 + {}^-8) + ({}^-4 + {}^-2) &= ({}^-1 + {}^-1) + ({}^-4 + {}^-2)\end{aligned}\right\} \quad \text{(I)}$$

Since the definition for addition of integers provides for uniqueness, it is to be expected that these sentences are true without reinforcing this with the verification. This leads us to make the following generalization.

Uniqueness Property of Addition for "Equals" Relation

For each $a, b, c \in I$, if $a = b$, then $a + c = b + c$.

Suppose that we begin with the above group of true sentences (I). Does it seem reasonable to expect that from each of them we can return to the sentence

$$(6 + {}^-8) = ({}^-1 + {}^-1)?$$

This suggests another generalization.

Cancellation Property of Addition for "Equals" Relation

For each $a, b, c \in I$, if $a + c = b + c$, then $a = b$.

Beginning again with the true mathematical sentence

$$(6 + {}^-8) = ({}^-1 + {}^-1),$$

verify the truth of the following sentences by performing the indicated sums and products.

$$\left.\begin{aligned}(6 + {}^-8) \times 3 &= ({}^-1 + {}^-1) \times 3\\ (6 + {}^-8) \times {}^-5 &= ({}^-1 + {}^-1) \times {}^-5\\ (6 + {}^-8) \times ({}^-2 + {}^-3) &= ({}^-1 + {}^-1) \times ({}^-2 + {}^-3)\end{aligned}\right\} \quad \text{(II)}$$

Here again we expect these sentences to be true because of the very nature of the definition of multiplication of integers. The observations made from these sentences are brought together in the following generalization.

Uniqueness Property of Multiplication for "Equals" Relation

For each $a, b, c \in I$, if $a = b$, then $a \times c = b \times c$.

As before, beginning with the group of true sentences (II), does it seem reasonable to expect that from each of them we can return to the sentence

$$(6 + {}^{-}8) = ({}^{-}1 + {}^{-}1)?$$

The generalization suggested by this is the following.

Cancellation Property of Multiplication for "Equals" Relation

For each $a, b, c \in I$, with $c \neq 0$, if $a \times c = b \times c$, then $a = b$.

The question now arises as to why the restriction was placed on c in stating this property. With c equal to any integer other than 0, say 2, we have a sentence,

$$15 \times 2 = (10 + 5) \times 2.$$

The cancellation property for multiplication allows us to write

$$15 = (10 + 5).$$

Now we shall consider the sentence

$$(2 + {}^{-}3) \times 0 = (10 + {}^{-}2) \times 0.$$

Verify that it is a true sentence. If we ignore the restriction, $c \neq 0$, and apply the cancellation property for multiplication, we would be forced to admit that $(2 + {}^{-}3) = (10 + {}^{-}2)$ is a true sentence. But it is not.

Let us now consider the uniqueness and cancellation properties of addition for the "less than" relation. We shall look for patterns in the following group of sentences from which generalizations may be made. Verify that the following sentences are true.

(1) If $({}^{-}8 + 6) < (3 + 2)$, then $({}^{-}8 + 6) + {}^{-}5 < (3 + 2) + {}^{-}5$.
(2) If $({}^{-}4 + {}^{-}1) < (0 + {}^{-}1)$, then $({}^{-}4 + {}^{-}1) + 6 < (0 + {}^{-}1) + 6$.
(3) If $({}^{-}8 + 6) + {}^{-}5 < (3 + 2) + {}^{-}5$, then $({}^{-}8 + 6) < (3 + 2)$.
(4) If $({}^{-}4 + {}^{-}1) + 6 < (0 + {}^{-}1) + 6$, then $({}^{-}4 + {}^{-}1) < (0 + {}^{-}1)$.

The generalization that follows from sentences (1) and (2) is:

Uniqueness Property of Addition for "Less than" Relation

For each $a, b, c \in I$, if $a < b$, then $a + c < b + c$.

The generalization that follows from sentences (3) and (4) is:

Cancellation Property of Addition for "Less than" Relation

For each $a, b, c \in I$, if $a + c < b + c$, then $a < b$.

After verification of the following sentences, generalizations concerning the uniqueness and cancellation properties of multiplication for the "less than" relation may be made.

(1) If $^{-}2 < 6$, then $^{-}2 \times 5 < 6 \times 5$.
(2) If $(8 + {}^{-}2) < 15$, then $(8 + {}^{-}2) \times 3 < 15 \times 3$.
(3) If $^{-}2 \times 5 < 6 \times 5$, then $^{-}2 < 6$.
(4) If $(8 + {}^{-}2) \times 3 < 15 \times 3$, then $(8 + {}^{-}2) < 15$.

Suppose that sentence (1) had been: "If $^{-}2 < 6$, then $^{-}2 \times {}^{-}5 < 6 \times {}^{-}5$." Would it have been true? In order for this sentence to become true, what order symbol should be placed between $(^{-}2 \times {}^{-}5)$ and $(6 \times {}^{-}5)$? We are now ready to make the following generalization:

Uniqueness Property of Multiplication for "Less than" Relation

For each $a, b, c \in I$, if $a < b$, then

(i) $a \times c < b \times c$ when $c > 0$;
(ii) $a \times c > b \times c$ when $c < 0$.

Suppose that sentence (3) had been: "If $2 \times {}^{-}5 < {}^{-}6 \times {}^{-}5$, then $2 < {}^{-}6$." Would it have been true? In order for this sentence to become true, what order symbol should be placed between 2 and $^{-}6$? We are now ready to make the following generalization:

Cancellation Property of Multiplication for "Less than" Relation

For each $a, b, c \in I$, if $a \times c < b \times c$, then

(i) $a < b$ when $c > 0$;
(ii) $a > b$ when $c < 0$.

The preceding properties are used when simplifying a mathematical sentence. The solution set of the simplified form is more easily determined. The following examples are illustrations.

Example 1 Change

$$p + {}^{-}3 = 17 + {}^{-}3$$

to a simpler sentence.

By the cancellation property of addition for =, a simpler sentence is

$$p = 17.$$

Example 2 Given $b + {}^{-}5 = 10$. Find a simpler sentence.

$b + {}^{-}5 = 10$	
$(b + {}^{-}5) + 5 = 10 + 5$	*Uniqueness of addition for =*
$b + ({}^{-}5 + 5) = 10 + 5$	*Associative property for +*
$b + 0 = 10 + 5$	*Additive inverses*
$b = 10 + 5$	*Additive identity*
$b = 15$	*Decimal notation*

The solution set of the simpler sentence, $b = 15$, is quite evident.

Example 3 Simplify

$$a \times (^{-}32 + {}^{-}8) = 15 \times (^{-}32 + {}^{-}8).$$

By the cancellation property of multiplication for =, a simpler sentence is

$$a = 15.$$

Example 4 Simplify

$$s \times (^{-}21 + 3) = {}^{-}36.$$

$s \times (^{-}21 + 3) = {}^{-}36$		
$s \times \quad {}^{-}18 = {}^{-}36$	*Renaming* $(^{-}21 + 3)$, *Definition* 8.2(c)
$s \times \quad {}^{-}18 = 2 \times {}^{-}18$	*Renaming* $^{-}36$, *Definition* 8.6(a)
$s = 2$	*Cancellation property of multiplication for* =

Example 5 Change $p + 3 < 17$ to a simpler sentence.

$p + 3 < 17$	
$(p + 3) + {}^{-}3 < 17 + {}^{-}3$	*Uniqueness of addition for* <
$p + (3 + {}^{-}3) < 17 + {}^{-}3$	*Why?*
$p + 0 < 17 + {}^{-}3$	*Why?*
$p < 17 + {}^{-}3$	*Why?*
$p < 14$	*Renaming* $17 + {}^{-}3$, *Definition* 8.2(a)

Is this sentence simpler than the original one? What is the greatest integer that will make this sentence true? Will it also make the original sentence true? Is there a least integer that will make both sentences true?

Example 6 Find the solution set of the sentence

$$a \times {}^{-}5 < 15.$$

$a \times {}^{-}5 < 15$	
$a \times {}^{-}5 < {}^{-}3 \times {}^{-}5$	*Renaming* 15, *Definition* 8.6(b)
$a > {}^{-}3$	*Cancellation property of multiplication for* <, *with* $c < 0$

Hence, the solution set is

$$\{a \in I : a > {}^{-}3\}.$$

8.8 SOLUTION SETS OF SENTENCES

In Section 6.3 only whole numbers were admitted to the solution sets of sentences. Now with the extension of the set of numbers, we are prepared to admit any integers. As observed in Section 8.7, finding the solution set of a sentence may be made easier by using the uniqueness and cancellation properties.

Example 1 Draw the graph on the number line of the set

$$A = \{b \in I : {}^-4 < b < 5\}.$$

From Definition 8.3(c),

$$^-4 < b < 5$$

means

$$^-4 < b \text{ and } b < 5.$$

Hence, the set A is the intersection of two sets. We shall represent the solution set of $^-4 < b$ on the number line with circles and that of $b < 5$ with dots:

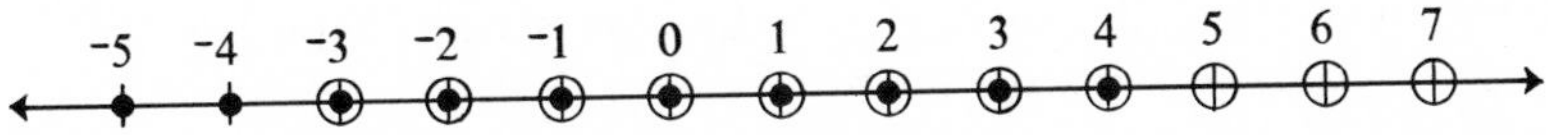

Notice that the least element in the solution set of $^-4 < b$ is $^-3$, but there is no greatest element. On the other hand, the greatest element in the solution set of $b < 5$ is 4, but there is no least element. We have

$$\begin{aligned} \{b \in I : {}^-4 < b < 5\} &= \{b \in I : {}^-4 < b\} \cap \{b \in I : b < 5\} \\ &= \{^-3, {}^-2, {}^-1, 0, 1, 2, 3, 4\}, \end{aligned}$$

and its graph consists of the points marked with both a circle and a dot.

Example 2 Find the solution set of

$$a + {}^-15 = 12, \qquad \text{where } a \in I.$$

We shall use the additive inverse of $^-15$.

$$\begin{aligned} a + {}^-15 &= 12 \\ (a + {}^-15) + 15 &= 12 + 15 && \textit{Uniqueness property of addition for} = \\ a + ({}^-15 + 15) &= 12 + 15 && \textit{Associative property for addition} \\ a + 0 &= 12 + 15 && \textit{Additive inverse} \\ a &= 12 + 15 && \textit{Additive identity} \\ a &= 27 && \textit{Renaming } (12 + 15) \end{aligned}$$

Hence, the solution set is $\{27\}$.

Example 3 Find the solution set of

$$|b| + 5 = 11, \qquad b \in I,$$

and draw the graph.

$$
\begin{aligned}
|b| + 5 &= 11 \\
|b| + 5 &= 6 + 5 && \textit{Renaming } 11 \\
|b| &= 6 && \textit{Cancellation property for addition}
\end{aligned}
$$

As integers are found that will make this sentence true, represent each on the number line with a dot (recall Definition 8.3(b)):

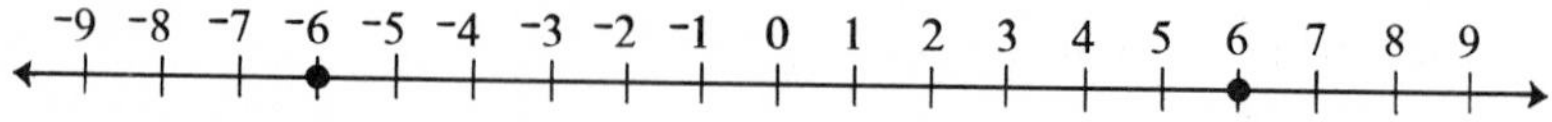

The solution set is $\{{}^-6, 6\}$.

Example 4 Make a graph of the solution set of

$$|x| < 6, \quad x \in I.$$

By selecting integers one at a time, we may find the solution set to be those integers represented in this graph:

The following discussion justifies the preceding result: By Definition 8.3(e), we have

$$
\begin{aligned}
&\text{if } x \geq 0,\ |x| = x; \\
&\text{if } x < 0,\ |x| = -x.
\end{aligned}
$$

Thus, for $|x| < 6$:

(1) If $x \geq 0$, then $|x| = x$. Hence, $x < 6$.
This may be written, $0 \leq x < 6$, by Definition 8.3(b) and 8.3(c).

(2) If $x < 0$, then $|x| = -x$. Hence, $-x < 6$.
This may be written, $x < 0$ and $-x < 6$ or,

$$x < 0 \text{ and } x > {}^-6, \text{ by Generalization 8.3,}$$

and converted to

$$x < 0 \text{ and } {}^-6 < x \text{ by Definition 8.3(b),}$$

then to

$${}^-6 < x < 0 \text{ by Definition 8.3(c).}$$

Combining ${}^-6 < x < 0$ and $0 \leq x < 6$, we have

$${}^-6 < x < 6.$$

The solution set of this sentence is

$$\{x \in I : {}^-6 < x\} \cap \{x \in I : x < 6\}.$$

Is this the same as the solution set of $|x| < 6$ shown above by the graph?

Example 5 Make a graph of the solution set of $|x| > 6$, $x \in I$. Integers belonging to the solution set may be found by trying one integer at a time to see if it makes the sentence true. The graph of the solution set is shown here:

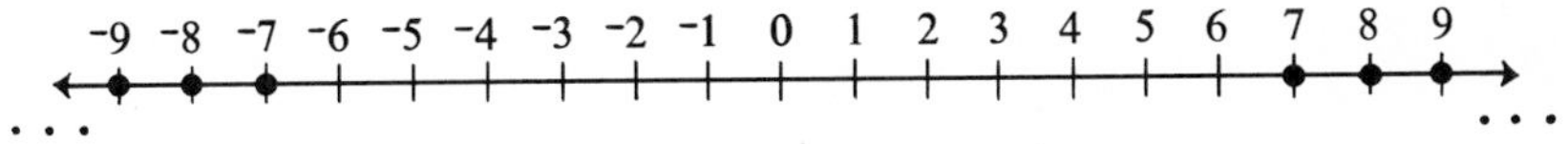

Verify that the solution set of the sentence

$$x > 6 \text{ or } x < {}^-6$$

is the same as the solution set of $|x| > 6$ given by the graph. Hence, the solution set of $|x| > 6$ may be written,

$$\{x \in I : x > 6\} \cup \{x \in I : x < {}^-6\}.$$

Observe that the solution set of $|x| < 6$ is the *intersection* of two sets; while the solution set of $|x| > 6$ is the *union* of two sets. What difference would there have been in the solution set of each, if the relations had been $\leq$ and $\geq$?

Example 6 Find the solution set of $|x + 2| < 5$, $x \in I$.

If you prefer, you may select integers, and test each to see if it will make the sentence true. What is the greatest integer which belongs to the solution set? the least integer? Can you determine whether the solution set is the intersection of two sets or the union? Did you find the solution set to be

$$\{{}^-6, {}^-5, {}^-4, {}^-3, {}^-2, {}^-1, 0, 1, 2\}?$$

Example 7 Find the solution set of $|x + 2| > 5$, $x \in I$.

Is there a greatest integer in the solution set? a least integer? Below is the graph of the solution set.

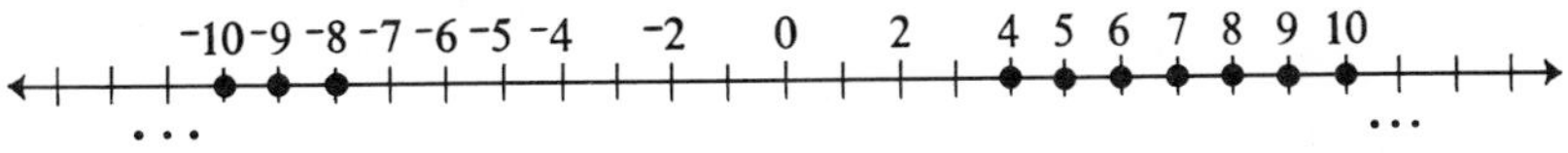

Is this a correct description of the solution set?

$$\{x \in I : x < {}^-7\} \cup \{x \in I : x \geq 4\}$$

Give other descriptions of the solution set.

EXERCISE 8.8

Find the solution set for each of the following, and draw the graph of each set.

1. $|m| < 4$, $m \in I$. (HINT: First consider $|m| = 4$, $m \in I$.)
2. $|m| > 4$, $m \in I$.
3. $|m| \leq 4$, $m \in I$.

4. $c > 15$ and $c < {}^{-}2, c \in I$.
5. $|x - 4| = {}^{-}9, x \in I$.
6. $|x + 4| \geq 6, x \in I$.
7. $|x + 4| > 6, x \in I$.
8. $|x + 4| < 6, x \in I$.
9. $|x + 4| \geq 6$ or $|x + 4| \leq 6, x \in I$.
10. $|x| > 3$ and $|x| > 5, x \in I$.
11. $|x| < 3$ or $|x| > 5, x \in I$.

8.9 THE SYSTEM OF INTEGERS

Recall that in Section 5.9 the basic description of a number system is given. Immediately following this, in Section 5.10, the *system of natural numbers* is described. The next mention of a number system is in Section 6.12, where the *system of whole numbers* is described.

In this chapter we have been developing the *system of integers*. Many of the properties mentioned are the same as those for the two systems previously studied. We have simply applied them to a new, enlarged set of numbers. The presence of an additive inverse for each element and the use of an additional operation, subtraction, distinguish the system of integers from the system immediately preceding it—the system of whole numbers. However, the presence of an additive inverse for each element makes it possible to define the operation of subtraction in terms of the addition operation. Hence, it is not necessary to regard subtraction as a new, independent operation with the same standing as the addition and multiplication operations. The inclusion of an additive inverse for each element is sufficient.

We can now describe the *system of integers*. This system consists of a set of elements,

$$I = \{\ldots, {}^{-}3, {}^{-}2, {}^{-}1, 0, 1, 2, 3, \ldots\},$$

the equivalence relation of equality, and two binary operations, addition and multiplication, defined for these elements, together with the following properties:

Commutative Properties

For each $a, b \in I, a + b = b + a$.
For each $a, b \in I, a \times b = b \times a$.

Associative Properties

For each $a, b, c \in I, (a + b) + c = a + (b + c)$.
For each $a, b, c \in I, (a \times b) \times c = a \times (b \times c)$.

Distributive Property for Multiplication over Addition

For each $a, b, c \in I$, $(a + b) \times c = (a \times c) + (b \times c)$;

$$c \times (a + b) = (c \times a) + (c \times b).$$

Identity Elements

For each $b \in I$, there exists a unique additive identity element $0 \in I$, such that $b + 0 = 0 + b = b$.
For each $b \in I$, there exists a unique multiplicative identity element $1 \in I$, such that $b \times 1 = 1 \times b = b$.

Additive Inverse

For each $b \in I$, there exists $-b \in I$, such that $b + -b = 0$.

Role of Zero in Multiplication

For each $a \in I$, $a \times 0 = 0 \times a = 0$.

Trichotomy Property

For each a, $b \in I$, $a = b$ or $a < b$ or $a > b$.

Uniqueness Properties for =

For each $a, b, c \in I$, if $a = b$, then $a + c = b + c$.
For each $a, b, c \in I$, if $a = b$, then $a \times c = b \times c$.

Cancellation Properties for =

For each $a, b, c \in I$, if $a + c = b + c$, then $a = b$.
For each $a, b, c \in I$ with $c \neq 0$, if $a \times c = b \times c$, then $a = b$.

Uniqueness Properties for <

For each $a, b, c \in I$, if $a < b$, then $a + c < b + c$.
For each $a, b, c \in I$, if $a < b$, then

(i) $a \times c < b \times c$ when $c > 0$;
(ii) $a \times c > b \times c$ when $c < 0$.

Cancellation Properties for $<$

For each $a, b, c \in I$, if $a + c < b + c$, then $a < b$.
For each $a, b, c \in I$, if $a \times c < b \times c$, then

(i) $a < b$ when $c > 0$;
(ii) $a > b$ when $c < 0$.

Now with subtraction an operation, one expects it to be included as an additional operational system. However, as pointed out in the preceding discussion, subtraction will not be given the same status as that given to addition and multiplication. The diagram that follows shows the two operational systems with the properties that each obeys. Compare the diagrams of the other systems given in Chapter 5 and 6, noting the additional features as each system was presented.

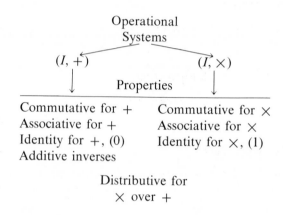

Operational
Systems

$(I, +)$ ← → $(I, \times)$

↓ Properties ↓

$(I, +)$	$(I, \times)$
Commutative for $+$	Commutative for $\times$
Associative for $+$	Associative for $\times$
Identity for $+$, (0)	Identity for $\times$, (1)
Additive inverses	

Distributive for
$\times$ over $+$

EXERCISE 8.9

1. Give other names for the following subsets of the integers.
   (a) Nonnegative integers (b) Positive integers
2. What names may be used for the pairs of integers: 8 and $^-8$; $^-12$ and 12?
3. In what way does the system of integers differ from the system of whole numbers? Which of the properties listed in Section 8.9 could have been listed as properties of the system of whole numbers or of the system of natural numbers?

9

The System of Rational Numbers

9.1 THE SET OF RATIONAL NUMBERS

Each time a new set of numbers has been introduced, it has been formed by annexing more numbers to the set already in use. This was done because of some new requirement that we wished to impose. To the first set, the *natural numbers,* was annexed the number zero in order to supply an additive identity, thus forming the set of *whole numbers*. Later, it was deemed desirable to extend the set of whole numbers so that subtraction could be defined as an operation. This was achieved by inventing a new set of numbers consisting of the additive inverse of each whole number except 0. The union of this new set and the set of whole numbers formed the set of *integers*.

As the systems of numbers were developed, we observed that each system retained most of the properties belonging to its predecessor while acquiring the new property that had motivated its formation. We know that

$$N \subset W \subset I$$

and that many properties applying to the elements of N also hold for W and I.

Since the inclusion of the additive inverses gave the integer system a decided advantage over the system of whole numbers, the introduction of further numbers and properties may be worth considering at this point. The presence of an additive inverse for each number in I suggests the idea of having *multiplicative inverses* also. We should expect these inverses to follow a pattern similar to that already established for the additive inverses.

The ordered pair consisting of an integer other than 0 and its additive inverse is matched under addition with the additive identity:

$$(a, -a) \xrightarrow{+} 0$$

This suggests that we should have a pair of numbers matched under multiplication with the multiplicative identity:

$$(a, *) \xrightarrow{\times} 1,$$

where * will be called the *multiplicative inverse* of a.

With this as a starting point, we shall demand that the set of integers I be increased to form a new set, Q', in such a way that each number in the set, except 0, will have a multiplicative inverse. We shall construct this new set from the numbers already at hand, the set of integers, by inventing for each integer c, not zero, a multiplicative inverse. The restriction on zero is justified in the following paragraph. The symbol

$$\frac{1}{c}$$

has through use become the commonly accepted way to write this multiplicative inverse of c. The definition follows.

Definition 9.1(a) For each $c \in I$, $c \neq 0$, there is a *multiplicative inverse*

$$\frac{1}{c}$$

such that

$$\left(c, \frac{1}{c}\right) \xrightarrow{\times} 1$$

or

$$c \times \frac{1}{c} = 1.$$

Why do we exclude 0 from the integers that have a multiplicative inverse? It has already been established in Section 6.6 that in the set of whole numbers, the product of zero and any number is zero:

$$a \in W, \quad (a, 0) \xrightarrow{\times} 0$$

This continued to hold with a an integer. In establishing the system of rational numbers, many of the properties of the earlier systems are included. If zero were to possess a multiplicative inverse (called M), then $0 \times M$ would be equal to 1 (Definition 9.1(a)). But this product is required to be 0, as pointed out above. Hence, there can be no number which when multiplied by zero is 1 and, thus, no multiplicative inverse of 0. Then $\frac{1}{0}$ has no meaning, and its exclusion when discussing rational numbers is an important point.

Definition 9.1(a) provides some new numbers, the multiplicative inverses of the integers except 0. The union of this set of numbers and the set of integers is the set we called Q'. With this set of numbers, Q', we take the first step toward constructing a number system. If this system is to fit into the orderly structure that has preceded it, the operations of addition and multiplication must be defined and the properties of the preceding systems included.

We choose to investigate multiplication first. With Q' consisting of elements such as

$$2, \frac{1}{2}, {}^{-}3, \frac{1}{{}^{-}3}, 4, \frac{1}{4}, 7, \frac{1}{7}, 0, 3, \frac{1}{3},$$

we note the presence of the integers. There will be no need to redefine multiplication for the integers, a subset of Q'. Our concern, then, becomes that of defining multiplication for pairs of elements such as

$$\left(2, \frac{1}{3}\right) \quad \text{and} \quad \left(\frac{1}{{}^{-}3}, \frac{1}{7}\right).$$

The definitions follow:

Definition 9.1(b) For each $a, b \in I, b \neq 0,$

$$a \times \frac{1}{b} = \frac{a}{b}.$$

Definition 9.1(c) For each $a, b \in I$, $ab \neq 0$,†

$$\frac{1}{a} \times \frac{1}{b} = \frac{1}{ab}.$$

It is immediately apparent that Definition 9.1(b) introduces other numbers. For example,

$$2 \times \frac{1}{3} = \frac{2}{3}.$$

But $\frac{2}{3}$ is not an integer or the multiplicative inverse of an integer. However, this number must be included. Why? Since we plan to accept multiplication as an operation, we are immediately forced to add to set Q' all of those numbers arising from

$$a \times \frac{1}{b}, \quad a, b \in I, b \neq 0.$$

This extended set of numbers will now be called set Q, the set of *rational numbers*.

9.2 THE MULTIPLICATION OPERATION IN THE SET OF RATIONAL NUMBERS

Since set Q includes all of the numbers of Q' and additional numbers such as

$$\frac{3}{7}, \frac{2}{5}, \frac{5}{{}^{-}7}, \frac{{}^{-}8}{2},$$

†ab is a short way of writing $a \times b$, and $ab \neq 0$ is a short way of saying that neither a nor b can be 0.

there is a need to extend multiplication to these. We shall use Definitions 9.1(b), 9.1(c), and properties for multiplication to guide us in forming a generalization for multiplying pairs of numbers such as

$$\left(\frac{2}{5}, \frac{3}{7}\right) \quad \text{or} \quad \left(\frac{5}{^-7}, \frac{^-8}{2}\right).$$

Example 1

$$\frac{2}{5} \times \frac{3}{7} = n$$

$$\frac{2}{5} \times \frac{3}{7} = \left(2 \times \frac{1}{5}\right) \times \left(3 \times \frac{1}{7}\right) \qquad \textit{Definition } 9.1\text{(b)}$$

$$= 2 \times \left(\frac{1}{5} \times 3\right) \times \frac{1}{7} \qquad \textit{Associative property for multiplication}$$

$$= 2 \times \left(3 \times \frac{1}{5}\right) \times \frac{1}{7} \qquad \textit{Commutative property for multiplication}$$

$$= (2 \times 3) \times \left(\frac{1}{5} \times \frac{1}{7}\right) \qquad \textit{Associative property for multiplication}$$

$$= (2 \times 3) \times \left(\frac{1}{5 \times 7}\right) \qquad \textit{Definition } 9.1\text{(c)}$$

$$= \frac{2 \times 3}{5 \times 7} \qquad \textit{Definition } 9.1\text{(b)}$$

Hence,

$$\frac{2}{5} \times \frac{3}{7} = \frac{2 \times 3}{5 \times 7}.$$

Using the plan shown in Example 1, find the products of the following:

(1) $\frac{^-5}{4} \times \frac{3}{8}$ (3) $\frac{5}{6} \times \frac{6}{5}$ (5) $\frac{0}{7} \times \frac{3}{5}$

(2) $\frac{8}{^-7} \times \frac{^-4}{9}$ (4) $\frac{^-9}{5} \times \frac{^-20}{^-27}$ (6) $\frac{a}{b} \times \frac{c}{d}$, $bd \neq 0$

From Problem (6) above is established a generalization for the multiplication of any two numbers in Q. Since it is possible to prove the generalization from previous definitions and properties, we shall call it a *theorem.*

Theorem I *For each $a, b, c, d \in I, bd \neq 0$,*

$$\frac{a}{b} \times \frac{c}{d} = \frac{a \times c}{b \times d} \quad \textit{or} \quad \frac{ac}{bd}.$$

By using Example 1 as a pattern, we may prove this theorem as follows:

$$\frac{a}{b} \times \frac{c}{d} = \left(a \times \frac{1}{b}\right) \times \left(c \times \frac{1}{d}\right) \qquad \textit{Why?}$$

$$= a \times \left(\frac{1}{b} \times c\right) \times \frac{1}{d} \qquad \textit{Why?}$$

$$= a \times \left(c \times \frac{1}{b}\right) \times \frac{1}{d} \qquad \textit{Why?}$$

$$= (a \times c) \times \left(\frac{1}{b} \times \frac{1}{d}\right) \qquad \textit{Why?}$$

$$= (a \times c) \times \left(\frac{1}{b \times d}\right) \qquad \textit{Why?}$$

$$= \frac{a \times c}{b \times d} \qquad \textit{Why?}$$

Hence,

$$\frac{a}{b} \times \frac{c}{d} = \frac{a \times c}{b \times d}, \quad a, b, c, d \in I, bd \neq 0,$$

and Theorem I has been proved.

Consider the next example.

Example 2

$$2 \times \frac{1}{2} = n$$

$$2 \times \frac{1}{2} = \frac{2}{2} \qquad \textit{Definition } 9.1(\text{b})$$

But

$$2 \times \frac{1}{2} = 1 \qquad \textit{Definition } 9.1(\text{a})$$

Because of the uniqueness that we demand of multiplication in set Q, $\frac{2}{2}$ and 1 name the same number. We then write

$$\frac{2}{2} = 1.$$

From Example 2 we make the following generalization.

Theorem II *For each $c \in I$, $c \neq 0$,*

$$\frac{c}{c} = 1.$$

Use Example 2 as a pattern in proving this theorem.

The above theorem indicates the many forms for the multiplicative identity in the system of rational numbers. Here we see that 1 has many representations, such as,

$$\frac{3}{3}, \frac{15}{15}, \frac{^-7}{^-7}, \frac{^-102}{^-102}.$$

The theorem excludes 0. Then $\frac{0}{0}$ cannot represent 1. The discussion showing that 0 has no multiplicative inverse, together with Example 2, should convince you that $\frac{0}{0}$ has no meaning.

Example 3

$$6 \times \frac{1}{1} = n$$

$$6 \times \frac{1}{1} = \frac{6}{1} \qquad \textit{Definition } 9.1(\text{b})$$

Since $6 \times 1 = 6$ by the multiplicative identity and $\frac{1}{1} = 1$ by Theorem II, if multiplication has uniqueness, $\frac{6}{1}$ and 6 must name the same number.

From Example 3 we make the following generalization.

Theorem III *For each $a \in I$,*

$$\frac{a}{1} = a.$$

Prove this theorem, using Example 3 as a pattern.

Example 4

$$\frac{5}{7} \times \frac{7}{5} = n$$

$$\frac{5}{7} \times \frac{7}{5} = \frac{5 \times 7}{7 \times 5} \qquad \textit{Theorem I}$$

$$= \frac{35}{35} \qquad \textit{Renaming } (5 \times 7) \textit{ and } (7 \times 5)$$

$$= 1 \qquad \textit{Theorem II}$$

Since

$$\left(\frac{5}{7}, \frac{7}{5}\right) \xrightarrow{\times} 1,$$

$\frac{7}{5}$ is the multiplicative inverse of $\frac{5}{7}$ by Definition 9.1(a).

From Example 4 we make the following generalization:

Theorem IV *For each $a, b \in I$, $ab \neq 0$,*

$$\frac{a}{b} \times \frac{b}{a} = 1$$

and $\frac{b}{a}$ is the multiplicative inverse of $\frac{a}{b}$.

Prove this theorem.

The numbers we have been using in the preceding examples belong to set Q. They consist of the integers, the multiplicative inverse of each integer except 0, and those additional numbers that were accepted because closure was demanded by the operation multiplication. As was predicted, it has now been established that every number in set Q, except 0, has its own multiplicative inverse.

Definition 9.2 The set of *rational numbers*, Q, consists of those numbers which may be expressed in the form

$$\frac{a}{b}, a, b \in I, b \neq 0.$$

The set of rational numbers Q and the multiplication operation form the binary operational system, $(Q, \times)$. The elements obey the commutative and associative properties for multiplication. The system has an identity number 1, and each element of Q, except 0, has a multiplicative inverse. And, 0 maintains the same roles as in the system of whole numbers. The name for 0 in Q may be expressed as

$$\frac{0}{b}, b \in I, b \neq 0.$$

EXERCISE 9.2

1. Using Theorem IV, find, when possible, the multiplicative inverse for each of the following.

   (a) $\frac{29}{11}$ (d) $\frac{^-4}{^-9}$ (g) $\frac{0}{2}$

   (b) $\frac{22}{33}$ (e) $\frac{1}{19}$ (h) 13

   (c) $\frac{^-2}{19}$ (f) $\frac{^-8}{^-8}$ (i) $\frac{21}{1}$

2. Use a theorem or definition to justify each of the following statements.

   (a) $^-12 = \frac{^-12}{1}$ (c) $\frac{1}{5} \times \frac{1}{3} = \frac{1}{5 \times 3}$

   (b) $\frac{2}{9} = 2 \times \frac{1}{9}$ (d) $\frac{13}{45} \times \frac{45}{13} = 1$

(e) $\frac{^-7}{^-7} = 1$ (g) $\frac{9}{^-10} \times \frac{11}{^-17} = \frac{9 \times 11}{^-10 \times {}^-17}$

(f) $^-5 \times \frac{1}{18} = \frac{^-5}{18}$ (h) $^-25 \times \frac{1}{^-25} = 1$

3. Justify the inclusion of each of the following in the set of rational numbers.

(a) 18 (c) 0 (e) $\frac{^-5}{2}$

(b) $\frac{1}{18}$ (d) $\frac{1}{^-5}$ (f) $\frac{3}{4}$

4. Find the product of each without using Theorem I. Instead, use Definitions 9.1(b), 9.1(c), Theorem II, and Theorem III when applicable to justify your procedure.

(a) $\frac{2}{5} \times \frac{3}{7}$ (d) $0 \times \frac{1}{4}$ (g) $\frac{36}{4} \times \frac{^-2}{^-9}$

(b) $\frac{1}{^-6} \times \frac{1}{3}$ (e) $\frac{3}{13} \times \frac{13}{3}$ (h) $\frac{0}{15} \times \frac{^-8}{5}$

(c) $8 \times \frac{1}{11}$ (f) $\frac{^-27}{14} \times \frac{5}{8}$

5. In case Theorems I, II, III, and IV have not been proved, use Definitions 9.1(a), 9.1(b), 9.1(c), and the properties assumed for the system of rational numbers to do this.

9.3 THE DIVISION OPERATION IN THE SET OF RATIONAL NUMBERS

Division was first introduced in this textbook for whole numbers. See Section 6.9. Was division accepted as an operation in the set of whole numbers? Why?

Now, we shall explore division in the set of rational numbers. As for the whole numbers, we relate division to multiplication as,

$$\frac{1}{12} \div \frac{1}{3} = n, \quad \text{if } \frac{1}{3} \times n = \frac{1}{12}.$$

The solution set is easily found to be $\{\frac{1}{4}\}$.

Adopting the definition for division of whole numbers, we define division for the rational numbers.

Definition 9.3 For each $a, b, n \in Q, b \neq 0$, if $b \times n = a$, then

$$a \div b = n.$$

Observe that now a, b, and n are elements of Q rather than W. The *quotient* is n, the *divisor* is b, and the *dividend* is a.

For each pair of rational numbers, is it possible to have an assignment under division? That is, is there an n for

$$(a, b) \xrightarrow{\div} n, \quad a, b, n \in Q, b \neq 0?$$

If so, then division can now be identified as an operation. What are the requirements for an operation?

We shall now consider the following examples of division in Q.

Example 1 Is there an n in Q such that $(12, 5) \overset{\div}{\rightarrow} n$? Are 12 and 5 rational numbers?

From Theorem III, 12 and 5 may be written $\frac{12}{1}$ and $\frac{5}{1}$. Hence, both are names for rational numbers.

$$12 \div 5 = n,$$

$$\text{if} \quad 5 \times n = 12. \qquad \textit{Definition } 9.3$$

$$\left(\frac{1}{5} \times 5\right) \times n = 12 \times \frac{1}{5} \qquad \textit{Uniqueness of multiplication}$$

$$1 \times n = 12 \times \frac{1}{5} \qquad \textit{Multiplicative inverse}$$

$$n = 12 \times \frac{1}{5} \qquad \textit{Multiplicative identity}$$

$$n = \frac{12}{5} \qquad \textit{Definition } 9.1\text{(b)}$$

Hence,

$$12 \div 5 = \frac{12}{5},$$

and $\{\frac{12}{5}\}$ is the solution set for the above sentence. Observe that the solution is an element of set Q.

Example 2 Justify each of the steps in this example.

$$\frac{2}{3} \div \frac{7}{13} = n,$$

$$\text{if} \quad \frac{7}{13} \times n = \frac{2}{3}.$$

$$\left(\frac{13}{7} \times \frac{7}{13}\right) \times n = \frac{2}{3} \times \frac{13}{7}$$

$$1 \times n = \frac{2}{3} \times \frac{13}{7}$$

$$n = \frac{2}{3} \times \frac{13}{7}$$

$$n = \frac{26}{21}$$

Therefore,

$$\frac{2}{3} \div \frac{7}{13} = \frac{26}{21},$$

and the solution set is $\{\frac{26}{21}\}$.

We shall now derive the procedure for dividing *each* pair of rational numbers, $\frac{p}{q} \div \frac{r}{s}$. We seek n in $\frac{p}{q} \div \frac{r}{s} = n$, for each $p, q, r, s \in I, n \in Q, qrs \neq 0$.

$$\frac{p}{q} \div \frac{r}{s} = n,$$

$$\text{if } \frac{r}{s} \times n = \frac{p}{q}. \qquad \textit{Definition 9.3}$$

$$\left(\frac{s}{r} \times \frac{r}{s}\right) \times n = \frac{s}{r} \times \frac{p}{q} \qquad \textit{Uniqueness of multiplication}$$

$$1 \times n = \frac{s}{r} \times \frac{p}{q} \qquad \textit{Multiplicative inverse}$$

$$n = \frac{s}{r} \times \frac{p}{q} \qquad \textit{Multiplicative identity}$$

$$n = \frac{sp}{rq} \qquad \textit{Theorem I}$$

We have established the following theorem.

Theorem V(a) *For each $p, q, r, s \in I$, $qrs \neq 0$,*

$$\frac{p}{q} \div \frac{r}{s} = \frac{sp}{rq}.$$

In the preceding development where

$$n = \frac{s}{r} \times \frac{p}{q},$$

by the commutative property for multiplication we write,

$$n = \frac{p}{q} \times \frac{s}{r}.$$

Hence,

$$\frac{p}{q} \div \frac{r}{s} = \frac{p}{q} \times \frac{s}{r}.$$

Observe that $\frac{s}{r}$ is the multiplicative inverse of $\frac{r}{s}$. From this, we state the following theorem.

Theorem V(b) *For each $p, q, r, s \in I$, $qrs \neq 0$,* $\frac{p}{q} \div \frac{r}{s}$ is equal to $\frac{p}{q}$ multiplied by the multiplicative inverse of $\frac{r}{s}$.

$$\frac{p}{q} \div \frac{r}{s} = \frac{p}{q} \times \frac{s}{r}$$

This justifies the *rule* which many of us learned: "To divide one fraction by another, invert the divisor and multiply." However, we now prefer to say: "To divide $\frac{p}{q}$ by $\frac{r}{s}$, multiply $\frac{p}{q}$ by the multiplicative inverse of $\frac{r}{s}$."

Example 3 Use the multiplicative inverse to find $\frac{7}{9} \div \frac{4}{5}$.

$$\frac{7}{9} \div \frac{4}{5} = \frac{7}{9} \times \frac{5}{4} \qquad \textit{$\tfrac{5}{4}$ is the multiplicative inverse of $\tfrac{4}{5}$}$$

$$= \frac{7 \times 5}{9 \times 4} \qquad \textit{Theorem I}$$

$$= \frac{35}{36} \qquad \textit{Renaming } (7 \times 5) \textit{ and } (9 \times 4)$$

It should be noted that the form $\frac{p}{q}$ may be considered to mean $p \div q$, as shown in the following example.

Example 4

$$3 \div 7 = n,$$

$$\text{if } 7 \times n = 3. \qquad \textit{Definition } 9.3$$

$$\left(7 \times \frac{1}{7}\right) \times n = 3 \times \frac{1}{7} \qquad \textit{Uniqueness of multiplication}$$

$$1 \times n = 3 \times \frac{1}{7} \qquad \textit{Multiplicative inverse}$$

$$n = 3 \times \frac{1}{7} \qquad \textit{Multiplicative identity}$$

$$n = \frac{3}{7} \qquad \textit{Definition } 9.1\text{(b)}$$

We see that $3 \div 7$ and $\frac{3}{7}$ represent the same number.

Using this interpretation we shall say that

$$\frac{\frac{2}{5}}{\frac{3}{7}} \quad \text{means} \quad \frac{2}{5} \div \frac{3}{7}.$$

By Theorem V(a)

$$\frac{2}{5} \div \frac{3}{7} = \frac{14}{15}.$$

But $\frac{14}{15}$ is a rational number by definition and hence

$$\frac{\frac{2}{5}}{\frac{3}{7}}$$

represents a rational number. From this we conclude that if a numeral can be changed to $\frac{a}{b}$, $a, b \in I$, $b \neq 0$, then the numeral is a name for an element of Q. This makes more significant the statement in the definition of a rational number (Definition 9.2) that a rational number is one that *may* be expressed in the form $\frac{a}{b}$, $a, b \in I$, $b \neq 0$.

In the same manner,

$$\frac{\frac{7}{3}}{\frac{7}{3}} \quad \text{means} \quad \frac{7}{3} \div \frac{7}{3},$$

and this may be simplified to 1. We conclude that the statement that has been in use throughout this chapter,

$$\frac{c}{c} = 1, c \in I, c \neq 0,$$

may be made to include all rational numbers except 0 and be written

$$\frac{c}{c} = 1, c \in Q, c \neq 0.$$

The preceding paragraph suggests another method for changing

$$\frac{\frac{2}{5}}{\frac{3}{7}}$$

to the form $\frac{a}{b}$, $a, b \in I, b \neq 0$. It follows that

$$\frac{\frac{2}{5}}{\frac{3}{7}} \times \frac{\frac{7}{3}}{\frac{7}{3}}$$

simply renames the original number, since multiplying by the multiplicative identity leaves the number unchanged. If we extend Theorem I,

$$\frac{a}{b} \times \frac{c}{d} = \frac{ac}{bd}, \quad a, b, c, d \in I, bd \neq 0,$$

to allow a, b, c, d to be rational numbers, then

$$\frac{\frac{2}{5}}{\frac{3}{7}} \times \frac{\frac{7}{3}}{\frac{7}{3}} = \frac{\frac{14}{15}}{1} \qquad \textit{$\tfrac{7}{3}$ chosen as the multiplicative inverse of $\tfrac{3}{7}$}$$

$$= \frac{14}{15}.$$

It is left to the reader to extend the statement $\frac{a}{1} = a, a \in I$ to include all rational numbers, that is, $\frac{a}{1} = a$, $a \in Q$. This is the reason for replacing $\frac{\frac{14}{15}}{1}$ by $\frac{14}{15}$.

EXERCISE 9.3

1. Use Definition 9.3 to show the relatedness of each division problem to multiplication.

   EXAMPLE $\frac{2}{3} \div \frac{5}{7} = n$ if $\frac{5}{7} \times n = \frac{2}{3}$

   (a) $\frac{5}{9} \div \frac{3}{4} = n$ (c) $\frac{4}{^-3} \div \frac{19}{7} = n$ (e) $\frac{^-12}{5} \div \frac{4}{^-7} = n$

   (b) $\left(18, \frac{2}{5}\right) \xrightarrow{\div} n$ (d) $\left(\frac{1}{3}, 2\right) \xrightarrow{\div} n$ (f) $\frac{3}{8} \div {}^-5 = n$

2. Use the multiplicative inverse to find the quotient of each.

(a) $\frac{21}{94} \div 7$ (d) $\frac{95}{^-7} \div \frac{15}{^-21}$ (g) $\frac{14}{5} \div 5$

(b) $\frac{^-23}{4} \div \frac{4}{23}$ (e) $\frac{4}{5} \div \frac{3}{7}$ (h) $\frac{^-3}{13} \div \frac{^-5}{^-26}$

(c) $\frac{7}{5} \div \frac{7}{5}$ (f) $\frac{^-27}{11} \div \frac{^-9}{^-22}$ (i) $\frac{46}{^-5} \div \frac{^-15}{23}$

3. Write each of the following as a division problem, then find the quotient.

(a) $\frac{\frac{3}{7}}{\frac{2}{5}}$ (d) $\frac{\frac{7}{15}}{\frac{1}{^-9}}$ (g) $\frac{16}{\frac{^-1}{2}}$

(b) $\frac{\frac{6}{11}}{\frac{17}{21}}$ (e) $\frac{\frac{2}{3}}{\frac{6}{9}}$ (h) $\frac{\frac{16}{1}}{\frac{2}{1}}$

(c) $\frac{\frac{^-19}{15}}{\frac{^-2}{3}}$ (f) $\frac{\frac{^-2}{5}}{\frac{^-3}{1}}$

4. Find the quotient for each part in Problem 3 above by using the multiplicative identity, $\frac{c}{c} = 1$, $c \in Q$, $c \neq 0$.

9.4 FRACTIONS; EQUIVALENT FRACTIONS

By Definition 9.2 every rational number can be expressed in the form $\frac{a}{b}$, where a is any integer and b is any nonzero integer. The numeral $\frac{a}{b}$ is called a *fraction,* with a the *numerator* and b the *denominator.*

Each number of set Q has many names of the form $\frac{a}{b}$. For instance, $\frac{1}{2}$, $\frac{2}{4}$, $\frac{3}{6}$, and $\frac{^-5}{^-10}$ name the same rational number. They are called *equivalent fractions.* The following examples present a further discussion of equivalent fractions.

Example 1 Show that $\frac{1}{2}$ and $\frac{2}{4}$ name the same rational number.

If a number is multiplied by the multiplicative identity, the result is that number. By Theorem II, some names for the multiplicative identity are $\frac{2}{2}$, $\frac{^-3}{^-3}$, $\frac{15}{15}$, and $\frac{21}{21}$. Multiplying $\frac{1}{2}$ by $\frac{2}{2}$ does not change the number named by $\frac{1}{2}$. Hence,

$$\frac{1}{2} = \frac{1}{2} \times \frac{2}{2} \quad \textit{Multiplicative identity, } \tfrac{2}{2}$$

$$= \frac{1 \times 2}{2 \times 2} \quad \textit{Theorem I}$$

$$= \frac{2}{4}. \quad \textit{Renaming } (1 \times 2) \textit{ and } (2 \times 2)$$

By using the multiplicative identity named $\frac{2}{2}$, it has been shown that $\frac{1}{2} = \frac{2}{4}$. Therefore we say that $\frac{1}{2}$ and $\frac{2}{4}$ are different names for the same rational number. These two names are equivalent fractions. The symbol $\simeq$ may be used, as $\frac{1}{2} \simeq \frac{2}{4}$, to denote equivalent fractions. However, we choose the more commonly used symbol, =, to say that the fractions name the same rational number.

Example 2 Use different names for the multiplicative identity to form a set of fractions equivalent to $\frac{^-2}{5}$.

The set consisting of five elements is

$$\left\{\frac{^-2}{5}, \frac{^-2}{5} \times \frac{^-3}{^-3}, \frac{^-2}{5} \times \frac{5}{5}, \frac{^-2}{5} \times \frac{^-2}{^-2}, \frac{^-2}{5} \times \frac{17}{17}\right\},$$

which may be written

$$\left\{\frac{^-2}{5}, \frac{6}{^-15}, \frac{^-10}{25}, \frac{4}{^-10}, \frac{^-34}{85}\right\}.$$

What name was used for the multiplicative identity to form the fraction $\frac{6}{^-15}$, which is equivalent to $\frac{^-2}{5}$? Form three other elements that belong to this set. Name the integer c used to form the multiplicative identity $\frac{c}{c}$ for each. By selecting different names for the multiplicative identity, one sees that each rational number has infinitely many fractions that represent it.

In the set of equivalent fractions discussed in Example 2, the fraction $\frac{^-2}{5}$ has an important role. It represents the *simplest* or the *basic* form of all the fractions in this set. Since $^-2$ and 5 are relatively prime, the fraction is considered to be the simplest. Are there other fractions in the set with numerator and denominator relatively prime? See Section 7.4 for the meaning of relatively prime. Which fraction is called the simpler, $\frac{^-2}{5}$ or $\frac{2}{^-5}$? We shall designate a fraction with the positive denominator as the simplest form.

Example 3 Change $\frac{3}{^-7}$ to an equivalent fraction having a positive denominator.

Multiply by the multiplicative identity $\frac{^-1}{^-1}$.

$$\frac{3}{^-7} \times \frac{^-1}{^-1} = \frac{^-3}{7}$$

Thus, $\frac{^-3}{7}$ is the required fraction.

Example 4 Change $\frac{^-3}{4}$ to an equivalent fraction having a positive numerator.

Try multiplying the given fraction by $\frac{^-1}{^-1}$.

Example 5 What is the basic form (the simplest fraction) used in constructing the set of equivalent fractions which contains $\frac{12}{15}$?

$$\frac{12}{15} = \frac{4 \times 3}{5 \times 3} \qquad \textit{Renaming } 12 \textit{ as } 4 \times 3 \textit{ and } 15 \textit{ as } 5 \times 3$$

$$= \frac{4}{5} \times \frac{3}{3} \qquad \textit{Theorem I}$$

$$= \frac{4}{5} \qquad \textit{Multiplicative identity } \tfrac{3}{3}$$

Since the numerator and denominator are relatively prime, $\frac{4}{5}$ is the simplest fraction of the set.

Since 12 and 15 in the fraction $\frac{12}{15}$ are not relatively prime, we say that the fraction is *reducible*. The fraction $\frac{4}{5}$ is *irreducible*, and is commonly referred to as the reduced form of $\frac{12}{15}$.

Example 6 Find a fraction with numerator and denominator relatively prime which is equivalent to $\frac{714}{1386}$.

Writing 714 and 1386 in completely factored form,

$$\frac{714}{1386} = \frac{17 \times 7 \times 3 \times 2}{11 \times 3 \times 7 \times 3 \times 2}.$$

This may be written as

$$\frac{714}{1386} = \frac{17}{11 \times 3} \times \frac{7 \times 3 \times 2}{7 \times 3 \times 2}.$$

Can you find a name for the multiplicative identity? Since $\frac{7 \times 3 \times 2}{7 \times 3 \times 2}$ is a name for 1,

$$\frac{714}{1386} = \frac{17}{11 \times 3}.$$

Hence, $\frac{17}{33}$ is the fraction we were seeking since 17 and 33 are relatively prime. The fraction $\frac{17}{33}$ is irreducible and may be thought of as the reduced form of $\frac{714}{1386}$.

Example 7 Are the fractions $\frac{9}{39}$ and $\frac{15}{65}$ equivalent?

Multiply each by the multiplicative identity.

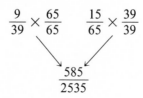

$$\frac{9}{39} \times \frac{65}{65} \qquad \frac{15}{65} \times \frac{39}{39}$$

$$\searrow \quad \swarrow$$

$$\frac{585}{2535}$$

Do you see that $\frac{9}{39}$ and $\frac{15}{65}$ name the same rational number? Hence, these two fractions are equivalent, and we write

$$\frac{9}{39} = \frac{15}{65}.$$

Example 8 Are the fractions $\frac{3}{7}$ and $\frac{4}{9}$ equivalent?

Observe the choice of the names for the multiplicative identity.

$$\frac{3}{7} \times \frac{9}{9} \qquad \frac{4}{9} \times \frac{7}{7}$$

$$\downarrow \qquad\qquad \downarrow$$

$$\frac{27}{63} \qquad\quad \frac{28}{63}$$

Here we see that $\frac{3}{7}$ and $\frac{4}{9}$ are names for different rational numbers. Hence, they are not equivalent fractions.

Can you find a pattern for selecting names for the multiplicative identity used in Examples 7 and 8? Compare the following pattern with the two examples:

$$\begin{array}{cc} \frac{a}{b} \times \frac{d}{d} & \frac{c}{d} \times \frac{b}{b} \\ \downarrow & \downarrow \\ \frac{a \times d}{b \times d} & \frac{c \times b}{d \times b} \end{array}$$

Since the denominators are the same, attention is given to the numerators, $a \times d$ and $c \times b$. If they are equal, we say that the fractions, $\frac{a}{b}$ and $\frac{c}{d}$, are equivalent. This will be used to define equivalent fractions.

Definition 9.4 For each $a, b, c, d \in I, bd \neq 0,$

$$\frac{a}{b} = \frac{c}{d}$$

if and only if

$$a \times d = b \times c.$$

Definition 9.4 is useful in testing the equivalence of fractions. Sometimes it is referred to as the *equivalence test*. Observe the use of this test in the following examples.

Example 9 Are $\frac{39}{15}$ and $\frac{52}{20}$ equivalent fractions?
Applying Definition 9.4, does $39 \times 20 = 15 \times 52$? Since each product is 780,

$$\frac{39}{15} = \frac{52}{20}.$$

Example 10 Apply the equivalence test on $\frac{5}{7}$ and $\frac{7}{9}$.
Does $5 \times 9 = 7 \times 7$? Since $45 \neq 49$,

$$\frac{5}{7} \neq \frac{7}{9}.$$

Example 11 Find S such that $\frac{5}{11} = \frac{30}{S}$.
If these two fractions are equivalent, we have:

$5 \times S = 11 \times 30$	*Definition* 9.4
$5 \times S = 330$	*Renaming* (11×30)
$5 \times S = 5 \times 66$	*Renaming* 330
$S = 66$	*Applying the cancellation property for multiplication (carried over from the system of integers)*

EXERCISE 9.4

1. Construct a set of fractions equivalent to each of the following such that each set contains six elements.

   (a) $\frac{2}{7}$ (b) $\frac{2}{1}$ (c) $\frac{^-21}{13}$ (d) $\frac{0}{4}$ (e) $\frac{^-4}{^-6}$

2. Name the $\frac{c}{c}$, c a nonzero integer, used in forming the following statements about fractions.

   (a) $\frac{1}{3} = \frac{7}{21}$ (c) $\frac{18}{9} = \frac{36}{18}$ (e) $\frac{^-54}{^-67} = \frac{54}{67}$

   (b) $\frac{5}{8} = \frac{^-10}{^-16}$ (d) $\frac{^-27}{18} = \frac{27}{^-18}$ (f) $\frac{7}{^-15} = \frac{35}{^-75}$

3. Find the simplest or basic form of each of these fractions.

   (a) $\frac{14}{21}$ (b) $\frac{13}{17}$ (c) $\frac{16}{^-18}$ (d) $\frac{108}{270}$ (e) $\frac{^-15}{^-21}$ (f) $\frac{^-84}{126}$

4. Show that the members of each of the following pairs are equivalent. Use Definition 9.4.

   (a) $\frac{35}{^-10}$ and $\frac{^-21}{6}$ (c) $\frac{1}{1}$ and $\frac{13}{13}$

   (b) $\frac{0}{7}$ and $\frac{0}{^-5}$ (d) $\frac{^-7}{^-1}$ and $\frac{49}{7}$

5. Find the fraction equivalent to each of the following such that the numerator and the denominator are relatively prime.

   (a) $\frac{288}{504}$ (b) $\frac{78}{42}$ (c) $\frac{75}{25}$

6. Which of the following are equivalent?

   $$\frac{403}{672}, \frac{30}{55}, \frac{42}{82}, \frac{78}{126}, \frac{84}{154}$$

7. This is a set of equivalent fractions.

   $$\left\{\frac{34}{14}, \frac{85}{35}, \frac{17}{7}, \frac{^-51}{^-21}, \frac{68}{28}, \frac{^-17}{^-7}\right\}$$

   Explain how this is known. Can you add other elements to this set? How many?

8. Use Definition 9.4 to find the solution set of each.

   (a) $\frac{x}{12} = \frac{15}{36}$ (c) $\frac{26}{^-11} = \frac{x}{^-5}$ (e) $\frac{x}{2} = \frac{2}{x}$ (g) $\frac{2}{x} = \frac{5}{(6 + x)}$

   (b) $\frac{7}{x} = \frac{35}{40}$ (d) $\frac{18}{^-7} = \frac{23}{x}$ (f) $\frac{(15 + x)}{8} = \frac{5}{2}$

9. Show that any integer a may be represented by $\frac{ak}{k}$, $k \in I$, $k \neq 0$.

10. Using the preceding problem give three fraction names for each of the following integers: 7, $^-2$, 0, 4, 100.

9.5 THE MEANING OF $\frac{-p}{q}$, $\frac{-p}{q}$, AND $-\left(\frac{p}{q}\right)$

Assuming that p and q are integers greater than 0, we shall find the sum of $\frac{^-p}{q}$ and $\frac{p}{q}$. Can you justify each of the steps?

$$\frac{-p}{q} + \frac{p}{q} = \frac{-p + p}{q} \qquad \textit{Why?}$$

$$= (-p + p) \times \frac{1}{q} \qquad \textit{Why?}$$

$$= 0 \times \frac{1}{q} \qquad \textit{Why?}$$

$$= 0 \qquad \textit{The role of } 0 \textit{ in multiplication}$$

Since the sum of $\frac{-p}{q}$ and $\frac{p}{q}$ is 0, one of the addends is the additive inverse of the other. That is, $\frac{-p}{q}$ is the additive inverse of $\frac{p}{q}$. Using the symbol for additive inverse, this is written

(i) $$\frac{-p}{q} = -\left(\frac{p}{q}\right).$$

Do you expect $\frac{-p}{q}$ also to be equal to $\frac{p}{-q}$? Consider the following:

$$\frac{-p}{q} \times \frac{^{-}1}{^{-}1} = \frac{-p \times {}^{-}1}{q \times {}^{-}1} \qquad \textit{Why?}$$

$$= \frac{p}{-q}$$

Since we have shown that $\frac{-p}{q}$ and $\frac{p}{-q}$ are equivalent, then we write,

(ii) $$\frac{-p}{q} = \frac{p}{-q}.$$

Applying the transitive property for equals to the statements given in (i) and (ii), we have

(iii) $$\frac{-p}{q} = \frac{p}{-q} = -\left(\frac{p}{q}\right).$$

Since the additive inverse of $\frac{5}{6}$ is $^{-}\left(\frac{5}{6}\right)$, it may be written as $\frac{^{-}5}{6}$ or as $\frac{5}{^{-}6}$. We identify as *positive rational numbers* those which can be written as $\frac{a}{b}$, where a and b are positive integers. The additive inverses of the positive rational numbers are the *negative rational numbers*. One rational number which is *neither positive nor negative is* 0.

EXERCISE 9.5

1. Use the preceding discussion to give two other names for each of the following:

   (a) $\frac{^{-}2}{3}$ (b) $\frac{7}{^{-}9}$ (c) $^{-}\left(\frac{3}{5}\right)$ (d) $^{-}\left(\frac{4}{4}\right)$ (e) $\frac{12}{^{-}12}$

2. The following are names for rational numbers. Identify those which are positive and those which are negative.

   (a) $-\left(\frac{^{-}7}{9}\right)$ (b) $-\left(\frac{7}{^{-}9}\right)$ (c) $-\left(\frac{^{-}7}{^{-}9}\right)$ (d) $\frac{^{-}7}{9}$ (e) $\frac{7}{^{-}9}$

3. Express each of the names given in Problem 2 in the simplest form.

4. Express each of the following in $\frac{a}{b}$ form, if a and b are integers, $b \neq 0$, and a and b are relatively prime.

(a) $-\left(\frac{^-9}{15}\right)$ (c) $\frac{2}{^-7} \times \frac{3}{^-14}$ (e) $\frac{2}{^-3} \times \frac{5}{6} \times \frac{1}{^-6}$

(b) $-\left(\frac{^-7}{^-14}\right)$ (d) $\frac{^-2}{7} \times \frac{^-3}{14}$ (f) $^-\left(\frac{2}{3}\right) \times \frac{^-5}{^-6} \times \frac{^-1}{6}$

5. Use examples to verify that $\frac{-a}{-b} = \frac{a}{b}$ for
   (a) $a > 0$ and $b > 0$
   (b) $a < 0$ and $b < 0$
   (c) $a > 0$ and $b < 0$
   (d) $a < 0$ and $b > 0$

   Does this constitute proof? From the examples, is a generalization evident? Try to prove this generalization.

9.6 THE ADDITION OPERATION IN THE SET OF RATIONAL NUMBERS

It was necessary to develop multiplication of rational numbers prior to addition. This sequence is unique to this set of numbers. In all other sets we developed addition first. You will understand the reason for this change as you study the examples that follow. In these we set out to justify the algorithms for addition that we use. Observe the important role of the properties, definitions, and theorems for multiplication in Q. The use of these would have been impossible if multiplication had not preceded addition. Bear in mind that all properties for addition in Q have been adopted from those in W and I. The distributive property for multiplication over addition is also included in this adoptive procedure and is applied to the numbers in Q. Also included is the additive identity $\frac{0}{b}$, $b \in I$, $b \neq 0$.

Addition of rational numbers meets the requirements of a binary operation; that is, (i) to each pair of rational numbers there is assigned *at least one* rational number under addition, and (ii) to each pair there is assigned *only one* rational number.

The following are examples showing the development of an algorithm for the addition operation. Carefully regard the justifications given in Example 1, then try to give them for Example 2.

Example 1

$\frac{2}{3} + \frac{5}{3} = \left(2 \times \frac{1}{3}\right) + \left(5 \times \frac{1}{3}\right)$	*Definition* 9.1(b)
$= (2 + 5) \times \frac{1}{3}$	*Distributive property for multiplication over addition*
$= 7 \times \frac{1}{3}$	*Renaming* (2 + 5)
$= \frac{7}{3}$	*Definition* 9.1(b)

Example 2

$$\frac{^-7}{12} + \frac{^-11}{12} = \left(^-7 \times \frac{1}{12}\right) + \left(^-11 \times \frac{1}{12}\right) \quad \textit{Why?}$$

$$= (^-7 + {}^-11) \times \frac{1}{12} \quad \textit{Why?}$$

$$= {}^-18 \times \frac{1}{12} \quad \textit{Why?}$$

$$= \frac{^-18}{12} \quad \textit{Why?}$$

$$= \frac{^-3 \times 6}{2 \times 6} \quad \textit{Why?}$$

$$= \frac{^-3}{2} \times \frac{6}{6} \quad \textit{Why?}$$

$$= \frac{^-3}{2} \quad \textit{Why?}$$

From the second step in each of these examples, we arrive at a generalization as expressed in the following theorem.

Theorem VI *For each $a, b, c \in I$, $c \neq 0$.*

$$\frac{a}{c} + \frac{b}{c} = \frac{a + b}{c}.$$

Use the procedure of Examples 1 and 2 to prove this theorem.

It should be noted that if two rational numbers are to be added, the denominators of their fractions must be the same if the preceding theorem is to be used. Now, we shall consider the addition of rational numbers when the denominators of their fractions are different.

Example 3

$$\frac{7}{50} + \frac{^-12}{25} = \frac{7}{50} + \left(\frac{^-12}{25} \times \frac{2}{2}\right) \quad \textit{Why?}$$

$$= \frac{7}{50} + \frac{^-24}{50} \quad \textit{Why?}$$

$$= \frac{7 + {}^-24}{50} \quad \textit{Why?}$$

$$= \frac{^-17}{50} \quad \textit{Why?}$$

In the second step of this example, $\frac{^-12}{25}$ is expressed as the equivalent fraction $\frac{^-24}{50}$. Now, the two fractions have the same denominator. This makes possible the application of Theorem VI.

Example 4 Show that $\frac{7}{18} + \frac{1}{6} = \frac{10}{18}$.
In order to apply Theorem VI, it is necessary to change $\frac{1}{6}$ to an equivalent

fraction having a denominator 18. This can be accomplished by multiplying by the multiplicative identity $\frac{3}{3}$.

The sum $\frac{10}{18}$ is reducible. It may be renamed as $\frac{(5 \times 2)}{(9 \times 2)}$. By Theorem I this may be written as $\frac{5}{9} \times \frac{2}{2}$. Since $\frac{2}{2}$ is the multiplicative identity 1, $\frac{10}{18}$ is equivalent to $\frac{5}{9}$. So,

$$\frac{7}{18} + \frac{1}{6} = \frac{5}{9}.$$

Thus the sum is expressed in a form in which the numerator and denominator are relatively prime. This is the simplest and, in general, the most acceptable form for an answer.

In Examples 3 and 4, there was little doubt as to the *common denominator* that should be used. In each case, it was the greater of the two denominators. Closer examination shows that 50 is a multiple of 25 and 18 is a multiple of 6. Then, it appears that the common denominator must be a multiple of each of the denominators. Obviously there are many multiples common to both denominators. Yet, the choice of a denominator is determined by selecting the *least common multiple.* Procedures for finding the least common multiple of two or more numbers were presented in Section 7.5. When one speaks of the least common multiple of two denominators, it is called the *least common denominator.* Application of the procedures used in finding the least common denominator will be given in the next chapter.

Follow the procedure shown in Example 5, and see if you can justify each step.

Example 5 Show that $\frac{1}{14} + \frac{5}{21} = \frac{13}{42}$.

$$\begin{aligned}
\frac{1}{14} + \frac{5}{21} &= \frac{1}{2 \times 7} + \frac{5}{3 \times 7} \\
&= \frac{1}{(2 \times 7)} \times \frac{3}{3} + \frac{5}{(3 \times 7)} \times \frac{2}{2} \\
&= \frac{1 \times 3}{(2 \times 7) \times 3} + \frac{5 \times 2}{(3 \times 7) \times 2} \\
&= \frac{3}{42} + \frac{10}{42} \\
&= \frac{3 + 10}{42} \\
&= \frac{13}{42}
\end{aligned}$$

EXERCISE 9.6

1. Use the procedure of examples 1 and 2 to find the sum of each of the following.

(a) $\frac{3}{4} + \frac{7}{4}$ (b) $\frac{^-5}{6} + \frac{1}{6}$ (c) $\frac{^-7}{13} + \frac{^-11}{13}$

2. Using examples 3 and 5 as a guide, find the sum of each.

(a) $\frac{1}{9} + \frac{2}{15}$ (c) $\frac{^-2}{15} + \frac{^-3}{10}$ (e) $\frac{4}{^-5} + \frac{3}{10}$

(b) $\frac{^-5}{12} + \frac{7}{16}$ (d) $\frac{^-7}{16} + \frac{3}{8}$ (f) $\frac{^-8}{^-9} + \frac{2}{^-27}$

3. Prove Theorem VI. Use Examples 1 and 2 as a guide for your procedure.

9.7 THE SUBTRACTION OPERATION IN THE SET OF RATIONAL NUMBERS

Since subtraction met all the requirements for an operation in the set of integers, it continues as such in set Q. Subtraction will be defined for rational numbers in the same manner that it is defined for integers (Definition 8.4).

Definition 9.7 For each $a, b, x \in Q$, if $b + x = a$, then

$$a - b = x.$$

Compare this definition and the examples that follow with those given for the integers in Section 8.4.

Example 1 What number is matched with $(\frac{9}{13}, \frac{6}{13})$ under subtraction?

$$\frac{9}{13} - \frac{6}{13} = n \quad \text{if} \quad \frac{6}{13} + n = \frac{9}{13}$$

$$\frac{9}{13} - \frac{6}{13} = \frac{3}{13} \quad \text{because} \quad \frac{6}{13} + \frac{3}{13} = \frac{9}{13}$$

The solution set is $\{\frac{3}{13}\}$.

Example 2 Find the solution set of $\frac{7}{12} - \frac{^-1}{4} = n$.

$$\frac{7}{12} - \frac{^-1}{4} = n \quad \text{if} \quad \frac{^-1}{4} + n = \frac{7}{12}$$

$$\frac{7}{12} - \frac{^-3}{12} = n \quad \text{if} \quad \frac{^-3}{12} + n = \frac{7}{12}$$

Why is it helpful to replace $\frac{^-1}{4}$ with an equivalent fraction? Now it is possible to replace n with $\frac{10}{12}$. Since this fraction is reducible, the solution set is $\{\frac{5}{6}\}$. Explain how you know that $\frac{10}{12}$ and $\frac{5}{6}$ are equivalent.

Generalization 8.4 was made for the subtraction of integers. It will now be given for rational numbers. As for integers, the additive inverse of the second number is added to the first number.

Generalization 9.7 For each $a, b, c, d \in I$, $bd \neq 0$.

$$\frac{a}{b} - \frac{c}{d} = \frac{a}{b} + {}^-\left(\frac{c}{d}\right).$$

This generalization is verified in the following example.

Example 3 Show that the same answer will be found for $\frac{5}{7} - \frac{2}{7}$ by using Definition 9.7 and Generalization 9.7.

Using Definition 9.7:

$$\frac{5}{7} - \frac{2}{7} = n$$

if

$$\frac{2}{7} + n = \frac{5}{7}$$

Hence,

$$\frac{5}{7} - \frac{2}{7} = \frac{3}{7}.$$

Using Generalization 9.7:

$$\frac{5}{7} - \frac{2}{7} = \frac{5}{7} + -\left(\frac{2}{7}\right)$$

$$= \frac{3}{7}$$

Example 4 Find n by applying Generalization 9.7.

$$\frac{2}{5} - \frac{3}{7} = n$$

$$\frac{2}{5} - \frac{3}{7} = \frac{2}{5} + -\left(\frac{3}{7}\right)$$

Find the least common denominator of 5 and 7. Explain how the sum may be written,

$$\frac{2}{5} + -\left(\frac{3}{7}\right) = \frac{14}{35} + -\left(\frac{15}{35}\right)$$

$$= \frac{^-1}{35}.$$

EXERCISE 9.7

1. Use Definition 9.7 to find n. Give n in the simplest form.

(a) $\frac{17}{2} - \frac{5}{2} = n$ (c) $\frac{4}{15} - \frac{7}{3} = n$ (e) $\frac{6}{7} - \frac{2}{3} = n$

(b) $\frac{^-3}{5} - \frac{^-8}{5} = n$ (d) $\frac{^-1}{9} - \frac{5}{18} = n$ (f) $\frac{5}{24} - \frac{^-11}{27} = n$

2. Use Generalization 9.7 to find n. The numerator and denominator of n should be relatively prime.

(a) $\frac{3}{4} - \frac{2}{5} = n$ (c) $\frac{^-3}{5} - \frac{^-8}{5} = n$ (e) $\frac{2}{15} - \frac{^-3}{3} = n$

(b) $\frac{2}{7} - \frac{^-5}{7} = n$ (d) $\frac{5}{18} - \frac{^-1}{9} = n$ (f) $\frac{^-4}{21} - \frac{7}{12} = n$

9.8 ORDER IN THE SET OF RATIONAL NUMBERS

Order is defined for rational numbers in the same way as it was done for integers; that is,

Definition 9.8 For each $a, b, c, d, p, q \in I$, $bdq \neq 0$ and $\frac{p}{q} > 0$,

$$\frac{a}{b} < \frac{c}{d} \quad \text{if and only if} \quad \frac{a}{b} + \frac{p}{q} = \frac{c}{d}.$$

The following examples will show how this definition helps in determining the order relation for each pair of rational numbers.

Example 1 Which is greater, $\frac{4}{17}$ or $\frac{5}{19}$?
First, express the two fractions with a common denominator:

$$\frac{4}{17} = \frac{4 \times 19}{17 \times 19}, \qquad \frac{5}{19} = \frac{5 \times 17}{19 \times 17}$$

Then $\frac{85}{323} > \frac{76}{323}$ because $\frac{85}{323} = \frac{76}{323} + \frac{9}{323}$ and $\frac{9}{323}$ is a positive rational number. Hence,

$$\frac{5}{19} > \frac{4}{17}.$$

Example 2 Which is greater, $\frac{^-9}{56}$ or $\frac{^-3}{20}$?
Again, express the two fractions with a common denominator. Write 56 and 20 in completely factored form:

$$56 = 2^3 \times 7$$
$$20 = 2^2 \times 5$$
$$\text{The lcm is } 2^3 \times 7 \times 5.$$

Then:

$$\frac{^-9}{56} = \frac{^-9}{2^3 \times 7} \times \frac{5}{5} \qquad \frac{^-3}{20} = \frac{^-3}{2^2 \times 5} \times \frac{2 \times 7}{2 \times 7}$$
$$= \frac{^-45}{280} \qquad\qquad = \frac{^-42}{280}$$

But

$$\frac{^-42}{280} = \frac{^-45}{280} + \frac{3}{280},$$

and we conclude that

$$\frac{^-9}{56} < \frac{^-3}{20}.$$

With the order of the rational numbers now established, we can say

$$\frac{2}{5} < \frac{3}{5} < \frac{4}{5}.$$

And we may speak of $\frac{3}{5}$ as being *between* $\frac{2}{5}$ and $\frac{4}{5}$. It is interesting to note that there is also a rational number between $\frac{2}{5}$ and $\frac{3}{5}$, one between $\frac{3}{5}$ and $\frac{4}{5}$, and so on.

Example 3 Find a rational number between $\frac{2}{5}$ and $\frac{3}{5}$.

$$\frac{2}{5} = \frac{2 \times 2}{5 \times 2} \qquad \frac{3}{5} = \frac{3 \times 2}{5 \times 2}$$
$$= \frac{4}{10} \qquad = \frac{6}{10}$$

Then

$$\frac{4}{10} < \frac{5}{10} < \frac{6}{10}.$$

Hence, $\frac{5}{10}$ is a rational number between $\frac{2}{5}$ and $\frac{3}{5}$.

Let us now consider the following example.

Example 4 Is there more than one rational number between $\frac{2}{5}$ and $\frac{3}{5}$?

$$\frac{2}{5} = \frac{2 \times 3}{5 \times 3} \qquad \frac{3}{5} = \frac{3 \times 3}{5 \times 3}$$
$$= \frac{6}{15} \qquad = \frac{9}{15}$$

Then

$$\frac{6}{15} < \frac{7}{15} < \frac{8}{15} < \frac{9}{15}.$$

Hence, $\frac{7}{15}$ and $\frac{8}{15}$ are both between $\frac{2}{5}$ and $\frac{3}{5}$. Is it possible to find even more rational numbers between $\frac{2}{5}$ and $\frac{3}{5}$? How many may be found?

Using the computations already made in Example 1, we can find a rational number between $\frac{4}{17}$ and $\frac{5}{19}$. Certainly, $\frac{80}{323}$ is between $\frac{76}{323}$ and $\frac{85}{323}$, and so

$$\frac{4}{17} < \frac{80}{323} < \frac{5}{19}.$$

Find others. From the results of Example 2, can you find several rational numbers between $\frac{^-9}{56}$ and $\frac{^-3}{20}$?

Examples 3 and 4 suggest an important statement concerning rational numbers.

The Property of Denseness

Between any two rational numbers a and b, $a < b$, there exists another rational number, c, such that

$$a < c < b.$$

Since the set of rational numbers has this property, it is said to be *everywhere dense*. Notice that this cannot be said of the set of integers. Find a counterexample which proves that the set of integers does not possess the property of denseness.

Another important statement made about pairs of integers may be made about pairs of rational numbers.

The Trichotomy Property

For each pair of rational numbers a, b,

$$a = b \quad \text{or} \quad a < b, \quad \text{or} \quad a > b.$$

EXERCISE 9.8

1. Verify each of the following by applying Definition 9.8.

(a) $3 > \frac{5}{2}$ (b) $\frac{^-7}{3} < \frac{1}{3}$ (c) $\frac{3}{2} > \frac{^-8}{3}$ (d) $\frac{^-5}{9} < \frac{^-3}{7}$

2. Replace "and" in each of the following with the appropriate symbol, =, <, or >. What property is being used?

(a) $\frac{4}{5}$ and $\frac{17}{20}$ (d) $\frac{17}{33}$ and $\frac{15}{28}$

(b) $\frac{^-3}{19}$ and $\frac{^-13}{57}$ (e) $\frac{875}{625}$ and $\frac{14}{10}$

(c) $\frac{2}{9}$ and $\frac{^-1}{3}$ (f) $\frac{3}{7}$ and $\frac{^-28}{28}$

3. Find one rational number between the two rational numbers given in parts (a), (b), and (e) of Problem 2.

4. Find four rational numbers between each of the following pairs.

(a) $\frac{2}{3}$ and $\frac{3}{5}$ (b) $\frac{^-4}{5}$ and $^-1$ (c) 0 and $\frac{1}{100}$

5. (a) Is there a least positive rational number?
(b) Is there a least nonnegative rational number?

9.9 THE NUMBER LINE

With the extension of the set of integers to the set of rational numbers, additional demands must be made on the number line. Now, we must consider a one-to-one correspondence between the elements of set Q and a set of points belonging to a line. The order of the rational numbers, established in Section 9.8, prescribes the sequence of the matching. The points of the line to be matched with the elements of Q are determined by a process of continuous subdividing of intervals, as illustrated in the figure at the top of the facing page.

/ 00455

D. C. Heath and Company

QUAN.	CATALOG NO.	DESCRIPTION
1	062562	INTRO TO MODERN MATHEMATICS 2ND E
		NO CHARGE

SENT TO YOU WITH THE COMPLIMENTS OF D.C. HEATH AND COMPANY.

IF YOU DESIRE FURTHER INFORMATION,
PLEASE CALL US TOLL FREE AT:
1-800-225-1388
(IN NYC 800-225-1388
IN MA. 1-800-842-1211)

OR CONTACT YOUR DISTRICT MANAGER:
JIM HAMANN
PREMIER BUILDING
9378 OLIVE BOULEVARD
ST. LOUIS, MISSOURI 63132

FORM 16-0152 REV. 11/71

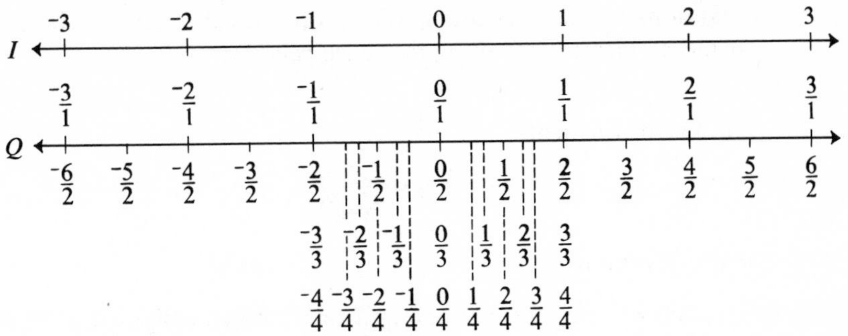

On line I is shown the matching of the integers $^-3$, $^-2$, $^-1$, 0, 1, 2, 3 with "equally spaced" points on the line. Line Q shows a few of the subdivisions of this interval and the matching of *some* of the rational numbers with points determined by the subdividing process.

Why are only some of the rational numbers so represented within the given interval? The discussion in Section 9.8 forces us to conclude that it would be impossible to represent all the rational numbers within this, or any other, interval. For, between any two rational numbers, no matter how "close," there is always another rational number. This is the property of denseness (Section 9.8). We also say that the set of points representing rational numbers on the line is *dense.*

Notice in the figure above that the rational numbers named by

$$\frac{0}{3}, \frac{1}{3}, \frac{2}{3}, \text{ and } \frac{3}{3}$$

have been matched with points that were located by subdividing the interval 0 to 1 into 3 intervals of equal length. The denominator of each of these fractions indicates the number of subdivisions of the interval, while the numerator of each indicates the number of these subdivisions.

EXERCISE 9.9

1. Make a fairly large (say 12-inch) number line, labeling points matching the following rational numbers and their opposites. (HINT: 2 is the opposite of $^-2$.)

$$^-2, \frac{^-7}{4}, \frac{^-5}{3}, \frac{^-3}{2}, \frac{^-4}{3}, \frac{^-5}{4}, \ ^-1, \frac{^-3}{4}, \frac{^-2}{3}, \frac{^-1}{2}, \frac{^-1}{3}, \frac{^-1}{4}, \ 0$$

2. After finishing Problem 1, further subdivide this number line, showing all the rational numbers named with denominators 6 and 12 between $^-2$ and 2.
3. Explain how the number line might be used in determining whether $a < b$ or $a > b$, $a, b \in Q$.

9.10 THE SYSTEM OF RATIONAL NUMBERS

We are now prepared to describe the *system of rational numbers.* As in the systems previously described, we identify the set of elements (numbers), the operations on these elements, the properties that govern these operations on the elements, and an equivalence relation.

The Set of Elements

$Q = \{q_1, q_2, q_3, \ldots\}$, where $q_i = \frac{a}{b}$, $a, b \in I$, $i \in N$, $b \neq 0$†

Two Operations

For each $q_1 + q_2$ there is one and only one q_3 assigned.
For each $q_1 \times q_2$ there is one and only one q_4 assigned.

An Equivalence Relation (=)

Order Relations (<, >)

Commutative Properties

The operations of addition and multiplication are commutative:

$$q_1 + q_2 = q_2 + q_1; \quad q_1 \times q_2 = q_2 \times q_1.$$

Associative Properties

The operations of addition and multiplication are associative:

$$(q_1 + q_2) + q_3 = q_1 + (q_2 + q_3),$$
$$(q_1 \times q_2) \times q_3 = q_1 \times (q_2 \times q_3).$$

Distributive Property

Multiplication is distributive over addition:

$$q_1 \times (q_2 + q_3) = (q_1 \times q_2) + (q_1 \times q_3),$$
$$(q_2 + q_3) \times q_1 = (q_2 \times q_1) + (q_3 \times q_1).$$

Identity Elements

The set Q contains an identity element for addition (0) and an identity element for multiplication (1):

$$q_1 + 0 = 0 + q_1 = q_1 \qquad \text{and} \qquad q_1 \times 1 = 1 \times q_1 = q_1.$$

Inverse Elements

Additive inverse. For each q_1 there is an element q_2 such that

†Here the elements of Q are represented by the symbols q_1, q_2, etc., where the small numerals are called *subscripts;* q_i, with $i \in N$, is a way of representing any one of the elements. In the properties below, q_1, q_2, and q_3 represent any three elements.

$$q_1 + q_2 = 0.$$

Multiplicative inverse. For each q_1 except zero there is an element q_2 such that

$$q_1 \times q_2 = 1.$$

The above statements characterize the system of rational numbers. Further properties which belong to this system, as well as to the preceding systems, are listed below for review and ready reference. They are included without proof.

Trichotomy Property

For each q_1, q_2,

$$q_1 = q_2 \quad \text{or} \quad q_1 < q_2 \quad \text{or} \quad q_1 > q_2.$$

This permits Q to be ordered.

Uniqueness Properties for =

If $q_1 = q_2$, then $q_1 + q_3 = q_2 + q_3$.
If $q_1 = q_2$, then $q_1 \times q_3 = q_2 \times q_3$.

Cancellation Properties for =

If $q_1 + q_3 = q_2 + q_3$, then $q_1 = q_2$.
If $q_1 \times q_3 = q_2 \times q_3$, $q_3 \neq 0$, then $q_1 = q_2$.

Uniqueness Properties for <

If $q_1 < q_2$, then $q_1 + q_3 < q_2 + q_3$.
If $q_1 < q_2$, then

(i) $q_1 \times q_3 < q_2 \times q_3$ if $q_3 > 0$;
(ii) $q_1 \times q_3 > q_2 \times q_3$ if $q_3 < 0$.

Cancellation Properties for <

If $q_1 + q_3 < q_2 + q_3$, then $q_1 < q_2$.
If $q_1 \times q_3 < q_2 \times q_3$, then

(i) $q_1 < q_2$ if $q_3 > 0$;
(ii) $q_1 > q_2$ if $q_3 < 0$.

The existence of additive inverses and multiplicative inverses makes possible the operations of subtraction and division. The set Q is closed under subtraction ($-$) and, except for division by zero, under division ($\div$). A property of Q that is not a property of the preceding number systems is the following.

The Property of Denseness

Between any two distinct elements q_1, q_2, there is another element, q_3, such that

$$q_1 < q_3 < q_2 \quad \text{or} \quad q_1 > q_3 > q_2.$$

The system of rational numbers is called a *field*. The requirements which have been met, are:

A set of elements (Q)
Two operations ($+$, $\times$)
An equivalence relation ($=$)
Associative and commutative properties for $+$ and $\times$
Identity elements for $+$ and $\times$
An inverse for $\times$ for each element except 0
An inverse for $+$ for each element
Distributivity for $\times$ over $+$

Since the system of rational numbers has an order relation, it is called an *ordered field*. Review the systems of natural numbers, whole numbers, and integers, determining in each case which properties are lacking that prevent the system from being classified as a field.

The following is a diagram showing the system of rational numbers described as a field.

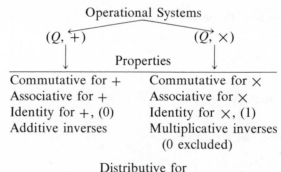

Operational Systems

$(Q, +)$	$(Q, \times)$
Properties ↓	↓
Commutative for $+$	Commutative for $\times$
Associative for $+$	Associative for $\times$
Identity for $+$, (0)	Identity for $\times$, (1)
Additive inverses	Multiplicative inverses (0 excluded)

Distributive for
$\times$ over $+$

This diagram should also assist you in comparing the operational systems of Q with others that have been studied. Notice that even though division is now an operation (except $\div$ by 0) it is not included as one of the operational systems. As you recall, it was defined in terms of multiplication.

Similarly, subtraction was defined in terms of addition in I. Hence, both are omitted.

Do you suppose that there are systems other than this one which can be called a field? Such systems may have a set of elements (S) other than numbers and operations (#, ⋇) other than (+, ×). Using these to describe a field we have the following: A set of elements (S), two operations (# and ⋇), and an equivalence relation. Both operations possess associativity, commutativity, and an identity element. For one operation, #, every element has an inverse. For the other operation, ⋇, every element has an inverse with one exception; that is, the identity for the # operation has no inverse with respect to the ⋇ operation. Also, the operation ⋇ is distributive over the operation #. With this description in mind, it will be possible to identify those systems that have the field properties.

With each number system evolving from the preceding one, you may ask if the rational number system is the last. It is not. In a later chapter you will find that there is a need for still other numbers. It will be interesting to see how they fit into our established scheme of developing new number systems.

10

Arithmetic of the Nonnegative Rational Numbers

10.1 INTRODUCTION

The study of the rational numbers in the elementary school usually consists of only the nonnegatives. At this level, the rationale for the multiplication, division, addition, and subtraction algorithms is not as rigorous as that presented in Chapter 9. Much of the procedure is based on models that give a concrete explanation of why the algorithm works. Hence, in this chapter we shall present some of these models and show how they are used in developing certain algorithms for the nonnegative rationals.

Nonnegative rationals written in decimal notation is another topic of elementary school mathematics. Since it is important that we understand why certain "rules" are used in the $\times$, $\div$, $+$, and $-$ algorithms, they shall be discussed in this chapter.

As the material in this chapter is studied, you should observe how it relates to the development given in Chapter 9.

10.2 MULTIPLICATION IN THE FORM $\frac{a}{b}$

An array constructed from a rectangular region is a very satisfactory model to use in developing the multiplication algorithm for nonnegative rational numbers. To construct an array from a rectangular region, we may proceed in the following manner:

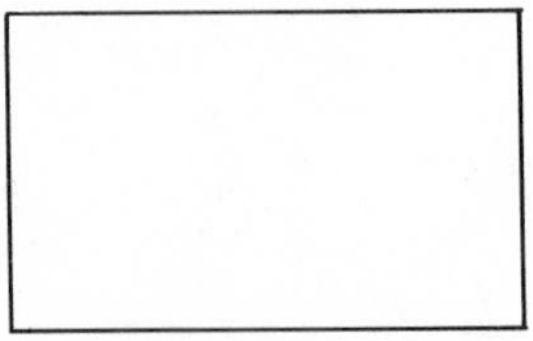

Rectangular Region

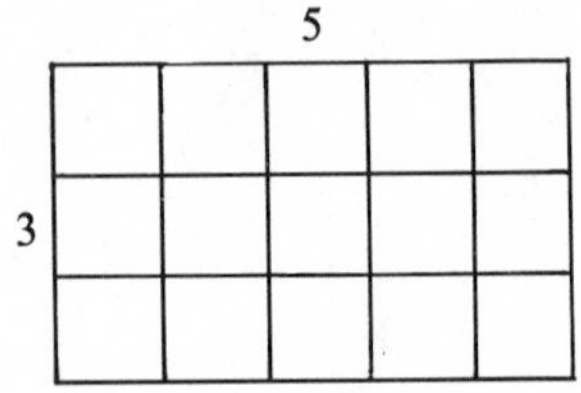

3×5 Array

As used in Chapter 7, a 3×5 array was used as a model for the multiplication $3 \times 5 = 15$.

The shaded subdivisions of the array shown next form another which is designated as a 2×4 array. Hence, $2 \times 4 = 8$.

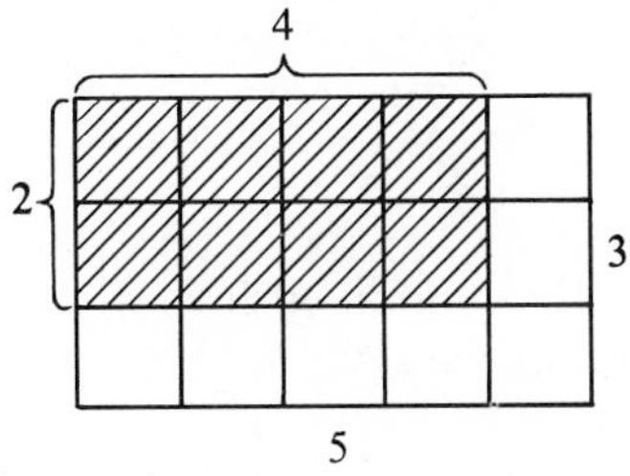

The shaded portion consists of 8 of the 15 in the total array. That is, the shaded portion is $\frac{8}{15}$ of the original array.

Now, we shall relate the number of rows and columns of the shaded array to those of the original array. The 2 rows is 2 of the 3 or $\frac{2}{3}$; the 4 columns is 4 of the 5 or $\frac{4}{5}$. So, in relation to the original array, we may describe the shaded portion as a $\frac{2}{3} \times \frac{4}{5}$ array. Do you see how we conclude from this discussion that

$$\frac{2}{3} \times \frac{4}{5} = \frac{2 \times 4}{3 \times 5}$$
$$= \frac{8}{15}?$$

From this we make the following generalization concerning the multiplication of nonnegative rationals in the form $\frac{a}{b}$:

Generalization 10.2

For each a, b, c, d nonnegative integers, $bd \neq 0$,

$$\frac{a}{b} \times \frac{c}{d} = \frac{a \times c}{b \times d}.$$

Is this the same as Theorem I given in Chapter 9? Cite any difference that you find.

Consider the next examples as the concept of the identity element is developed.

Example 1 Show that $1 \times \frac{2}{3} = \frac{2}{3} \times 1 = \frac{2}{3}$.

We begin with a 1×1 region called the unit region. It is subdivided as shown.

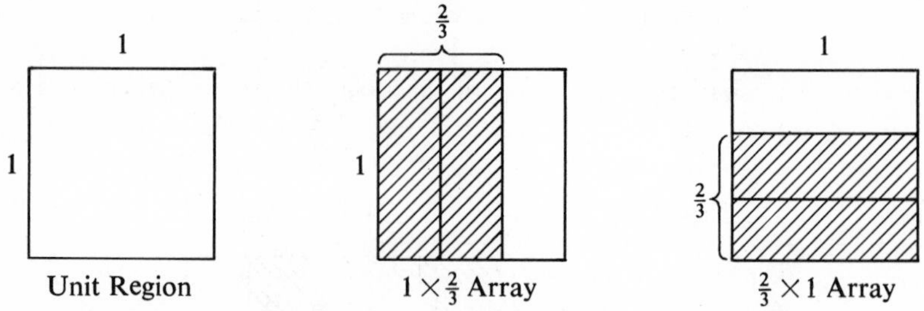

Since the shaded portion in each consists of 2 of the 3 subdivisions of the original unit region, we say,

$$1 \times \frac{2}{3} = \frac{2}{3} \times 1 = \frac{2}{3}.$$

Hence, 1 is the identity element.

Example 2 Show that $\frac{2}{2}$ and $\frac{3}{3}$ are names for the multiplicative identity 1 by using an array.

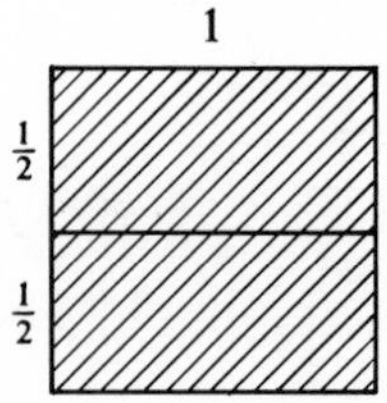

The unit region has been subdivided into 2 congruent regions. These subdivisions may be described as 2 of the 2 in the unit region, or as $\frac{2}{2}$. Hence, the shaded portion forms a $\frac{2}{2} \times 1$ array. But, another name for this array is 1×1. Thus, $\frac{2}{2}$ is obviously another name for 1, the multiplicative identity.

Use this diagram to explain that $\frac{3}{3}$ is still another name for 1.

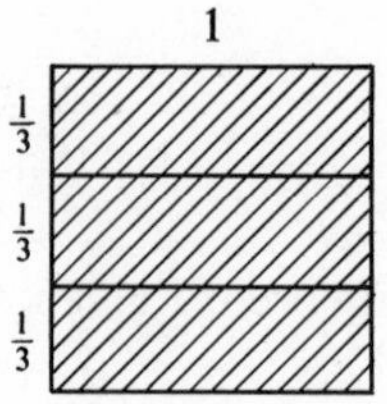

From this example, we are prepared to say that $\frac{5}{5}$, $\frac{9}{9}$, $\frac{12}{12}$, and $\frac{c}{c}$, $c \neq 0$, are names for the multiplicative identity 1.

The next example shows the construction of an array that is a model for the multiplication of two particular nonnegative rational numbers.

Example 3 Use an array to find the product of $\frac{3}{4}$ and $\frac{2}{5}$.

A $\frac{3}{4} \times \frac{2}{5}$ array is needed. This is a case in which it is more practical to begin with a rectangular region rather than the unit region. Subdivide the region into 4 congruent regions. Shade 3 of them as shown with ////. Next, subdivide the region into 5 congruent regions. Shade 2 of them as shown with \\\\. Observe that the two sets of congruent subdivisions overlap. That portion marked with cross hatches represents the $\frac{3}{4} \times \frac{2}{5}$ array.

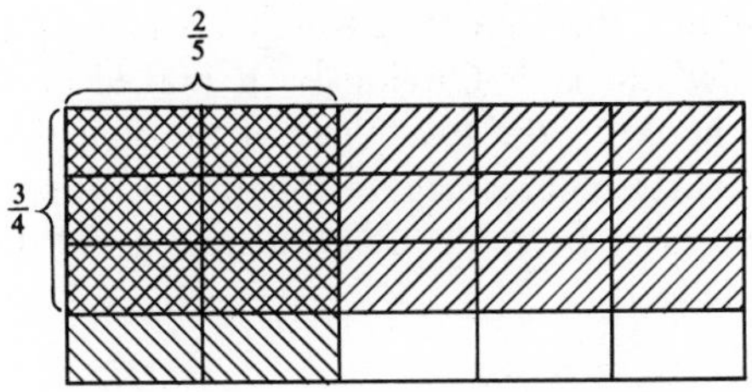

From Generalization 10.2 and the array, we have:

$$\frac{3}{4} \times \frac{2}{5} = \frac{3 \times 2}{4 \times 5}$$
$$= \frac{6}{20}$$
$$= \frac{3 \times 2}{10 \times 2}$$
$$= \frac{3}{10} \times \frac{2}{2}$$

Hence,

$$\frac{3}{4} \times \frac{2}{5} = \frac{3}{10}$$

Justification for the steps in the preceding may be found in Section 9.2.

Another model for multiplication of rational numbers is the *stretcher–shrinker machine*. The machine shown here will be used to multiply non-negative rationals. First, the part on the left does the "stretching." The section hooked up to it then takes over and does the "shrinking." Each section must be programmed as to the amount of stretch and shrink that is to take place, as shown in the examples that follow.

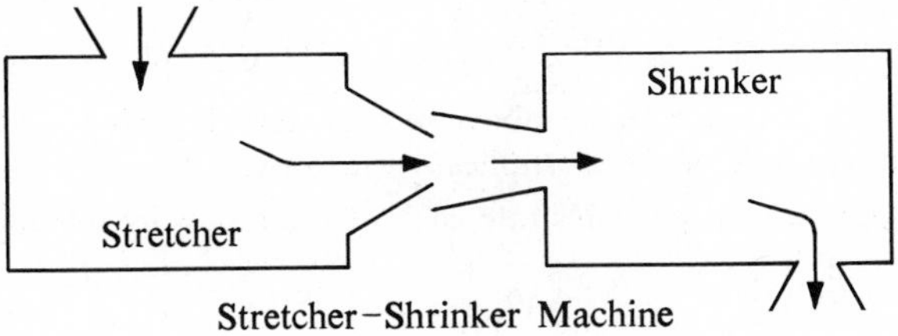

Stretcher–Shrinker Machine

Example 4 The number three is sent through a machine that stretches by multiplying by 2, then shrinks by multiplying by $\frac{1}{6}$. What number comes out?

When 3 is dropped into the stretcher part, it is multiplied by 2. The 6 then goes into the shrinker part where it is multiplied by $\frac{1}{6}$. The number that comes out is 1. Is this the same number you would get by multiplying 3 by $\frac{2}{6}$?

Now interchange the shrinker and the stretcher in the above machine. Send 3 through the shrinker first. It would shrink to $\frac{3}{6}$ when multiplied by $\frac{1}{6}$. Next, send $\frac{3}{6}$ through the stretcher which multiplies by 2. The number which comes out is 1. Obviously, it makes no difference which sequence is used to hook up the two parts of the machine.

From this stretcher–shrinker machine it appears that to multiply 3 by $\frac{2}{6}$, one "stretches" 3 by multiplying it by 2 and "shrinks" the result by multiplying it by $\frac{1}{6}$.

Example 5 Program the machine to stretch by multiplying by 5 and shrink by multiplying by $\frac{1}{7}$. What is the *output* if the *input* is $\frac{2}{3}$?

As $\frac{2}{3}$ goes through the stretching machine, it is made 5 times as large. The shrinking machine takes the $\frac{10}{3}$ and multiplies it by $\frac{1}{7}$; $\frac{10}{21}$ drops out. Do you see that this is the same as multiplying $\frac{2}{3}$ by $\frac{5}{7}$? The input was $\frac{2}{3}$, the machine performed a stretching then a shrinking, and the output was $\frac{10}{21}$. Would the output be the same if the stretcher and shrinker were interchanged?

EXERCISE 10.2

1. Make an appropriate array to be used in finding the product of each.

   (a) $\frac{2}{3} \times 1$ (c) $\frac{2}{3} \times \frac{1}{4}$ (e) $\frac{2}{5} \times \frac{3}{4}$

   (b) $1 \times \frac{3}{4}$ (d) $\frac{1}{5} \times \frac{2}{3}$ (f) $\frac{2}{6} \times \frac{2}{3}$

2. Illustrate how the stretcher-shrinker machine may be used to find the product of each of the following. Explain the programming necessary for each machine.

   (a) $5 \times \frac{2}{3}$ (b) $4 \times \frac{8}{9}$ (c) $\frac{3}{7} \times \frac{2}{5}$

3. Use Generalization 10.2 to find the product of each.

   (a) $\frac{2}{7} \times \frac{3}{3}$ (c) $\frac{2}{3} \times \frac{1}{4}$ (e) $\frac{2}{6} \times \frac{2}{3}$

   (b) $\frac{3}{8} \times \frac{5}{9}$ (d) $\frac{3}{7} \times \frac{2}{15}$ (f) $\frac{5}{9} \times \frac{6}{10}$

4. Show how each of the above answers may be expressed in irreducible form by making use of the multiplicative identity.

5. Draw an array to illustrate each of the following sentences.

   (a) $\frac{3}{2} \times \frac{2}{2}$ (b) $\frac{3}{2} \times \frac{1}{2}$ (c) $\frac{3}{2} \times \frac{3}{4}$

10.3 DIVISION IN THE FORM $\frac{a}{b}$

The multiplicative inverse and the multiplicative identity will have an important role in this discussion of division of nonnegative rational numbers. Refer to Chapter 9, Definition 9.1(a) and Theorem II, for multiplicative inverse and multiplicative identity respectively.

Observe the pattern found in the following exploration:

$$\begin{array}{lcl} 6 \div 3 = 2 & \text{and} & 6 \times \frac{1}{3} = 2 \\ 18 \div 2 = 9 & \text{and} & 18 \times \frac{1}{2} = 9 \\ 10 \div 5 = 2 & \text{and} & 10 \times \frac{1}{5} = 2 \\ 24 \div 6 = 4 & \text{and} & 24 \times \frac{1}{6} = 4 \end{array}$$

You should know that the pairs $(3, \frac{1}{3})$, $(2, \frac{1}{2})$, $(5, \frac{1}{5})$, and $(6, \frac{1}{6})$ are multiplicative inverses.

From this exploration and Theorem V(b) in Chapter 9, we make the following generalization concerning the division of nonnegative rationals.

Generalization 10.3

For each nonnegative rational number $\frac{a}{b}$ and $\frac{c}{d}$, $bdc \neq 0$,

$$\frac{a}{b} \div \frac{c}{d} = \frac{a}{b} \times \frac{d}{c}.$$

Example 1 Find the quotient of $\frac{7}{9} \div \frac{5}{8}$.

Since $\frac{5}{8} \times \frac{8}{5} = 1$, the multiplicative inverse of $\frac{5}{8}$ is $\frac{8}{5}$. Using Generalization 10.3, we write,

$$\begin{aligned} \frac{7}{9} \div \frac{5}{8} &= \frac{7}{9} \times \frac{8}{5} \\ &= \frac{56}{45}. \end{aligned}$$

As was shown in Chapter 9, $\frac{3}{5} \div \frac{7}{8}$ may be written $\dfrac{\frac{3}{5}}{\frac{7}{8}}$. We shall also interpret $(\frac{2}{3} + \frac{1}{2}) \div (\frac{1}{6} + \frac{1}{3})$ as $\dfrac{\frac{2}{3} + \frac{1}{2}}{\frac{1}{6} + \frac{1}{3}}$. The following examples illustrate how these may be expressed as nonnegative rational numbers, using the multiplicative identity.

Example 2 Use the multiplicative identity to find a simpler name for

$$\frac{\frac{3}{5}}{\frac{7}{8}}.$$

Multiplying by the multiplicative identity $\frac{40}{40}$, we have

$$\frac{\frac{3}{5}}{\frac{7}{8}} \times \frac{40}{40} = \frac{24}{35}.$$

Do you see the reason for selecting the name $\frac{40}{40}$ for the multiplicative identity?

Example 3 Show how the multiplicative identity may be used to find a simpler name for

$$\frac{\frac{2}{3} + \frac{1}{2}}{\frac{1}{6} + \frac{1}{3}}.$$

Do you agree that $\frac{6}{6}$ would be an appropriate name for the multiplicative identity?

$$\begin{aligned} \frac{\frac{2}{3} + \frac{1}{2}}{\frac{1}{6} + \frac{1}{3}} \times \frac{6}{6} &= \frac{(\frac{2}{3} + \frac{1}{2}) \times 6}{(\frac{1}{6} + \frac{1}{2}) \times 6} \\ &= \frac{(\frac{2}{3} \times 6) + (\frac{1}{2} \times 6)}{(\frac{1}{6} \times 6) + (\frac{1}{3} \times 6)} \\ &= \frac{4 + 3}{1 + 2} \\ &= \frac{7}{3} \end{aligned}$$

You should be able to justify each of the steps in the above algorithm.

You have probably discovered from the preceding examples that the choice of a name for the multiplicative identity, $\frac{c}{c}$, is made by making c the least common multiple of the denominators.

EXERCISE 10.3

1. Use Generalization 10.3 in finding the quotient of each. Identify the multiplicative inverse used in each problem.

   (a) $\frac{7}{9} \div \frac{5}{4} =$ (d) $\frac{2}{3} \div \frac{1}{4} =$ (g) $\frac{1}{4} \div \frac{2}{3} =$

   (b) $\frac{4}{3} \div \frac{7}{2} =$ (e) $\frac{2}{7} \div 3 =$ (h) $\frac{8}{9} \div 11 =$

   (c) $\frac{5}{3} \div \frac{8}{7} =$ (f) $3 \div \frac{2}{7} =$ (i) $11 \div \frac{8}{9} =$

2. Use the multiplicative identity in finding a simpler name for the following. The numerator and denominator of the answer should be relatively prime.

   (a) $\dfrac{\frac{5}{12}}{\frac{2}{3}}$ (c) $\dfrac{\frac{2}{3} + \frac{1}{2}}{\frac{11}{6} + 2}$ (e) $\dfrac{\frac{2}{3} + \frac{1}{4}}{\frac{11}{12} + \frac{2}{3}}$ (g) $\dfrac{5 + \frac{1}{4}}{3 + \frac{1}{3}}$

   (b) $\dfrac{\frac{3}{5}}{\frac{4}{3}}$ (d) $\dfrac{\frac{7}{8} + \frac{1}{5}}{\frac{1}{10} + 6}$ (f) $\dfrac{\frac{1}{3} + \frac{1}{2}}{\frac{2}{6} + \frac{5}{10}}$ (h) $\dfrac{2 + \frac{1}{5}}{\frac{1}{3} + \frac{1}{20}}$

10.4 ADDITION AND SUBTRACTION IN THE FORM $\frac{a}{b}$

An appropriate model for the addition and subtraction of nonnegative rational numbers is that portion of the number line representing only those numbers equal to or greater than 0. See Section 9.9. The following examples are used to discover generalizations for adding and subtracting nonnegative rationals when written in the form $\frac{a}{b}$.

Example 1 Find the sum of $\frac{3}{8}$ and $\frac{4}{8}$.

Line segments going in the same direction with measures of $\frac{3}{8}$ and $\frac{4}{8}$ have been placed end to end along the number line.

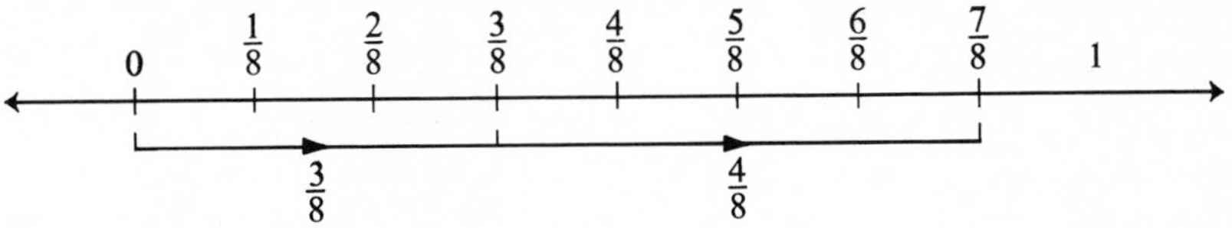

With one end of the first line segment on 0 and the terminal end of the second one on $\frac{7}{8}$ of the number line, we say:

$$\frac{3}{8} + \frac{4}{8} = \frac{7}{8}.$$

That is:

$$\frac{3}{8} + \frac{4}{8} = \frac{3 + 4}{8}.$$

This example verifies Theorem VI, Chapter 9, which we choose to refer to in this chapter as a generalization for adding nonnegative rational numbers.

Generalization 10.4(a)

For each of the nonnegative rational numbers $\frac{a}{c}$ and $\frac{b}{c}$, $c \neq 0$,

$$\frac{a}{c} + \frac{b}{c} = \frac{a + b}{c}.$$

With subtraction defined in terms of addition, the number line also serves as a model for developing a generalization for the subtraction of nonnegative rationals. Directed line segments will be used as in Example 1 with the agreement that the direction for nonnegative rationals is always to go to the right.

Example 2 Find the solution set for $\frac{5}{8} - \frac{2}{8} = n$.

Applying Definition 9.7,

$$\text{if } \frac{2}{8} + n = \frac{5}{8}, \text{ then } \frac{5}{8} - \frac{2}{8} = n.$$

To find n, we ask, "What must be added to $\frac{2}{8}$ to equal $\frac{5}{8}$?"

From the number line, we are able to determine n by counting the one-eighths.

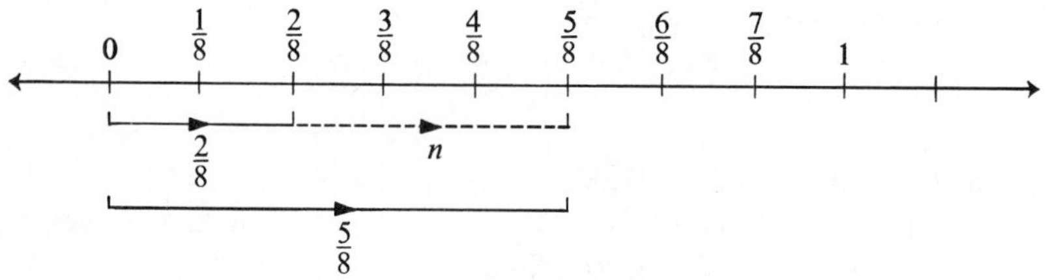

We find n to be $\frac{3}{8}$, so

$$\frac{5}{8} - \frac{2}{8} = \frac{3}{8}.$$

But,

$$\text{since } \frac{3}{8} = \frac{5-2}{8}, \quad \text{then } \frac{5}{8} - \frac{2}{8} = \frac{5-2}{8}.$$

Example 3 Find the solution set of $\frac{1}{4} - \frac{3}{4} = n$.

The definition for subtraction demands that we find a number which when added to $\frac{3}{4}$ equals $\frac{1}{4}$. The number line is used as in Example 2. An appropriate directed line segment is shown for $\frac{3}{4}$. Is it possible to find a line segment

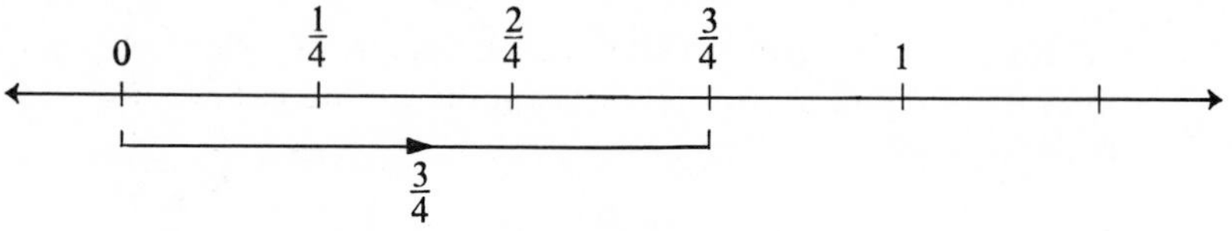

directed to the right which may be joined to the end of the one shown, and will have its terminal end point at $\frac{1}{4}$? Obviously not. Hence, we conclude that there is no number in the set of nonnegative rationals which when added to $\frac{3}{4}$ is equal to $\frac{1}{4}$. The solution set of $\frac{1}{4} - \frac{3}{4} = n$ is the empty set. In what set of numbers will there be a nonempty solution set? See Section 9.7.

This leads to a generalization for the subtraction of nonnegative rational numbers with the necessary restriction.

Generalization 10.4(b)

For nonnegative rational numbers $\frac{a}{c}$ and $\frac{b}{c}$, $c \neq 0$, $\frac{a}{c} > \frac{b}{c}$,

$$\frac{a}{c} - \frac{b}{c} = \frac{a-b}{c}.$$

In the preceding examples of this section, the denominators of each pair of fractions were the same. Is it possible to add and subtract numbers having different denominators by using Generalizations 10.4(a) and 10.4(b)?

Example 4 Find the sum of $\frac{2}{3}$ and $\frac{1}{2}$.

Since the denominators are not the same, it appears that Generalization 10.4(a) is not applicable. However, $\frac{2}{3}$ and $\frac{1}{2}$ each have many different names. Is it possible to find names for both which have the same denominators?

Using the multiplicative identity, we are able to find a set of equivalent fractions for each.

$$\frac{2}{3}\colon \left\{\frac{2}{3}, \frac{2}{3}\times\frac{2}{2}, \frac{2}{3}\times\frac{3}{3}, \frac{2}{3}\times\frac{4}{4}, \ldots\right\}$$

$$\left\{\frac{2}{3}, \frac{4}{6}, \frac{6}{9}, \frac{8}{12}, \ldots\right\}$$

$$\frac{1}{2}\colon \left\{\frac{1}{2}, \frac{1}{2}\times\frac{2}{2}, \frac{1}{2}\times\frac{3}{3}, \frac{1}{2}\times\frac{4}{4}, \ldots\right\}$$

$$\left\{\frac{1}{2}, \frac{2}{4}, \frac{3}{6}, \frac{4}{8}, \ldots\right\}$$

The elements in the two sets having the same, but least, denominator are $\frac{4}{6}$ and $\frac{3}{6}$, names for $\frac{2}{3}$ and $\frac{1}{2}$, respectively.

Our example now becomes:

$$\begin{aligned}\frac{2}{3}+\frac{1}{2} &= \left(\frac{2}{3}\times\frac{2}{2}\right)+\left(\frac{1}{2}\times\frac{3}{3}\right)\\ &= \frac{4}{6} + \frac{3}{6}\\ &= \frac{4+3}{6} \qquad \textit{Generalization } 10.4(\text{a})\\ &= \frac{7}{6}\end{aligned}$$

It is possible to find the denominator 6 by finding the least common multiple of the denominators, 3 and 2. See Section 7.5. Since our discussion involves denominators, we shall call 6 the *least common denominator* of $\frac{2}{3}$ and $\frac{1}{2}$. The least common denominator determines the selection of appropriate names for the multiplicative identity used in converting $\frac{2}{3}$ and $\frac{1}{2}$ to fractions with the denominator 6, as

$$\frac{2}{3}\times\frac{2}{2}=\frac{4}{6} \quad\text{and}\quad \frac{1}{2}\times\frac{3}{3}=\frac{3}{6}.$$

By applying the same procedure, Generalization 10.4(b) may be used to find $\frac{2}{3}-\frac{1}{2}$.

Example 5 Find the sum of $\frac{5}{6}$ and $\frac{2}{15}$.

Another procedure for finding the least common denominator is given in Section 7.5. The prime factors of 6 and 15 are found.

$$\begin{array}{ccc} 6 & \qquad & 15 \\ \swarrow \ \searrow & & \swarrow \ \searrow \\ 3 \times 2 & & 3 \times 5 \end{array}$$

Since the denominator must be a multiple of both 6 and 15, yet it must be the least one, the factors are selected as shown.

$$\begin{array}{c} 6 \\ \swarrow \ \searrow \\ 2 \times 3 \times 5 = 30 \\ \nwarrow \ \nearrow \\ 15 \end{array}$$

The diagram shows that 30 is the least multiple of both 6 and 15. Hence, 30 is the least common denominator of the fractions $\frac{5}{6}$ and $\frac{2}{15}$. See Section 9.6.

The least common denominator having been determined, we are now prepared to select names for the multiplicative identity that may be used to convert $\frac{5}{6}$ and $\frac{2}{15}$ to fractions having 30 as the denominator. Since

$$\frac{5}{6} \times \frac{5}{5} = \frac{25}{30} \quad \text{and} \quad \frac{2}{15} \times \frac{2}{2} = \frac{4}{30},$$

the names for the multiplicative identity are $\frac{5}{5}$ and $\frac{2}{2}$. Now, we are prepared to find the sum of $\frac{5}{6}$ and $\frac{2}{15}$ using Generalization 10.4(a). You should be able to complete this problem as was done in the preceding example.

EXERCISE 10.4

1. Use Generalizations 10.4(a) and 10.4(b) to give a simpler name for each when possible. For those not possible explain why.

   (a) $\frac{3}{5} + \frac{1}{5}$ (c) $\frac{4}{9} + \frac{5}{9}$ (e) $\frac{12}{23} - \frac{8}{23}$

   (b) $\frac{7}{13} + \frac{2}{13}$ (d) $\frac{7}{8} - \frac{2}{8}$ (f) $\frac{2}{11} - \frac{5}{11}$

2. Find the least common denominator and give the names for the multiplicative identity which must be used to convert each fraction to one having the lcd as denominator.

   (a) $\frac{7}{8} + \frac{9}{5}$ (c) $\frac{3}{10} + \frac{4}{25}$ (e) $\frac{1}{11} + \frac{1}{3}$

   (b) $\frac{5}{12} + \frac{7}{16}$ (d) $\frac{5}{6} + \frac{3}{7}$ (f) $8 + \frac{2}{3}$

3. Express each as a nonnegative rational number.

   (a) $\frac{2}{15} + \frac{4}{3}$ (c) $\frac{6}{14} + \frac{4}{7}$

   (b) $\frac{5}{19} + \frac{2}{3}$ (d) $\frac{12}{25} + \frac{7}{15}$

4. Find the nonnegative rational number named by each of the following if there is one.

(a) $\left(\frac{1}{2} + \frac{1}{3}\right) + \frac{1}{5}$ (f) $\frac{0}{7} + \frac{3}{8}$

(b) $\frac{1}{2} + \left(\frac{1}{3} + \frac{1}{5}\right)$ (g) $4 - \left(\frac{2}{3} - \frac{5}{6}\right)$

(c) $\left(\frac{1}{2} - \frac{1}{3}\right) - \frac{1}{5}$ (h) $10 + \frac{3}{4}$

(d) $\frac{1}{2} - \left(\frac{1}{3} - \frac{1}{5}\right)$ (i) $\frac{4}{5} - 2$

(e) $\frac{0}{7} - \frac{3}{8}$ (j) $\frac{4}{3} + \left(\frac{7}{12} - \frac{4}{21}\right)$

10.5 NONNEGATIVE RATIONAL NUMBERS IN DECIMAL NOTATION

The expanded form for writing numerals in several different bases was discussed in Chapter 3. Each of these related to a particular place value chart. Base ten related to a decimal place value chart, shown in Section 3.11. A portion of it is shown here.

$10 \times 10 \times 10$	10×10	10	1

Observe that each successive position going from right to left is 10 times the preceding one. And each successive position going from left to right is $\frac{1}{10}$ the preceding one. Now, suppose that we wish to extend the chart by taking $\frac{1}{10}$ of the last position, 1, and continue to take $\frac{1}{10}$ of each successive position going to the right. In so doing, we extend the preceding place value chart. The following diagram shows the extended chart. Each of the rows gives different forms for naming the positions. Observe the use of negative exponents in the last row. A discussion of these and the 0 exponent follows.

PLACE VALUE CHART

Thousand								*Thousandth*		
Hundred	*Ten*	*One*	*Hundred*	*Ten*	*One*	*One Tenth*	*One Hundredth*	*One*	*Ten-*	*Hundred-*
100,000	10,000	1,000	100	10	1	0.1	0.01	0.001	0.0001	0.00001
. . .	. . .	$10 \times 10 \times 10$	10×10	10	1	$\frac{1}{10}$	$\frac{1}{10 \times 10}$	$\frac{1}{10 \times 10 \times 10}$	$\frac{1}{\ldots}$	$\frac{1}{\ldots}$
10^5	10^4	10^3	10^2	10^1	10^0	$\frac{1}{10^1}$	$\frac{1}{10^2}$	$\frac{1}{10^3}$	$\frac{1}{10^4}$	$\frac{1}{10^5}$
10^5	10^4	10^3	10^2	10^1	10^0	10^{-1}	10^{-2}	10^{-3}	10^{-4}	10^{-5}

In Section 3.8 we introduced the symbol called the *exponent.* We shall now justify the definition of 10^0 (or any natural number with the exponent 0) and the definition of negative exponents.

We begin with

$$10^2 = 10 \times 10 \quad \text{and} \quad 10^3 = 10 \times 10 \times 10,$$

and find $10^2 \times 10^3$:

$$\begin{aligned} 10^2 \times 10^3 &= (10 \times 10) \times (10 \times 10 \times 10) \\ &= 10^5 \\ &= 10^{2+3}. \end{aligned}$$

This reminds us of the familiar generalization used in algebra:

$$a^n \times a^m = a^{n+m}, \qquad a \in N, \qquad n, m \in I.$$

Using this generalization and the multiplicative identity, we shall demonstrate the reason for choosing 10^0 as another name for 1:

$$10^2 \times 1 = 10^2 \qquad \textit{Multiplicative identity}$$

and

$$\begin{aligned} 10^2 \times 10^0 &= 10^{2+0} && a^n \times a^m = a^{n+m}, \quad a \in N, \quad n, m \in I \\ &= 10^2 && \textit{Additive identity} \end{aligned}$$

Thus, 10^0 and 1 must name the same number by the uniqueness of multiplication.

Now let us investigate 10^{-1}:

$$10^1 \times \frac{1}{10^1} = 1 \qquad \textit{Multiplicative inverse}$$

and

$$\begin{aligned} 10^1 \times 10^{-1} &= 10^{1+-1} && a^n \times a^m = a^{n+m}, \quad a \in N, \quad n, m \in I \\ &= 10^0 && \textit{Additive identity} \\ &= 1 \end{aligned}$$

Thus, 10^{-1} and $\frac{1}{10^1}$ must name the same number by the uniqueness of multiplication.

Do not mistake the preceding discussion for proof. The discussion merely points out that we have been consistent in using 10^0 as 1 and 10^{-1} as $\frac{1}{10^1}$. Hence, we give the following definitions.

Definition 10.5(a) For each $p \in Q, p \neq 0$,

$$p^0 = 1.$$

Definition 10.5(b) For each $p \in Q, p \neq 0, n \in I, n > 0$,

$$p^{-n} = \frac{1}{p^n}.$$

Notice that in Definition 10.5(b), $n > 0$. It is beyond the scope of this chapter to consider n other than a positive integer.

Now, let us use the extended place value table shown here to discuss the representation of rational numbers.

	10^2	10^1	10^0	$\frac{1}{10^1}$	$\frac{1}{10^2}$	$\frac{1}{10^3}$
(a)				5	6	
(b)	1	2	3	7		

Suppose that digits are placed in the place value chart as shown in (a). Does this represent a rational number? If so, how is it named? Using expanded notation we write:

$$\left(5 \times \frac{1}{10}\right) + \left(6 \times \frac{1}{10^2}\right) = \frac{5}{10} + \frac{6}{100}$$
$$= \frac{50}{100} + \frac{6}{100}$$
$$= \frac{56}{100}$$

Since $\frac{56}{100}$ is a numeral in the form $\frac{a}{b}$, $a, b \in I$, $b \neq 0$, it is a name for a rational number. Recall that a numeral in this form is called a fraction (Section 9.3).

Another name for (a) in the table is given by

$$0.56.$$

When the name is given in this form, it is said to be in *decimal notation,* while the $\frac{a}{b}$ form is called the *fraction form.* The former is also frequently referred to as the *decimal fraction form.* Attention is called to the symbol ".", the *decimal point.* This dot separates the 10^0 (or 1) and the $\frac{1}{10}$ positions of the digits in the numeral. The expanded notation used in the preceding paragraph directs us in reading the numeral 0.56. Since $\frac{5}{10} + \frac{6}{100}$ is $\frac{56}{100}$, we read 0.56 as "fifty-six hundredths."

We shall now investigate the meaning of the digits in (b) of the table. Using expanded notation, we write:

$$(1 \times 10^2) + (2 \times 10^1) + (3 \times 10^0) + \left(7 \times \frac{1}{10}\right)$$
$$= 100 + 20 + 3 + \frac{7}{10}$$
$$= 123 + \frac{7}{10}$$

The numeral $123 + \frac{7}{10}$ is formed by two addends. The addend 123 is called the *integral part,* and $\frac{7}{10}$ is called the *fractional part* of the numeral. We shall adopt the convention of writing this as $123\frac{7}{10}$, omitting the + symbol. We may refer to a numeral in this form as being in the *mixed form.* This numeral is read "one hundred twenty-three *and* seven tenths."

The question now is: "Does $123\frac{7}{10}$ represent a rational number?" Since $123\frac{7}{10}$ is the same as $123 + \frac{7}{10}$, we have:

$$\begin{aligned} 123\frac{7}{10} &= 123 + \frac{7}{10} \\ &= \left(\frac{123}{1} \times \frac{10}{10}\right) + \frac{7}{10} \\ &= \frac{1230}{10} + \frac{7}{10} \\ &= \frac{1237}{10} \end{aligned}$$

Since $\frac{1237}{10}$ is a numeral in the form $\frac{a}{b}$, $a, b \in W$, $b \neq 0$, we have shown that $123\frac{7}{10}$ represents a rational number.

Another name for the rational number $123\frac{7}{10}$ is

$$123.7,$$

with the decimal point separating the integral and fractional parts of the numeral. Again, the expanded notation form helps in reading this numeral. Since

$$123.7 = 100 + 20 + 3 + \frac{7}{10},$$

the numeral is read "one hundred twenty-three and seven tenths." Notice that there is no difference in reading the two numerals, $123\frac{7}{10}$ and 123.7.

Decimal notation is easily written from expanded notation. Consider the following examples.

Example 1 Express in decimal notation

$$\left(5 \times \frac{1}{10}\right) + \left(3 \times \frac{1}{10^2}\right) + \left(2 \times \frac{1}{10^3}\right).$$

Using the place value chart, each of the numbers 5, 3, and 2 is placed in the proper position.

$\frac{1}{10}$	$\frac{1}{10^2}$	$\frac{1}{10^3}$
5	3	2

Reading from the chart, the expanded form in decimal notation is .532.

Example 2 Express the expanded form

$$(3 \times 1) + (2 \times 10^{-1}) + (7 \times 10^{-2})$$

in decimal notation.

Placing the 3, 2, and 7 in the proper positions in the place value chart, we have:

1	10^{-1}	10^{-2}
3	2	7

The expanded form in decimal notation is 3.27.

EXERCISE 10.5

1. Write each of the following in decimal notation.
   (a) $8\left(\frac{1}{10}\right) + 0\left(\frac{1}{10^2}\right) + 5\left(\frac{1}{10^3}\right)$
   (b) $2(10^3) + 0(10^2) + 1\left(\frac{1}{10^1}\right) + 2\left(\frac{1}{10^2}\right)$
   (c) $3\left(\frac{1}{10^2}\right) + 4\left(\frac{1}{10^3}\right) + 3\left(\frac{1}{10^4}\right)$
   (d) $4(10^1) + 6(10^0) + 8(10^{-1}) + 5(10^{-2})$
   (e) $7(10^2) + 9(10^1) + 0(10^0) + 1(10^{-1}) + 2(10^{-2})$
   (f) $5(10^{-1}) + 2(10^{-2}) + 6(10^{-3})$
2. Write each of the following in expanded notation:
   (a) 2.71 (b) 0.0207 (c) 0.10201 (d) 40.005
3. Write in decimal notation:
   (a) Six hundred ninety-one thousandths
   (b) Six hundred and ninety-one thousandths
   (c) Nine hundred seven and sixty-two ten-thousandths
   (d) Seventy and seven hundredths
4. Give another name for each of the following.
   (a) 5^0 (d) 7^{-2} (g) 5×1^0 (i) 5×1^{-1}
   (b) 8^{-1} (e) $5^1 \times 5^2$ (h) 5×1^1 (j) $2^{-2} \times 2^{-3}$
   (c) 12^0 (f) $2^2 \times 2^3$
5. Change each of the following mixed forms to the fraction form. (HINT: $5\frac{2}{9} = 5 + \frac{2}{9}$. Now find the sum.)
   (a) $3\frac{2}{5}$ (c) $106\frac{3}{17}$ (e) $145\frac{1}{5}$
   (b) $17\frac{2}{9}$ (d) $234\frac{5}{7}$ (f) $1682\frac{2}{3}$

10.6 ADDITION AND MULTIPLICATION ALGORITHMS FOR NONNEGATIVE RATIONAL NUMBERS WRITTEN IN DECIMAL NOTATION

The algorithms for whole numbers have been discussed in Section 7.6. A review of the addition and multiplication algorithms in that section will remind you of the role of the properties as each algorithm was developed.

The development of the addition algorithm for nonnegative rationals written in decimal notation is shown in Examples 1, 2, and 3. You should be able to justify each step in the development by citing the appropriate property.

Example 1 Find the sum of 23.6 and 15.12.

Express each addend in expanded notation.

$$23.6 = (2 \times 10) + (3 \times 1) + \left(6 \times \frac{1}{10}\right)$$

$$15.12 = (1 \times 10) + (5 \times 1) + \left(1 \times \frac{1}{10}\right) + \left(2 \times \frac{1}{100}\right)$$

$$\text{Sum} = (2 + 1) \times 10 + (3 + 5) \times 1 + (6 + 1) \times \frac{1}{10} + \left(2 \times \frac{1}{100}\right)$$

$$= (3 \times 10) + (8 \times 1) + \left(7 \times \frac{1}{10}\right) + \left(2 \times \frac{1}{100}\right)$$

(i) $$= 30 + 8 + \frac{7}{10} + \frac{2}{100}$$

(ii) $$= 38.72$$

The partial sums given in line (i) may be arranged in columns as shown in the following algorithm. The shortened form needs no explanation. Observe the location of line (ii) in each of the algorithms.

$$\begin{array}{r} 23.6 \\ \underline{15.12} \\ 30. \\ 8. \\ 0.7 \\ \underline{0.02} \\ 38.72 \end{array} \qquad \text{or} \qquad \begin{array}{r} 23.6 \\ \underline{15.12} \\ 38.72 \end{array}$$

Example 2 Find the sum of 0.09 and 0.02.†

Writing each addend in expanded notation,

†It is considered good practice to use the zero in the one's place when writing a number less than one in decimal notation.

$$
\begin{aligned}
0.09 + 0.02 &= \left(9 \times \frac{1}{100}\right) + \left(2 \times \frac{1}{100}\right) \\
&= \frac{9}{100} + \frac{2}{100} \\
&= \frac{9+2}{100} \\
&= \frac{11}{100}.
\end{aligned}
$$

Renaming 11 by decimal numeration, we have:

$$
\begin{aligned}
&= \frac{10+1}{100} \\
&= \frac{10}{100} + \frac{1}{100} \\
&= \frac{1}{10} + \frac{1}{100} \\
&= 0.11
\end{aligned}
$$

Do you see that this justifies the algorithm shown?

$$
\begin{array}{r}
0.09 \\
\underline{0.02} \\
0.01 \\
\underline{0.1} \\
0.11
\end{array}
$$

Example 3 Find the sum of 3.8 and 0.734.

This development is quite long, but since it justifies "carrying," it should be carefully studied.

$$
\begin{aligned}
3.8 &= (3 \times 1) + \left(8 \times \frac{1}{10}\right) \\
0.734 &= (0 \times 1) + \left(7 \times \frac{1}{10}\right) + \left(3 \times \frac{1}{100}\right) + \left(4 \times \frac{1}{1000}\right) \\
\hline
\text{Sum} &= (3+0) \times 1 + (8+7) \times \frac{1}{10} + \left(3 \times \frac{1}{100}\right) + \left(4 \times \frac{1}{1000}\right) \\
&= (3 \times 1) + \left(15 \times \frac{1}{10}\right) + \left(3 \times \frac{1}{100}\right) + \left(4 \times \frac{1}{1000}\right) \\
&= 3 + \frac{15}{10} + \frac{3}{100} + \frac{4}{1000} \\
&= 3 + \frac{10+5}{10} + \frac{3}{100} + \frac{4}{1000}
\end{aligned}
$$

$$
\begin{aligned}
&= 3 + \left(\frac{10}{10} + \frac{5}{10}\right) + \frac{3}{100} + \frac{4}{1000} \\
&= 3 + \left(1 + \frac{5}{10}\right) + \frac{3}{100} + \frac{4}{1000} \\
&= (3 + 1) + \frac{5}{10} + \frac{3}{100} + \frac{4}{1000} \\
&= 4 + \frac{5}{10} + \frac{3}{100} + \frac{4}{1000} \\
&= 4.534
\end{aligned}
$$

Compare the following algorithm with the preceding development. Take particular notice of that part where "carry 1" is justified.

$$
\begin{array}{r}
3.8 \\
\underline{0.734} \\
3. \\
1.5 \\
0.03 \\
\underline{0.004} \\
4.534
\end{array}
\qquad \text{or} \qquad
\begin{array}{r}
\textcircled{1} \\
3.8 \\
\underline{0.734} \\
4.534
\end{array}
$$

We see that Examples 1, 2, and 3 demonstrate that the addition algorithm for nonnegative rational numbers written in decimal notation follows that for the whole numbers. Likewise, the algorithm for subtraction is the same as the one developed in Section 7.6.

The multiplication algorithm developed for whole numbers is the same as the one that is used for nonnegative rationals written in decimal notation. The procedure for locating the decimal point presents some difficulty. The following examples justify the procedure commonly used. The development is quite detailed, but necessary to give an adequate explanation.

Example 4 Find the product of 0.1 and 0.01.

Writing each factor in expanded notation we have

$$
\begin{aligned}
0.1 \times 0.01 &= \frac{1}{10} \times \frac{1}{100} \\
&= \frac{1}{1000}.
\end{aligned}
$$

Written in decimal notation, this is

$$= 0.001.$$

Example 5 Find the product of 0.03×0.002.

$$0.03 \times 0.002 = \left(3 \times \frac{1}{100}\right) \times \left(2 \times \frac{1}{1000}\right)$$

(i) $$= (3 \times 2) \times \left(\frac{1}{100} \times \frac{1}{1000}\right)$$

$$= 6 \times \frac{1}{100{,}000}$$

$$= 0.00006$$

Observe line (i). Does it suggest an algorithm?

Example 6 Find the product of 4.3 and 0.02.

$$4.3 \times 0.02 = \left[(4 \times 1) + \left(3 \times \frac{1}{10}\right)\right] \times \frac{2}{100}$$

$$= \left(4 + \frac{3}{10}\right) \times \frac{2}{100}$$

$$= \frac{43}{10} \times \frac{2}{100}$$

(i) $$= (43 \times 2) \times \left(\frac{1}{10} \times \frac{1}{100}\right)$$

$$= \frac{86}{1000}$$

$$= \frac{80 + 6}{1000}$$

$$= \frac{80}{1000} + \frac{6}{1000}$$

$$= \frac{8}{100} + \frac{6}{1000}$$

$$= 0.086$$

Does line (i) reinforce the suggestion for an algorithm observed in Example 5? Do each of these examples justify the familiar "rule" for locating the decimal point in the product—that is, that the number of places to the right of the decimal point in the product is equal to the sum of the numbers of places to the right of the decimal point in the factors?

The algorithm for the division of nonnegative rational numbers written in decimal notation will be developed in Section 10.7.

EXERCISE 10.6

1. Use expanded notation to justify each of the following.
   (a) $1.95 + 4.372 = 6.322$
   (b) $0.007 + 26.32 = 26.327$
   (c) $836.43 + 471.8 = 1308.23$
   (d) $197.68 + 0.7364 = 198.4164$

2. Find the sums in Problem 1 by arranging in columns and showing the partial sums. Then, give the short form of the algorithm.
3. Find the product of each of the following, using the algorithm suggested by Examples 5 and 6.

(a) 0.2×0.3	(d) 21.005×0.014	(g) 0.258×10
(b) 18.1×3.1	(e) 42.02×12.6	(h) 1.6×10^2
(c) 68.02×15.05	(f) 0.0258×10	(i) 1.6×10^3

4. Use the standard algorithm to find a simpler name for each of the following.

(a) $8.7 + 2.1$	(c) $203.96 + 5.24$	(e) $86.07 + 7.018$
(b) $8.7 - 2.1$	(d) $203.96 - 5.24$	(f) $86.07 - 7.018$

10.7 CONVERSION OF FRACTIONS TO DECIMAL NOTATION AND VICE VERSA

At times it is convenient or even necessary to convert from the fraction form of a rational number to decimal notation, or from decimal notation to the fraction form.

Making use of expanded notation as in Section 10.5, we are able to convert from decimal notation to the fraction form. Let us consider one more example of this process.

Example 1 Express 4.032 in the fraction form. Using expanded notation, we have:

$$\begin{aligned} 4.032 &= (4 \times 1) + \left(0 \times \frac{1}{10}\right) + \left(3 \times \frac{1}{100}\right) + \left(2 \times \frac{1}{1000}\right) \\ &= 4 + \frac{0}{10} + \frac{3}{100} + \frac{2}{1000} \\ &= \frac{4000}{1000} + \frac{0}{1000} + \frac{30}{1000} + \frac{2}{1000} \\ &= \frac{4032}{1000} \end{aligned}$$

Alternatively, we may write this in the mixed form:

$$\begin{aligned} 4.032 &= 4 + \frac{0}{10} + \frac{3}{100} + \frac{2}{1000} \\ &= 4 + \left(\frac{0}{1000} + \frac{30}{1000} + \frac{2}{1000}\right) \\ &= 4 + \frac{32}{1000} \\ &= 4\frac{32}{1000} \end{aligned}$$

Converting from the fraction form to decimal notation requires a different approach. Let us consider the following examples.

Example 2 Express $\frac{3}{5}$ in decimal notation.

$$\begin{aligned}\frac{3}{5} &= \frac{3}{5} \times \frac{2}{2}\\ &= \frac{3 \times 2}{5 \times 2}\\ &= \frac{6}{10}\\ &= 6 \times \frac{1}{10}\\ &= 0.6\end{aligned}$$

Here the name of the multiplicative identity was chosen so as to make the new denominator 10.

Example 3 Convert $\frac{5}{8}$ to decimal notation.

$$\begin{aligned}\frac{5}{8} &= \frac{5}{2 \times 2 \times 2} \times \frac{5 \times 5 \times 5}{5 \times 5 \times 5}\\ &= \frac{5 \times 5 \times 5 \times 5}{10 \times 10 \times 10}\\ &= \frac{625}{1000}\\ &= \frac{600}{1000} + \frac{20}{1000} + \frac{5}{1000}\\ &= \frac{6}{10} + \frac{2}{100} + \frac{5}{1000}\\ &= \left(6 \times \frac{1}{10}\right) + \left(2 \times \frac{1}{100}\right) + \left(5 \times \frac{1}{1000}\right)\\ &= 0.625\end{aligned}$$

In the first step, why was the name of the multiplicative identity chosen to be

$$\frac{5 \times 5 \times 5}{5 \times 5 \times 5}?$$

Instead of using $\frac{5 \times 5 \times 5}{5 \times 5 \times 5}$ or $\frac{125}{125}$ as a name for the multiplicative identity, we shall use $\frac{1000}{1000}$.

$$
\begin{aligned}
\frac{5}{8} &= \frac{5}{8} \times \frac{1000}{1000} \\
&= \frac{5}{8} \times \left(1000 \times \frac{1}{1000}\right) \\
&= \left(\frac{5}{8} \times 1000\right) \times \frac{1}{1000} \\
&= \frac{5000}{8} \times \frac{1}{1000} \\
&= 625 \times \frac{1}{1000} \\
&= \frac{625}{1000} \\
&= \frac{600}{1000} + \frac{20}{1000} + \frac{5}{1000} \\
&= \frac{6}{10} + \frac{2}{100} + \frac{5}{1000} \\
&= \left(6 \times \frac{1}{10}\right) + \left(2 \times \frac{1}{100}\right) + \left(5 \times \frac{1}{1000}\right) \\
&= 0.625
\end{aligned}
$$

Is it possible to use the methods suggested in Examples 2 and 3 for converting all fractions to the decimal notation? That is, is it possible to find for every fraction an equivalent fraction such that the new denominator is a power of ten?

Let us consider the fraction $\frac{2}{3}$. Using $\frac{10}{10}$ as the multiplicative identity, we have

$$\frac{2}{3} = \frac{2}{3} \times \frac{10}{10}.$$

By steps similar to those used in the second solution of Example 3, this becomes

$$\frac{2}{3} = \frac{20}{3} \times \frac{1}{10}.$$

Using the division algorithm, we find that $20 \div 3$ has a remainder; that is,

$$20 = (3 \times 6) + 2.$$

Would the result have been different had another name for the multiplicative identity been chosen? Try $\frac{100}{100}$, $\frac{1000}{1000}$, and others. Obviously, we must find another method for converting fractions such as $\frac{2}{3}$ to decimal notation. Since

$\frac{2}{3}$ may be interpreted as $2 \div 3$, we shall use Definition 9.3 and write

$$2 \div 3 = c \quad \text{if} \quad 3 \times c = 2.$$

We first ask whether there is a *whole number c* such that when it is multiplied by 3, the product is 2. Since there is none, we shall rename 2 as $\frac{20}{10}$ (explain how this is done). Then, $2 \div 3$ is written $\frac{20}{10} \div 3$ and

$$\frac{20}{10} \div 3 = c \quad \text{if} \quad 3 \times c = \frac{20}{10}.$$

Obviously, there is no $c \in W$ that satisfies this sentence either. However, is there a $\frac{t}{10}$, $t \in W$, such that

$$3 \times \frac{t}{10} = \frac{20}{10}?$$

If t is 6, then $3 \times \frac{t}{10}$ is less than $\frac{20}{10}$, but if t is 1 more, 7, then $3 \times \frac{t}{10}$ is greater than $\frac{20}{10}$. Applying the test for the *largest acceptable partial quotient* developed in Section 7.6, we have:

$$\left(3 \times \frac{t}{10}\right) \leq \frac{20}{10} < \left(3 \times \frac{t+1}{10}\right),$$

$$\left(3 \times \frac{6}{10}\right) \leq \frac{20}{10} < \left(3 \times \frac{6+1}{10}\right).$$

Hence, $\frac{t}{10}$ is $\frac{6}{10}$, the largest acceptable *partial quotient in tenths*. Explain why the symbol "$\leq$" is used.

The following algorithm presents the preceding discussion:

$$\begin{array}{r|c|l} 3 & \frac{20}{10} & \\ & \frac{18}{10} & \frac{6}{10}, \qquad \text{since } \left(3 \times \frac{6}{10}\right) \leq \frac{20}{10} < \left(3 \times \frac{6+1}{10}\right). \\ \hline & \frac{2}{10} & \end{array}$$

When $\frac{6}{10}$ is used for the missing factor c, there is a remainder, $\frac{2}{10}$. Having exhausted the tenths, we rename $\frac{2}{10}$ as $\frac{20}{100}$ and seek $\frac{h}{100}$ as near to, but not greater than $\frac{20}{100}$. That is:

$$\left(3 \times \frac{h}{100}\right) \leq \frac{20}{100} < \left(3 \times \frac{h+1}{100}\right),$$

$$\left(3 \times \frac{6}{100}\right) \leq \frac{20}{100} < \left(3 \times \frac{6+1}{100}\right).$$

Hence, $\frac{h}{100}$ is $\frac{6}{100}$, the largest acceptable *partial quotient in hundredths.*

Beginning with $2 \div 3$, the developing algorithm is shown here. Compare this with the division algorithm for whole numbers in Section 7.6.

$$
\begin{array}{l|l|l}
3 \,\overline{)\,2} & & \\
\quad 0 & 0 & (3 \times 0) \leq 2 < (3 \times [0 + 1]) \\
\hline
2 = \frac{20}{10} & & \\
\quad \frac{18}{10} & \frac{6}{10} & \left(3 \times \frac{6}{10}\right) \leq \frac{20}{10} < \left(3 \times \frac{6+1}{10}\right) \\
\hline
\quad \frac{2}{10} = \frac{20}{100} & & \\
\qquad \frac{18}{100} & \frac{6}{100} & \left(3 \times \frac{6}{100}\right) \leq \frac{20}{100} < \left(3 \times \frac{6+1}{100}\right) \\
\hline
\qquad \frac{2}{100} = \frac{20}{1000} & &
\end{array}
$$

With the remainder always being the digit 2, it is obvious that the process continues. The quotient is the sum of the partial quotients, which is

$$0 + \frac{6}{10} + \frac{6}{100} + \cdots .$$

In decimal notation this is:

$$0.66 \ldots .$$

Converting the fraction forms in the preceding algorithm to decimal notation, we have the algorithm shown here:

$$
\begin{array}{l|l}
3 \,\overline{)\,2} & \\
\quad 0 & 0 \\
\hline
\quad 2.0 & \\
\quad 1.8 & 0.6 \\
\hline
\quad 0.20 & \\
\quad 0.18 & 0.06 \\
\hline
\quad 0.020 & \\
\quad 0.018 & 0.006 \\
\hline
\quad 0.0020 & \vdots
\end{array}
$$

Compare this with the algorithm using fraction forms. Comparing the second step in each, we note that $\frac{20}{10}$ is equivalent to 2.0, $\frac{18}{10}$ is equivalent to 1.8, and $\frac{6}{10}$ is equivalent to 0.6.

Since each remainder continues to be a digit 2, there is no end to the partial quotients having a digit 6.

$$
\begin{aligned}
2 \div 3 &= 0 + 0.6 + 0.06 + 0.006 + \cdots \\
&= 0.666 \ldots
\end{aligned}
$$

Observe that 6 continues to repeat in the quotient. We call such a numeral a *repeating decimal.* It may be written

$$0.\overline{6},$$

a bar being customarily drawn over the digit (or digits) that repeat in a numeral. Writing $23.7\overline{214}$ means that the next three digits are 214 and all subsequent ones fall in that pattern: 23.7214214214 . . . .

Use the division algorithm to express $\frac{1}{7}$ in decimal notation. You should find that the result has a repeating pattern with six different digits: $0.\overline{142857}$. It could have been predicted that the repeating pattern would have six or less digits, since division by 7 could never involve remainders other than the digits 1, 2, 3, 4, 5, 6. Notice that in converting $\frac{2}{3}$, only one digit appeared as a remainder and the repeating pattern contained only one digit.

Example 4 Express $\frac{1}{4}$ in decimal notation.

This may be done easily by using the procedures given in Example 2 and Example 3. What names would you suggest for the multiplicative identity? For comparison we now use the division algorithm as developed in the preceding discussion, dividing 1 by 4:

$$
\begin{array}{r|l|l}
4 & 1 & \\
 & 0 & 0 \\
\hline
 & 1.0 & \\
 & 0.8 & 0.2 \\
\hline
 & 0.20 & \\
 & 0.20 & 0.05 \\
\hline
 & & 0.25
\end{array}
$$

Hence,

$$\frac{1}{4} = 1 \div 4 = 0.25.$$

Since in the last step of Example 4 the remainder is 0, the quotient 0.25 is called a *terminating decimal.* Of course, by continuing the algorithm, we may write the quotient as

$$0 + 0.2 + 0.05 + 0.000 + 0.0000 + 0.00000 + \cdots = 0.25000\ldots = 0.25\overline{0}.$$

From this we could say that $0.25\overline{0}$, or 0.25, may be regarded as a repeating decimal, since the 0 repeats. So it is quite correct to refer to terminating decimals as repeating decimals also. However, they are usually called terminating decimals.

From the preceding discussion, we see that one method of converting a fraction to decimal notation is to use division. Each of the examples involved the fraction $\frac{a}{b}$, $a, b \in N$, $a < b$. Is it possible to convert the fraction to a decimal numeral if $a, b \in Q$, $b \neq 0$ with a and b written in decimal notation? Converting such a fraction to a decimal numeral will be shown in the following three examples.

Example 5 Convert $\frac{0.2}{0.3}$ to decimal notation.
Using the multiplicative identity $\frac{10}{10}$.

$$\frac{0.2}{0.3} = \frac{0.2}{0.3} \times \frac{10}{10}$$
$$= \frac{2}{3}$$

Using the division algorithm previously demonstrated,

$$= 0.\overline{6}$$

Example 6 Convert $\frac{0.75}{0.6}$ to decimal notation.

$$\frac{0.75}{0.6} = \frac{0.75}{0.6} \times \frac{10}{10}$$
$$= \frac{7.5}{6}$$

Since this may be written $7.5 \div 6$, we use the division algorithm as shown.

$6 \overline{)7.5}$		
6	1	$(6 \times 1) \leq 7.5 < (6 \times [1 + 1])$
1.5		
1.2	0.2	$(6 \times .2) \leq 1.5 < (6 \times [0.2 + 0.1])$
0.30		
0.30	0.05	$(6 \times 0.05) \leq 0.30 < (6 \times [0.05 + 0.01])$
	1.25	

Example 7 Convert $\frac{0.00644}{0.07}$ to decimal notation.
Why is $\frac{100}{100}$ used to name the multiplicative identity?

$$\frac{0.00644}{0.07} = \frac{0.00644}{0.07} \times \frac{100}{100}$$
$$= \frac{0.644}{7}$$

Writing $\frac{0.644}{7}$ as $0.644 \div 7$, use the division algorithm to show

$$\frac{0.644}{7} = 0.092.$$

In these examples you should observe that by choosing a particular power of 10 to name the multiplicative identity, each fraction written in division form became:

$$0.2 \div 0.3 = 2 \div 3$$
$$0.75 \div 0.6 = 7.5 \div 6$$
$$0.00644 \div 0.07 = 0.644 \div 7$$

A generalization is made from this.

Generalization 10.7

For $a, b, c \in Q, n \in N, b \neq 0$, if $a \div b = c$, then

$$(a \times 10^n) \div (b \times 10^n) = c.$$

By applying this generalization, it is possible to convert each divisor to a natural number. This makes the division much easier. Consequently, in any division problem, both dividend and divisor should be multiplied by that power of 10 which will convert the divisor to a natural number. That is:

$$\begin{aligned} 5.7325 \div 43.023 &= (5.7325 \times 10^3) \div (43.023 \times 10^3) \\ &= 5732.5 \div 43023 \end{aligned}$$

Example 8 Use the preceding generalization to find the quotient of $5.23 \div 0.16$.

$$\begin{aligned} 5.23 \div 0.16 &= (5.23 \times 10^2) \div (0.16 \times 10^2) \\ &= 523 \div 16 \end{aligned}$$

The division algorithm follows:

$$\begin{array}{r|r|r} 16 & 523 & \\ & 480 & 30 \\ \hline & 43 & \\ & 32 & 2 \\ \hline & 11 & 32 \end{array}$$

In the set of whole numbers, the result is written

$$523 = (16 \times 32) + 11.$$

In the set of rationals, the quotient may be written

$$32 + \tfrac{11}{16} \quad \text{or} \quad 32\tfrac{11}{16}.$$

The quotient consists of an integral and a fractional part. If the algorithm is continued as in Example 6, the quotient is 32.6875, which is in decimal notation. Here, we have three different ways of expressing the quotient.

In our daily experiences, we rarely have need for a great number of places in a decimal numeral, such as 0.666666 . . . or 83.7218452. The number of digits that is needed depends on the nature of the problem. A method of dropping off digits that are considered unnecessary is called *rounding*.

The following are examples of rounding:

If 0.364 is rounded to the nearest hundredth, it becomes 0.360.
If 0.364 is rounded to the nearest tenth, it becomes 0.400.
If 126.3 is rounded to the nearest one, it becomes 126.0.

If 126.3 is rounded to the nearest ten, it becomes 130.0.
If 126.3 is rounded to the nearest hundred, it becomes 100.0.

In the following examples, techniques for rounding are shown.

Example 9 Round 32.6749 (i) to the nearest tenth, and (ii) to the nearest hundredth.

(i) Place a ray underneath the digits that are to be dropped, as

$$32.6\underrightarrow{749}.$$

Since the digit 7 immediately over the end point of the ray is greater than 5, the digit 6 is increased by 1 when 7, 4, and 9 are dropped. Thus, 32.6749 is rounded to 32.7000. We say that 32.6749 rounded to the nearest tenth is 32.7000.

(ii) Place a ray under the digits to be dropped, as

$$32.67\underrightarrow{49}.$$

Since the digit 4 is less than 5, the digit 7 is not changed when 4 and 9 are dropped:

$$32.67\underrightarrow{49} \text{ is rounded to } 32.6700$$

We say that 32.6749 rounded to the nearest hundredth is 32.6700.

Example 10 Round 59,846 (i) to the nearest hundred, and (ii) to the nearest thousand.

(i) Using the ray to indicate the digits to be dropped, we have

$$59{,}8\underrightarrow{46}.$$

Since the digit 4 is less than 5, then 59,846 rounded to the nearest hundred is 59,800.

(ii) Show that 59,846 rounded to the nearest thousand is 60,000.

In the preceding examples, the first "drop off" digit was either less than or greater than 5. What happens when this digit is equal to 5? For example, how are 7.45 and 0.75 rounded to the nearest tenth? There is really no common agreement as to how this should be done. The two accepted methods are: (1) The digit preceding the "drop off" digit 5 is increased by 1, or (2) the digit preceding the "drop off" digit 5 must be even. If it is not even, it is made so by adding 1. By (2), 7.45 rounded to the nearest tenth is 7.4 and 0.75 is 0.80. This may be called the *even rule* for rounding.

The results of Examples 2, 3, and 4 lead us to conclude that all rational numbers may be written in decimal notation either terminating or repeating. When a rational number is written in fraction form, is it possible to determine whether it can be expressed as a repeating decimal or as a terminating one without going through the process of converting? Consider the following three rational numbers,

$$\frac{1}{16}, \frac{6}{40}, \text{ and } \frac{14}{1750}.$$

They may be represented by terminating decimals. Verify this. In the simplest fraction form each may be expressed with its denominator written as a product of primes:

$$\frac{1}{16} = \frac{1}{2^4}$$

$$\frac{6}{40} = \frac{3}{20} = \frac{3}{2^2 \times 5}$$

$$\frac{14}{1750} = \frac{1}{125} = \frac{1}{5^3}$$

Notice that the prime factors of each of the denominators are powers of 2 or 5. If a fraction can be changed to a form having the denominator as a power of 10, it may then be written in terminating decimal form. What $k \in N$ must be used for $\frac{k}{k}$ (the multiplicative identity) to change each of the fractions

$$\frac{1}{2^4}, \frac{3}{2^2 \times 5}, \frac{1}{5^3}$$

to ones having a power of 10 as denominator? (See Examples 2 and 3.)

It should be apparent that if prime factors other than 2 or 5 are present in the denominator, it cannot be changed to a power of 10. Investigate these fractions:

$$\frac{4}{9}, \frac{7}{15}, \frac{11}{70}.$$

We draw the conclusion that the condition for changing a fraction in simplest form to a terminating decimal form is that its denominator have prime factors that are powers of 2 and/or 5, and that no different prime factors be present.

We have now learned how to convert a terminating decimal to the fraction form, and to convert the fraction form to either a terminating or a repeating decimal. It seems appropriate at this time to consider a procedure for changing a repeating decimal to the fraction form. Consider the following examples.

Example 11 Express $0.\overline{3}$ in the fraction form.

Let n represent the rational number expressed by $0.\overline{3}$.

$n = 0.\overline{3} = 0.333\ldots$	
$10 \times n = 10 \times 0.333\ldots$	*Uniqueness of multiplication*
$10n = 3.333\ldots$	*Renaming* (10 × 0.333 . . .)
$9n + n = 3 + 0.333\ldots$	*Renaming* 10 *and* 3.33 . . . *and distributive property*
$n = 0.333\ldots$	*Given*
$9n = 3$	*Cancellation for addition*
$\frac{1}{9} \times 9n = \frac{1}{9} \times 3$	*Uniqueness of multiplication*
$n = \frac{1}{3}$	*Explain*

Since the solution set is $\{\frac{1}{3}\}$, then $0.\overline{3} = \frac{1}{3}$.

Example 12 Change $7.0\overline{3}$ to the fraction form. Justify each step.

$$n = 7.033\ldots$$
$$10n = 70.333\ldots$$
$$100n = 703.333\ldots$$
$$10n + 90n = 70.333\ldots + 633$$
$$90n = 633$$
$$\frac{1}{90} \times 90n = \frac{1}{90} \times 633$$
$$n = \frac{633}{90}$$
$$n = \frac{211}{30}$$

The solution set is $\{\frac{211}{30}\}$. Hence, $7.0\overline{3} = \frac{211}{30}$.

We may now conclude that every positive rational number may be expressed in either a terminating decimal form or a repeating decimal form. If we allow the terminating form to assume the repeating form by annexing $\overline{0}$, then the statement is: *Every rational number may be expressed in the repeating decimal form.* From Examples 11 and 12 it seems reasonable to expect that *every number expressed in the repeating decimal form is a rational number.*

EXERCISE 10.7

1. Show that the following numerals represent rational numbers by putting each in fraction form.

   (a) $1.\overline{3}$ (c) $113.\overline{2}$ (e) $2.\overline{07}$ (g) $0.0\overline{134}$
   (b) $11.\overline{3}$ (d) $3.\overline{14}$ (f) $0.\overline{134}$ (h) $0.00\overline{134}$

2. Express each of the numerals in Problem 1 in the mixed form when possible.

3. Use expanded notation to show how each of these numerals is read.

(a) 0.73 (d) 1.0201 (g) 0.10201 (i) 2407.3
(b) 9.271 (e) 10.201 (h) 405.055 (j) 0.0030
(c) 52.06 (f) 102.01

4. Use the multiplicative identity to change each of the following to decimal notation.

(a) $\frac{2}{5}$ (c) $\frac{17}{25}$ (e) $\frac{69}{20}$ (g) $\frac{2}{16}$ (i) $\frac{381}{250}$

(b) $\frac{3}{2}$ (d) $\frac{207}{50}$ (f) $\frac{3}{8}$ (h) $\frac{54}{200}$ (j) $\frac{3}{500}$

5. Use the division algorithm to convert each of the following to decimal notation. Use the bar symbol over the digits which repeat.

(a) $\frac{3}{4}$ (c) $\frac{1}{7}$ (e) $\frac{1}{3}$ (g) $\frac{1}{9}$ (i) $\frac{1}{999}$

(b) $\frac{5}{8}$ (d) $\frac{25}{7}$ (f) $\frac{2}{3}$ (h) $\frac{1}{99}$

6. Indicate which of the following fractions would be terminating decimals if converted into decimal form. After deciding on which are terminating, convert to decimal notation. In the case of those that are repeating, give the pattern of the repeating digits.

(a) $\frac{1}{16}$ (c) $\frac{21}{45}$ (e) $\frac{36}{666}$ (g) $\frac{23}{20}$ (i) $\frac{15}{75}$

(b) $\frac{10}{18}$ (d) $\frac{27}{143}$ (f) $\frac{52}{78}$ (h) $\frac{24}{20}$

7. Use the procedure presented in Examples 11 and 12 to convert the following repeating decimals to the fraction form.

(a) $0.\overline{6}$ (d) $2.\overline{5}$ (g) $0.\overline{2}$ (j) $0.1\overline{2}$
(b) 0.444 . . . (e) $2.0\overline{5}$ (h) $0.0\overline{2}$ (k) $0.\overline{12}$
(c) 0.1313 . . . (f) $2.00\overline{5}$ (i) $0.\overline{02}$ (l) $1.\overline{12}$

8. Round each of the following (i) to the nearest tenth and (ii) to the nearest hundredth. Use the even rule where applicable.

(a) 0.073 (c) 63.726 (e) 0.651
(b) 0.487 (d) 0.354 (f) 809.984

9. Round each of the following to (i) the nearest ten (ii) to the nearest hundred. Use the even rule where applicable.

(a) 3268.0 (c) 126.8 (e) 435
(b) 9737.0 (d) 765.0 (f) 29,764.4

EXAMPLE Convert 0.333 . . . to the fraction form, using the result of part (g) of Problem 5. You should have found that $\frac{1}{9}$ was equivalent to the decimal form 0.111 . . . . We proceed as follows:

$$\frac{1}{9} = 0.111\ldots$$

$$0.333\ldots = 3 \times 0.111\ldots$$

Using $\frac{1}{9}$ instead of the decimal form 0.111 . . . , we have:

$$0.333\ldots = 3 \times \frac{1}{9}$$
$$= \frac{3}{9}$$
$$= \frac{1}{3}$$

10. Given the following information, use a method suggested by the solution of the example below to convert each of the repeating decimals to the fraction form.

$$\frac{1}{9} = 0.\overline{1}, \qquad \frac{1}{99} = 0.\overline{01}, \qquad \frac{1}{999} = 0.\overline{001}$$

(a) $0.\overline{2}$ (b) $0.\overline{02}$ (c) $0.\overline{002}$ (d) $0.\overline{32}$ (e) $0.\overline{321}$

(f) $0.3\overline{21}$ (HINT: $0.32121\ldots = 0.3 + 0.021\overline{21} = 0.3 + (\frac{1}{10} \times 0.21\overline{21})$.)

EXAMPLE Find the sum of $0.\overline{3}$ and $0.\overline{6}$.
These numerals may be written as 0.333 . . . and 0.666 . . . . It is convenient to "line up" the digits and add in this manner:

$$\begin{array}{l} 0.333\ldots \\ \underline{0.666\ldots} \\ 0.999\ldots \end{array}$$

That is,

$$0.\overline{3} + 0.\overline{6} = 0.\overline{9}.$$

Show that $0.\overline{3} = \frac{1}{3}$ and $0.\overline{6} = \frac{2}{3}$. What is the sum of $\frac{1}{3}$ and $\frac{2}{3}$?
From the information developed in this example, is it possible to say that $0.\overline{9} = 1$? Explain. Can you show in another way that $0.\overline{9} = 1$?

11. Express each sum in repeating decimal form.
    (a) $0.\overline{12} + 0.\overline{43}$ (c) $25.\overline{3} + 12.\overline{4}$ (e) $0.\overline{9} + 1$
    (b) $3.\overline{1} + 0.\overline{8}$ (d) $5.\overline{0} + 0.\overline{01}$ (f) $52.\overline{3} + 100.\overline{3}$
12. Verify each of the following.
    (a) $0.\overline{5} + 0.\overline{6} = 1.\overline{2}$ (c) $0.\overline{9} + 0.\overline{01} = 1.\overline{01}$ (e) $0.\overline{23} + 0.\overline{78} = 1.\overline{02}$
    (b) $0.\overline{9} + 1 = 2$ (d) $1.\overline{7} + 1.\overline{6} = 3.\overline{4}$
13. Change each of the following division problems to a form that will make the use of the division algorithm simpler. For parts (e) through (h), find the quotient of each, giving the complete algorithm as demonstrated in Example 6.
    (a) $18 \div 0.6$ (d) $18 \div 0.0006$ (g) $17.5 \div 0.005$
    (b) $18 \div 0.06$ (e) $17.5 \div 0.5$ (h) $17.5 \div 0.0005$
    (c) $18 \div 0.006$ (f) $17.5 \div 0.05$ (i) $100 \div 0.9$
14. Express the quotient of each in the mixed form consisting of an integral and a fractional part.
    (a) $395 \div 17$ (b) $7.36 \div 0.21$ (c) $1.183 \div 0.042$

11

The System of Real Numbers

11.1 THE SET OF REAL NUMBERS

As was predicted in Chapter 9 there is another system of numbers that we need to investigate. We shall construct new numbers to satisfy new requirements and then join these numbers to the set of rational numbers, to form a new set, the set of real numbers.

The discussion that follows brings out the need for numbers other than the rational numbers which we have at hand and shows the manner in which certain of them are discovered. To understand this development we need first to explore briefly some notions of exponents, followed by a more detailed discussion of finding squares and square roots of numbers.

Remember the use of exponents in the expanded form of a numeral. Thus, we have

$$10^2 = 10 \times 10 = 100, \quad 7^3 = 7 \times 7 \times 7 = 343, \qquad \text{and so on.}$$

We say that "ten *squared*," or "the *square* of ten," is 100, and that "seven *cubed*," or "the *cube* of seven," is 343. Consider the following question:

What number has 9 as its square?

We know that $(3)^2 = 9$; also $(^-3)^2 = 9$, by Definition 8.6(b). The question may be restated as:

What is the solution set for the mathematical sentence $x^2 = 9$?

Thus, x may be 3 or $^-3$, and the solution set is $\{3, {}^-3\}$.

A quick check will establish that the solution set for $x^2 = \frac{64}{121}$ is $\{\frac{8}{11}, \frac{^-8}{11}\}$.

A generalization may be stated.

Generalization for Squares of Numbers

For each rational number $q \geq 0$, if $q^2 = n$, then $(-q)^2 = n$, $n \geq 0$.

Let us list the integers between $^-5$ and 4:

$$M = \{a : a \in I \text{ and } {}^-5 < a < 4\},$$
$$= \{{}^-4, {}^-3, {}^-2, {}^-1, 0, 1, 2, 3\}.$$

Next, list the squares corresponding to the elements in M. There will be repetitions, since by the generalization for squares of numbers, the squares of $^-3$, $^-2$, and $^-1$ are the same as the squares of 3, 2, and 1. Thus, only nonnegative integers are squares of integers, and so $^-16$, for example, is not the square of any integer. The squares of the integers are:

$$S = \{0, 1, 4, 9, 16, 25, 36, 49, 64, 81, \ldots\}$$

The generalization for squares of numbers makes it clear that the squares of all rational numbers are nonnegative.

Find, when possible, the solution sets for the following open sentences.

(1) $x^2 = 81$ (2) $x^2 = \frac{4}{25}$ (3) $x^2 = \frac{169}{9}$ (4) $x^2 = {}^-64$ (5) $x^2 = 2$

The preceding discussion makes it evident that Problem (4) has no solution in any set studied heretofore, since the square of any rational number is nonnegative. No solution is apparent for Problem (5) even though 2 is nonnegative. In Section 11.3 we shall discuss the truth of the statement that there is no rational number whose square is 2.

If we wish $x^2 = 2$ to have a solution and if our prediction is true that there is no rational number whose square is 2, then it will be necessary to invent some new numbers. Our task in this chapter is to create a new set of numbers whose elements will make possible the solution of sentences such as $x^2 = 2$. When this new set is joined to the set of rational numbers, we shall have a set called the *set of real numbers, R.* It is not beyond our powers of imagination to suspect that there may be a system that offers solutions for sentences such as Problem (4). However, that exploration will be left for study beyond this course.

11.2 SQUARES AND SQUARE ROOTS

Writing 324 in its completely factored form, we have:

$$324 = 2 \times 2 \times 3 \times 3 \times 3 \times 3$$
$$= 2^2 \times 3^4$$
$$= (2 \times 3^2) \times (2 \times 3^2)$$

Hence,

$$324 = (2 \times 3^2)^2, \text{ or } (18)^2.$$

This tells us that 324 is the square of 18. We say (compare Section 5.7) that 324 is a *perfect square,* meaning now that it is the square of an integer, 18. From the generalization for squares of numbers, we know that $(^{-}18)^2 = 324$ also. The numbers 18 and $^{-}18$ are called the *square roots* of 324. Formally, we have:

Definition 11.2(a) A *nonnegative integer n* is said to be a *perfect square* if it can be written as the product of two equal factors:

$$n = k \times k, \quad \text{where} \quad k \in I, k \geq 0$$

We say that k and $-k$ are the *square roots* of n.

If each k is written in completely factored form, it is clear that any prime factor of k must appear twice in n. Hence, every prime factor of n must appear an even number of times for n to be a perfect square. In the example above, 3 appears four times and 2 twice.

Next, we consider whether the rational number $\frac{25}{81}$ may be called a perfect square. Writing the numerator and the denominator in completely factored form, we have:

$$\frac{25}{81} = \frac{5 \times 5}{3 \times 3 \times 3 \times 3}, \quad \text{or} \quad \frac{5^2}{3^4}$$

$$= \left(\frac{5}{3^2}\right) \times \left(\frac{5}{3^2}\right), \quad \text{or} \quad \left(\frac{5}{3^2}\right)^2$$

$$= \left(\frac{5}{9} \times \frac{5}{9}\right), \quad \text{or} \quad \left(\frac{5}{9}\right)^2$$

We now state:

Definition 11.2(b) A nonnegative rational number $\frac{a}{b}$, a and b relatively prime, is a perfect square if $a = m^2$, $m > 0$, and $b = n^2$, $n > 0$. Then $\frac{m}{n}$ and $-(\frac{m}{n})$ are the square roots of $\frac{a}{b}$.

Thus, $\frac{25}{81}$ is a perfect square, and $\frac{5}{9}$ and $^{-}(\frac{5}{9})$ are its square roots.

11.3 IRRATIONAL NUMBERS

We now return to the open sentence $x^2 = 2$, which is Problem (5) in Section 11.1. Since 2 is already in its completely factored form and no prime factors occur an even number of times, 2 is not the square of a positive integer. Is it, perhaps, the square of some rational number $\frac{a}{b}$ where $a > 0$, $b > 0$, $a, b \in I$ and a and b are relatively prime? Then $(\frac{a}{b})^2$ would be equal to 2.

Let us assume that this is true:

$$\left(\frac{a}{b}\right)^2 = 2$$

$$\frac{a}{b} \times \frac{a}{b} = 2$$

$$\frac{a^2}{b^2} = 2$$

$$\left(a^2 \times \frac{1}{b^2}\right) \times b^2 = 2b^2$$

$$a^2 \times \left(\frac{1}{b^2} \times b^2\right) = 2b^2$$

$$a^2 \times 1 = 2b^2$$

$$a^2 = 2b^2$$

Now a^2 is a positive integer whose complete factorization is composed of even powers of primes. In the same manner b^2 is composed of prime factors each occurring an even number of times. But none of the factors in a^2 and b^2 are the same, since a and b were assumed to be relatively prime. Moreover, since the two integers a^2 and $2b^2$ are presumed equal, their completely factored forms would be the same (the fundamental theorem of arithmetic, Section 7.3). Now a^2 is made up of a set of prime factors, each occurring an even number of times. But the presence of the prime factor 2 in $2b^2$ ensures an odd number of factors—the even number of factors of b^2 plus the one extra factor, 2, makes an odd number of factors. Because of this, there is no possibility that a^2 does equal $2b^2$, and we are forced to conclude that the original assumption, $(\frac{a}{b})^2 = 2$, is false. Hence, there is no rational number $\frac{a}{b}$ whose square is 2.

Since there is no rational number whose square is 2, we shall invent a number such that its square will be 2 and represent it by $\sqrt{2}$ (read "square root of 2"). To distinguish it from the rational numbers, we say it is an *irrational number.* It is understood that $\sqrt{2}$ is positive, and so another number whose square is 2 is $^-\sqrt{2}$. Thus, the solution set of

$$x^2 = 2 \quad \text{is} \quad \{\sqrt{2}, {}^-\sqrt{2}\}.$$

The convention that the symbol $\sqrt{\ }$ represents a positive number holds no matter what numeral appears with it; that is, $\sqrt{9} = 3$, although both 3 and $^-3$ are solutions of $x^2 = 9$.

In an exactly comparable manner we speak of *perfect cubes* and *cube roots* and, in general, *perfect* nth *powers* and nth *roots,* where n is a natural number, $n > 1$. For example, a solution of the sentence $x^3 = 8$ is 2. (Is there another?) Since $2^3 = 8$, we write $\sqrt[3]{8} = 2$. Similarly, $\sqrt[5]{31}$ (read "fifth root of 31") is a number such that $(\sqrt[5]{31})^5 = 31$; it is a solution of the sentence $x^5 = 31$.

Numerals using the symbol $\sqrt{\ }$ are said to be written in *radical form.* In the general symbol for the radical form, $\sqrt[n]{x}$, n is called the *index of the root* and x is the *radicand.* The index is a natural number greater than 1. When the index is not written, it is understood to be 2. If n is even, the symbol has meaning in the real number system *only if x is positive.* $\sqrt[4]{16} = 2$, since $2^4 = 16$. We know that $(^-2)^4 = 16$ also, hence, $^-2$ is *a* 4th root of 16, but positive 2 is called the *principal* 4th root of 16. However, $\sqrt[4]{^-16}$ has no meaning in the real number system since an even power of a real number cannot be a negative number. There is no real number that raised to the 4th power gives $^-16$. However, if n is odd, then a positive radicand yields a positive number; a negative one gives a negative number. Both roots are real numbers. For example, $\sqrt[3]{27} = 3$, and $\sqrt[3]{^-27} = {^-3}$. Natural numbers for n are used in this text.

There are other irrational numbers besides those that arise from roots of numbers that are not perfect powers. One such irrational number has a special name, π (Greek letter, pronounced "pi"). It is the name for the ratio of the circumference of any circle to its diameter. (The value of π is only approximately equal to $\frac{22}{7}$.) There are other irrational numbers, some with special names, but we shall not include them in this discussion.

EXERCISE 11.3

1. Write the complete factorization for these numbers and show that they are perfect squares.
   (a) 1521 (d) 9025
   (b) $225a^4b^6$, a and b primes (e) 2025
   (c) $\frac{864}{150}$
2. Prove that no prime number is a perfect square.
3. Formulate a rule for identifying integers that are perfect cubes. (HINT: Compare Definition 11.2(a).)
4. Show that $^-64$ is not a perfect square, but is a perfect cube.
5. Find:
   (a) $\sqrt{81}$ (e) $\sqrt{9 \times 16}$
   (b) $\sqrt[3]{\frac{27}{125}}$ (f) $\sqrt{9} \times \sqrt{16}$
   (c) $\sqrt{\frac{72}{200}}$ (g) $\sqrt{9 + 16}$
   (d) $\sqrt[3]{^-216}$ (h) $\sqrt{9} + \sqrt{16}$
6. Discuss the solution sets for the following sentences.
   (a) $x^2 = 49$ (d) $x^3 + 27 = 0$
   (b) $x^2 - 81 = 0$ (e) $x^4 + 16 = 0$
   (c) $x^2 - 5 = 0$ (f) $x^3 + 7 = 0$

11.4 THE NUMBER LINE AND $\sqrt{2}$; COMPLETENESS

Some time before the fourth century B.C., the Pythagoreans, a group of Greek mathematicians who were disciples of Pythagoras, made the, to them, shocking discovery that $\sqrt{2}$ is not a rational number. Prior to that time the

Pythagoreans had held to the belief that the measures of all things in the universe could be given by rational numbers greater than 0. In the light of this philosophy, it was most disturbing to find proof that there is no rational number whose square is 2.

This problem may have arisen in connection with a theorem that is still called the *Pythagorean theorem.* This theorem may be described by applying it to a particular problem. Consider the following illustration showing a right triangle the measures of whose sides are 3, 4, and 5, with a square drawn

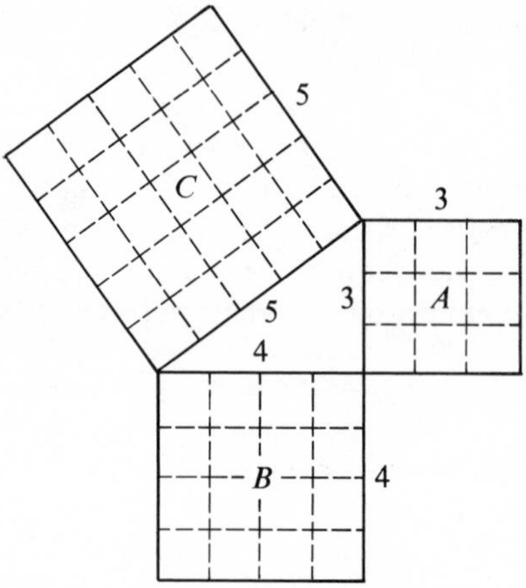

on each of the sides. If cutouts of the squares were made, we could completely cover square C with the small squares drawn in squares A and B. This implies that the area of square C is the same as the combined area of square A and square B. Writing the area of each of the squares, we have:

$$\begin{aligned}(5 \times 5) &= (3 \times 3) + (4 \times 4)\\ 5^2 &= 3^2 + 4^2\\ 25 &= 9 + 16\end{aligned}$$

The Pythagorean theorem states that the relationship

$$c^2 = a^2 + b^2$$

holds for all right triangles, the measure of whose sides are a, b, c, elements of the set of real numbers, where a and b are measures of the *legs* and c is the measure of the *hypotenuse.* The identification of hypotenuse and legs is made in the illustration below. Notice the location of the hypotenuse and the legs in relation to the right angle.

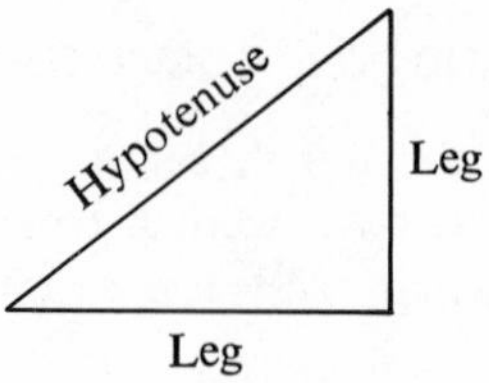

The theorem may be stated as follows:

Pythagorean Theorem *The square of the measure of the hypotenuse of a right triangle is equal to the sum of the squares of the measures of the two legs.*

By applying this theorem, we may find the measure of any one of the sides if the other two are given. Also, the relationship $c^2 = a^2 + b^2$ may be used as a test in determining whether the triangle is a right triangle; this applies the reverse statement (called a *converse*) of the Pythagorean theorem. This theorem and its converse are proved in most high school geometry courses.

Now, we shall proceed to find the measure of the hypotenuse of the triangles given in the examples which follow.

Example 1 Given the right triangle the measures of whose sides are 5 and 12, find the measure of the hypotenuse, c.

Applying the Pythagorean theorem, we have:

$$c^2 = 5^2 + 12^2$$
$$c^2 = 25 + 144$$
$$c^2 = 169$$

Is there a rational number whose square is 169? If so, what is it?

The situation that completely jolted the Pythagoreans was an example such as the one following.

Example 2 Given a square whose sides each measure 1, find the measure of its diagonal.

With a right triangle involved, we apply the Pythagorean theorem:

$$c^2 = 1^2 + 1^2$$
$$c^2 = 1 + 1$$
$$c^2 = 2$$

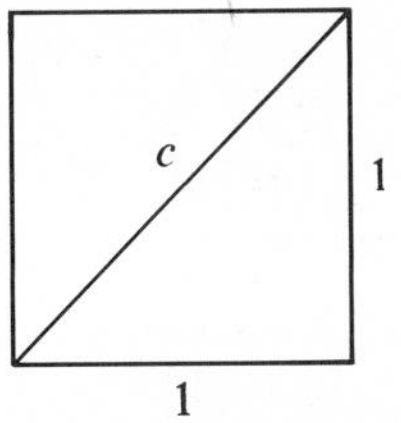

The Pythagoreans expected to find a rational number whose square is 2. The search for such a number brought about the discovery that there is no such number. A proof of this was given in Section 11.3. Thus, in our notation,

$$c = \sqrt{2}.$$

Our next consideration is: Does the irrational number $\sqrt{2}$ have a place in the ordering scheme of the rational numbers? Can we determine the location of a point on the number line to which $\sqrt{2}$ may be matched? Recall the development of the number line with each extended set of numbers serving to fill it in more fully, until the set of rational numbers was reached. Then we found that we could always insert a rational number between any two, and the matching points on the number line could be packed as closely as we pleased. At that point we spoke of the number line as being dense. But did all the points have matching numbers? We shall demonstrate that $\sqrt{2}$ may be matched with a point, and we know that this point is not already matched with a rational number. This implies that for each rational number there corresponds a unique point, but there are still some points not included in this correspondence; that is, there are some "holes" in the number line. When these "holes" are "filled in" with the irrational numbers, the number line is then said to be *complete*.

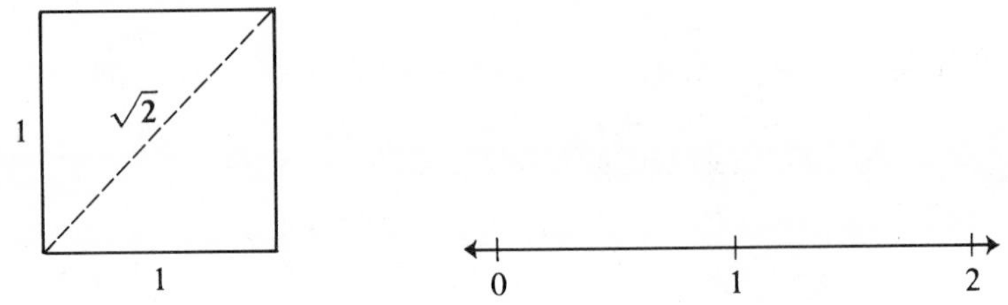

Figure 11–1

We shall now demonstrate the use of a model to locate $\sqrt{2}$ on the number line. Draw a square the measure of each of whose sides is 1. Draw a number line, making the unit intervals the same length as a side of the square (Figure 11–1). Make a cutout of the square, placing it on the number line as shown in Figure 11–2. Now pivot the square about the vertex located at 0 until the diagonal fits along the number line, as in Figure 11–3.

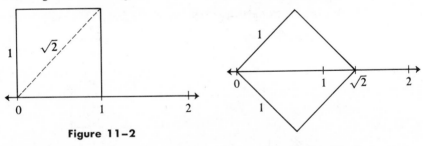

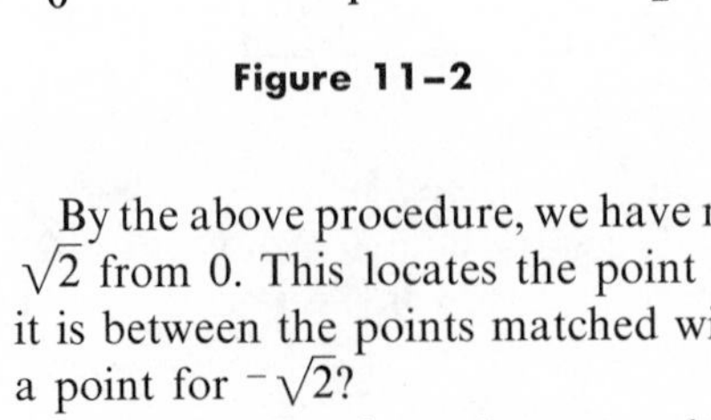

Figure 11–2

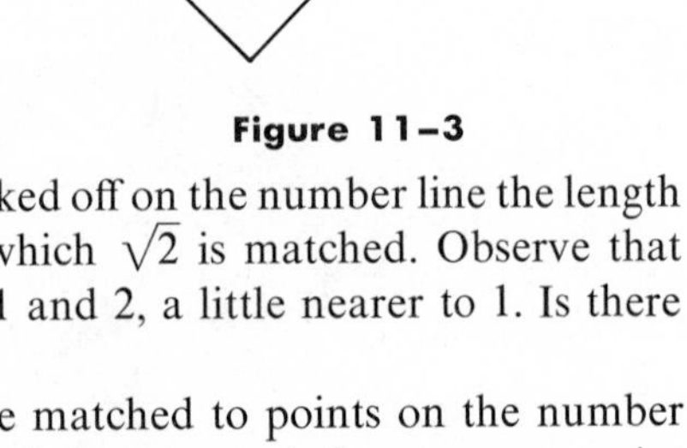

Figure 11–3

By the above procedure, we have marked off on the number line the length $\sqrt{2}$ from 0. This locates the point to which $\sqrt{2}$ is matched. Observe that it is between the points matched with 1 and 2, a little nearer to 1. Is there a point for $^{-}\sqrt{2}$?

Other irrational numbers can also be matched to points on the number line. In this informal, intuitive manner it is suggested that to every point on the number line there corresponds a number, rational or irrational, that is a real number. The proof of this property of *completeness* involves the concept of *limit,* which is beyond the scope of this book. We may now say that *the set of real numbers, R, is the set of numbers such that for every point of the number line there is a corresponding number and for every number of R there is a corresponding point on the number line.* This is essentially the *property of completeness.* Recall the denseness property of the rational numbers. Now this new set of numbers, the irrationals, take their place among the rationals to make the number line complete.

11.5 DECIMAL APPROXIMATIONS OF IRRATIONAL NUMBERS

In the preceding chapter it was shown that every rational number could be expressed in a repeating decimal (including the terminating decimal) form, and that every such decimal could be changed to the fraction form (Section 10.7). It is not possible to express irrational numbers exactly in the decimal form, for, if this could be done, they would be rational numbers. Nevertheless, it is often desirable to find a rational number which is very close to a given irrational number. The following discussion illustrates one procedure.

We seek an approximate value for $\sqrt{5}$. By definition the irrational number $\sqrt{5}$ is a solution of the open sentence

$$x^2 = 5.$$

Hence our aim will be to find a number whose square is as close to 5 as seems useful. Since $2^2 < 5$ and $3^2 > 5$, it follows that

(a) $$2 < \sqrt{5} < 3.$$

If we try successive tenths between 2 and 3, we find that $(2.2)^2 < 5$ and $(2.3)^2 > 5$, and hence

(b) $$2.2 < \sqrt{5} < 2.3.$$

A trial of successive hundredths (2.21, 2.22, 2.23, etc.) shows that $(2.23)^2 < 5$ and $(2.24)^2 > 5$, and hence

(c) $$2.23 < \sqrt{5} < 2.24.$$

The successive intervals within which $\sqrt{5}$ is included are shown on the drawing of the number line below.

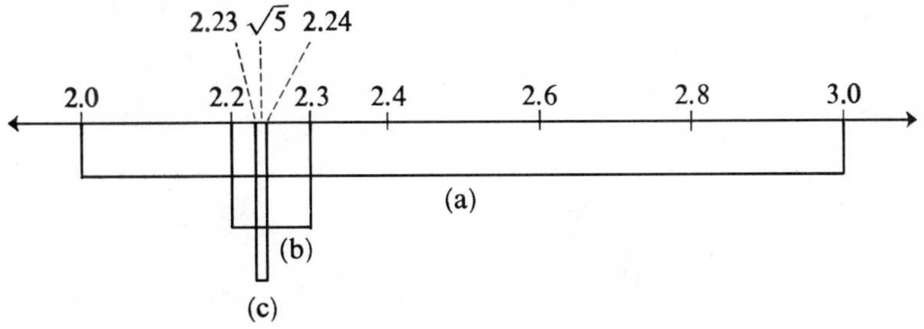

Sentences such as (a), (b), and (c) are called *inequalities*. These inequalities may be combined into one sentence:

$$2 < 2.2 < 2.23 < \sqrt{5} < 2.24 < 2.3 < 3$$

If further refinement is desired, the process may be continued. We already know that our quest will never be completely successful, since if it were that would mean that $\sqrt{5}$ was a rational number. It is this process of continued approximation which makes us aware that our approach to irrational numbers differs markedly from that used previously in constructing new sets of numbers.

Another method that is useful for approximating square roots of natural numbers which are not perfect squares is called the *divide-and-average method.* Consider an example.

We seek a number whose square is 41, that is, some k such that

$$k \times k = 41.$$

We might guess that this number is 6. If this were correct, then $41 \div 6$ would be 6. However, $41 \div 6 \approx 6.8$. Evidently 6 is too small and 6.8 is too large to be used as k. Since k is somewhere between 6 and 6.8, let us use their average,

$$\frac{6.0 + 6.8}{2} = 6.4,$$

as a new approximation and repeat the process. If 6.4 were $\sqrt{41}$, then 6.4×6.4 would equal 41, or $41 \div 6.4$ would be 6.4. Actually it is very close:

$$41 \div 6.4 \approx 6.406$$

If we wish greater accuracy, the average

$$\frac{6.400 + 6.406}{2} = 6.403$$

will be an even closer approximation. Does

$$6.403 \times 6.403 = 41?$$

No, but 6.403 is $\sqrt{41}$ correct to four digits.

Another example shows that when using a second method, it is not even necessary to start with a very good guess. To find an approximation for $\sqrt{185}$, start with 10 as a first guess, even though it is evidently far from the best one. Since $185 \div 10 = 18.5$, it is true that

$$10 < \sqrt{185} < 18.5.$$

The average of 10 and 18.5,

$$\frac{10.0 + 18.5}{2} = 14.3,$$

will be our new approximation. After dividing 185 by 14.3, the next approximation will be

$$\frac{14.30 + 12.94}{2} = 13.62.$$

Repeating the process, we divide 185 by 13.62 and find

$$\frac{13.620 + 13.583}{2} = 13.601$$

to be the new approximation. It is possible by squaring or by a new division to discover that $\sqrt{185}$ falls between 13.601 and 13.602. Then, $\sqrt{185}$, *correct to 2 decimal places,* is 13.60. If precision to 3 decimal places is desired, it is possible to repeat the process—that is, to divide again and examine the new average. However, it may be easier to square 13.601 and 13.602, see if 185 falls between these two squares, and discover which square is closer to 185. In this case 13.601 is the required approximation.

When it is necessary to use irrational numbers in computation, their approximate values may be adequate. At other times it may be convenient to use a radical form. We call 2.23 or 2.24 an *approximate decimal form* for the square root of 5, while $\sqrt{5}$ is its *exact* or *radical form.*

EXERCISE 11.5

1. Estimate between two successive hundredths each of the following irrational numbers. Write the results as inequalities.
   (a) $\sqrt{7}$ (Use 2.6 as a first estimate.)
   (b) $\sqrt{300}$ (Use 17 as a first estimate.)
   (c) $\sqrt{21}$
2. (a) Find $\sqrt[3]{19}$. That is, find a number, x, such that $x^3 = 19$. Obviously, $2 < x < 3$ since $2^3 = 8$ and $3^3 = 27$. Try 2.7 as a first approximation.
   (b) $\sqrt[3]{100}$ (c) $\sqrt[3]{-19}$ (d) $\sqrt[3]{-9}$
3. Use the "divide and average" method to find each of the following, expressing the approximation correct to the nearest .001 and also to the nearest .01.
   (a) $\sqrt{3779}$ (b) $\sqrt{285.2}$ (c) $-\sqrt{23}$

11.6 THE SYSTEM OF REAL NUMBERS

The set of real numbers, R, is the union of two *disjoint* sets, the set of *rational* numbers, Q, and the set of *irrational* numbers. It is the set of numbers such that every element corresponds to a point on the number line and also every point on the number line corresponds to an element of the set.

The *system of real numbers* is composed of

a set of elements, R,
two binary operations, $+$ and $\times$,
an equivalence relation, $=$,
an order relation, $<$,

together with certain properties which make the system a *complete ordered field:*

Commutative Properties

For each $a, b \in R$,

$$a + b = b + a.$$

For each $a, b \in R$,

$$a \times b = b \times a.$$

Associative Properties

For each $a, b, c \in R$,

$$(a + b) + c = a + (b + c).$$

For each $a, b, c \in R$,

$$(a \times b) \times c = a \times (b \times c).$$

Distributive Property of Multiplication over Addition

For each $a, b, c \in R$,

$$a \times (b + c) = (a \times b) + (a \times c).$$

$$(b + c) \times a = (b \times a) + (c \times a).$$

Identity Elements

For each $a \in R$,

$$a + 0 = 0 + a = a.$$

For each $a \in R$,

$$a \times 1 = 1 \times a = a.$$

Inverse Elements

For each $a \in R$,

$$-a \in R \qquad \text{and} \qquad a + -a = 0.$$

For each $a \in R, a \neq 0$,

$$\frac{1}{a} \in R \qquad \text{and} \qquad a \times \frac{1}{a} = 1.$$

The properties listed above are called *field properties.*

Trichotomy Property

For each $a, b \in R$, one and only one of the following sentences is true:

$$a < b \quad \text{or} \quad a = b \quad \text{or} \quad a > b.$$

Properties stated next involve the equals relation, $=$.

Uniqueness Properties for $=$

For each $a, b, c \in R$, if $a = b$, then

$$a + c = b + c.$$

For each $a, b, c \in R$, if $a = b$, then

$$a \times c = b \times c.$$

Cancellation Properties for $=$

For each $a, b, c \in R$, if $a + c = b + c$, then

$$a = b.$$

For each $a, b, c \in R$, $c \neq 0$, if $a \times c = b \times c$, then

$$a = b.$$

Properties stated below involve the order relation, $<$.

Uniqueness Properties for $<$

For each $a, b, c \in R$, if $a < b$, then

$$a + c < b + c.$$

For each $a, b, c \in R$, if $a < b$, then

(i) $\quad a \times c < b \times c \quad$ if $\quad c > 0$;
(ii) $\quad a \times c > b \times c \quad$ if $\quad c < 0$.

Cancellation Properties for $<$

For each $a, b, c \in R$, if $a + c < b + c$, then

$$a < b.$$

For each $a, b, c \in R$, $c \neq 0$, if $a \times c < b \times c$, then

(i) $\quad a < b \quad$ if $\quad c > 0$;
(ii) $\quad a > b \quad$ if $\quad c < 0$.

Now that the system of real numbers has been described in the preceding summary, we shall identify it as a field through the following diagram. See Section 9.10 for field requirements.

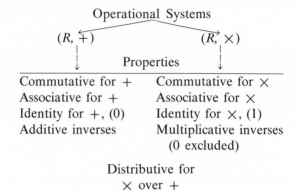

Operational Systems

$(R, +)$ $(R, \times)$

Properties

Commutative for $+$	Commutative for $\times$
Associative for $+$	Associative for $\times$
Identity for $+$, (0)	Identity for $\times$, (1)
Additive inverses	Multiplicative inverses (0 excluded)

Distributive for
$\times$ over $+$

The discussion of this system closes our study of the development of the real number system. We have traced this development in the hope that students will be able to assign numbers to their proper places in this hierarchy. The following diagram shows how the sets of numbers are related.

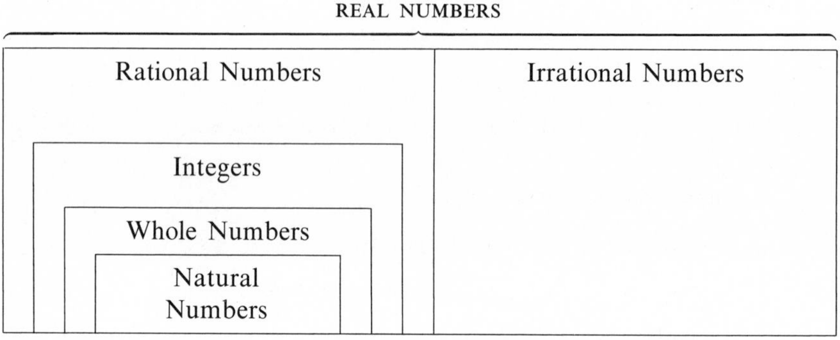

EXERCISE 11.6

1. List the classifications to which each of the following belongs.
   (a) 13 (Natural, Whole, Integer, Rational, Real)
   (b) $\sqrt[3]{5}$ (Irrational, Real)
   (c) $\frac{2}{7}$
   (d) 0
   (e) $^{-}(\frac{3}{5})$
   (f) $^{-}51$
   (g) 7.3
   (h) π
   (i) $11.\overline{27}$
   (j) 2.020020002 . . .
   (k) $-\sqrt{16}$
   (l) $\sqrt[4]{-16}$
   (m) $\sqrt[3]{(\frac{27}{64})}$
2. Make a chart tracing the structure of the real number system, showing the elements, operations, and properties which distinguish each succeeding system.

12

Metric Geometry

12.1 DISCRETE AND CONTINUOUS SETS; COUNTING AND MEASURING

We frequently find it necessary to answer such questions as "How many persons belong to your family?" or "How many chairs are in this room?" The answer to each of these questions is a whole number. Each element of these sets is a separate, whole object. Such sets are called *discrete* sets, and their elements can be *counted.*

Other questions that confront us daily are such questions as:

"How long is the room?"
"How tall is Joe?"
"How wide is the picture?"

"How long?" and "How tall?" and "How wide?" are questions about something considered to be all in one piece without any separations, or, in more mathematical language, *continuous.* The answers to these questions cannot be determined by counting separate objects or elements of a set, but require the *measuring* process.

12.2 NONMETRIC AND METRIC GEOMETRY

The various sets of points studied in Chapter 4, such as line and plane, are examples of continuous sets of points. In that chapter we were not dealing with measure, and the type of work done there is called *nonmetric geometry.* When measure is associated with sets of points, the resulting study is called *metric geometry.* Thus, if the union or the intersection of two line segments is being discussed, we are dealing with a nonmetric situation; if the length of a line segment is under consideration, we have a metric situation.

In this chapter, we shall study the process of measuring. By understanding the measuring process, one comprehends more fully the meaning of length, area, and volume.

12.3 CONGRUENCE; MEASUREMENT OF A LINE SEGMENT

If we write $\overline{AB} = \overline{AB}$, the symbol "=" indicates that the same set of points has been named. Measurement is not considered.

When two sets of points are not identical, we wish to have some means of comparing them. Let us consider two line segments, $\overline{CD}$ and $\overline{AB}$:

A comparison of these two line segments may be made by imagining a "copy" of $\overline{CD}$ to be placed on $\overline{AB}$. Such a copy may be represented by placing an edge of paper along $\overline{CD}$, marking the point C and the point D on the edge, and then placing the edge of the paper along $\overline{AB}$ with C coinciding with A:

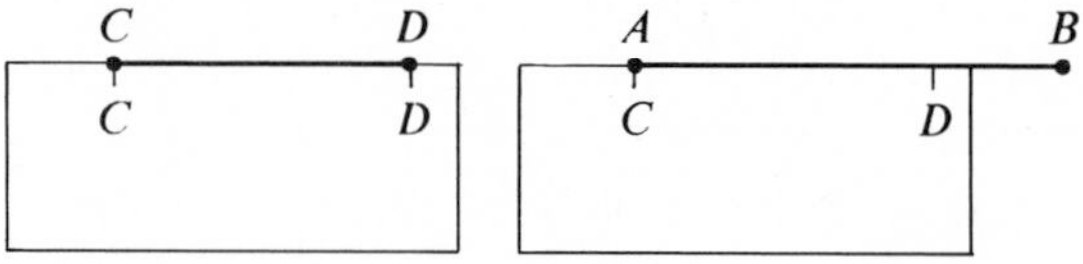

Where does D fall? Is it between A and B? This leads us to say that $\overline{CD}$ *is shorter than* $\overline{AB}$.

Let us compare $\overline{CD}$ with $\overline{XY}$ below. Here we find that D does not fall between X and Y, and so we say that $\overline{CD}$ *is longer than* $\overline{XY}$:

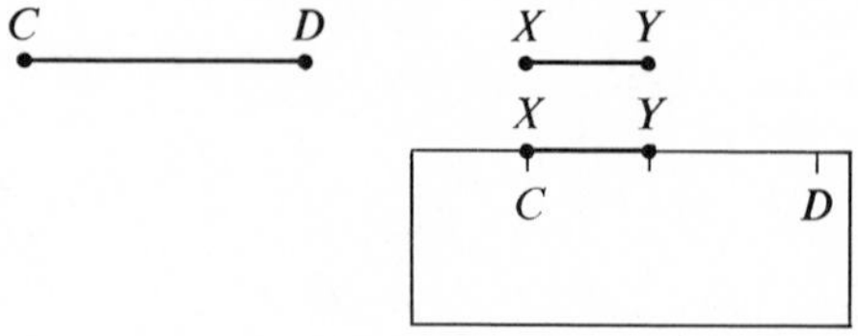

By comparing $\overline{CD}$ with $\overline{PQ}$ below, we find that when C coincides with P, D falls on Q:

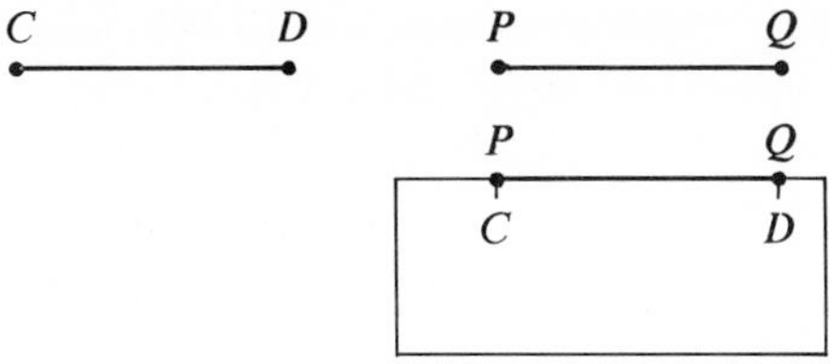

When this occurs, we say that $\overline{CD}$ *is congruent to* $\overline{PQ}$. Informally, we may say that $\overline{CD}$ "just fits" $\overline{PQ}$. The symbol for "is congruent to" is "$\cong$." Using this symbol, we write

$$\overline{CD} \cong \overline{PQ}.$$

Note that it would be incorrect to write $\overline{CD} = \overline{PQ}$. Why? Are $\overline{CD}$ and $\overline{PQ}$ identical sets of points?

The *measuring process* is based on our ability to lay off a specified unit of measurement along a continuous set of points. This idea is used when the farmer says that his horse is 15 "hands" high or a gate is 45 "paces" from his barn.

In measuring a line segment, a *unit* line segment is selected and this is laid off along the given segment in such a way that consecutive copies of the unit segment have only an end point in common. The copies of the unit segment are, of course, congruent segments. We may also describe this as "rubber-stamping" a line segment with a unit line segment. Suppose that we wish to rubber-stamp $\overline{XY}$ with the line segment u. That is, copies of u are marked end-to-end along $\overline{XY}$. Here $\overline{XY}$ is *subdivided* exactly by the unit u:

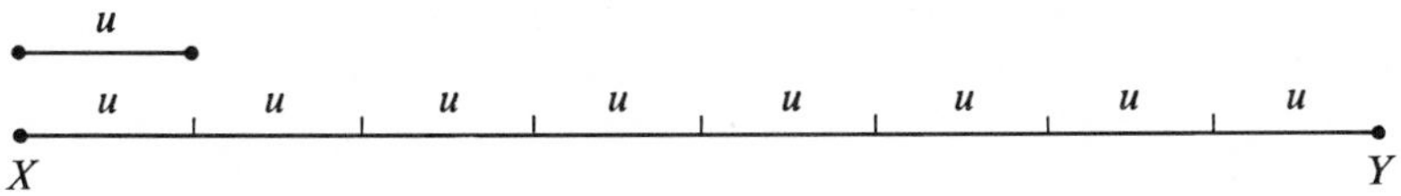

Counting the number of congruent copies of u, we find that there are 8. We say that the *measure* of $\overline{XY}$ is 8, and we write

$$m\overline{XY} = 8.$$

When "m" is written in front of $\overline{XY}$, it denotes that a measure is being related to the line segment. The segment is no longer being considered merely as a set of points as was done in Chapter 4. When we write "$m\overline{XY}$," a number is being associated with the continuous set of points $\overline{XY}$, just as when we write "N(A)," a number is being associated with the discrete set A.

When the measurement of $\overline{XY}$ (or any line segment) is discussed, the following distinctions must be made:

8, the *number,* is the *measure;*
u is the *unit of measurement;*
$8u$ is the *length* or *measurement.*

Measurement of lengths is called *linear measurement,* and units of length are called *linear units.*

Now suppose that $\overline{XY}$ is subdivided by the unit t. We lay off t on $\overline{XY}$ and see that it "fits on" $\overline{XY}$ 6 times, but with some of $\overline{XY}$ left over:

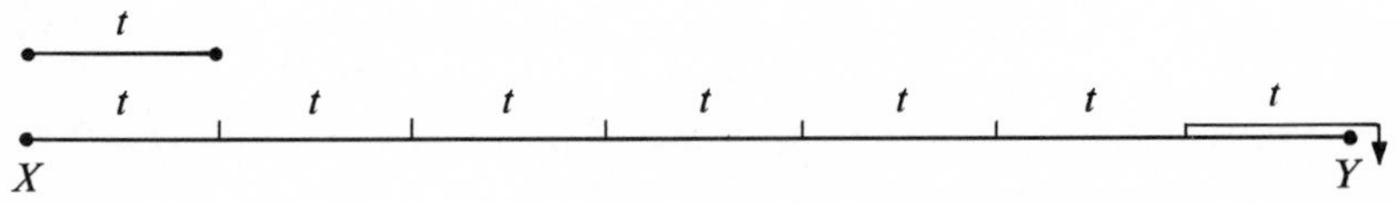

Since the leftover portion is almost another unit, the measure of $\overline{XY}$ is nearer 7 than 6. That is, the measure of $\overline{XY}$ is 7 *to the nearest unit t.* To indicate that the measure is very near 7, we use the symbol "$\approx$" to mean "is *approximately* equal to" and write

$$\mathrm{m}\overline{XY} \approx 7.$$

We might also write

$$6 < \mathrm{m}\overline{XY} < 7.$$

To say that the measurement, or length, of $\overline{XY}$ *is* $7t$ means that the measurement is given *to the nearest unit t.*

Suppose that the unit of measurement is changed to a smaller unit, h. This new unit h is laid off on $\overline{XY}$ as shown:

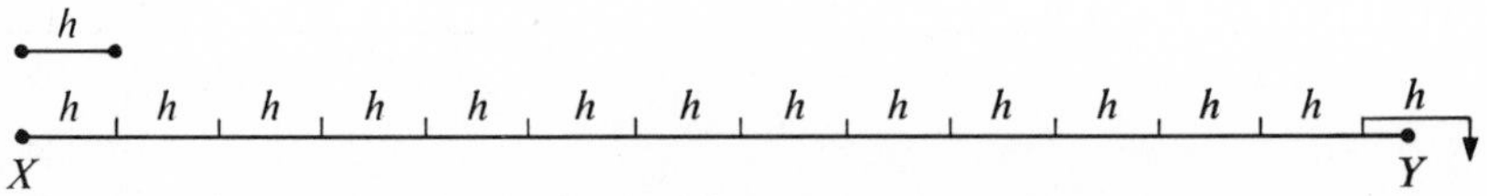

The measure of $\overline{XY}$ may be determined by counting the number of h units. The h unit fits on $\overline{XY}$ 13 times with a portion of $\overline{XY}$ left over. The measure of $\overline{XY}$ is nearer 13 than it is to 14. We write

$$\mathrm{m}\overline{XY} \approx 13 \quad \text{and} \quad 13 < \mathrm{m}\overline{XY} < 14.$$

The measurement of $\overline{XY}$ may be given as $8u$ or $7t$ or $13h$, depending on the unit used. Since the unit h is smaller than the unit t, we say that the measurement 13*h* is more *precise* than the measurement 7*t*. That is, the use of a smaller unit gives a more precise measurement than does the larger unit.

Even though the process of measurement is an approximate one, for every line segment there can be assigned a positive real number. If these numbers are the same for two line segments, then we say that the measures are the same and these line segments are congruent. On the other hand, if two line segments are congruent, then their measures are the same.

EXERCISE 12.3

1. Use the method of comparing line segments as in Section 12.3 to determine whether $\overline{AP}$ is longer than, shorter than, or congruent to the line segments given in the list below. Use an edge of a sheet of paper (or a string) to make your comparisons.

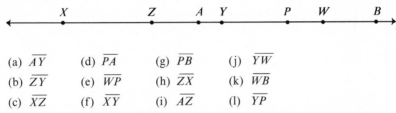

(a) $\overline{AY}$ (d) $\overline{PA}$ (g) $\overline{PB}$ (j) $\overline{YW}$
(b) $\overline{ZY}$ (e) $\overline{WP}$ (h) $\overline{ZX}$ (k) $\overline{WB}$
(c) $\overline{XZ}$ (f) $\overline{XY}$ (i) $\overline{AZ}$ (l) $\overline{YP}$

2. In Problem 1(d), above, did you see the need for making the comparison with the edge of the sheet of paper as was done in the other problems? Why?
3. Lay off the unit p along $\overline{PQ}$.

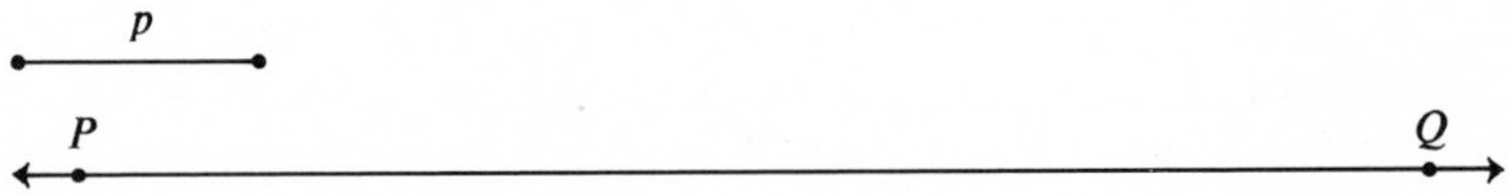

(a) How many whole units fit into $\overline{PQ}$? Is there any portion of $\overline{PQ}$ left over?
(b) What is the best measure to give for $\overline{PQ}$, using p as a unit? Explain.
(c) The measure of $\overline{PQ}$ is between what two whole numbers? Answer this by filling the blanks with the correct numerals:

$$\_\_\_\_ < m\overline{PQ} < \_\_\_\_$$

(d) What is the length of $\overline{PQ}$ to the nearest unit p?
(e) Would the use of a unit smaller than p give a more precise measure? Demonstrate. What is the measure of $\overline{PQ}$ when this smaller unit is used? Give the measure to the nearest unit. Is this measure the same as that given in part (b) above? What is the length of $\overline{PQ}$ to the nearest new unit?

12.4 STANDARD UNITS OF LENGTH

Unless the same unit of measurement was used by all members of the class to measure $\overline{PQ}$ in Problem 3(e) of Exercise 12.3, a difficulty will arise if the results are compared. An agreement on a unit of measurement is necessary if comparisons are to be made. When many people agree on a certain unit for measuring—that is, when it is accepted nationally—that unit is said to be a *standard unit of measurement* for that country.

Some of the earliest attempts at developing standard units were made by the Egyptians along the Nile as early as 6000 B.C. They turned to the human body for their units of measurement. The *cubit,* the length from the tip of the middle finger to the elbow, was used by the Egyptians in making measurements needed in the construction of their pyramids. They also used the hand and the fingers as units. The Romans used the thumb for the unit we call the *inch.*

In order that a country might have standard units for measuring, its king issued decrees proclaiming what these units should be. From fact and fable, interesting stories have come to us concerning these decrees. King Edward II of England proclaimed that three barleycorns placed end to end constituted an *inch,* the *foot* was the heel to toe measurement of King Alfred's foot, while King Henry I decreed that the *yard* was the length from the tip of his nose to the end of his thumb when his arm was outstretched. Of interest is the account of the establishment of the *rod* at some time during the sixteenth century. Sixteen men coming out of church on a certain Sunday were lined up. The sum of the lengths of the left feet of these sixteen men determined the rod. The inch, the foot, the yard, and the rod are linear units of the so-called *English system* of measurement.

The *metric system,* having the *meter* as the basic unit of length, was invented by a group of scientists. The idea was presented in 1791 to the French government by a committee of the Academy of Sciences, and the standard units were accepted in 1799 by a conference made up of representatives from many countries. The objective of these men was to establish a system of measurement that they hoped would be accepted by all nations. They felt the need of a standard international system, particularly when it was necessary to compare scientific data resulting from measuring. The acceptance of this system met with much opposition. You will be interested in reading about the struggle that took place in France before this system was finally adopted. Even though the metric system has been legalized in the United States (in 1866), the English system is used more extensively except in scientific work.

A bar made of platinum and iridium kept by the National Bureau of Standards in Washington, D.C., has our national *standard meter* marked on it. Originally the meter was defined by the French as one ten-millionth of the earth's quadrant passing through Paris. In 1960 it was redefined as the length equal to 1,650,763.73 wavelengths in a vacuum of the orange-red radiation of krypton 86. An instrument has been invented which is capable of measuring wavelengths with great precision. Now, the meter may be accurately determined by means of this instrument.

12.5 RULERS AND SYSTEMS OF LINEAR MEASUREMENT

In order to make more precise measurements, it is necessary to use smaller units (Section 12.3). Thus, a foot is subdivided into 12 inches, indicated on a *foot ruler* by the numerals 1, 2, . . . , 12 located at one-inch intervals. Thus a foot ruler represents a portion of the number line. When the one-inch intervals, representations of line segments, are subdivided into congruent intervals, we have still smaller units. Only a few of the subdivisions have been shown in the following drawing:

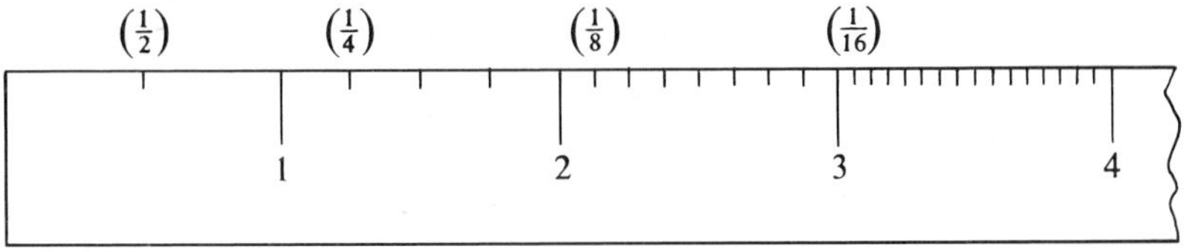

The first interval is subdivided into two congruent intervals (half inches); hence, each subdivision here is a half inch. The other intervals show an inch subdivided into $\frac{1}{4}$-, $\frac{1}{8}$-, and $\frac{1}{16}$-inch intervals. Look at your foot ruler to get a complete picture of these subdivisions. Each mark has a *rational number* assigned. The system of designating units on a ruler is called a *scale.*

The *yardstick* is used to measure cloth, the length of a room, and objects for which a foot ruler is not practical. The yard is subdivided into three foot-units. Since one yard is the same length as three feet, we choose to use the symbol "=" (the same as) to express this relation and others shown below. The relations between the more commonly used units of length in the English system are as follows:

$$1 \text{ yard} = 3 \text{ feet}$$
$$1 \text{ foot} = 12 \text{ inches}$$
$$1 \text{ yard} = 3 \text{ feet} = 36 \text{ inches}$$

Other linear units of measurement in the English system may be found in any junior high school mathematics textbook.

When a pilot reports the altitude the airplane is flying, he gives it as a number of feet above sea level. The people of Denver speak of their city as "the mile-high city." Is it possible for a pilot to fly over Denver at an altitude of 4000 feet? It is necessary for the pilot to know that a mile is equivalent to 5280 feet in order for him to adjust his flight plans.

Many of the rulers that we use today are also subdivided into *centimeter* units. This is a subdivision of the basic metric unit of length, the *meter.*

A *centi*meter is one of the 100 subdivisions of a meter, or $\frac{1}{100}$ meter. This is a very convenient unit for measuring short lengths. Other units smaller than the meter are the *decimeter* and the *millimeter*. As you no doubt surmise, a *deci*meter is one of the 10 subdivisions of a meter and a *milli*meter is one of 1000 subdivisions.

This drawing represents a meter which has been greatly reduced. Each of the ten subdivisions represents $\frac{1}{10}$ of a meter or a decimeter.

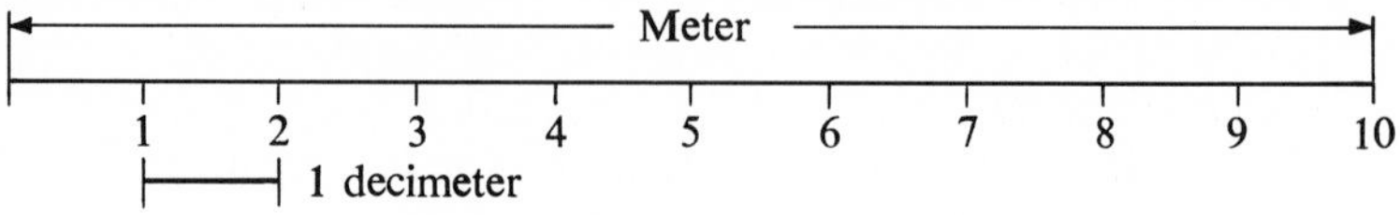

We now "remove" the decimeter from the preceding drawing and enlarge it. It is then subdivided into 10 centimeters. One of the centimeters is shown subdivided into 10 millimeters.

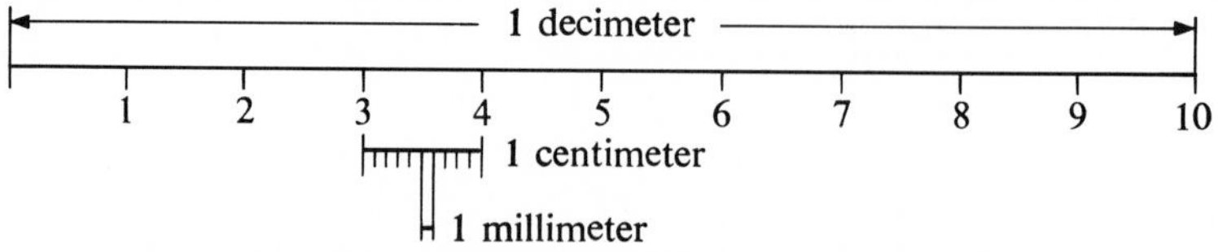

Summarizing the relationship among these units, we have:

$$
\begin{aligned}
1 \text{ meter} &= 10 \text{ decimeters}\\
1 \text{ decimeter} &= 10 \text{ centimeters}\\
1 \text{ centimeter} &= 10 \text{ millimeters}
\end{aligned}
$$

$$1 \text{ meter} = 10 \text{ decimeters} = 100 \text{ centimeters} = 1000 \text{ millimeters}$$

The abbreviations for these units are: m for meter, dm for decimeter, cm for centimeter, and mm for millimeter.

The Latin prefixes *deci-*, *centi-*, and *milli-* attached to the word *meter* are used in designating $\frac{1}{10}$, $\frac{1}{100}$, and $\frac{1}{1000}$, respectively, of a meter. The Greek prefixes *deca-*, *hecto-*, and *kilo-* attached to the word *meter* are used to designate 10, 100, and 1000 times a meter. If a meter is placed end to end 10 times, the measurement is called a *decameter* (dkm); for 100 times, a *hectometer* (hm); for 1000 times, a *kilometer* (km).

The following diagram indicates the relatedness of each of the units to the meter. The direction of the arrows is significant in interpreting the relationship. Observe the two arrows connecting the meter and the centimeter. One arrow starts at meter and goes to centimeter. This is read, "One meter is 100 centimeters". The returning arrow is interpreted as, "One centimeter is $\frac{1}{100}$ of a meter." You should be able to interpret the other arrows.

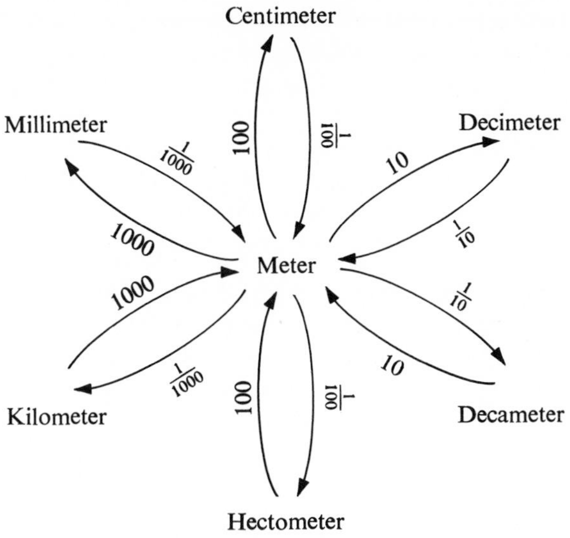

Observe the relationship between the metric system of measurement and the decimal numeral (Section 10.5). For example, consider the measurement 5.326 meters. Using expanded notation, we can write:

$$5.326 = (5 \times 1) + \left(3 \times \frac{1}{10}\right) + \left(2 \times \frac{1}{100}\right) + \left(6 \times \frac{1}{1000}\right)$$

$$= \quad 5 \quad + \quad \frac{3}{10} \quad + \quad \frac{2}{100} \quad + \quad \frac{6}{1000}$$

But $\frac{3}{10}$ of a meter is 3 dm, $\frac{2}{100}$ of a meter is 2 cm, and $\frac{6}{1000}$ of a meter is 6 mm. Hence, 5.326 meters may be identified as 5 meters, 3 decimeters, 2 centimeters, and 6 millimeters.

This is useful in case one wishes to identify a measurement in meters in terms of the three subdivision units of the meter. However, usually measurements need to be converted to a single unit as shown in the examples which follow.

Example 1 Convert 5.3 meters to (a) centimeters, and (b) decimeters.

(a) Since 1 meter = 100 centimeters,

$$5.3 \text{ m} = 5.3 \times 100 = 530 \text{ cm}.$$

(b) Since 1 meter = 10 decimeters,

$$5.3 \text{ m} = 5.3 \times 10 = 53 \text{ dm}.$$

Example 2 Convert 850 millimeters to meters.

Since 1 millimeter = $\frac{1}{1000}$ meter,

$$850 \text{ mm} = 850 \times \tfrac{1}{1000} = 0.850 \text{ m}.$$

Example 3 Convert 8.2 decameters to centimeters.
Since 1 decameter $= \frac{1}{10}$ meter, then

$$8.2 \text{ dkm} = 8.2 \times \tfrac{1}{10} = 0.82 \text{ m}.$$

Since 1 m = 100 cm, then

$$0.82 \text{ m} = 0.82 \times 100 = 82 \text{ cm}.$$

Hence,

$$8.2 \text{ dkm} = 82 \text{ cm}.$$

Example 4 How many millimeters are there in 1 hectometer?
Since 1000 mm = 1 m, and $\frac{1}{100}$ hm = 1 m, then

$$\tfrac{1}{100} \text{ hm} = 1000 \text{ mm}.$$

Hence,

$$1 \text{ hm} = 100 \times 1000 = 100{,}000 \text{ mm}.$$

Observe the role of the meter in converting the units in each of the preceding examples. The meter serves as the base. However, it is possible to become sufficiently skilled to convert from one unit to another by bypassing the meter.

Since in the United States we continue to use both the English and the metric systems, we should be able to express measurements interchangeably in either system. In the United States the legal length of the yard is $\frac{3600}{3937}$ of a meter. Thus, the meter in our country is 39.37 inches. This is how many inches longer than the yard? In Great Britain and France the meter has been established as 39.37079 inches.

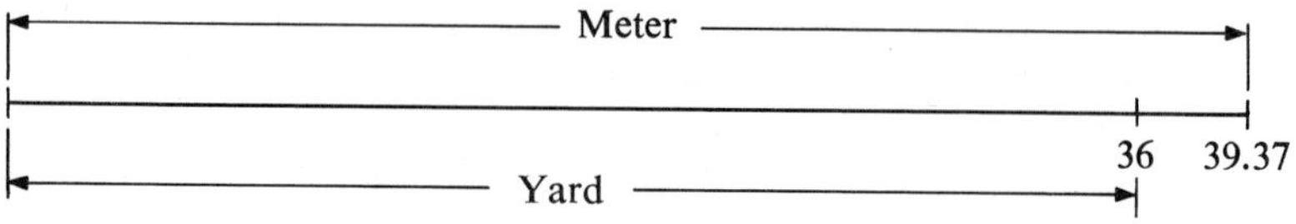

The following gives the approximate relation between linear units in the metric system and the inch in the English system, as used by the United States:

$$1 \text{ meter} \approx 39.37 \text{ inches}$$
$$100 \text{ centimeters} \approx 39.37 \text{ inches}$$
$$1 \text{ centimeter} \approx .3937 \text{ inch}$$

Since

$$1 \text{ cm} \approx 0.3937 \text{ in.},$$

then

$$1 \text{ in.} \approx \tfrac{1}{0.3937} \text{ cm} \approx 2.540005 \text{ cm}.$$

In the United States, the inch is defined as 2.54 centimeters. The centimeter is slightly over $\frac{1}{3}$ of an inch as shown in the following diagram.

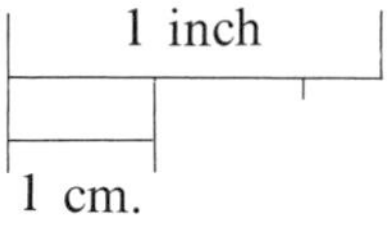

Another useful relationship is that of the mile and the kilometer.

$$1 \text{ mile} \approx 1.61 \text{ kilometers}$$

Observe that the symbol for approximately equal to ($\approx$) is used to show that the conversion factors from one system to another are approximate. Some may prefer to use the equals ($=$) instead, but keep in mind that the result obtained in making the conversion is approximate.

The following examples show conversions from one system to another.

Example 5 A highway sign in France gives the distance to Paris as 84 km. Interpret this in terms of miles for an American driver.

Since 1 mi $\approx$ 1.61 km, then

$$1 \text{ km} \approx \tfrac{1}{1.61} \text{ mi} \approx 0.621 \text{ mi}.$$

Hence,

$$84 \text{ km} \approx (84 \times 0.62) \text{ mi} \approx 52.08 \text{ mi}.$$

Again you are reminded that this result is approximate, but is quite adequate knowledge for the American driver.

Example 6 A person running the 100 yard dash wishes to compare it with the distance a person runs in a 100 meter race.

Since 1 in. $\approx$ 2.54 cm,

$$1 \text{ yd} \approx (36 \times 2.54) \text{ cm} \approx 91.44 \text{ cm},$$
$$100 \text{ yd} \approx (100 \times 91.44) \text{ cm} \approx 9144 \text{ cm}.$$

Since 1 cm $= \tfrac{1}{100}$ m,

$$9144 \text{ cm} = (9144 \times \tfrac{1}{100}) \text{ m} = 91.44 \text{ meters}.$$

Hence, 100 yds $\approx$ 91.44 m, which is 8.56 m less. How many yards less is it?

EXERCISE 12.5

1. Construct a 6-inch ruler with inch subdivisions on one edge and centimeter subdivisions on the other. Begin marking the subdivisions from the same end.
   (a) Use the ruler to give the nearest centimeter measurement for 1 inch, 2 inches, $\frac{1}{2}$ inch, $3\frac{1}{4}$ inches.
   (b) Using 1 centimeter as .3937 inch, one inch is approximately equal to how many centimeters? Give your answer to the nearest hundredth.
2. By observation identify each of the points given on line p with the numerals from both the inch scale and the centimeter scale. Give each to the nearest subdivision on the scales.

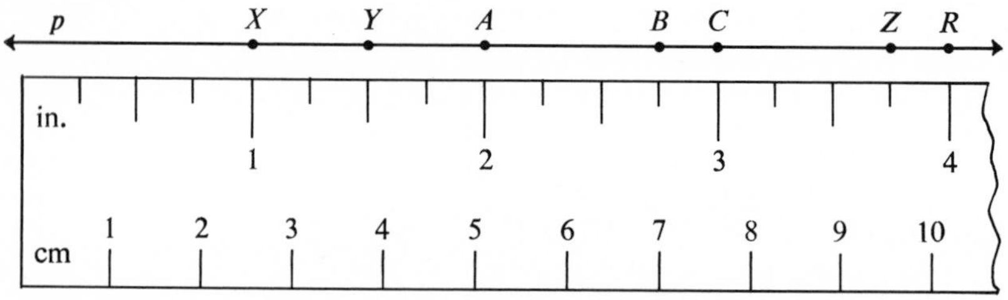

3. Use one inch as approximately equal to 2.54 cm to verify the results found in Problem 2. For instance, point Y is matched with $1\frac{1}{2}$ on the inch scale. Using 1 inch $\approx$ 2.54 cm, $1\frac{1}{2}$ inches $\approx (1\frac{1}{2} \times 2.54)$ cm. This is 3.81 cm. Hence, point Y is matched with about 3.81 on the centimeter scale. Since the subdivision is a centimeter, Y is nearest to 4 on the centimeter scale.
4. Using the figure in Problem 2, find the length of each of the following in inches, in centimeters, then in millimeters. Give the centimeter measurement to the nearest hundredth of a centimeter and the millimeter measurement to the nearest tenth of a millimeter.

   (a) $\overline{XY}$ (c) $\overline{AB}$ (e) $\overline{ZC}$ (g) $\overline{YA}$
   (b) $\overline{XA}$ (d) $\overline{AR}$ (f) $\overline{YB}$ (h) $\overline{XB}$
5. Since $\overline{XB}$ in the figure in Problem 2 is equal to $(\overline{XY} \cup \overline{YA}) \cup \overline{AB}$, show how the length of $\overline{XB}$ may be developed from this fact. Compare the length obtained with that using the ruler.
6. Convert the following measurements to the units indicated.
   (a) 10 meters to centimeters
   (b) 10 centimeters to meters
   (c) 2.7 centimeters to millimeters
   (d) 2.7 millimeters to centimeters
   (e) 27 millimeters to decimeters
   (f) 0.068 centimeter to meters
   (g) $1\frac{1}{2}$ decimeters to meters
   (h) 3 meters to millimeters
7. Express each of the following measurements in terms of all four units: meters, decimeters, centimeters, and millimeters.

   EXAMPLE. 2.310 meters is 2 meters, 3 decimeters, 1 centimeter, and 0 millimeters.

   (a) 8.263 meters
   (b) 0.58 meter
   (c) 1.002 meters
   (d) 203.023 centimeters
   (e) 6803.000 millimeters
   (f) 29.505 decimeters
8. Convert the following measurements to the units indicated.
   (a) 5 in. to cm
   (b) 35 cm to in.
   (c) 3.4 mm to in.
   (d) 7.5 ft to m
   (e) 256 mi to km
   (f) 583 yd to hm
   (g) 12 dm to ft
   (h) 21 km to yd
9. Use 1 in. $\approx$ 2.54 cm to show that 1 mi $\approx$ 1.61 km.
10. Which of the following record the longer measurement?
    (a) 1 mi or 1 km
    (b) $1\frac{2}{3}$ yd or 1.5 m
    (c) 2 mm or $\frac{2}{10}$ in.
    (d) $\frac{3}{4}$ yd or 75 cm
    (e) 2 ft or $\frac{2}{3}$ m
    (f) 6 in. or 15 cm

11. A hiking club hiked 12 miles in 4 hours. About how many kilometers did they hike in 6 hours, if they maintained the same rate?
12. An American-made car is being driven at the rate of 50 mi per hour in Italy. Is the speed limit being exceeded if it is posted as 75 km per hour?

12.6 ACCURACY OF MEASUREMENT

A recorded measurement under the most ideal circumstances can at best be only an approximation. The use of precision measuring instruments by skilled technicians still produces approximate results.

Yet, we do know that of two measurements taken, the one using the smaller unit gives a more precise measurement. That is, as smaller units for measuring are chosen, closer approximations to the actual length are obtained.

Let us consider measuring the line segment *AB* with the units of an inch, a half inch, and a fourth inch as shown in the diagram.

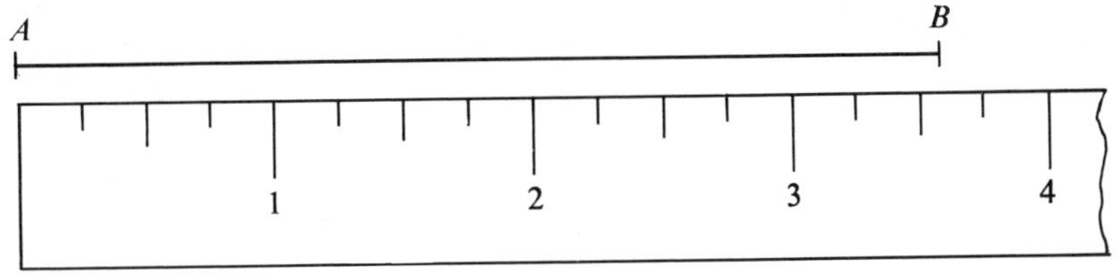

We find

$$m\overline{AB} \text{ to the nearest inch to be } 4;$$
$$m\overline{AB} \text{ to the nearest } \tfrac{1}{2} \text{ inch to be } 3\tfrac{1}{2};$$
$$m\overline{AB} \text{ to the nearest } \tfrac{1}{4} \text{ inch to be } 3\tfrac{3}{4}.$$

What is the *greatest possible error* that could have been made in assigning these measures to $\overline{AB}$? This is determined by finding $\frac{1}{2}$ of the unit used in measuring. The greatest possible error is found for each in the following examples.

Example 1 The $m\overline{AB}$ to the nearest inch is 4. What is the greatest possible error? The unit used is an inch, so the

$$\textit{greatest possible error} \text{ is } \tfrac{1}{2} \times 1 = \tfrac{1}{2} \text{ inch.}$$

Example 2 Since $m\overline{AB}$ to the nearest $\frac{1}{2}$ inch is $3\frac{1}{2}$, what is the greatest possible error? The unit used is $\frac{1}{2}$ inch, so the

$$\textit{greatest possible error} \text{ is } \tfrac{1}{2} \times \tfrac{1}{2} = \tfrac{1}{4} \text{ inch.}$$

Example 3 Since $m\overline{AB}$ to the nearest $\frac{1}{4}$ inch is $3\frac{3}{4}$, what is the greatest possible error?

The unit used is $\frac{1}{4}$, so the

greatest possible error is $\frac{1}{2} \times \frac{1}{4} = \frac{1}{8}$ inch.

From these examples we see that the smaller the unit selected for measuring, the smaller is the greatest possible error. This in turn produces a closer approximation of the actual length.

The next question might be: What is the percent of error for each of the preceding measurements? The percent of error is determined by dividing the greatest possible error by the measure, then converting this number to percent. The data necessary for computing the greatest possible error and the percent of error for each of the measurements of $\overline{AB}$ is given in the following table. The last two columns show the procedure for computing these.

Measurement	*Unit*	*Greatest Possible Error*	*Percent of Error*
5 in.	$1 = 1$ in.	$\frac{1}{2} \times 1 = 0.5$ in.	$\frac{0.5}{5} = 0.10 = 10\%$
$4\frac{1}{2}$ in.	$\frac{1}{2} = 0.5$ in.	$\frac{1}{2} \times 0.5 = 0.25$ in.	$\frac{0.25}{4.5} = 0.0\bar{5} = 5.\bar{5}\%$
$4\frac{3}{4}$ in.	$\frac{1}{4} = 0.25$ in.	$\frac{1}{2} \times 0.25 = 0.125$ in.	$\frac{0.125}{4.75} = 0.026 = 2.6\%$

We shall now identify the *accuracy of a measurement* with the percent of error. That is, the smaller the percent of error of two or more measurements, the more accurate is that measurement. Using this criterion, $4\frac{3}{4}$ in. is the most accurate of the three measurements given for $\overline{AB}$.

It is important to note that in order to compare the accuracy of several measurements, one must know what unit was used. This is simple if one does his own measuring, but more difficult to determine when recorded measurements are to be interpreted by someone else. The following examples show how units are determined for recorded measurements.

Example 4 What unit was used in measuring a line segment whose recorded length is 3.57 cm.?

Find the last digit, from left to right, in the measure. It is 7 and in the hundredths place. The unit used in measuring is 0.01 cm.

Example 5 What unit is used in measuring 263 yards?

Since the last digit is in the ones place, the unit used is 1 yard.

Example 6 Two measurements reported for the height of Al, the basketball center, are 6 ft. $10\frac{1}{4}$ in. and 6 ft. $10\frac{1}{8}$ in. Which measurement is more precise? Which is more accurate?

The following data are necessary in answering the two questions.

	6 ft $10\frac{1}{4}$ in.	6 ft $10\frac{1}{8}$ in.
Unit	$\frac{1}{4}$ in.	$\frac{1}{8}$ in.
Greatest Possible Error	$\frac{1}{8} = 0.125$	$\frac{1}{16} = 0.0625$
Percent of Error	$\frac{0.125}{82.25} = 0.0015 = 0.15\%$	$\frac{0.0625}{82.125} = 0.00076 = 0.076\%$

Since the unit for the measurement 6 ft $10\frac{1}{8}$ in. is smaller than that for 6 ft $10\frac{1}{4}$ in., the former is more precise. Comparing the percent of error, $0.076\% < 0.15\%$; hence 6 ft $10\frac{1}{8}$ in. is the more accurate measurement.

EXERCISE 12.6

1. Given the following measurements:

   (a) 3 cm (b) 2.4 cm (c) 0.06 m (d) 7.148 dm (e) 23.39 mm

   Find (i) the unit of measurement; (ii) the greatest possible error; (iii) the percent of error of each.
2. The distance between two towns is reported to be 18.7 miles. Later, it is reported to be 19 miles. Which measurement is more precise? What is the greatest possible error for each? What is the percent of error for each? Which is the more accurate measurement? Why?
3. Which is the more accurate measurement, $4\frac{5}{11}$ or $4\frac{8}{9}$? Explain.

12.7 PERIMETER AND CIRCUMFERENCE

Problems that frequently come to our attention involve finding the *perimeter* of a polygon or the *circumference* of a circle. A typical problem that requires finding the perimeter of a rectangle is the following: How much fencing is needed to enclose a 75 ft × 165 ft lot? The circumference of a circle must be found to solve the following problem: How far does a pony travel in one complete revolution on a 50 ft radius merry-go-round?

Since a number is assigned to each perimeter and to each circumference, we call the number the measure. The measure and the linear unit together give the measurement.

The following examples should direct you in developing formulas for finding the perimeter of polygons.

Example 1 How much fencing is needed to enclose a 75 ft × 165 ft rectangular lot?

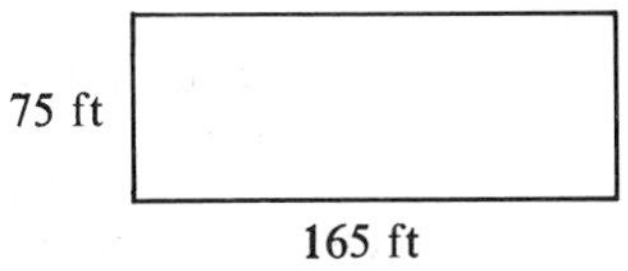

Let P represent perimeter, then

$$\begin{aligned} m(P) &= 75 + 165 + 75 + 165 \\ &= 480. \end{aligned}$$

Hence, the amount of fencing required is 480 feet.

Other ways to find the measure of the perimeter are as follows:

(i) $m(P) = (2 \times 75) + (2 \times 165)$

(ii) $m(P) = 2 \times (75 + 165)$

Equating (i) and (ii),

$$(2 \times 75) + (2 \times 165) = 2 \times (75 + 165),$$

an example of the distributive property for multiplication over addition.

Consider a rectangle having b the measure of the base and a the measure of the altitude.

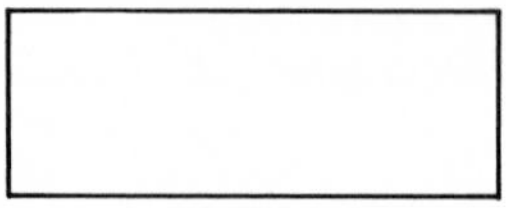

Then, $m(P) = 2a + 2b$ or $2(a + b)$.

Here is a parallelogram having b and c measures of the sides as shown.

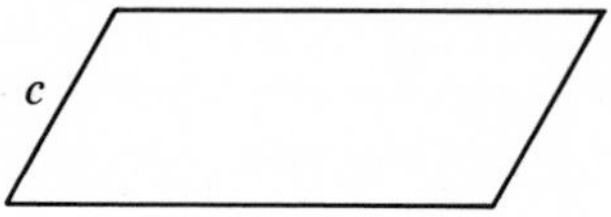

Then, $m(P) = 2b + 2c$ or $2(b + c)$.

Example 2 What is the length of a rope used to outline a play area as shown?

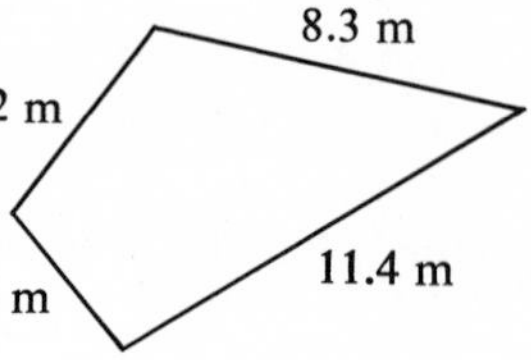

$$\begin{aligned} m(P) &= 8.3 + 11.4 + 5.8 + 6.2 \\ &= 31.7 \end{aligned}$$

Hence, the rope is 31.7 meters or 31 m and 7 dcm.

From this, it follows that for a polygon having $b, c, d, e, \ldots$ measures of the sides, the measure of the perimeter is the sum of all the sides.

$$m(P) = b + c + d + e + \cdots$$

How is the measure of the circumference of a circle determined? By the circumference of a circle is meant the boundary of the circular region. Its measure certainly cannot be found in the same way that the measure of the perimeter of a polygonal region was found. One procedure is to roll the circle along a line for one complete revolution. The measure of how far it rolls is the measure of the circumference of the circle. See the diagram.

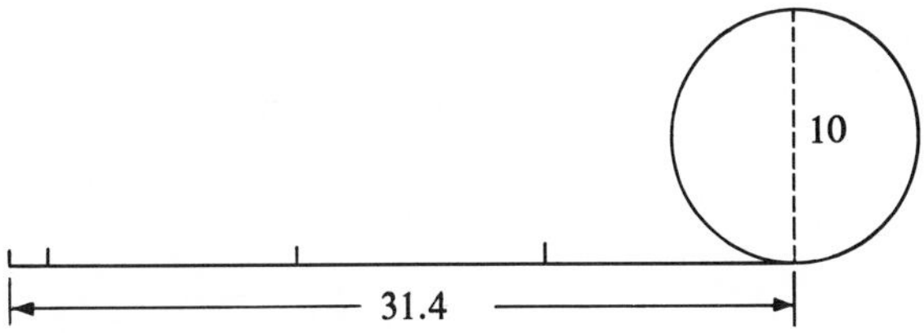

The measure of the circumference of the circle shown in the diagram is approximately 31.4. The measure of the diameter (a line segment passing through the center and having its end points on the circle) of this circle is 10. In a similar manner, the measures of the circumferences of four other circles were found and the data recorded in the following table:

	$m(\text{C})$	$m(\text{D})$	$m(\text{C}) \div m(\text{D})$
Circle 1	31.4	10	3.14
Circle 2	12.54	4	3.135
Circle 3	19.46	6.2	3.138
Circle 4	15.72	5	3.144
Circle 5	26.39	8.4	3.141

Observe the data in the last column. Rounding to the nearest hundredth, each becomes 3.14. This relationship between the measure of the circumference and the measure of the diameter ($c \div d$) is a number approximately equal to 3.14. Actually, it is the irrational number, 3.14159 . . . . Historically, this number is called π. For most problems the approximate decimal notation, 3.14, or the fraction $\frac{22}{7}$ is used.

From the preceding, we write

$$c \div d = \pi.$$

Using the definition for $\div$,

$$c = d \times \pi.$$

Hence, c and d represent the measures of the circumference and the diameter, respectively.

Now, the formula

$$c = \pi d$$

may be used instead of the rolling and measuring process to find the measure of the circumference of a circle. The example which follows illustrates this.

Example 3 How far does a pony travel in one revolution on a 50 ft merry-go-round?

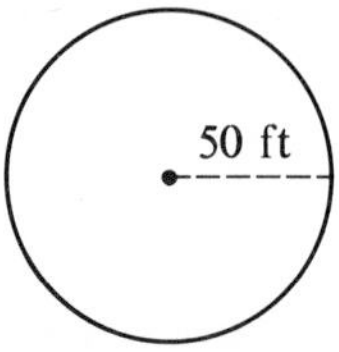

A 50 ft merry-go-round means that its radius (a line segment having one end point on the center and the other on the circle) has a measurement of 50 ft. The measure of the diameter is 2×50 or 100.

Using the formula

$$c = \pi d$$

with 100 for the measure of the diameter and 3.14 for π

$$\begin{aligned} c &= 100 \times 3.14 \\ &= 314. \end{aligned}$$

Hence, the pony travels 314 ft in one revolution of the merry-go-round.

EXERCISE 12.7

1. Find the measure of the perimeter of each of the following:

(a)

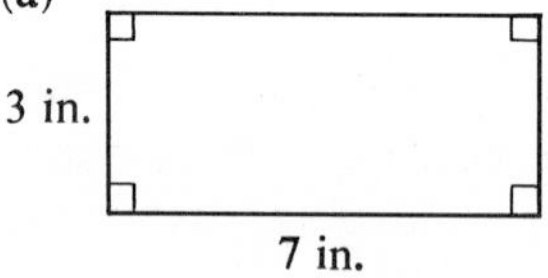

(b)

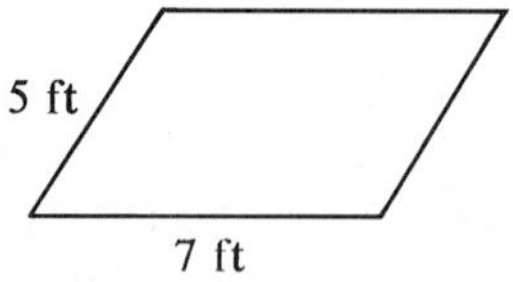

(c)

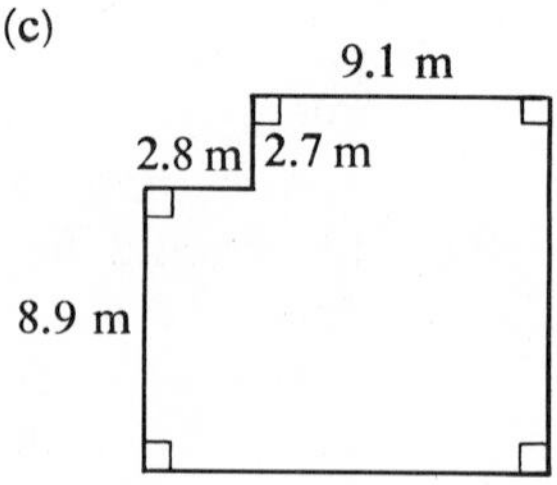

(d)

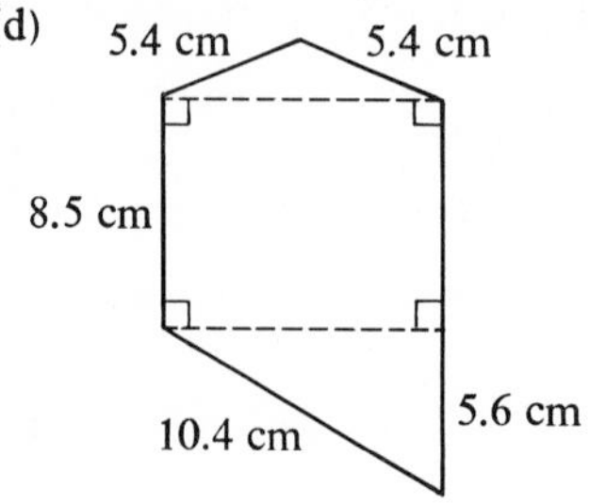

(e)

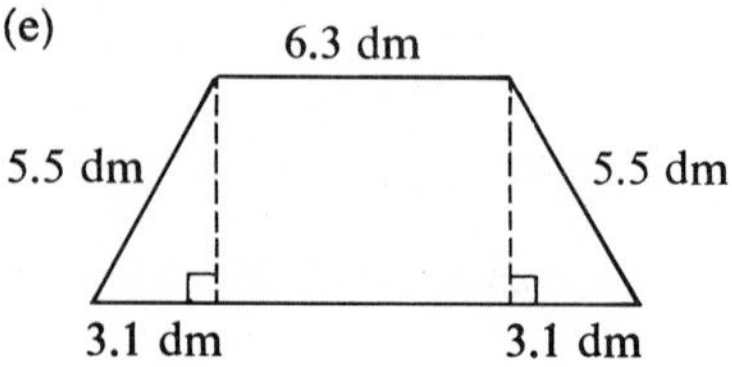

2. Find the measure of the circumference of each of the following:

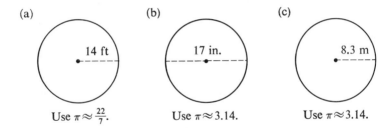

(a)

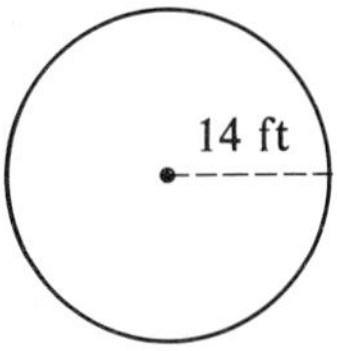

Use $\pi \approx \frac{22}{7}$.

(b)

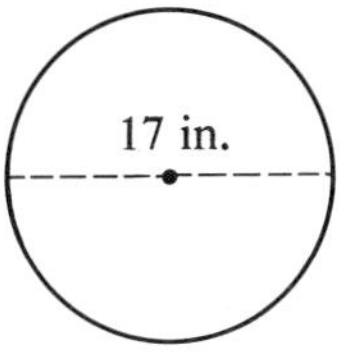

Use $\pi \approx 3.14$.

(c)

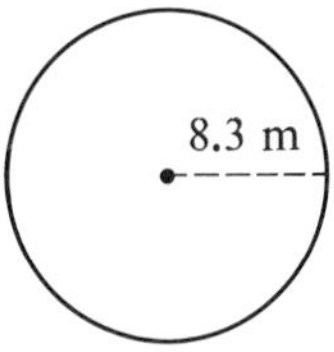

Use $\pi \approx 3.14$.

12.8 MEASUREMENT OF A REGION

Since it is possible to assign a measure to each line segment, one expects the same for a region. A line segment is subdivided by using a line segment as a unit. The measure is determined by counting the congruent subdivisions. If a measure is to be assigned to a region, then it must be subdivided into congruent subdivisions that may be counted. Obviously, the unit must be a region.

The region that follows has been subdivided by a square-shaped region.

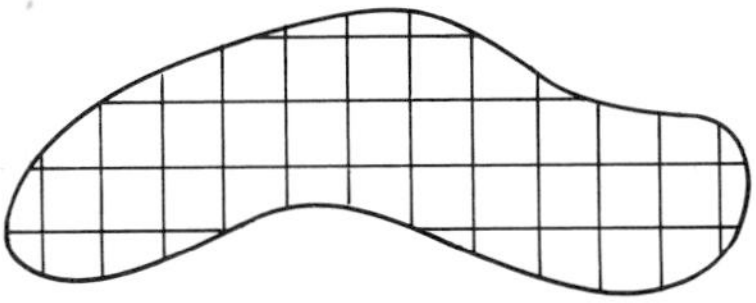

What is the difficulty in counting the subdivisions? Would the region ever be completely filled in by making the unit smaller and smaller? Your answer, no doubt, is that it would not. However, the smaller the unit, the more precise is the measure. That is, the smaller the unit, the closer would be the approximation to the actual "size" of the region.

The difficulty in measuring a region such as this is evident. This leads us to consider the measure of *polygonal regions:* those regions bounded by polygons.

We shall begin with *rectangular region R*. Since a *square region* is a very convenient unit to use, we shall select s as the unit of measurement. The following diagram shows R subdivided by the square unit s.

R

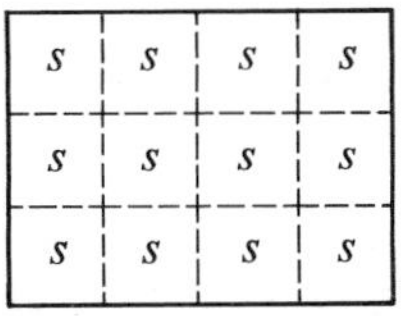

How many subdivisions are there in R? The number of subdivisions is the measure of region R. The measure of a region is called *area*. In other words, the area of a region is a *number*. Since the area is the number of subdivisions, it will vary as the unit is changed. The following can be said concerning rectangular region R which has been measured by unit s:

12 is the *area* (measure) of R;
s is the *unit of measurement;*
$12s$ is the *measurement*.

The area of region R may be determined by counting the subdivisions. However, since the subdivisions form a 3×4 array, the area may be determined by finding the product of 3 and 4.

As the rectangular region R was being subdivided, so were the line segments (the sides) that form the rectangle. Consequently, the numbers, 3 and 4, are the measures of the line segments as shown.

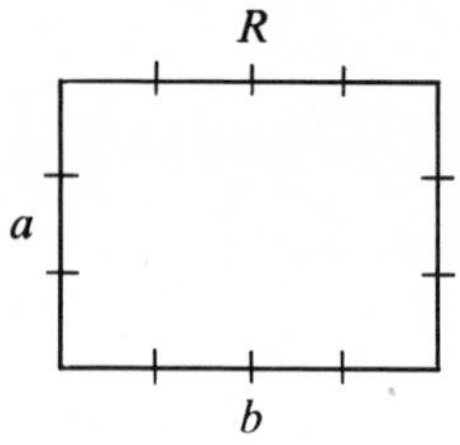

If b is the measure of one side called the *base,* and a is the measure of a connecting side called the *altitude,* the area A is expressed by this formula:

$$A_{\square} = b \times a$$

This information will now be used to find the area of other polygonal regions. In the following examples, formulas will be developed for areas of a few of these regions.

Example 1 Find the formula which may be used to determine the area of a region bounded by a parallelogram.

Let b and a be the measures of the base and the altitude, respectively. See Figure 12–1. Observe that the altitude is not one of the sides as it was in the rectangle. *The altitude is the perpendicular line segment between two parallel sides.* It is often referred to as the *height.*

By making a "cut" along the altitude, followed by a translation (a slide) along the base, the triangular region is relocated as shown in Figure 12–2.

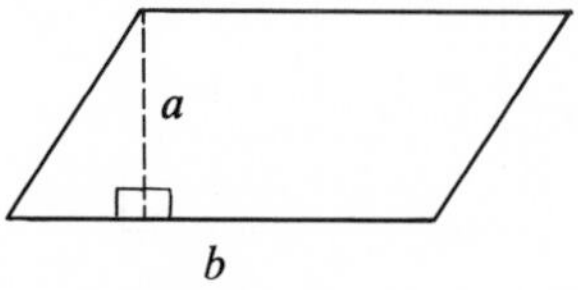

Figure 12–1

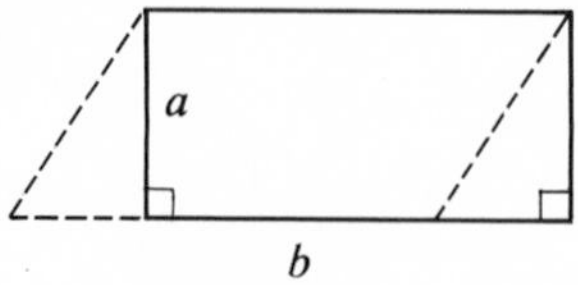

Figure 12–2

A rectangular region has been formed by the rearrangement of these parts of the parallelogram-region.

Now that the parallelogram has been converted into a rectangle with no loss of any of the original region, the area of the parallelogram-region is given in terms of the rectangular region—that is,

$$A_{\square} = b \times a,$$

where $A_{\square}$ is the area of the parallelogram-region, b is the measure of the base, and a the measure of the altitude.

From now on, when speaking of the area of a rectangular region we shall refer to it as the area of a rectangle. Corresponding names will be used for all other polygonal regions. In so doing, one must realize that area certainly is not the measure of a polygon, but of its region.

Example 2 How may a formula be developed for the area of a triangle?

This development will begin with a parallelogram-region, *PQRS*, Figure 12–3. "Cutting" this region along diagonal *d* forms two triangular regions *PQS* and *RSQ*, shown in Figure 12–4.

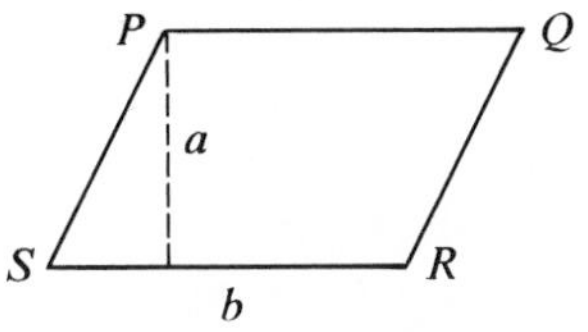

Figure 12–3

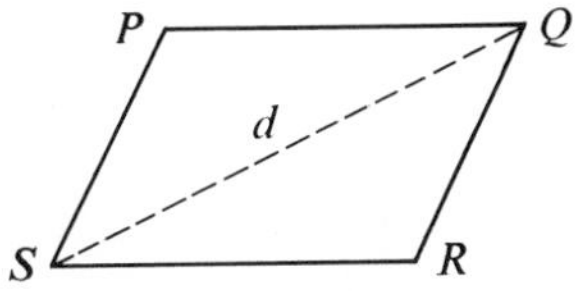

Figure 12–4

Make a $\frac{1}{2}$ turn of triangle *PQS* about *P*. See Figure 12–5. Now, by a translation, $\triangle PQS$ slides onto $\triangle RSQ$, shown in Figure 12–6.

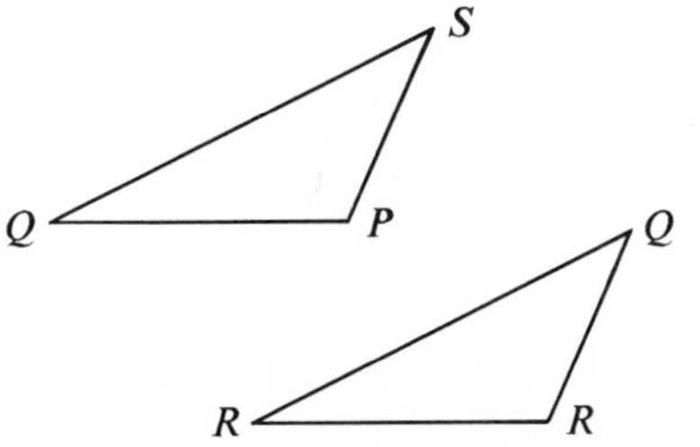

Figure 12–5

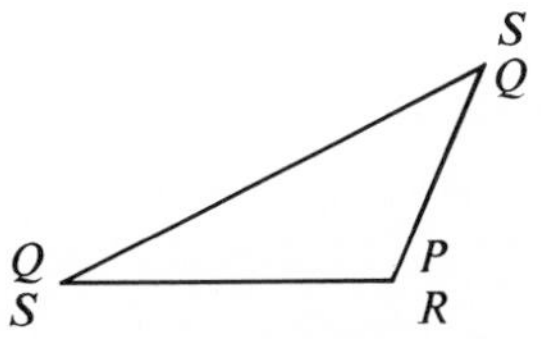

Figure 12–6

Thus, these two triangular regions are congruent. This leads to the following observation: The triangular region *QRS* is $\frac{1}{2}$ of the region *PQRS*. Since the formula for the area of a parallelogram is

$$A_{\square} = b \times a,$$

the formula for the area of a triangle may be written

$$A_{\triangle} = \tfrac{1}{2} \times b \times a.$$

A nonpolygonal region that has special interest for us is the circular region. Our experience with the region, discussed at the beginning of this section, leads us to wonder if it is possible to find the area of a circular region. Let us consider the following example.

Example 3 Find a formula for the area of a circular region.
We begin with a circular region with r the measure of the radius and d the measure of the diameter (Figure 12–7).

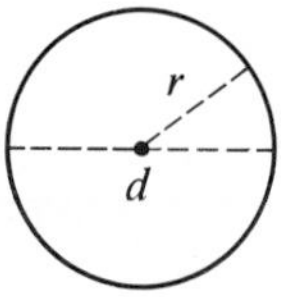

Figure 12–7

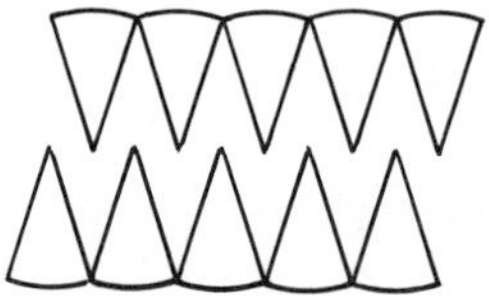

Figure 12–8

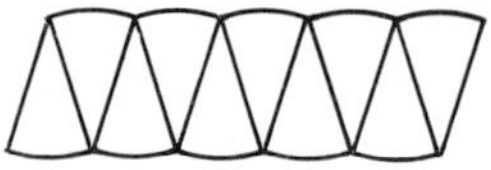

Figure 12–9

The circular region in Figure 12–7 is "cut" along the diameter making two parts. Each of these parts is "cut" into sectors (pie shapes) as in Figure 12–8. Then the two parts are assembled as shown in Figure 12–9. Put together in this manner these regions are the same in total area as the original circular region.

If the number of sectors is increased without limit, the scalloped effect in Figure 12–9 will approach two parallel line segments. This makes Figure 12–9 begin to look like Figure 12–10. Do you see that a rectangular region is approached, with a and b approaching r and $\frac{1}{2}c$, respectively?

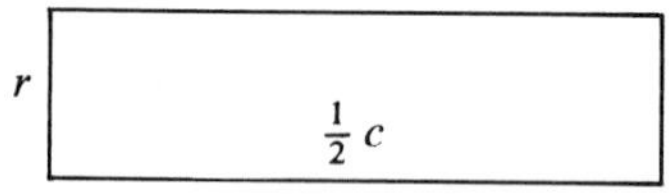

Figure 12–10

The area of this region is,

$$A_{\square} = r \times \tfrac{1}{2}c.$$

But, the circumference, c, of a circle is πd. Hence,

$$A_{\square} = r \times \tfrac{1}{2}\pi d.$$

Since $\quad A_{\square} = A_{\odot}$ and $d = 2r$,

then, $\quad A_{\odot} = \pi r^2.$

Other useful area formulas may be found in a handbook of mathematical tables and formulas. The few given here have been developed to show you how formulas are created, and with the hope that you will become interested in creating others.

EXERCISE 12.8

1. Use the preceding method to develop a formula for finding the area of each of the following regions. Can you identify each of the regions?

(a)

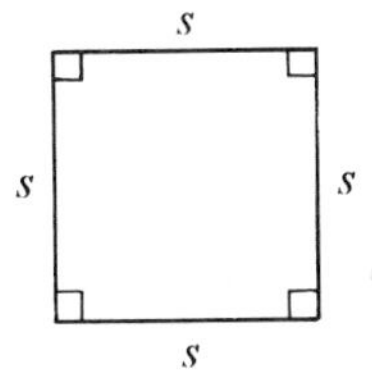

(c)

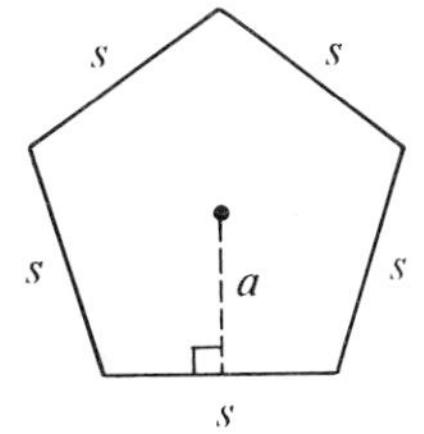

(b)

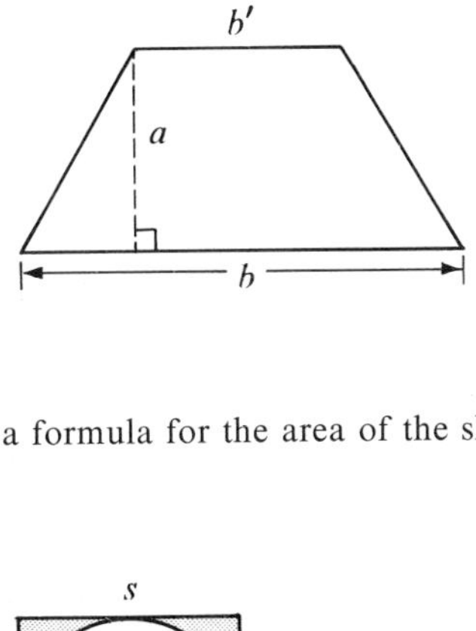

(d)

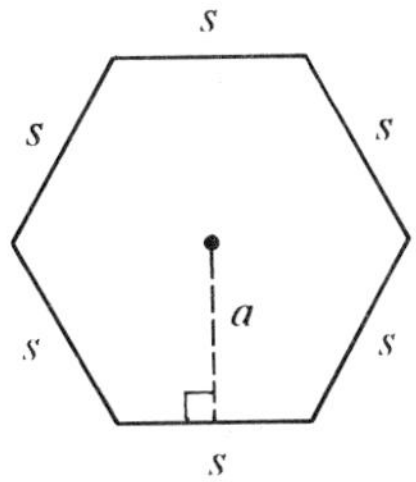

2. Develop a formula for the area of the shaded portion of each of the following regions.

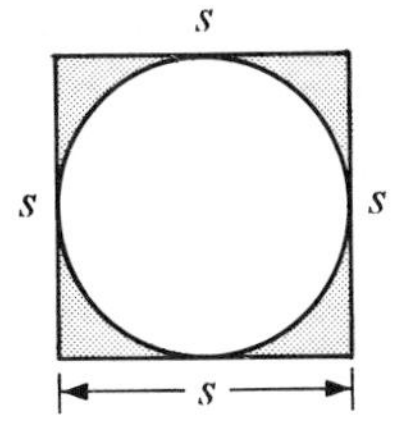

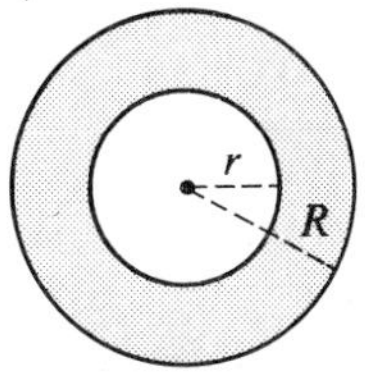

12.9 STANDARD UNITS FOR MEASURING REGIONS

Since a unit of measurement for the line segment was an inch in the English system and a centimeter in the metric system, the selection of a *square inch* and a *square centimeter* seems natural as the standard units for measuring a rectangular region. A square inch is a square region whose sides measure one inch; a square centimeter is defined similarly.

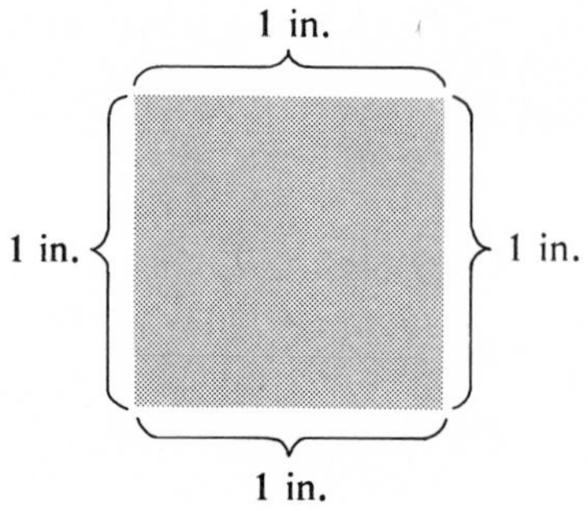

1 Square Inch

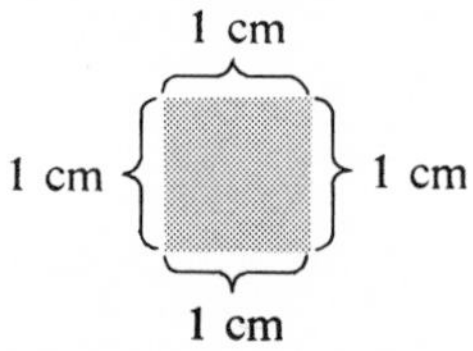

1 Square Centimeter

Other standard units of the English system used to measure regions are the square foot, the square yard, the square rod, and the square mile; while the square millimeter, the square decimeter, and the square meter are a few of the standard units of the metric system. The relationship of the square inch (sq in.) to the square foot (sq ft), and that of the square millimeter (sq mm) to the square centimeter (sq cm) are shown in the two diagrams. Other relationships follow.

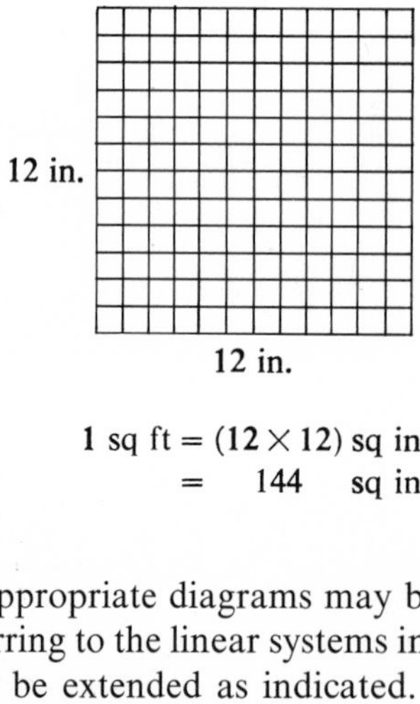

$$1 \text{ sq ft} = (12 \times 12) \text{ sq in.} = 144 \text{ sq in.}$$

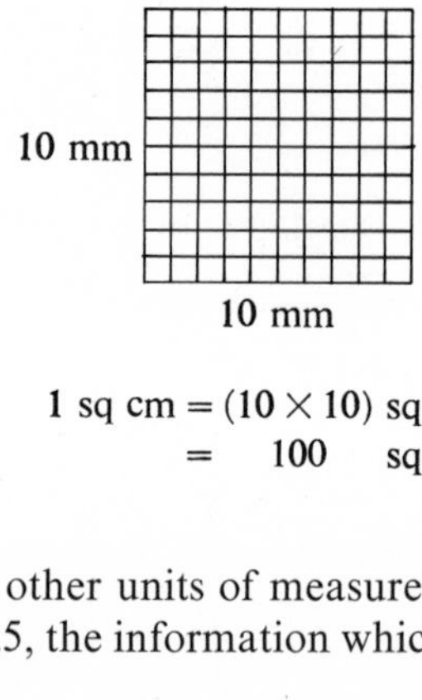

$$1 \text{ sq cm} = (10 \times 10) \text{ sq mm} = 100 \text{ sq mm}$$

Appropriate diagrams may be made for other units of measurement. By referring to the linear systems in Section 12.5, the information which follows may be extended as indicated.

Since 1 ft = 12 in.,

$$\text{then } 1 \text{ sq ft} = (12 \times 12) \text{ sq in.} = 144 \text{ sq in.}$$

Since 1 yd = 3 ft,

$$\text{then } 1 \text{ sq yd} = (3 \times 3) \text{ sq ft} = 9 \text{ sq ft.}$$

Since 1 cm = 10 mm,

$$\text{then } 1 \text{ sq cm} = (10 \times 10) \text{ sq mm} = 100 \text{ sq mm.}$$

Since 1 m = 10 dcm,

then 1 sq m = (10 × 10) sq dcm = 100 sq dcm.

Using the linear relationship between the inch and the centimeter, we have:

Since 1 in. ≈ 2.54 cm,

then 1 sq in. ≈ (2.54 × 2.54) sq cm ≈ 6.45 sq cm.

From this,

$$1 \text{ sq cm} \approx \frac{1}{6.45} \text{ sq in.} \approx 0.16 \text{ sq in.}$$

The accompanying drawing illustrates the comparison of these two square units of measurement.

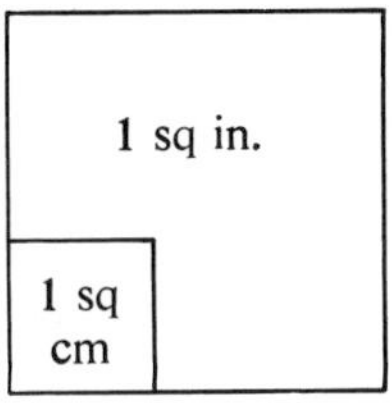

Relationships between other square units may be determined from these as shown in the examples.

Example 1 How many square centimeters are there in a square meter?

Since 100 cm = 1 m, then (100 × 100) sq cm = 1 sq m. That is, 10,000 sq cm = 1 sq m.

Example 2 One square centimeter is what fractional part of a square meter?

Referring to Example 1, 10,000 sq cm = 1 sq m. Hence:

$$1 \text{ sq cm} = \frac{1}{10{,}000} \text{ sq m} = 0.0001 \text{ sq m}$$

Example 3 Determine the number of square centimeters in a square foot.

Since 1 sq in. ≈ 6.45 sq cm and 1 sq ft = 144 sq in., then

1 sq ft ≈ (144 × 6.45) sq cm.

That is, 1 sq ft ≈ 928.8 sq cm.

Example 4 A square foot is what fractional part of a square meter?

In Example 3 we found that 1 sq ft ≈ 928.8 sq cm, and from Example 2 we see that 1 sq cm = 0.0001 sq m. Hence:

$$1 \text{ sq ft} \approx (0.0001 \times 928.8) \text{ sq m},$$
$$1 \text{ sq ft} \approx 0.09 \text{ sq m}.$$

Precision and accuracy of measurement are determined for square units in the same manner as for linear units. See Section 12.6, then study the next example.

Example 5 Which gives the greater accuracy of measurement, 32.7 sq cm or 32.74 sq cm?

The data necessary for our investigation are given in the table.

	32.7 sq cm	*32.74 sq cm*
Unit	0.1 sq cm	0.01 sq cm
Greatest Possible Error	$\frac{1}{2} \times 0.1 = 0.05$	$\frac{1}{2} \times 0.01 = 0.005$
Percent of Error	$\frac{0.05}{32.7} = 0.0015 = 0.15\%$	$\frac{0.005}{32.74} = 0.00015$ $= 0.015\%$

Since the unit for the measurement of 32.7 sq cm is 0.1 sq cm and 0.01 sq cm for the measurement 32.74 sq cm, the latter is more precise. Comparing the percent of error, $0.015\% < 0.15\%$; hence 32.74 sq cm is the more accurate measurement.

EXERCISE 12.9

1. Make use of the linear units of measurement to convert each of the following to square units.
   (a) 1 sq cm to sq dm
   (b) 1 sq dm to sq cm
   (c) 1 sq mm to sq m
   (d) 1 sq m to sq mm
   (e) 1 sq ft to sq yd
   (f) 1 sq yd to sq ft
   (g) 1 sq mm to sq dm
   (h) 1 sq dm to sq mm
2. Use the relationship 1 sq in. ≈ 6.45 sq cm in making the following conversions. Also, make use of any relationships that you have previously developed.
   (a) 1 sq in. to sq mm
   (b) 1 sq ft to sq mm
   (c) 1 sq mm to sq in.
   (d) 1 sq mm to sq ft
3. Find the area of each of the regions. Since the linear units must be the same in each region, it may be necessary to convert before using the appropriate formula.

(a)

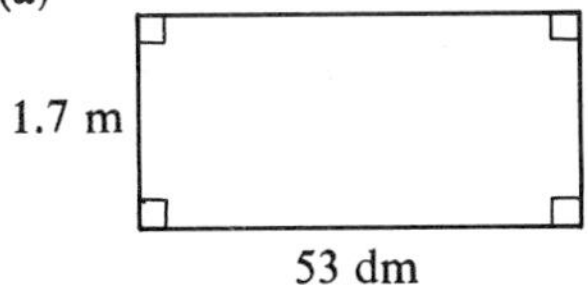

(b)

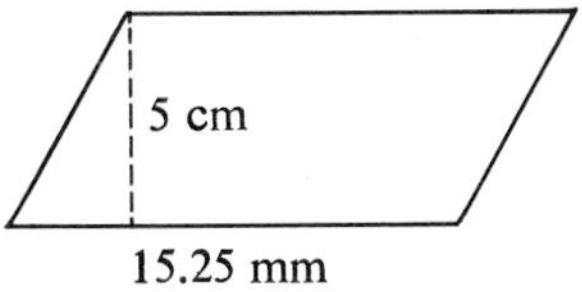

(c)

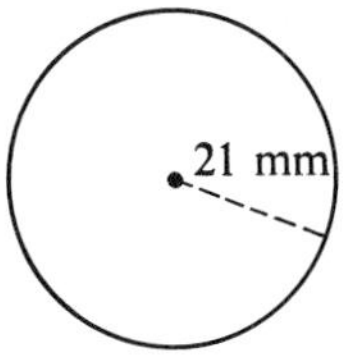

(d)

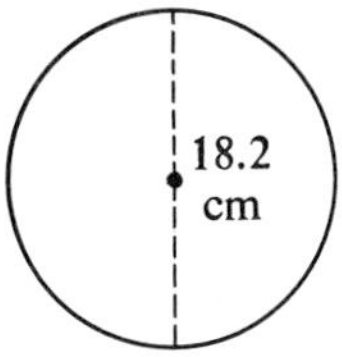

4. Find the area of each of the following regions by adding the areas of the subdivisions. Some regions have been subdivided while others will need subdividing.

(a)

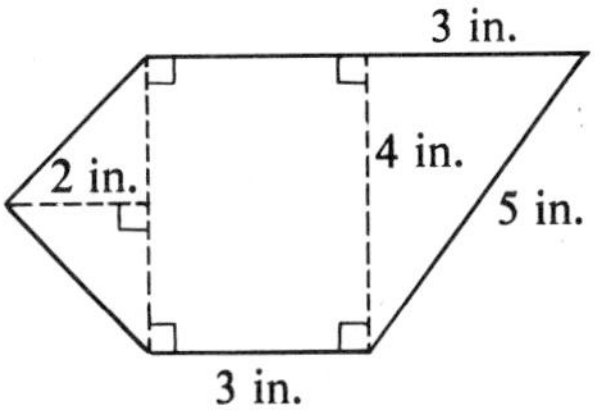

(b)

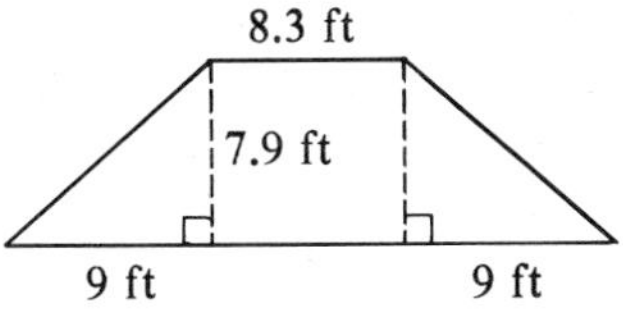

(c)

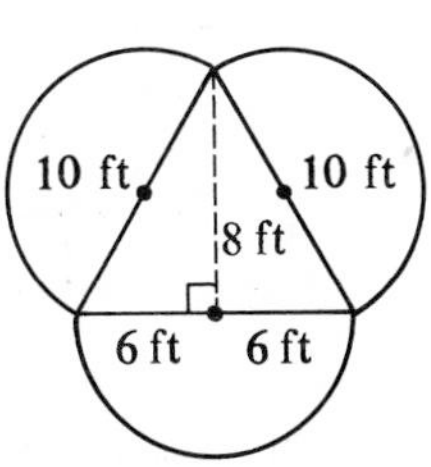

(d)

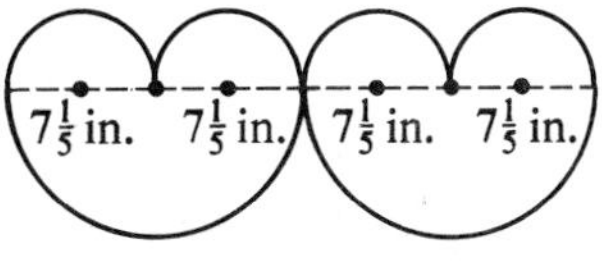

(e)

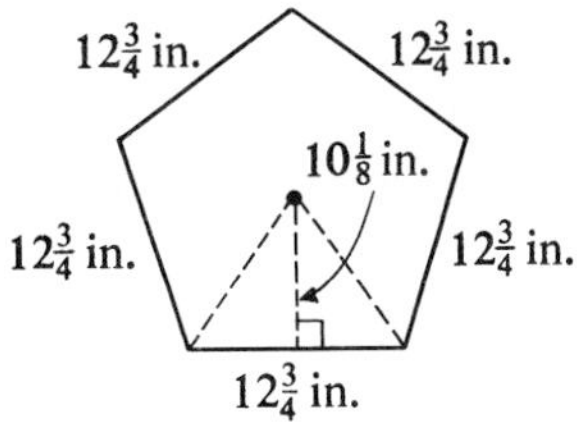

(f)

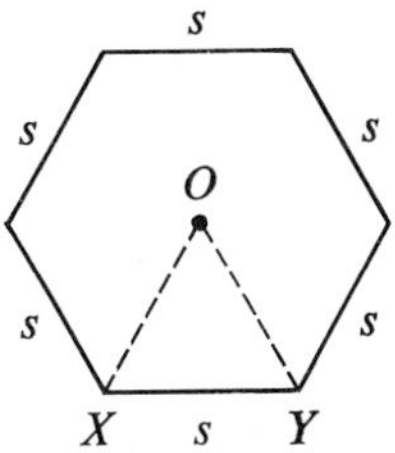

Area of $\triangle XOY$ is 18.7 sq cm

12.10 MEASUREMENT OF SPACE FIGURES

A *space figure* is usually referred to as a *solid figure*. Hence, these two terms will be used interchangeably throughout this discussion. Some of the more

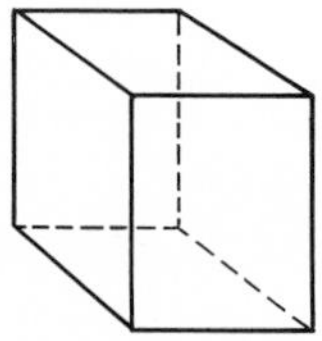

Rectangular Prism

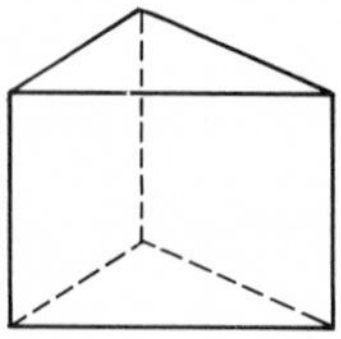

Triangular Prism

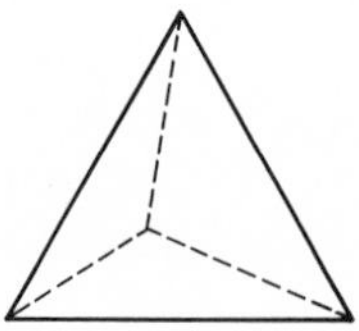

Triangular Pyramid

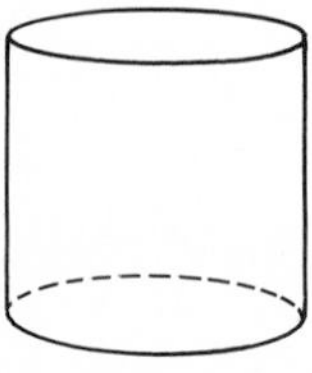

Circular Cylinder

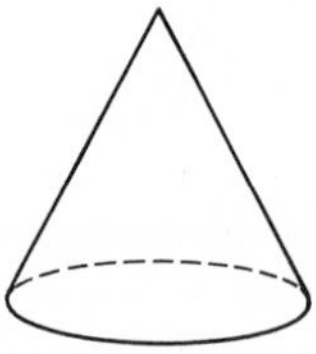

Circular Cone

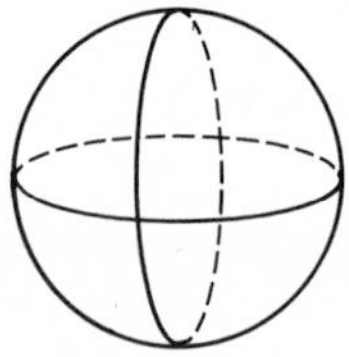

Sphere

common space figures with their names are illustrated here. The shape of the *base* of prisms, pyramids, cylinders, and cones is indicated by the name given to the solid figure. A triangular prism means a prism having a triangular-shaped base; a circular cylinder has a circular-shaped base. If a picture of a rectangular pyramid is drawn, it will look much like the ancient Egyptian pyramids with their rectangular-shaped bases. The solid figures shown here, except the sphere, are *right* solid figures. An informal characterization is to say that they do not "lean."

Is it possible to select a unit that may be used to measure the interior of these figures? Since a line segment was the unit used to measure a line segment and a region to measure a region, one expects that a solid will be used to measure a solid figure. A solid in the shape of a *cube* will be selected as the unit. A cube is a rectangular prism whose edges have the same measure. See the diagram.

The interior of a rectangular prism is shown subdivided by a cube. The

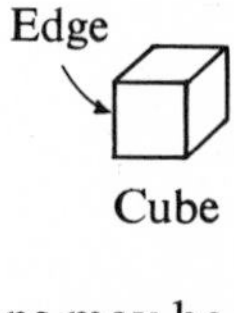

Cube

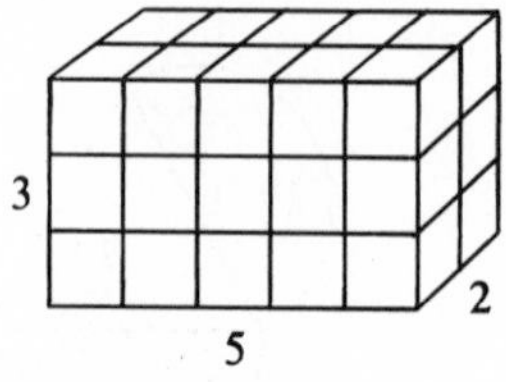

subdivisions may be counted to find the measure, which is 30. This measure is a number and is called *volume*. Another method of determining the measure of the rectangular prism is to multiply the number of units in one layer by the number of layers. Since the units in each layer are arranged in an array, multiplying 5 by 2 gives the number of units in one layer. This number is then multiplied by the number of layers, 3. Hence, the measure is found by finding the product, $(5 \times 2) \times 3$.

Observe that subdividing the interior of the rectangular prism also subdivided the edges. Thus, 5, 2, and 3 become measures of the edges. Generalizing, if a, b, and h are measures of the edges of a rectangular prism, then the volume is found by

$$V = a \times b \times h.$$

(NOTE: A right rectangular prism is used throughout the discussion.)

The advantages of standardizing the unit of measurement are evident from previous discussions. A standard unit of volume in the English system is the *cubic inch,* while in the metric system it is the *cubic centimeter*. Some

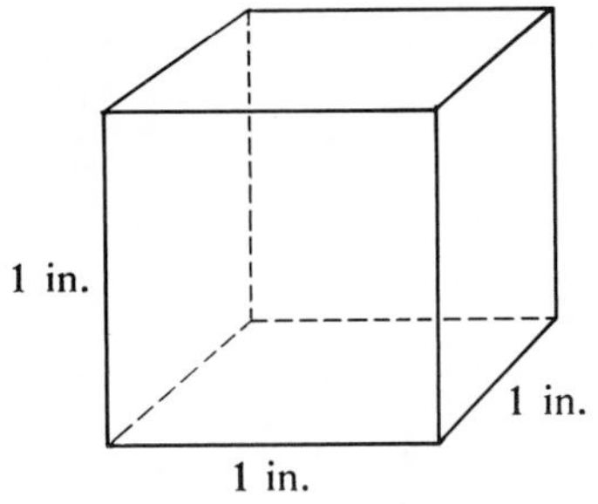

1 Cubic Inch
(cu in.)

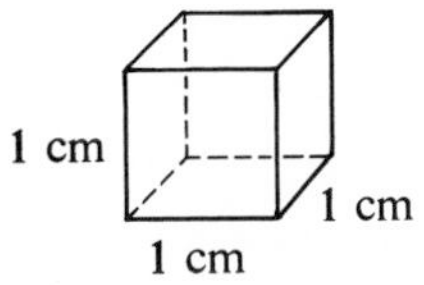

1 Cubic Centimeter
(cc)

other standard units in the English system used to measure volume are the cubic foot, the cubic yard, and the cubic rod; while the cubic millimeter, the cubic meter, and the cubic decimeter are a few of the standard units of the metric system.

Relationships between the different standard units of measurement of volume are developed in a manner similar to those for area.

$$\begin{aligned}
\text{Since 1 ft} &= 12 \text{ in.,}\\
\text{then 1 cu ft} &= (12 \times 12 \times 12) \text{ cu in.}\\
&= 1728 \text{ cu in.}\\
\text{Since 1 ft} &= \tfrac{1}{3} \text{ yd,}\\
\text{then 1 cu ft} &= (\tfrac{1}{3} \times \tfrac{1}{3} \times \tfrac{1}{3}) \text{ cu yd}\\
&= \tfrac{1}{27} \text{ cu yd}
\end{aligned}$$

$$\begin{aligned}
\text{Since 1 cm} &= 10 \text{ mm,}\\
\text{then 1 cc} &= (10 \times 10 \times 10) \text{ cu mm}\\
&= 1000 \text{ cu mm}\\
\text{Since 1 dm} &= \tfrac{1}{10} \text{ m,}\\
\text{then 1 cu dm} &= (\tfrac{1}{10} \times \tfrac{1}{10} \times \tfrac{1}{10}) \text{ cu m}\\
&= \tfrac{1}{1000} \text{ cu m}
\end{aligned}$$

Other relationships between cubic measurements may be developed from linear measurements as has been demonstrated.

Using the relationship of an inch to a centimeter, we are able to find corresponding cubic measurements.

Since 1 in. $\approx$ 2.54 cm,

$$1 \text{ cu in.} \approx (2.54 \times 2.54 \times 2.54) \text{ cc} \approx 16.39 \text{ cc.}$$

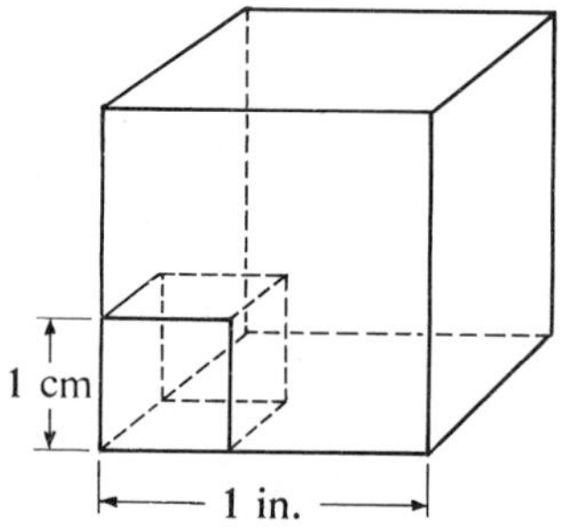

The accompanying model helps in visualizing the comparison of the two cubic units.

Tables of other equivalent units of volume may be constructed when needed. Also, a handbook of mathematical tables is recommended for a more extensive use of standard units of measurement.

Precision and accuracy of cubic measurements are determined in the same manner as for both linear and square units. See Sections 12.6 and 12.9.

Returning to measuring the volume of a (right) rectangular prism, we observe that in the formula

$$V = a \times b \times h$$

$(a \times b)$ gives the area of the base of the prism. Naming this B, the formula becomes

$$V = B \times h.$$

This formula may now be used to find the volume of a prism and also a cylinder. The method used to find the area of the base will depend on its shape. You may need to review the formulas for finding the area of a circle, a triangle, a square, and other polygons.

In the following experiments, formulas will be developed for finding the volume of a pyramid and a cone.

Experiment I

Material: Containers in the shape of a (right) prism and a (right) pyramid having congruent bases and heights the same measure; container of water.

Procedure: Use the pyramid to fill the prism with water.

Question: How many times was the water in the pyramid emptied into the prism?

Conclusion: Since 3 pyramids of water filled the prism, 1 pyramid filled how much of the prism?

Hence, the volume of the pyramid is $\frac{1}{3}$ the volume of the prism. And,

since $$V_{\text{Prism}} = B \times h,$$
then $$V_{\text{Pyramid}} = \tfrac{1}{3}(B \times h).$$

Since each pyramid has a prism as its counterpart, the volume of any pyramid is found by the preceding formula.

Experiment II

Material: Containers in the shape of a (right) circular cylinder and a (right) circular cone having congruent bases and heights the same measure; container of water.

Follow the plan of Experiment I. What is your conclusion?

since $$V_{\text{Cylinder}} = B \times h,$$
then $$V_{\text{Cone}} = \tfrac{1}{3}(B \times h).$$

The two experiments show that the formulas for the pyramid and cone are the same,

$$V = \tfrac{1}{3}(B \times h).$$

The difference in the calculation again is found in determining the area of the different shaped bases.

The development of the formula for the volume of a sphere is not given in this textbook. The formula is

$$V_{\text{Sphere}} = \tfrac{4}{3}\pi r^3.$$

EXERCISE 12.10

1. Each of the listed space figures has a base (B) 18 sq in. and height (h) 15 in. Find the volume of each.
   (a) Prism (b) Pyramid (c) Cylinder (d) Cone
2. The measure of each edge of a right rectangular prism is 6 cm
   (a) Find the volume, rounding to the nearest tenth.
   (b) Compare the percent of error of the rounded volume to that of the volume before rounding.
   (c) What special name can be given to this prism?
3. Find the volume of a sphere if
   (a) the radius is 3 ft (use $\pi = 3.14$).
   (b) the diameter is 5 mm.
   (c) the radius is 21 in. (use $\pi = \frac{22}{7}$).
4. Find the volume of each of these right solid figures.

(i)

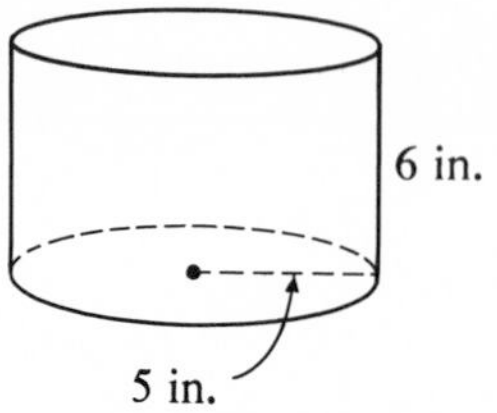

(iii)

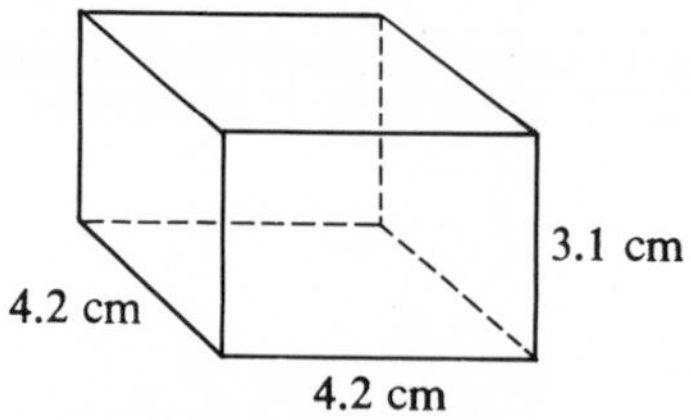

A square base

(ii)

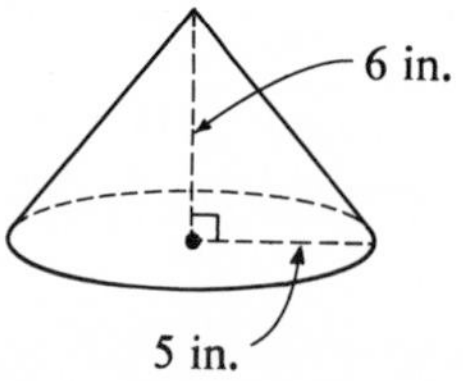

(iv)

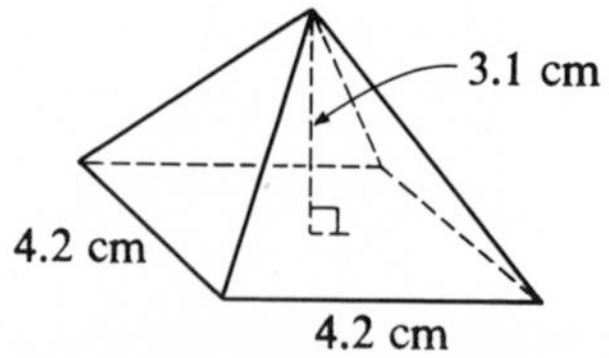

A square base

(a) Give the volume of (i) and (ii) in cubic inches, then convert to the nearest tenth of a cubic centimeter.

(b) Give the volume of (iii) and (iv) to the nearest tenth of a cubic centimeter, then convert the rounded volume to the nearest hundred cubic millimeters. How many cubic millimeters were lost in this rounding process? Compare the percent of error of the rounded volume in cubic millimeters with the volume had it not been rounded.

12.11 MEASUREMENT OF THE INTERIOR OF AN ANGLE

The unit used in measuring a line segment was a line segment. It seems quite natural, then, to choose an angle as the unit for measuring an angle. The part of the angle that we measure is its *interior*. We measure it by laying off *congruent angles* in such a way that consecutive angles have a side in common.

It is sufficient for us to think of congruent angles in very much the same way that we thought of congruent line segments. Suppose that we consider these two angles:

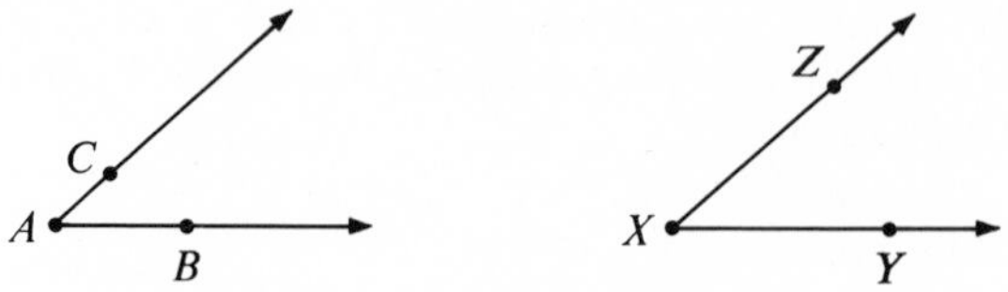

We shall imagine placing one upon the other with the idea of matching the two sets of points. Vertex A is matched with vertex X, $\overrightarrow{AB}$ is matched with $\overrightarrow{XY}$, and $\overrightarrow{AC}$ is matched with $\overrightarrow{XZ}$. If it is possible to make such a matching, we say that these two angles are *congruent* and write

$$\angle CAB \cong \angle ZXY.$$

Since these angles are not identical sets of points, we do not write $\angle CAB = \angle ZXY$ (see Section 4.6). We shall speak of congruent angles in an intuitive manner as those angles that are "just alike" and "fit." We think of one angle as being a copy of the other. It is important for us to know that for every angle there exists an angle congruent to it.

We shall now direct our attention to laying off copies of a unit angle, u, upon the interior of $\angle PQR$. The measure is determined by the number of copies of the unit angle that have been made:

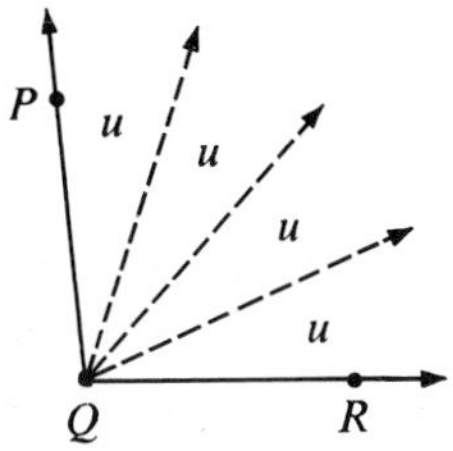

Here, $\angle PQR$ is exactly subdivided by the unit u, and so the measure of $\angle PQR$ in this case is 4. This may be written

$$m\ \angle PQR = 4,$$

which is read, "The measure of angle PQR is 4." Thus, we have:

4 is the *measure;*
u is the *unit of measurement;*
$4u$ is the *measurement.*

As in the case of a line segment, a unit may not always exactly subdivide the interior of an angle. Observe the measurement of $\angle RBT$ shown here:

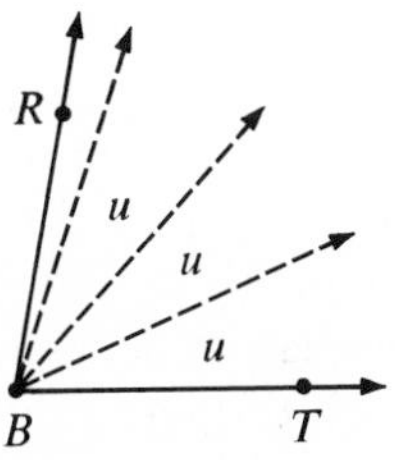

In this case a portion of $\angle RBT$ is left over, and the measure is given to the nearest number of units used. The measure of this angle with u as the unit is approximately 3, and we write

$$m\ \angle RBT \approx 3.$$

It would certainly be possible to subdivide the unit *u* into congruent angles. If it is subdivided into 4 congruent angles, then each new unit, which we shall call *t*, will be $\frac{1}{4}$ of the original unit *u*. When this unit is fitted into the interior, there is still some left over. Even though the measurement is still an approximation, it is considered to be more "precise" than when the larger unit *u* was applied.

As with line segments, to every angle there can be assigned a positive real number. If the numbers assigned to two angles are the same, we say that the angles are congruent, and if two angles are congruent, then their measures are the same.

The difficulties encountered when the units of measurement are not standardized have already been explored in the discussion of line segments. Thus, we shall now look for a standard unit that may be used in measuring the interior of an angle. This standard unit of measurement of an angle is called a *degree,* and the symbol denoting this unit is "°." If the measurement of an angle is 45 degrees, we may write it as 45°.

How is the standard angle unit determined? One should be aware that any unit may be arbitrarily selected. It becomes the standard unit only when people agree to accept it as such. The unit that has been agreed upon is one of the 180 congruent angles formed by 181 rays drawn from point *P* as illustrated (due to the difficulty in drawing 181 rays, only 19 were drawn here):

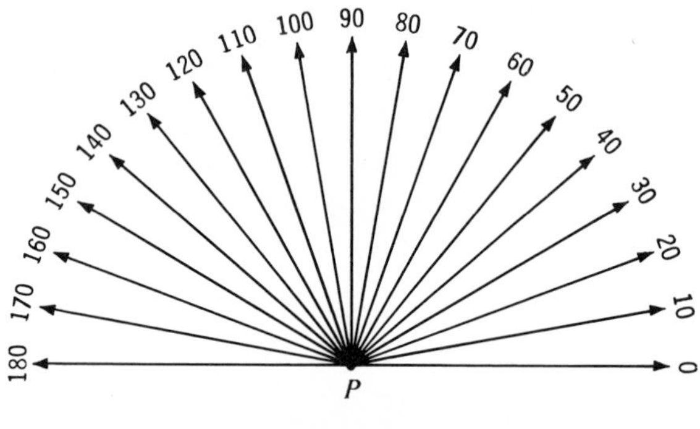

Beginning at the right, a whole number is assigned in order to each ray. That is, to the first ray is assigned 0, to the second 1, etc. (In our drawing, the numbers assigned are 0, 10, 20, etc.) Note that the union of the ray marked 0 and the one marked 180 is a line.

For more precise measurement, the *degree* angle is subdivided into 60 congruent angles. Each of these subdivisions is called a *minute.* Also each of these minute angles is subdivided into 60 congruent angles. Each of these subdivisions is called a *second.* That is:

$$1 \text{ degree} = 60 \text{ minutes}$$

or in symbols:

$$1° = 60'$$
$$1 \text{ minute} = 60 \text{ seconds}$$

or in symbols:

$$1' = 60''$$

$$1 \text{ degree} = 60 \text{ minutes} = (60 \times 60) \text{ seconds}$$

$$1^\circ = 60' = 3600''$$

A *protractor* is an instrument used to measure an angle. It has marked on it the scale shown in the preceding diagram, and is usually subdivided to show degrees:

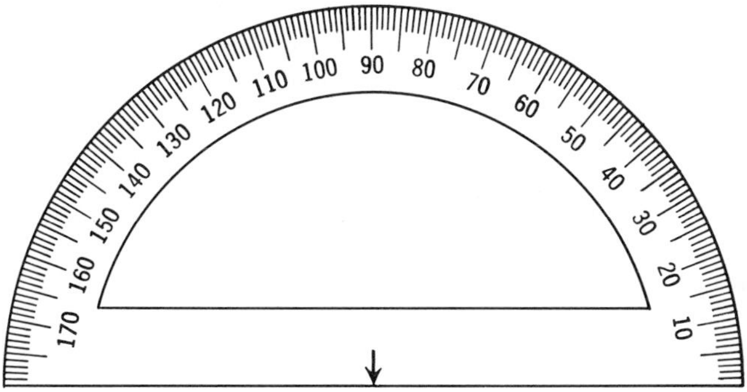

A commercially produced protractor usually has a second scale also, going from left to right, which is convenient for measuring angles in some positions.

To measure $\angle ABC$, we proceed in the following manner:

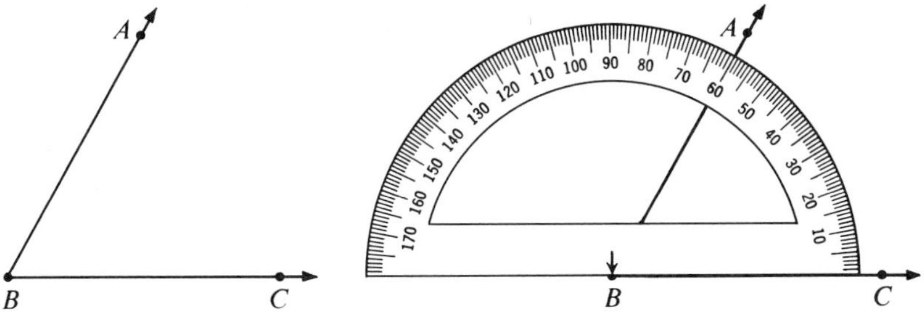

Place the arrow on the protractor at the vertex of the angle, B, with the edge of the protractor lying along $\overrightarrow{BC}$. The measure of $\angle ABC$ is determined by the position on the scale where it is crossed by $\overrightarrow{BA}$. As accurately as we can read the scale, m $\angle ABC$ is 60. A more precise instrument might have given the measurement as 59° 30′, which is read "fifty-nine degrees and thirty minutes." This latter measurement might also be given as $59\frac{1}{2}°$.

The measurement of an angle is precise to the least unit mentioned. For instance, the measurement 3 degrees 4 minutes 15 seconds is precise to the nearest second. It may be written

$$3°4'15'' = \left(3 + \frac{4}{60} + \frac{15}{60 \times 60}\right)^{\circ}$$

$$= 3\frac{255°}{3600}.$$

An angle whose measurement is 90° is called a *right angle.* The rays forming a right angle are *perpendicular*. Each of the following angles is a right angle. Their sides are perpendicular.

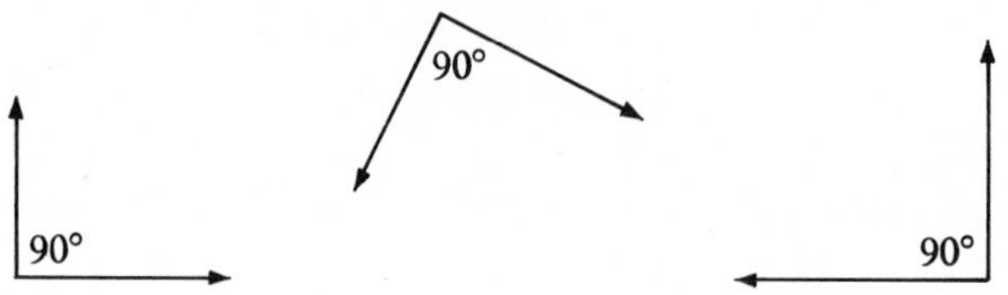

EXERCISE 12.11

1. Describe the "size" of the unit angle by giving what it is called and how it is determined.
2. This is a drawing showing a protractor placed with its arrow on the common vertex, Z, of several angles.

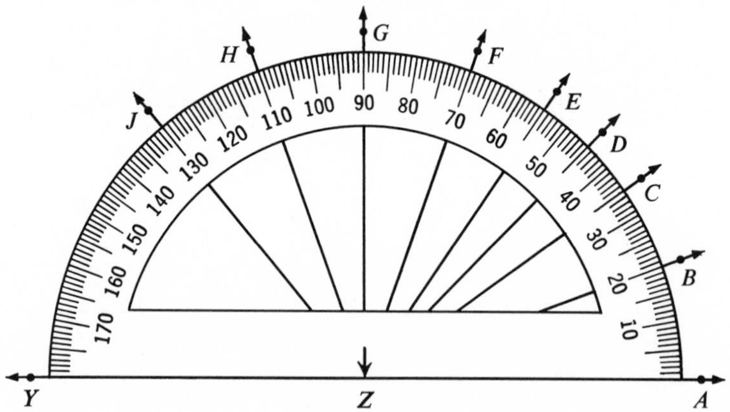

(a) Even though the edge of the protractor does not fall along one ray of each angle, is it possible to find the measure from the scale? Explain.
(b) What is the difference between the measurement of $\angle AZB$ and its measure?
(c) Using the scale on this protractor, what is the most precise measurement which may be given for any of the angles?
(d) Give the measure of each of the angles in the following list which are represented in the drawing above.

EXAMPLE. m $\angle AZF = 70$.

(1) $\angle AZC$ (4) $\angle FZH$ (7) $\angle DZE$

(2) $\angle AZG$ (5) $\angle JZE$ (8) $\angle CZH$

(3) $\angle CZE$ (6) $\angle CZD$

(e) Using the drawing, identify an angle such that
   (1) its measure is the same as that of $\angle EZC$.
   (2) its measure is twice the measure of $\angle DZF$.
   (3) its measure is $\frac{1}{2}$ of m$\angle GZA$.
   (4) it is congruent to $\angle AZD$.

3. Give the following measurements in degree units.

(a) $25^\circ\ 10'$ (c) $7^\circ\ 30''$ (e) $4^\circ\ 5'\ 10''$

(b) $36^\circ\ 60'$ (d) $18^\circ\ 15'\ 20''$ (f) $45^\circ\ 12'\ 27''$

4. Change $18^\circ\ 20'\ 45''$ to

(a) degree units. (b) minute units. (c) second units.

5. Find the percent of error for each of these measurements.

(a) 90° (c) $89^\circ\ 59'$ (e) $89^\circ\ 43'\ 2''$

(b) $89^\circ\ 16'$ (d) $89^\circ\ 10'\ 45''$

13

Relations and Functions

13.1 INTRODUCTION

In previous chapters, the term *relation* was used in referring to equality (=) and order (< or >). These symbols were used to connect two elements such as: $a = b$ and $x < y$, which established a correspondence or relation, between the pairs of the elements. With an agreement established as to the way these elements are connected, they may be written as an *ordered pair* such as: (a, b), (x, y). The way in which the components of the ordered pair are connected is usually called the *rule*.

We shall now consider some examples that explore further the concept of relation.

Example 1 Given the two sets:

$$S = \{\text{Colorado, Illinois, New York, California}\},$$
$$C = \{\text{Springfield, Albany, Denver, Baton Rouge, Sacramento}\}.$$

Construct the set of ordered pairs which relates the elements of the two sets by the phrase, "is the capital of."

The phrase "is the capital of" directs the assignments made in the arrow diagram. Each capital city in set C has an assignment (image) of a state in set S.

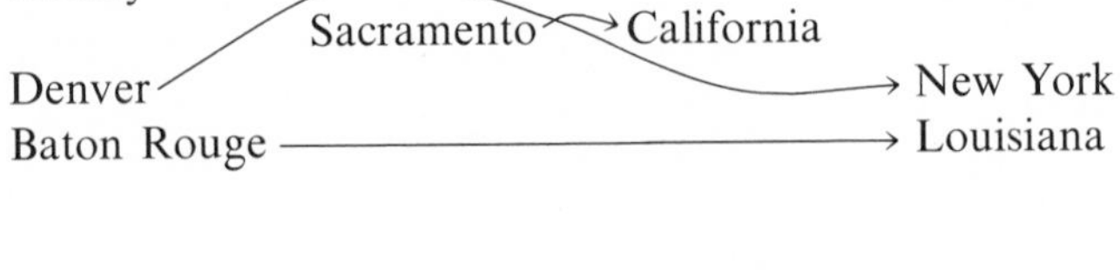

The result of this pairing is given by the set of ordered pairs,

{(Springfield, Ill.), (Albany, N.Y.), (Denver, Colo.), (Baton Rouge, La.), (Sacramento, Calif.)}.

Each of the ordered pairs belongs to the relation "is the capital of." The ordered pair (Springfield, Illinois) may be written,

Springfield "is the capital of" Illinois.

The ordered pair (Illinois, Springfield) does not belong to "is the capital of." Neither does (Memphis, Tennessee) belong to this set, as Memphis is not the capital of Tennessee.

Example 2 The elements of set $A = \{4, 9, 16, 25\}$ and set $B = \{2, {}^-2, 4, {}^-3, {}^-5, {}^-4, 5, 3\}$ have been paired as indicated by the arrows. Identify the rule that appears to have been used in pairing the elements of the two sets.

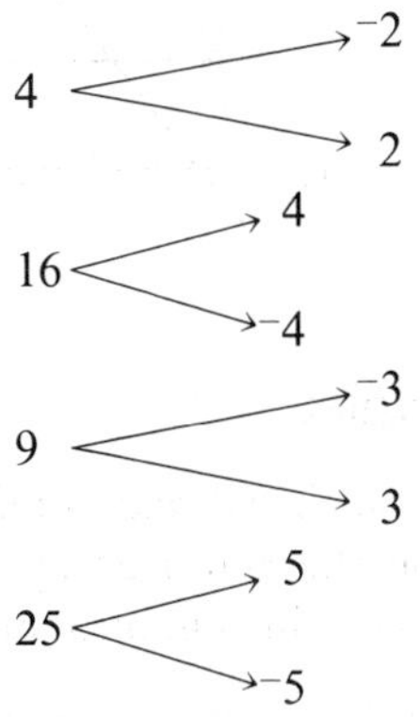

The rule for the pairing appears to be "is the square of." That is, 4 "is the square of" $^-2$, 4 "is the square of" 2, and 16 "is the square of" 4 bear this out. The set of ordered pairs

$$\{(4, {}^-2), (4, 2), (16, 4), (16, {}^-4), (9, {}^-3), (9, 3), (25, {}^-5), (25, 5)\}$$

are members of the set described by this phrase.

Definition 13.1 A relation is a set of ordered pairs (x, y) whose components are connected by a descriptive phrase R, such that $x\,R\,y$ is true for each $x \in U$ and $y \in U$.

The phrase that describes the way in which the components of the ordered pair are connected is used to identify the relation. It is quite common to speak of the connective phrase as the relation. That is, "is the capital of" is referred to as the relation.

EXERCISE 13.1

1. Identify the connective phrase that apparently has been used to relate the two components of the ordered pairs in each of the following sets:
   (a) $\{(1, 0), (2, 1), (3, 2), (4, 3)\}$
   (b) $\{(2, 4), (3, 9), (4, 16), (5, 25)\}$
   (c) $\{(6, 72°), (7, 74°), (8, 80°), (9, 89°), (10, 93°), (11, 98°), (12, 102°)\}$
   (d) {(Africa, Nile), (North America, Mississippi), (Asia, Yangtze), (Europe, Danube)}
   (e) {(Nile, Africa), (Mississippi, North America), (Yangtze, Asia), (Danube, Europe)}
2. Given $U = \{1, 3, 5\}$ and R representing "is greater than," list the ordered pairs of the set that will make the sentence $x\ R\ y$ true.
3. R represents "is parallel to." Construct a set of ordered pairs that belong to $x\ R\ y$.

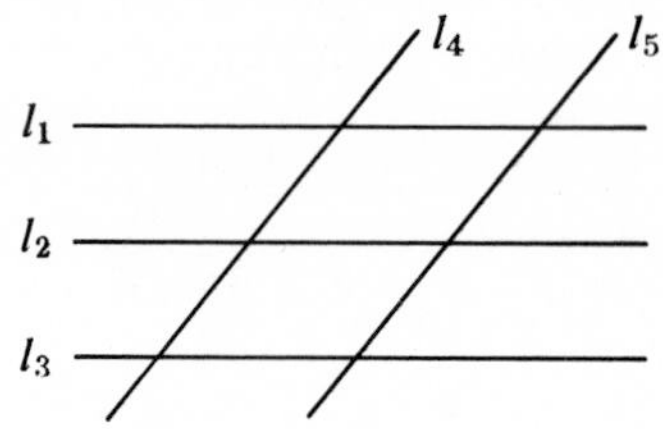

13.2 PROPERTIES OF RELATIONS

As an example in which we may study the properties of relations, let us consider the relation characterized by "has the same color hair as" for the set of all girls in a certain schoolroom. It might look like this:

{(Mary, Julia), (Elizabeth, Peggy), (Jane, Mary), (Jane, Julia), (Susan, Susan), . . .}

While (Julia, Rebecca) is an ordered pair, it will not belong in this relation if Julia is a blonde and Rebecca is a brunette. We shall now introduce an abbreviation and write for the ordered pair (Mary, Julia)

Julia K Mary

to mean, in this case, "Julia has the same color hair as Mary."

In our study of relations, we shall look for certain properties which the relations may or may not possess. These are the reflexive, symmetric, and transitive properties, which we met in Section 6.11. In the following statements of these properties, K should be understood to represent the rule which enables us to select the relation that may be under consideration, and a, b, and c are elements of the set to which K is applied.

Reflexive property: For each a, $a\ K\ a$.

Translated to the above example, this means that each girl in the room has the same color hair as herself. The element (Susan, Susan) is an illustration.

Symmetric property: For each a and b, if $a\,K\,b$, then $b\,K\,a$.

For our schoolgirls this says: If Peggy has the same color hair as Elizabeth, then Elizabeth has the same color hair as Peggy; that is, if (Elizabeth, Peggy) is an element of the set, then (Peggy, Elizabeth) must be also.

Transitive property: For each a, b, c, if $a\ K\ b$ and $b\ K\ c$, then $a\ K\ c$.

It holds true that if Julia has the same color hair as Mary, and Mary has the same color hair as Jane, then Julia has the same color hair as Jane.

Hence, the relation "has the same color hair as" applied to the set of girls in a given schoolroom is reflexive, symmetric, and transitive.

Now, let us examine some other relations and determine if any of these three properties is satisfied.

Example 1 Which properties does the "less than or equal", ($\leq$), relation for the set of integers have?

(1) Is the relation reflexive? For each a, $a \leq a$ since $a = a$ for all a. Hence, the relation is reflexive.
(2) Is the relation symmetric? For each a, b, if $a \leq b$, it does not follow that $b \leq a$. A counterexample: $7 \leq 9$, but $9 \nleq 7$. While the property of symmetry holds for *some* elements, that is, $7 \leq 7$, it does not hold for *all* cases. Hence, the relation is not symmetric.
(3) Is the relation transitive? For each a, b, c, if $a \leq b$ and $b \leq c$, then $a \leq c$. This is a true statement.

Hence, the relation $\leq$ is reflexive and transitive, but not symmetric.

Example 2 Which properties does the "is not equal to," ($\neq$), relation for the set of integers have?

(1) For each a, "$a \neq a$" is not a true statement. Hence, the reflexive property is not satisfied.
(2) For each a, b, "if $a \neq b$, then $b \neq a$" is a true statement. Hence, the symmetric property holds.
(3) For each a, b, c, "if $a \neq b$ and $b \neq c$," it does not necessarily follow that $a \neq c$. A single counterexample: $9 \neq 7$ and $7 \neq 9$, but $9 = 9$.

Hence, only the symmetric property is satisfied.

Example 3 Which properties does the "is a descendant of" relation for the set of all people have?

(1) For each a, "a is a descendant of a" is not valid.
(2) For each a, b, if a is a descendant of b, then b *is not* a descendant of a.

(3) For each a, b, c, if a is a descendant of b and b is a descendant of c, then a is a descendant of c.

Hence, this relation is neither reflexive nor symmetric, but is transitive.

Relations which satisfy all three of these properties are called *equivalence relations*. Such a relation separates or *partitions* the set of elements involved into disjoint subsets called *equivalence classes*.

Referring to the earlier example concerning the set of girls, we note that those with black hair fall into one class or subset, and those with red hair into another, etc. Each girl belongs in one subset and no girl is in more than one. The classes are disjoint sets and their union is the total set of elements to which the relation applies.

This diagram shows how an equivalence relation partitions a set of elements into disjoint subsets.

U = set of all girls in a schoolroom

R: "the same color hair as"

EXERCISE 13.2

1. Discuss the reflexive, symmetric, and transitive properties for each of the relations described by the following.
   (a) "Is the brother of" for the set of all people; for the set of all men and boys.
   (b) "Votes the same way as" for the set of all voting citizens of the United States of America.
   (c) "Differs in weight by not more than $\frac{1}{2}$ lb." for the set of all people.
   (d) "Is in the same grade as" for the set of all 10-year-old schoolboys in a single city.
2. (a) Given the relation

$$\{(2, 3), (3, 2), (2, 2), (3, 3)\},$$

   verify that it is both reflexive and symmetric.
   (b) What elements must be added to the relation

$$\{(1, 5)\ (4, 4), (3, 5), \ldots\}$$

   to make it both reflexive and symmetric? (HINT: Five more ordered pairs are necessary.)

13.3 A RELATION AS A SUBSET OF A CARTESIAN PRODUCT

We have been considering rather informally certain sets of ordered pairs with the first component coming from some given set and the second component from the same set or a different set. In general, we shall now consider a set of ordered pairs (x, y) with x taken from set A and y from set B. All such ordered pairs are elements of the Cartesian product set, $A \times B$ (Section 5.5). Hence, we know that *any relation with elements* (x, y) *is a subset of* $A \times B$.

Contrary to the convention used in Chapter 5 for the grids displaying elements of a Cartesian product, we shall now agree, unless otherwise specified, to show the first component on the horizontal line and the second component on the vertical line.

Example 1 Given $A = \{1, 2, 3, 4, 5\}$ and $B = \{3, 4, 5\}$, the set $A \times B$ may be displayed as:

B						
5	(1, 5)	(2, 5)	(3, 5)	(4, 5)	(5, 5)	
4	(1, 4)	(2, 4)	(3, 4)	(4, 4)	(5, 4)	
3	(1, 3)	(2, 3)	(3, 3)	(4, 3)	(5, 3)	
	1	2	3	4	5	A

The following set of ordered pairs has been selected from the above array, $A \times B$:

$$M = \{(2, 3), (3, 3), (3, 4), (4, 3), (4, 4), (4, 5), (5, 3), (5, 4), (5, 5)\}$$

It is not always obvious why a certain set has been selected. However, in this particular case it is possible to show that each pair (x, y) fits the rule

$$x + 1 \geq y.$$

Check that this is true.

We have two ways of describing this particular relation. We may use the roster notation as above or the set builder notation as follows:

$$M = \{(x, y), x \in A, y \in B : x + 1 \geq y\}$$

This description of M is the counterpart for ordered pairs of the set builder notation introduced in Section 2.4. M could also be described as

$$M = \{(x, y) \in (A \times B) : x + 1 \geq y\}.$$

This last description emphasizes the fact that the relation M is a subset of $A \times B$.

It would be of interest to construct the graph of the above relation. However, let us first recall the method discussed in Section 6.13 for making the graph of the solution set of sentences such as $2 + n = 5$, $1 + 3 > y$, etc. The graph of the solution set of such sentences is constructed by establishing a one-to-one correspondence between the elements of the set and points in the number line.

NOTE: Frequently, different names are given to sentences; those having the equals (=) symbol are called *equations,* while those having an order (< or >) symbol are called *inequations* (or inequalities). The place holders (n or y) are called *variables.*

Making a graph on the number line suggests a procedure for constructing a graph of a set of ordered pairs, a relation. We choose to establish a correspondence between the elements of such a set and points in a plane, called the *number plane.*

The elements of a relation form a subset of the Cartesian product of two given sets of numbers. For example,

$$P = \{(2, 1), (3, 2), (5, 4)\}$$

is a subset of the Cartesian product $W \times W$, W the set of whole numbers; while

$$Q = \{(^-2, 3), (3, ^-3)\}$$

is a subset of $I \times I$, I the set of integers. The Cartesian product $W \times W$ is called the *universal set,* U, of set P, and $I \times I$ is the universal set of set Q.†

Let us now establish a one-to-one correspondence between the elements of $I \times I$ and points in a plane. Using $I \times I$ as the universal set U, we list a few of the elements. Obviously, it is impossible to list them all.

$$U = I \times I = \{\ldots (^-1, 0), (^-1, 1), (^-1, 2), \ldots, (0, 0), (0, 1), (0, 2), \ldots, (1, 0), (1, 1), (1, 2), \ldots\}$$

It is possible to match each element of U with a point in a plane called the *number plane.* Establishing this correspondence is similar to locating a town on a map (Section 5.2). We begin with a pair of perpendicular number

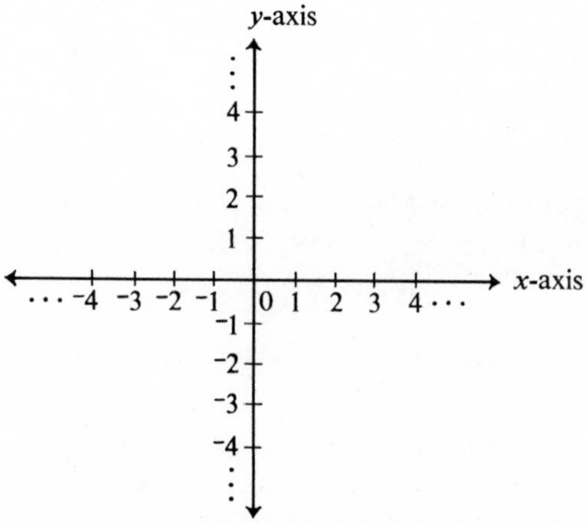

Figure 13–1

†The universal set may consist of a set of ordered pairs or a set of elements that make up those ordered pairs.

lines as reference lines; see Figure 13–1. They are called *axes*. The horizontal line is the *x-axis* and the vertical is the *y-axis*. The intersection of the two axes is at the point (0, 0), or O, called the *origin*. With the universal set $I \times I$, the coordinates on the number lines (the axes) are integers.

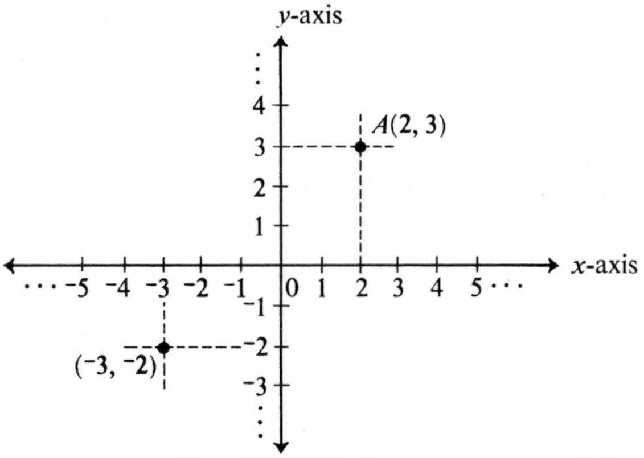

Figure 13–2

To locate a point in the number plane which corresponds to the ordered pair $(2, 3) \in I \times I$, begin at the origin, O, count to 2 on the x-axis. Draw a dotted line (Figure 13–2) perpendicular to the x-axis at 2. Next, count from O to 3 on the y-axis. Draw a perpendicular at point 3 to the y-axis. Since the point of intersection of these two lines is 2 units along the x-axis from the origin and 3 units along the y-axis, the ordered pair that locates point A is (2, 3). The ordered pair (2, 3) is called the coordinate of point A.

Use the same procedure to locate the point associated with $(^-3, ^-2)$. Remember that the first component $(^-3)$ is located along the x-axis and the second component $(^-2)$ along the y-axis. Then, the point of intersection of the lines perpendicular to the x-axis at $^-3$ and to the y-axis at $^-2$ locates the ordered pair $(^-3, ^-2)$.

The ordered pairs $(x, y) \in I \times I$ are associated with the points in the *lattice* displayed in Figure 13–3. Using this array of lattice points, each point has

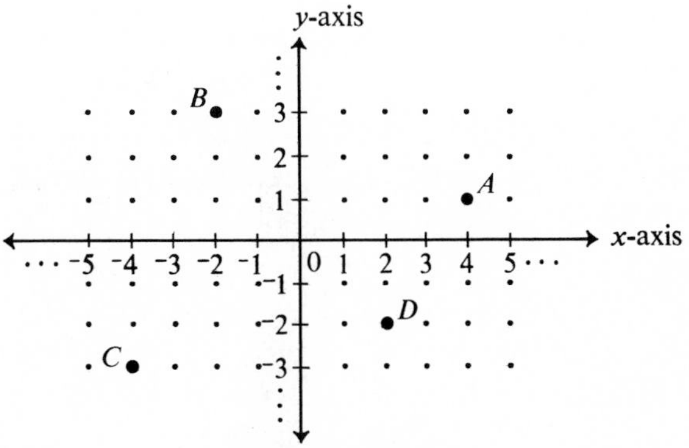

Figure 13–3

an ordered pair belonging to $I \times I$ associated to it. Find the ordered pairs associated with the points A, B, C, and D.

The number plane for $R \times R$, R the set of real numbers, consists of *all* points of the plane instead of lattice points as in $I \times I$. This is due to the completeness of R.

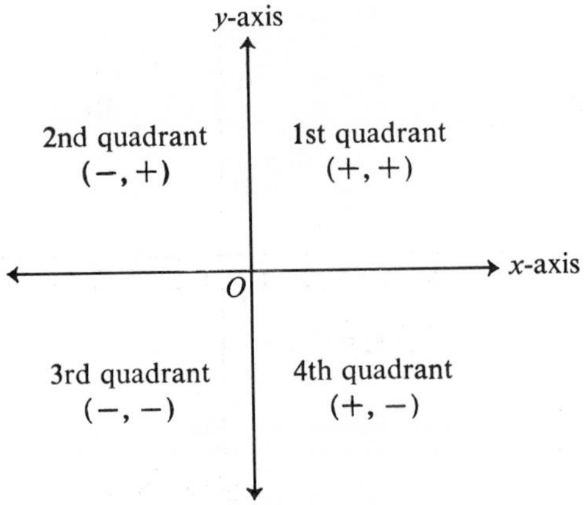

Figure 13–4

Observe that the number plane may be thought of as being subdivided into four quadrants as designated in Figure 13–4. The ordered pairs whose first and second components are positive are located in the first quadrant; those whose first component is negative and second component positive, in the second quadrant. Locate the ordered pairs (x, y), $x < 0$, $y < 0$; and (x, y), $x > 0$, $y < 0$.

Now that the elements of $I \times I$ have been matched with points in the number plane, we are prepared to represent the relation $P = \{(2, 1), (3, 2), (5, 4)\}$, a subset of $I \times I$, as in Figure 13–5. The assignment of the elements

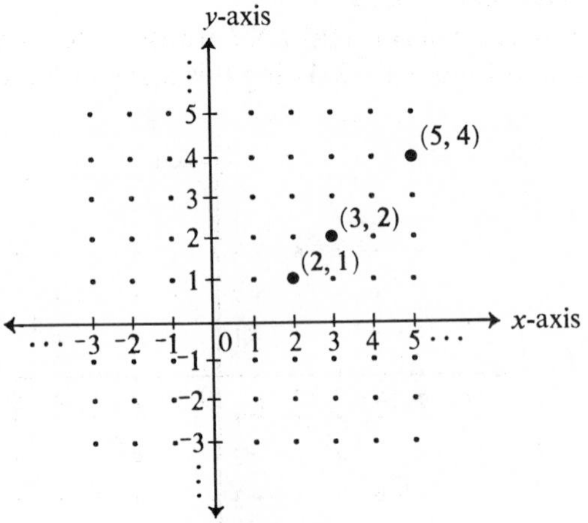

Figure 13–5

of P to points in the number plane having the same ordered pairs is the *graph* of relation P. The relation P is a subset of $I \times I$.

In Example 1, the elements (x, y) of the relation were selected using the sentence (rule) $x + 1 \geq y$. This sentence has *two variables*, x and y. The solution set is the relation

$$M = \{(2, 3), (3, 3), (3, 4), (4, 3), (4, 4), (4, 5), (5, 3), (5, 4), (5, 5)\},$$

a subset of $A \times B$. Figure 13–6 shows $A \times B \subset I \times I$ of the number plane and the graph of the relation M. Notice that in this example $A \times B$ is U.

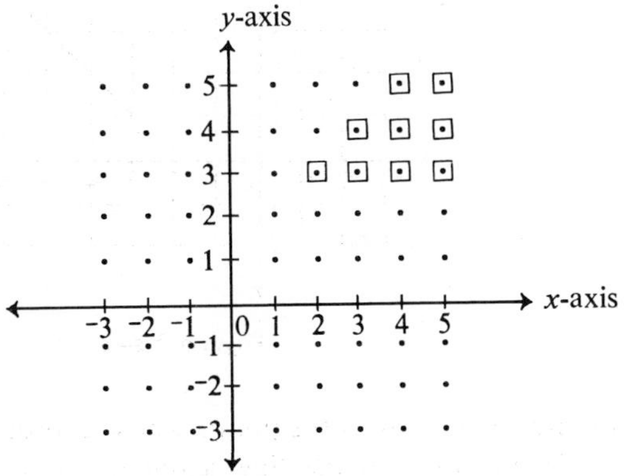

Figure 13–6

Example 2 Given $U = \{1, 2, 3, 4, 5, 6\}$, that is, $U = \{z \in I : 1 \leq z \leq 6\}$, and

$$S = \{(x, y), x \in U, y \in U : x - 2 = y\},$$

verify that

$$S = \{(3, 1), (4, 2), (5, 3), (6, 4)\}.$$

Let $U \times U$, a subset of $I \times I$, be represented by the "dots" in the first quadrant, Figure 13–7. The *graph* of the relation S is indicated by the large

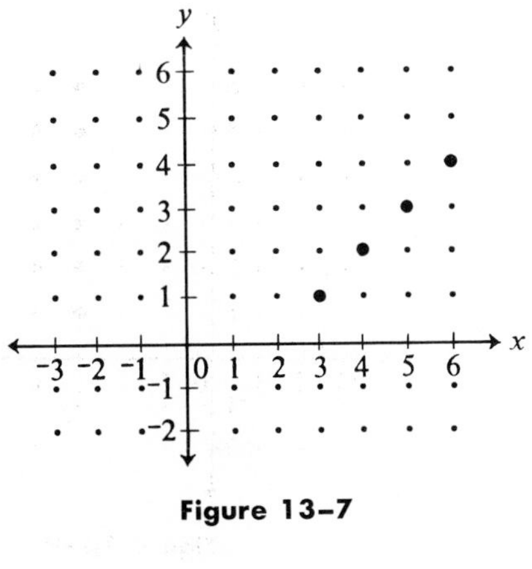

Figure 13–7

black dots. If on the other hand U is changed to the subset of R, the set of *real numbers,* then x (the first component) and y (the second component) would take on all real values beginning with 1 and ending with 6.

$$U = \{z \in R : 1 \leq z \leq 6\}$$

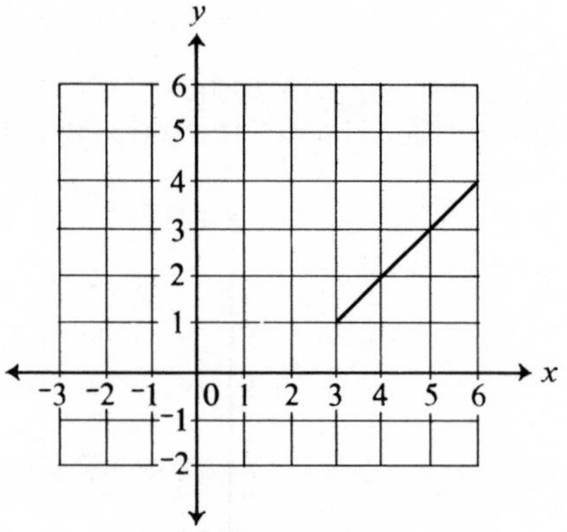

Figure 13–8

Then Figure 13–8 pictures the points whose coordinates are elements of the relation. Every point on the line segment, the coordinates of whose end points are (3, 1) and (6, 4) has coordinates taken from the real numbers allowed in U. Remember the completeness of the real number system. Obviously, the ordered pairs cannot be displayed in roster notation.

Example 3 Given $U = \{1, 2, 3, 4, 5, 6\}$, $U = \{z \in I : 1 \leq z \leq 6\}$,

$$P = \{(x, y), x \in U, y \in U : x - 2 \leq y\}.$$

Check to see if each point in Figure 13–9 represents an ordered pair belonging to P.

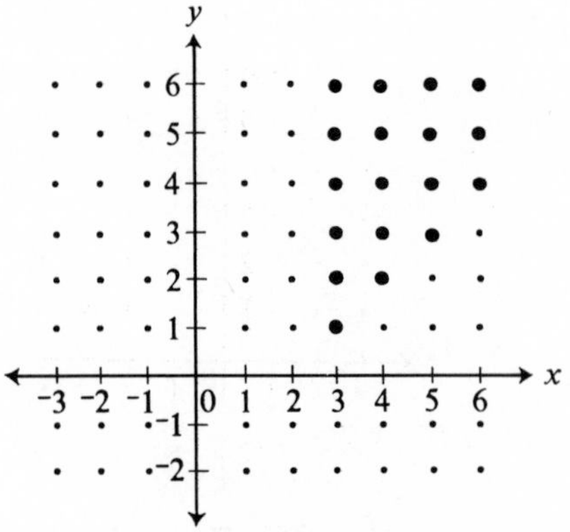

Figure 13–9

Now let U be changed to

$$U = \{z \in R : 1 \leq z \leq 6\}.$$

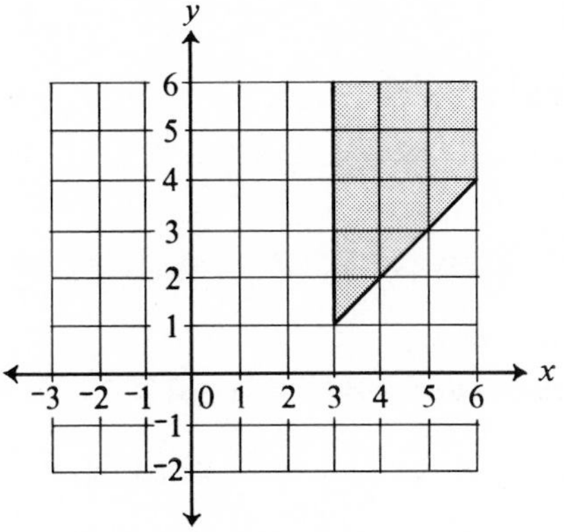

Figure 13–10

Is (5.9, 5.01) an element of the new relation? Check $(\sqrt{11}, \sqrt{30})$, $(\frac{14}{3}, \frac{8}{3})$, and (5, 3). How would Figure 13–9 be changed? Verify that this new relation (new because U has been changed) is shown graphically in Figure 13–10. If U is again changed, this time to include only the *rational numbers,*

$$U = \{z \in Q : 1 \leq z \leq 6\},$$

what has been deleted from the preceding relation? Explain how the change from real numbers to rational numbers will affect the graph. Is it possible to picture the deletion of the irrational numbers?

Example 4 Given $U = R$, the set of real numbers, and

$$M = \{(x, y) \in R \times R : y = |x|\}.$$

Some members of M would be

$$(3, 3), (^-\sqrt{5}, \sqrt{5}), (0, 0), (^-3, 3), (\tfrac{-7}{3}, \tfrac{7}{3}).$$

List five other members. Verify that M would appear graphically as in Figure 13–11.

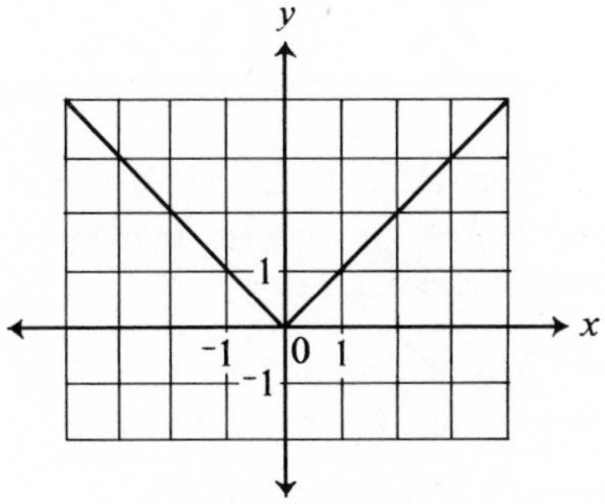

Figure 13–11

From the preceding examples, the graph of a relation $(x, y) \in I \times I$ requires a lattice grid, while $(x, y) \in R \times R$ requires a line grid. This applies also for any $U \times U$ that is a subset of either.

EXERCISE 13.3

1. If U is the set of integers and $M = \{(x, y) \in I \times I : |x| + 1 = |y|\}$, which of the pairs

$$(4, 5), (3, {}^-4), (6.7, {}^-7.7), ({}^-6, 5), ({}^-7, 8), ({}^-1, 0), (1, 2)$$

are elements of M?

2. Given the relation $K = \{(x, y) \in R \times R : x^2 + y^2 = 25\}$, which of the pairs

$$(3, 4), ({}^-5, 0), (2, \sqrt{21}), (\sqrt{27}, {}^-\sqrt{2}), (4.1, 2.9), (2, {}^-\sqrt{21}), (3, {}^-4), (0, 5)$$

are elements of K?

3. Write out the set of ordered pairs that make up the relation

$$M = \{(x, y), x \in N, y \in N : y > x\}$$

considered on a subset

$$A = \{1, 2, 3, 4\}, (A = \{z \in N : 1 \leq z \leq 4\}),$$

of the set of natural numbers. Graph the relation M.

4. In Problem 3, change the subset A of the set of natural numbers to a subset, B, of the set of real numbers:

$$B = \{z \in R : 1 \leq z \leq 4\}$$

Write at least five ordered pairs which are elements of this relation:

$$M' = \{(x, y), x \in B, y \in B : y > x\}$$

Is $(\sqrt{7}, \sqrt{15})$ an element of M'? Is $(7, \sqrt{21})$? Can you now write out the complete set? Graph the relation.

5. Using the same permitted values for x and y, first as in Problem 3 and then as in Problem 4, consider the relation

$$N = \{(x, y) : y = 2x\}.$$

Can you write five elements of the relation in each case? Graph the relation.

6. Assuming that $U = R$, the set of real numbers, find an ordered pair which belongs to each relation described below.

(a) $3x + y = 13$ (c) $|x| + |y| = 0$

(b) $5 - x = y$ (d) $4 - |y| \leq 2x$

7. Given the graphs of six relations.

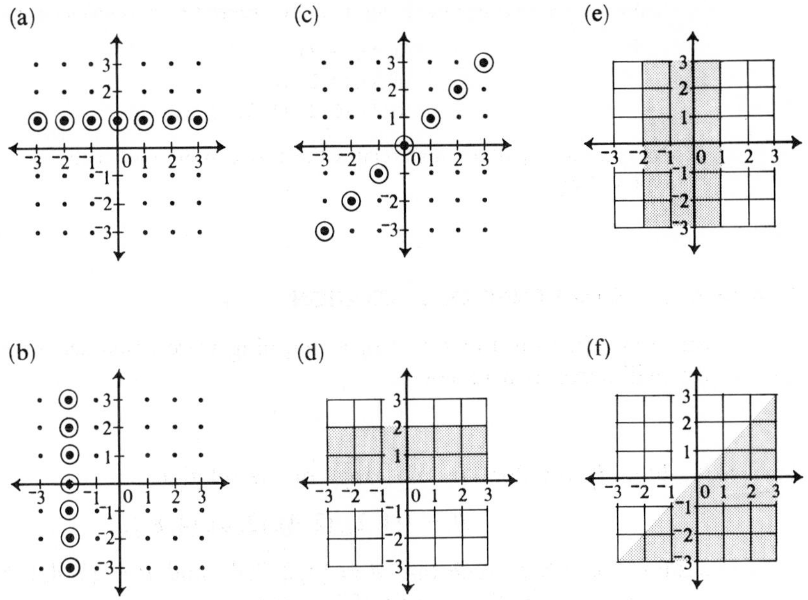

(i) Find 5 ordered pairs that belong to each relation.
(ii) Describe each relation in set builder notation.

8. Here the elements of a relation have been graphed for $U = N \times N$. Write the

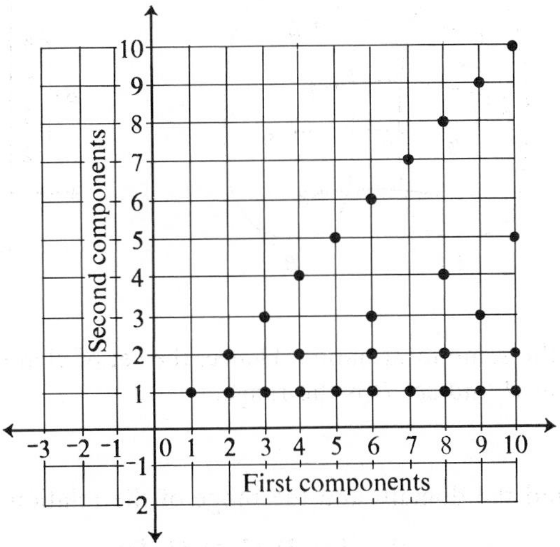

set of ordered pairs that is the relation represented in this graph. Can you detect the pattern? If so, supply the next five rows and columns. That is, extend the coordinate axes to include 11, 12, 13, 14, 15, and then locate the points whose ordered-pair representations follow the same pattern. Which of the pairs

$$(11, 1), (12, 3), (14, 7), (15, 10), (15, 5), (4, 12)$$

belong to this new set? HINT: Study these ordered pairs selected from the relation.

$$(3, 1), (3, 3)$$
$$(4, 1), (4, 2), (4, 4)$$
$$(6, 1), (6, 2), (6, 3), (6, 6)$$

Do you see a pattern that relates the first component to the second? What is R in $x\ R\ y$?

13.4 DOMAIN AND RANGE OF A RELATION

Since the domain and range of a mapping were discussed in Section 2.7, this will serve as a review.

Example 1 Identify the domain and range of the relation

$$R = \{(1, 2), (2, 4), (3, 6), (4, 8)\},$$

a subset of $A \times B$, such that $A = \{1, 2, 3, 4\}$ and $B = \{2, 4, 6, 8\}$.

The relation R, "is one half of," is represented by an arrow diagram and graph. The arrow diagram shows that each element of set A has an assignment in set B, and that each element in B has been assigned. The graph

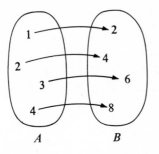

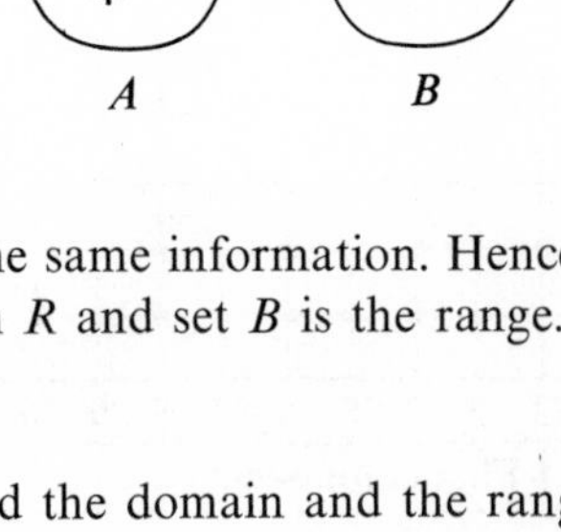

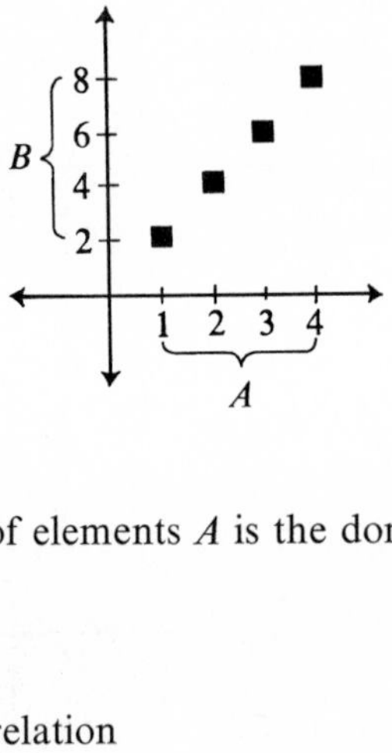

gives the same information. Hence, the set of elements A is the domain of relation R and set B is the range.

Example 2 Find the domain and the range of the relation

$$R = \{(2, 3), (2, 4), (3, 4)\},$$

a subset of

$$A \times B = \{(2, 3), (2, 4), (3, 3), (3, 4), (4, 3), (4, 4)\}.$$

From $A \times B, A = \{2, 3, 4\}$ and $B = \{3, 4\}$. An arrow diagram and a graph show the relation R, "is less than."

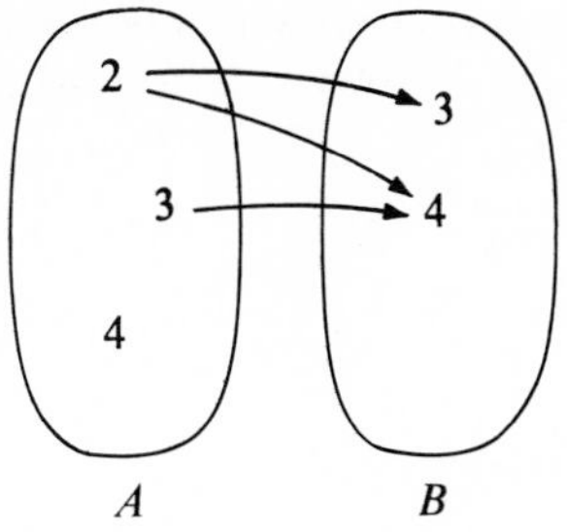

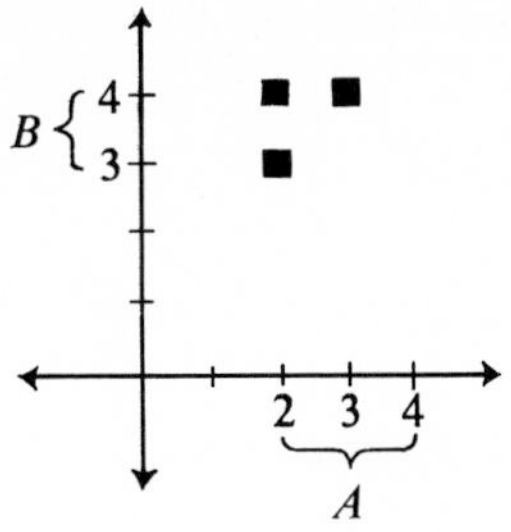

The domain is {2, 3} and the range is $B = \{3, 4\}$. The element 4 of set A does not belong to the domain of the relation, since it does not have an assignment in set B.

Example 3 Determine the domain and the range of the relation represented by the accompanying arrow diagram and graph.

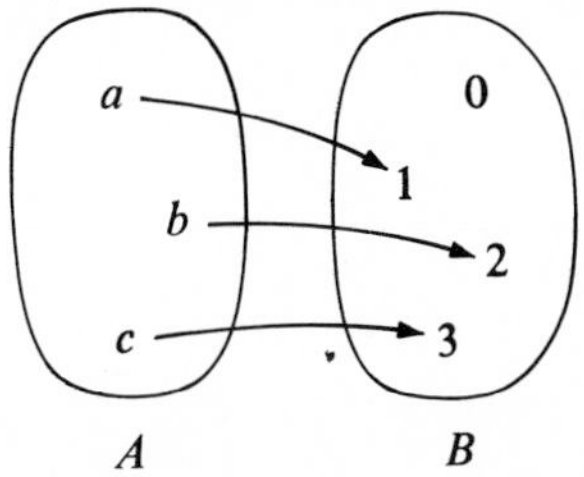

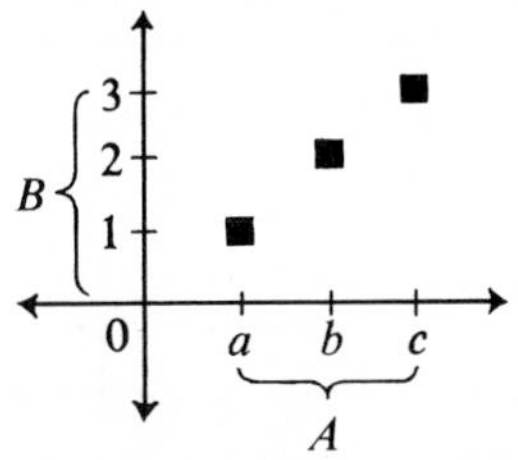

The relation $R = \{(a, 1), (b, 2), (c, 3)\}$, a subset of $A \text{ X } B$. Each element of A has an assignment in B, but each element in B has not been assigned. Hence, set A is the domain and {1, 2, 3} is the range of the relation R.

We are now prepared to give a definition for the domain and range of a relation.

Definition 13.4 The set of first components of the ordered pairs which belong to a relation is called the *domain* of the relation; the set of second components of these ordered pairs is called the *range* of the relation.

EXERCISE 13.4

1. Using $U = \{1, 2, 3, 4, 5\}$, write the following relations on $U \text{ X } U$ in roster notation. Then give the domain and the range for each relation.
   (a) $S_1 = \{(x, y) : xy \text{ is an odd number}\}$
   (b) $S_2 = \{(x, y) : x - y = 4\}$
   (c) $S_3 = \{(x, y) : y = x + 5\}$
   (d) $S_4 = \{(x, y) : x = 2 \text{ and } y > 3\}$

2. Make a graph of each of the following relations. Identify the domain and range of each. Refer to Section 2.5 for a discussion of "and" and "or."
   (a) $\{(x, y) \in N \times N : x = 4 \text{ and } y = 3\}$
   (b) $\{(x, y) \in N \times N : x = 4 \text{ or } y = 3\}$
   (c) $\{(x, y) \in R \times R : x < 4 \text{ and } y < 3\}$
   (d) $\{(x, y) \in R \times R : x < 4 \text{ or } y < 3\}$

13.5 FUNCTIONS

Let us now study a particular kind of relation. Consider these two relations:

$$A = \{(0, 2), (1, 3), (^-1, 1), (2, 4), (^-2, 0)\}$$
$$B = \{(1, 3), (1, 2), (2, 1), (3, ^-2)\}.$$

Observe that the domain of relation A is $\{0, 1, ^-1, 2, ^-2\}$. To each element of this domain, there is assigned one and only one element of the range, $\{2, 3, 1, 4, 0\}$. This is not the case in relation B. There is an element in the domain which has two elements assigned to it—that is, to 1 is assigned both 3 and 2.

The assignment of the elements in each relation is shown by an arrow diagram and by a graph.

Relation A

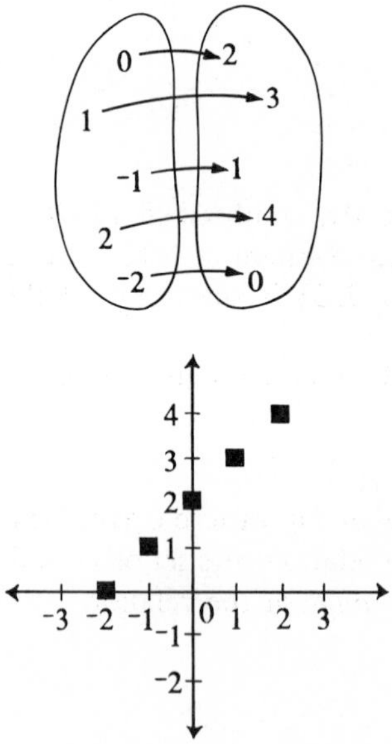

Relation B

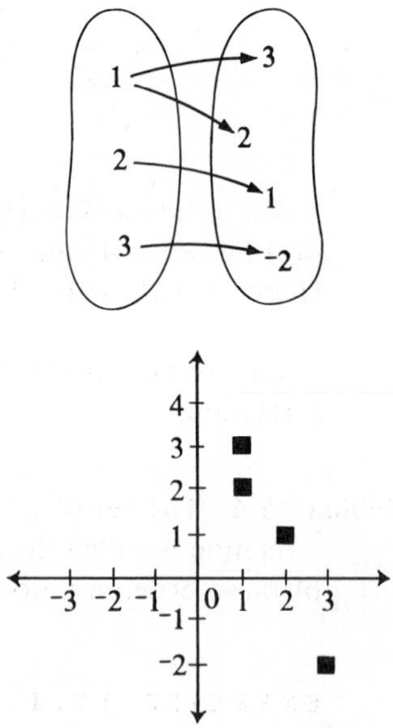

Our interest is in relation A. Do you recall that in Chapter 7 this was called a *mapping?* We shall now call such a relation a *function.* Thus, that which distinguishes a function from a relation is the limiting of *exactly one* element of the range to each element of the domain. It is evident that all functions are relations, but not all relations are functions. Formally we define a function in the following manner.

Definition 13.5 If, in a relation, to every element of a set called the domain there is assigned exactly one element of a set called the range, then that relation is called a *function*.

Example 1 Is the relation

$$S = \{(x, y) \in N \text{X} N : y = x^2 \text{ and } 1 \leq x \leq 5\}$$

a function?

In roster notation this relation is

$$S = \{(1, 1), (2, 4), (3, 9), (4, 16), (5, 25)\}.$$

To each element of the domain, $\{1, 2, 3, 4, 5\}$, is there exactly one assignment in the range, $\{1, 4, 9, 16, 25\}$? The answer is "yes." Since the necessary conditions have been fulfilled, we conclude that relation S is a function.

Example 2 Which graphs of the relations represent functions?

(a) $U \subset W \text{X} W$; Domain $= \{1, 2, 3, 4, 5\}$; Range $= \{1, 2, 3\}$

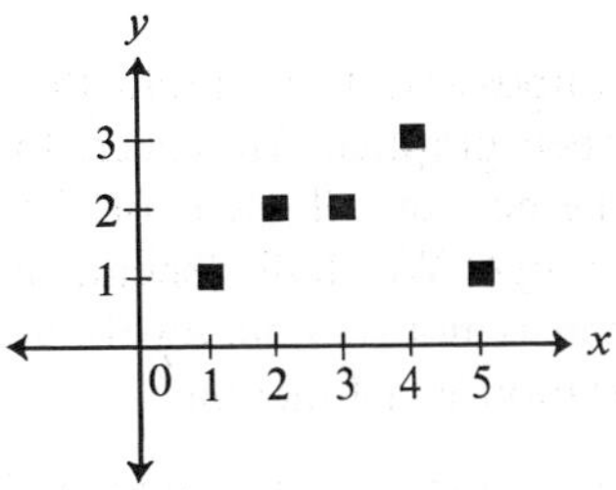

List the ordered pairs (x, y) represented by the dots. What is the domain? What is the range? Can you tell the domain and range by inspecting the graph of the relation? Does any element in the domain have assigned to it more than one element in the range? Since the answer is "no," this relation satisfies the requirement for a function. It is permissible to have one element in the range be an assignment (image) of more than one element in the domain. Observe that 1 in the range is assigned to both 1 and 5 in the domain. Find another example in the graph.

(b) $U \subset N \text{X} N$; Domain $= \{1, 2, 3, 4\}$; Range $= \{1, 2, 3\}$

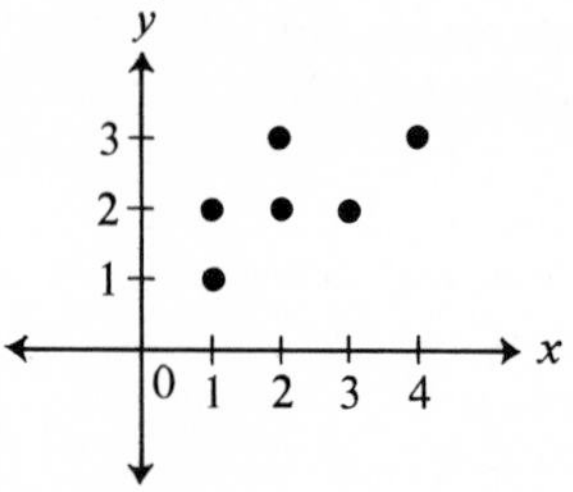

By listing the ordered pairs associated with the points shown in the graph, you are able to conclude that this relation is not a function. Which elements of the domain have more than one element of the range assigned? Explain how this information may be found by inspecting the graph.

(c) $U \subset R \times R$; Domain $= \{x \in R: 0 \leq x \leq 5\}$
Range $= \{y \in R: 0 \leq y \leq 2.5\}$

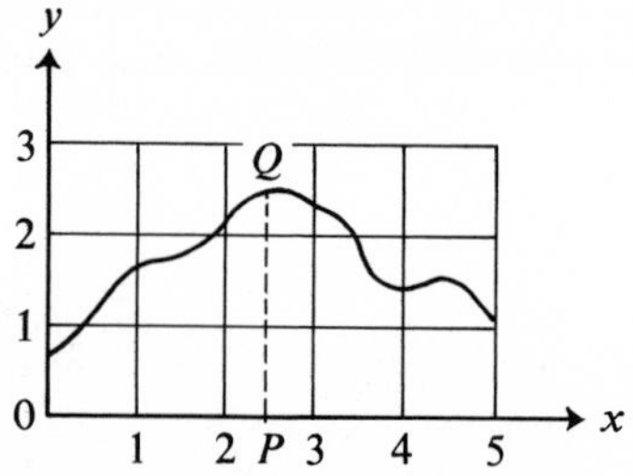

Obviously, it is impossible to write all the ordered pairs describing the relation that has been graphed. However, to investigate this relation, we must determine whether any element of the domain has more than one assignment in the range. We shall depend on a counterexample to prove that a given relation is not a function. If no such example is found, we assume that the relation is a function.

The following procedure is suggested as a way to look for counterexamples. Let the point P on the x-axis correspond to some number in the domain. The vertical line through P (indicated by the dotted line) contains all the points whose x-component is the same as that of P. The point Q then has the same x-component as P. But, point Q is the intersection of the vertical line and the curve. Hence, the ordered pair associated with Q belongs to the relation. As P takes on all possible values of the domain, a different vertical line corresponds to each. Thus, by "sliding" the vertical line back and forth, one can determine whether it intersects the curve in more than one point. In case that it does, then for some x there are multiple values for y, and the relation $(x, y) \in R \times R$ is not a function. This "vertical line" test is an easy way to determine whether a relation is not a function if the domain of its graph is represented on the horizontal axis.

(d) $U \subset R \times R$; Domain $= \{x \in R: 0 \leq x \leq 3\}$
Range $= \{y \in R: {}^-3 \leq y \leq 3\}$

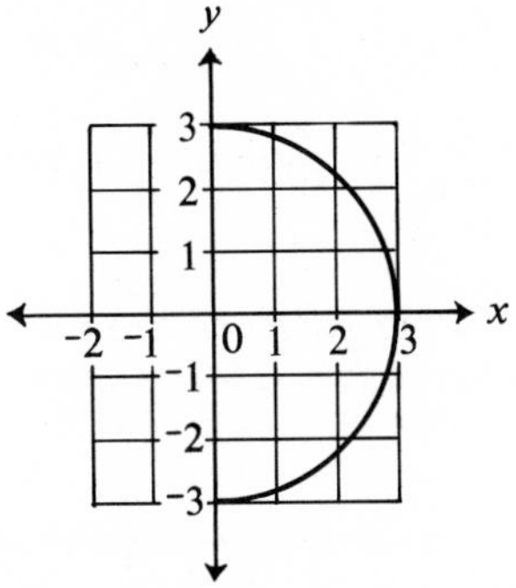

Apply the "vertical line" test to the graph. Why do you conclude that the relation is not a function?

Example 3 Use the graph of the relation S_1 to determine whether it is a function.

$$S_1 = \{(x, y) \in R \times R: x = 3 \text{ and } x = y\}$$

Some ordered pairs which belong only to $x = 3$ are

$$(3, 0), (3, 1), (3, 0.5), (3, {}^-1), (3, 2), (3, {}^-9).$$

Notice that the first component of each ordered pair is 3. Since no restrictions are placed on the second component in $x = 3$, y is every real number. The graph of $x = 3$ is the line indicated in the diagram.

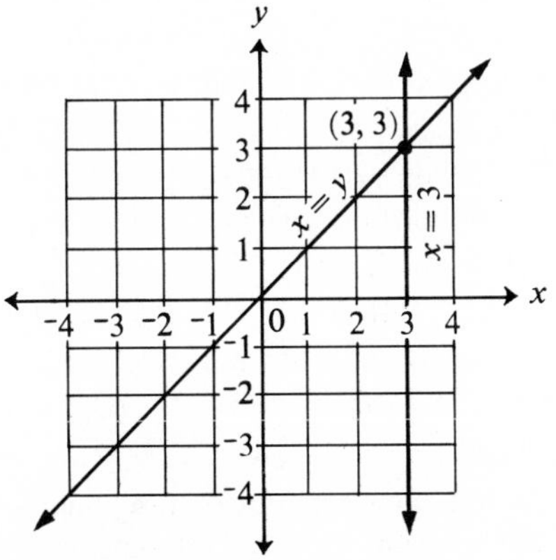

Assigning real numbers to x in $x = y$, corresponding real values of y are found. If x is $^-2$, the corresponding value of y is $^-2$. The graph of the ordered pairs of real numbers belonging to $x = y$ is the line indicated in the diagram.

Since S_1 specifies $x = 3$ *and* $x = y$, then the ordered pairs belonging to the relation are found to be those points in the intersection of the two graphs.

The diagram shows only one point in the intersection. Hence,

$$S_1 = \{(3, 3)\}.$$

Do you see that relation S_1 is a function?

Example 4 Use the graph of the relation S_2 to determine whether it is a function.

$$S_2 = \{(x, y) \in R \times R : x \leq 3 \text{ and } x \leq y\}$$

Verify that this is the graph of the relation S_2.

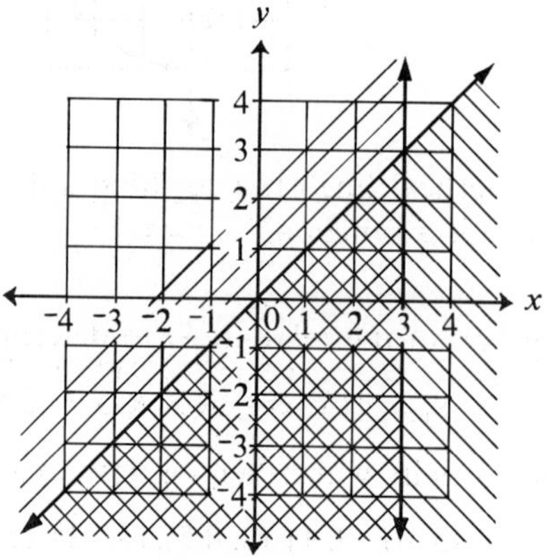

Since the cross-hatched portion is the graph of the relation S_2, is S_2 a function? Are (3, 1), (3, 2), (3, $^-4$) elements of S_2?

If $x \leq 3$ and $x \leq y$ is changed to $x < 3$ and $x < y$, what difference would there be in the graph?

EXERCISE 13.5

1. Consider the following relations.
   (i) Give the domain and the range of each.
   (ii) Identify the functions and explain why the others are not functions.
      (a) $\{(7, 3), (4, 9), (7, 6), (3, ^-7), (4, 0), (2, 1), (7, ^-3)\}$
      (b) $\{(^-2, 7), (^-2, 4), (^-2, 3), (^-2, ^-1), (^-2, 6)\}$
      (c) $\{(7, ^-2), (4, ^-2), (3, ^-2), (^-1, ^-2), (6, ^-2)\}$
      (d) $\{(5, 7), (^-2, 1), (2, 3), (8, 1), (3, ^-2), (2, \sqrt{9})\}$
      (e) $\{(2, ^-5)\}$
2. The following relations are presented without the formal set builder notation. Tell which are functions. Be sure to designate the domain and the range of each.
   (a) y is the uncle of x.
   (b) y is the perimeter measure in inches of triangle x.
   (c) y is the average temperature on x day.
   (d) y is the day of the week on which x was the average temperature.
3. Graph each of these relations:

$$A = \{(1, 2), (2, 4), (5, 5), (4, 4)\}$$
$$B = \{(1, 2), (1, 5), (3, 4), (3, 6)\}.$$

Apply the "vertical line" test to see whether each is a function.

4. Make a graph of the relation the Cartesian product $M \times S \subset I \times I$, where $M = \{1, 3, 5\}$ and

$$S = \{y \in M : 2 \leq y \leq 5\}.$$

Be careful that the graph includes *only* the points required.

5. Use the "vertical line" test to discover which of the relations graphed below are functions.

(a)

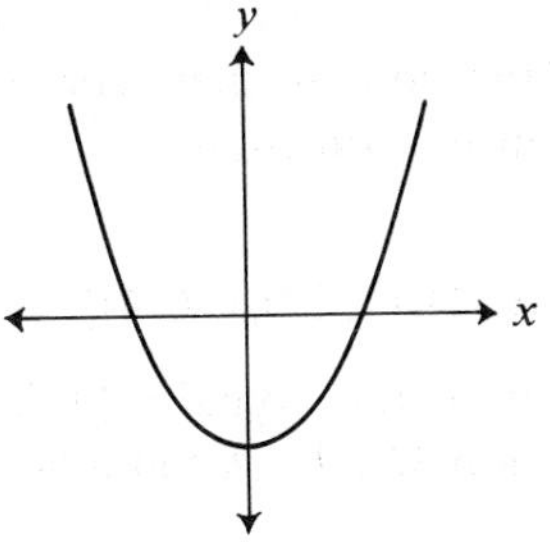

(b)

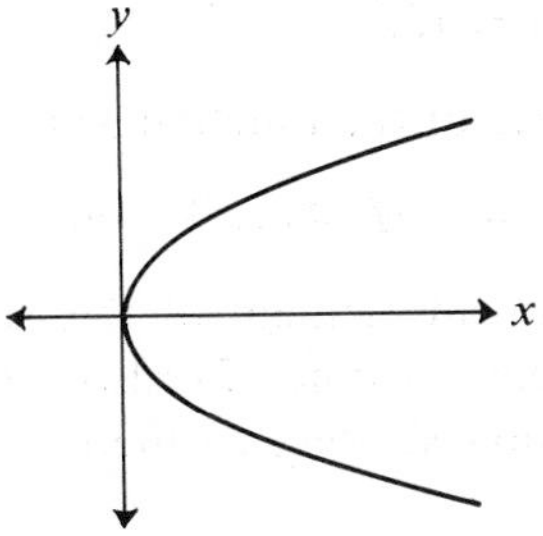

(c)

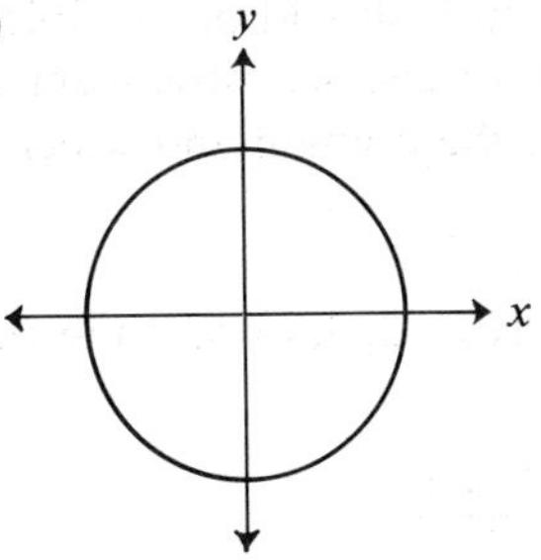

(d)

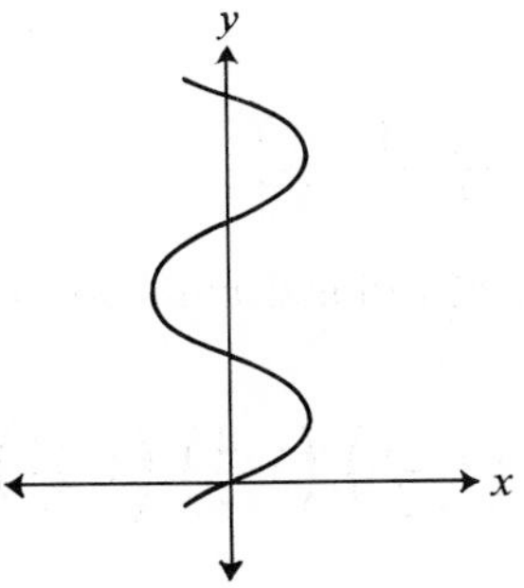

(e)

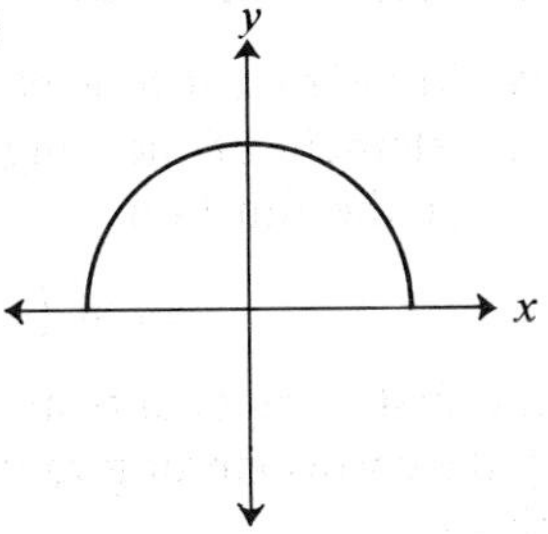

(f)

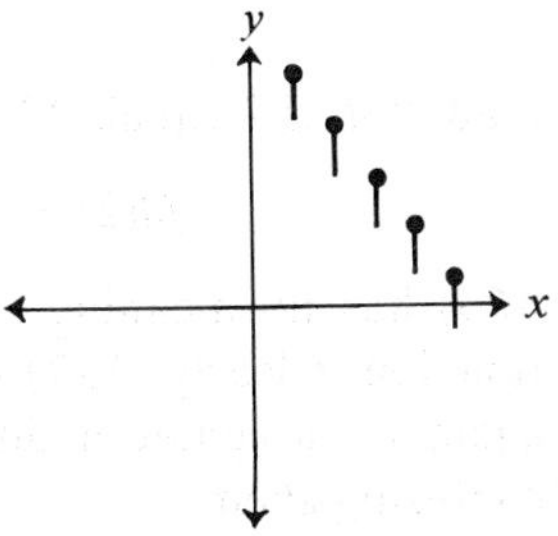

(g)

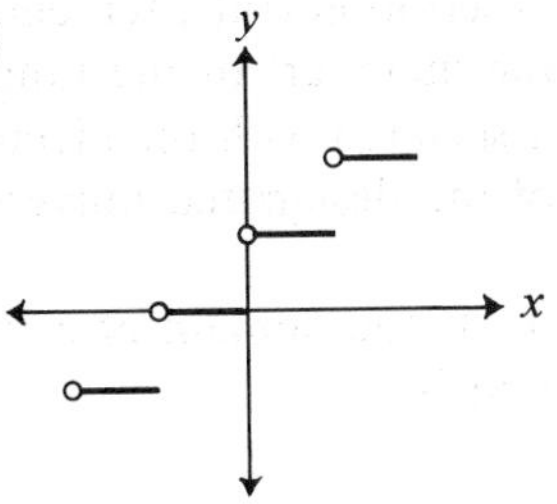

NOTE: The hollow dot indicates no point.

(h)

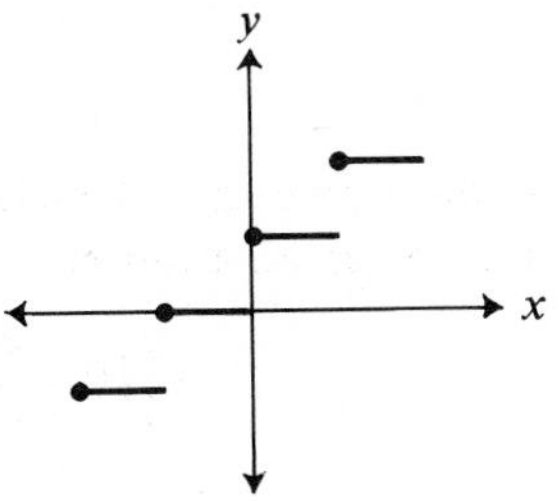

6. Make a graph of the following relations. Determine which are functions.
   (a) $A = \{(x, y) \in R \times R: x + y = 4 \text{ and } y = 2\}$
   (b) $B = \{(x, y) \in R \times R: x + y = 4 \text{ or } y = 2\}$
   (c) $C = \{(x, y) \in R \times R: x + y \leq 4 \text{ and } y \leq 2\}$
   (d) $D = \{(x, y) \in R \times R: x + y \leq 4 \text{ or } y \leq 2\}$
   (e) $E = \{(x, y) \in R \times R: 0 \leq x \leq 3 \text{ and } x + 1 \leq y\}$

13.6 FUNCTIONAL NOTATION

We know that if to each element of the domain there corresponds exactly one element of the range, the relation is a function.

For the function

$$M = \{(1, 4), (2, {}^-3), (6, 1), (7, 12), ({}^-4, 4)\},$$

the first components are the elements of the domain, $\{1, 2, 6, 7, {}^-4\}$. For 7, the corresponding value in the function M is 12. A common way of writing this is

$$M(7) = 12,$$

read, "M of 7 equals 12" or "the value of the function M at 7 is 12." Then

$$M(2) = {}^-3, \quad M(1) = 4, \quad M({}^-4) = 4, \quad \text{etc.}$$

$M(3)$ has no meaning, since 3 is not an element of the domain for this function. Clearly, $M(7)$ does not mean M times 7. Also, M may be used again to designate a different function so long as there is no chance of confusing them.

If, for a function F, we are given the universal set, the domain, and a description (rule, sentence, graph), then it should be possible to determine the members of the range. They are, of course, those elements which are associated with the given elements of the domain and which are members of the designated universal set.

Example 1 If the domain of a function F is $\{0, 1, 2, 3, 4\}$ and the rule of association is

$$y = \frac{x^2 - 5}{3},$$

the range of F is

$$\left\{\frac{{}^-5}{3}, \frac{{}^-4}{3}, \frac{{}^-1}{3}, \frac{4}{3}, \frac{11}{3}\right\}$$

if the universal set includes the set of rational numbers. The function in the roster notation is

$$F = \left\{\left(0, \frac{{}^-5}{3}\right), \left(1, \frac{{}^-4}{3}\right), \left(2, \frac{{}^-1}{3}\right), \left(3, \frac{4}{3}\right), \left(4, \frac{11}{3}\right)\right\}.$$

What is $F(2)$? What is $F(5)$?

It is sometimes necessary to restrict the domain. Certain elements of the domain may not produce number values for the function which are admissible. If the function in Example 1 had been limited to the set of integers, then the elements listed in the domain would not be admissible, since, while they are integers, they produce values of y that are not integers.

Example 2 Consider the function described by

$$f(x) = \frac{4}{2 - x}$$

in the set $R \times R$. There is one member of the set of real numbers for which the corresponding function value is not defined. Consequently, that number is not an element of the domain. You should recognize that this number is $x = 2$. Give a reason for this. Are there any other real numbers that should be omitted from the domain? Write the domain of this function in set builder notation. Give the values of $f(1.9)$, $f(2.1)$, $f(102)$, and $f(^-98)$. Some readers may know how to draw the graph of this function, but that is not necessary for our present work.

EXERCISE 13.6

1. Given the function described by $F(x) = 2|x|$ over $R \times R$, find the values of $F(3)$, $F(^-2)$, $F(0)$, and $F(\frac{^-1}{2})$. State the greatest possible domain and range for F.
2. Give in roster notation the domain and the range of

$$f = \{(2, 1), (3, 5), (7, 27), (4, 5), (6, 15)\}.$$

   What are the values of $f(3)$, $f(4)$, $f(6)$, and $f(27)$?
3. Given $G = \{(x, y) \in R \times R : y = 2x + 1\}$, find $G(24)$, $G(^-4)$, $G(\frac{3}{4})$, and $G(\frac{^-1}{2})$.
4. Given $g = \{(x, y) \in Q \times Q : xy = 1\}$, complete each of the following.

   (a) $g(4) =$ _____ (d) $g($_____$) = 5$ (g) $g($_____$) = \frac{5}{9}$
   (b) $g(\frac{1}{3}) =$ _____ (e) $g($_____$) = \frac{^-1}{10}$ (h) $g($_____$) = 0$
   (c) $g(0) =$ _____ (f) $g(\frac{3}{4}) =$ _____ (i) $g(100) =$ _____
   (j) Domain of $g =$ __________
   (k) Range of $g =$ __________
5. The domain of the function H is the set of real numbers. $H(x) = 7$. What is the range of H?
6. If the function f is defined by

$$f(x) = 2x$$

   on $I \times I$, and the domain of f is the set of integers, what is the range of f?
7. A function F is defined by

$$F(x) = \frac{6}{3 + x^2}$$

on $R \times R$. For what real values of x is $F(x)$ defined? Is $F(x)$ zero for any value of x? Is $F(x)$ less than zero for any value of x? For what value of x is $F(x)$ the greatest? What is the range of the function? Illustrate with a graph.

8. Two functions on $R \times R$ are described as follows:

$$F_1(x) = 2|x|, \qquad F_2(x) = 4 - |x|$$

(a) What are the admissible values of x for F_1?
(b) What are the admissible values of x for F_2?
(c) Discuss the corresponding range for each function.

14

Mathematical Systems

14.1 EXPLORING MATHEMATICAL SYSTEMS

Number systems have been developed throughout this textbook. Beginning with the natural numbers, there was a progressive development of number systems going to the whole numbers, to the integers, to the rational numbers, and culminating with the real numbers. Each of these systems consists of two operational systems—namely, $(S, +)$ and $(S, \times)$, where S is an *infinite set of numbers.*

Is it possible for an operational system to consist of a *finite set of numbers?* A familiar finite set is the set of numbers of the clock used in telling time. Is it possible to define an operation on these numbers? Consider, for example: A plane departs in 3 hours. It is now 11 o'clock. What is the departure time of the plane? Obviously, this is some type of addition problem. Yet, to write $11 + 3 = 2$ looks a bit strange. Hence, the symbol $\oplus$ will be invented to designate this new "addition," and $11 \oplus 3 = 2$ is read, "11 circle plus 3 is equal to 2."

The clock will help to interpret the symbol $\oplus$. Suppose that the hour hand is at 11. Where will it be 3 hours from now? The hour hand moves clockwise for 3 successive intervals, stopping at 2. The symbol $\oplus$ is used to designate a clockwise turning of the hour hand.

For further exploration, a 5-minute clock will be used instead of the

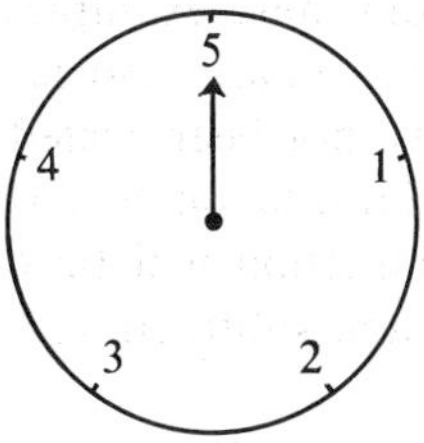

conventional one. Place the clock hand on 5. We are now ready to add using clock arithmetic. To "add" 2 to 1, move the hand from the 5 position one interval in a clockwise direction to the 1 position. Follow with a clockwise movement of 2 intervals, stopping at 3. This represents the "sum" of 1 and 2. Similarly, to "add" 1 to 2, the hand is moved from 5 clockwise through 2 intervals, followed with a clockwise move of 1 interval. The terminal location is again 3. Therefore, either $1 \oplus 2$ or $2 \oplus 1$ is the same as moving the hand clockwise 3 intervals from 5. This relation, "is the same as," an equivalence relation, is designated by the symbol, $=$.

Hence, $$1 \oplus 2 = 3$$
and $$2 \oplus 1 = 3.$$

Interpreting 5-minute clock addition in the preceding manner, the following "circle plus" grid is constructed.

⊕	1	2	3	4	5
1	2	3	4	5	1
2	3	4	5	1	2
3	4	5	1	2	3
4	5	1	2	3	4
5	1	2	3	4	5

Does the grid give convincing evidence that $\oplus$ is an operation, and that commutativity holds? Explain. Is there an identity element? Observe the 5-row and the 5-column. Since 5 in $\oplus$ acts like 0 in $+$, from now on 5 will be replaced by the symbol 0. That is, the identity element for $\oplus$ will be represented by 0.

Check the grid to see whether there is a 5 (0) sum in each row and column. If so, this means that each number of the set has a $\oplus$-inverse. For example, $2 \oplus 3 = 0$, so 2 is the $\oplus$-inverse of 3 and 3 is the $\oplus$-inverse of 2.

Use the 5-minute clock to find each of the following sums:

$(2 \oplus 1) \oplus 3$ $\quad$ $(4 \oplus 2) \oplus 3$ $\quad$ $(3 \oplus 2) \oplus 1$
$2 \oplus (1 \oplus 3)$ $\quad$ $4 \oplus (2 \oplus 3)$ $\quad$ $3 \oplus (2 \oplus 1)$

From these examples, does it seem appropriate to say that 5-minute clock numbers are associative under $\oplus$? Even though all triple sets of numbers have not been tested, it is quite possible to do so in a finite set.

Subtraction has been previously defined in terms of addition. Clock subtraction will be defined in a similar manner. Using the symbol $\ominus$ to denote subtraction,

$$3 \ominus 2 = x,$$

which means we seek a number x, an element of $\{0, 1, 2, 3, 4\}$, such that

$$2 \oplus x = 3.$$

The number x may be found by looking in the $\oplus$ grid along the 2-row for the sum 3. The number of the column where the 3 was found is the value for x, namely 1. Hence,

$$2 \oplus 1 = 3 \qquad \text{and} \qquad 3 \ominus 2 = 1.$$

As a second example, find the solution set for

$$1 \ominus 4 = x.$$

By the definition of $\ominus$, we seek the solution set for

$$4 \oplus x = 1.$$

The grid supplies 2 as the missing number, and so

$$1 \ominus 4 = 2.$$

Evidently, "subtraction" on the clockface means moving the hand backward, or counterclockwise. Hence, moving 1 space forward from 0 and then 4 spaces backward brings the hand to the place marked 2.

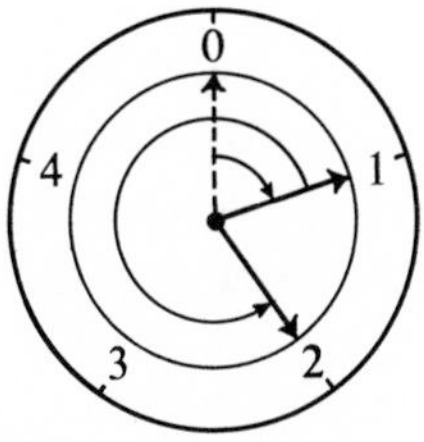

We stop here to make the following observations concerning the preceding discussion. First, although the set of elements used here, $\{0, 1, 2, 3, 4\}$, is a subset of the set of whole numbers, $W = \{0, 1, 2, 3, \ldots\}$, this new set possesses all the properties for $\oplus$ which W has for $+$, and some besides. The set $\{0, 1, 2, 3, 4\}$ and the operation $\oplus$ as defined by the grid together possess the properties of commutativity, associativity, and an identity element. Besides, there is an "additive" inverse for every element, and this permits an operation of "subtraction." The last two are properties that the system of whole numbers lacks.

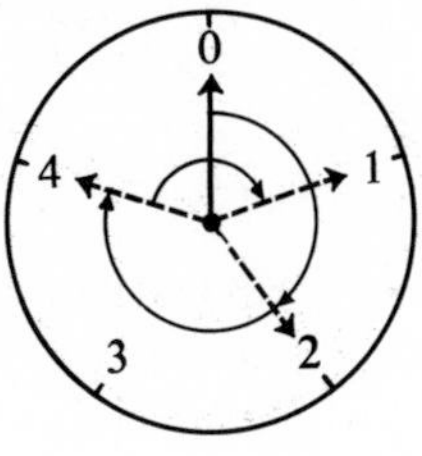

Consider 5-minute clock multiplication. The symbol $\otimes$ will be read, "circle times," and $3 \otimes 2$ will be interpreted as 3 clockwise turns of 2 intervals each. This is done by beginning with the hand pointing to 0. The hand is turned clockwise through 2 intervals, followed with two more such turns, making 3 turns of 2 each. The hand will stop on 1. Hence,

$$3 \otimes 2 = 1.$$

Using the clock to find $2 \otimes 0$, the hand is placed at 0. With 2 turns of 0 intervals, the hand remains at 0. Hence,

$$2 \otimes 0 = 0.$$

A "circle times" grid may be constructed.

$\otimes$	0	1	2	3	4
0	0	0	0	0	0
1	0	1	2	3	4
2	0	2	4	1	3
3	0	3	1	4	2
4	0	4	3	2	1

Is the grid symmetric with respect to the main diagonal from the upper left-hand corner to the lower right? This is evidence that $\{0, 1, 2, 3, 4\}$ is commutative under $\otimes$.

What evidence is there in the grid that indicates that this set of numbers has an identity element under $\otimes$? Since 1 is the multiplicative identity element for all previously studied number systems, one expects it to be the circle times identity element also.

Does each number have a $\otimes$-inverse? If so, each row and column in the grid should contain the identity element, 1. This occurs except in the 0-row and the 0-column. Consequently, each number except 0 has a $\otimes$-inverse. This is the case for what other number system? Since

$$3 \otimes 2 = 1, \quad 1 \otimes 1 = 1, \quad 4 \otimes 4 = 1,$$

then 3 is the $\otimes$-inverse of 2, while 1 and 4 are their own inverses.

Division in $\{0, 1, 2, 3, 4\}$ is defined in the same way as was done in previous number systems. Using $\oplus\!\!\!\!\div$ to denote division in 5-minute clock numbers, we see that

$$3 \,⊕\!\!\!\!\div\, 2 = x, \quad \text{if} \quad 2 \otimes x = 3$$

gives the relatedness of "circle divide" and "circle times." The number for x is found in the $\otimes$ grid to be 4. Obtaining a $⊕\!\!\!\!\div$ result from the $\otimes$ grid is much simpler than by using the clock as a model.

The following examples are used to investigate whether $\otimes$ is distributive over $\oplus$. Find the value of each of these expressions.

$$\begin{array}{ll} 3 \otimes (2 \oplus 1) & (3 \otimes 2) \oplus (3 \otimes 1) \\ 2 \otimes (4 \oplus 3) & (2 \otimes 4) \oplus (2 \otimes 3) \\ 4 \otimes (0 \oplus 2) & (4 \otimes 0) \oplus (4 \otimes 2) \end{array}$$

These few examples certainly do not constitute proof; however, since no counterexample has been found, distribution of $\otimes$ over $\oplus$ is accepted.

The number system described in the preceding discussion is summarized by the following diagram. The set of elements will be denoted by C_5, the set of 5-minute clock numbers.

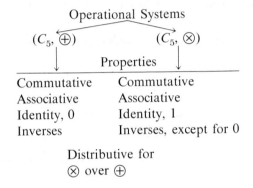

Operational Systems

$(C_5, \oplus)$	$(C_5, \otimes)$
Properties	
Commutative	Commutative
Associative	Associative
Identity, 0	Identity, 1
Inverses	Inverses, except for 0

Distributive for
$\otimes$ over $\oplus$

Suppose that we have a set of elements that are not numbers, a binary operation on these elements, an equivalence relation, and properties that these elements obey under the operation. Obviously, this cannot be identified as a number system. Instead, it is called a *mathematical system.*

Definition 14.1 A *mathematical system* consists of one or more operational systems,† an equivalence relation, and properties that the elements obey under the operation(s).

The number system consisting of $(C_5, \oplus, \otimes)$, the equivalence relation $=$, and the properties that the elements of C_5 obey under the operations $\oplus$ and $\otimes$ may also be called a mathematical system.

The following examples are tested by Definition 14.1 to determine whether each is a mathematical system.

Example 1 Given the set of elements $\{x, y, z, t\}$, and $\odot$ defined by the grid, is $\odot$ an operation as defined?

†Recall that an operational system consists of a set of elements and an operation defined on those elements (Definition 2.5(c)).

⊙	x	y	z	t
x	z	t	x	y
y	t	x	y	z
z	x	y	z	t
t	y	z	t	m

Since (t, t) has an assignment, m, that does not belong to the given set, then ⊙ is not an operation. This conclusion would be the same if (t, t) had no assignment. Since an operational system does not exist, the requirements for a mathematical system have not been met.

Example 2 Given $\{a, b, c\}$ and # defined by the grid.

#	a	b	c
a	b	a or c	a
b	c	b	b or c
c	a or b	c	a

Why is # , as defined by the grid, not an operation?
Do all pairs of elements have exactly one assignment? Since the set of elements, $\{a, b, c\}$, and # do not form an operational system, an essential requirement of a mathematical system does not exist.

Example 3 Consider $\{k, p, m, n\}$ and ⊡ defined by the grid.

⊡	k	p	m	n
k	n	m	p	k
p	m	k	n	p
m	p	n	k	m
n	k	p	m	n

An examination of the grid shows that $\{k, p, m, n\}$ and ⊡ form an operational system. Hence, one requirement for a mathematical system has been met. We shall now examine for properties also necessary.

To investigate whether the commutative property holds, we look at the main diagonal (upper left to lower right). As was the case for the natural numbers, the grid is symmetric with respect to this diagonal. This is evidence of the *commutative property*. Why? To establish that $\boxdot$ is associative requires many different problems in the search for a counterexample. Does

$$(p \boxdot m) \boxdot k = p \boxdot (m \boxdot k)?$$

Try several other examples. If no counterexample is found, one may assume *associativity*. This is not a proof. However, since this is a finite system (has a finite number of elements), all possibilities may be investigated. This would then constitute a proof.

Now look for a row which repeats the elements exactly as given at the top of the columns. This is the row headed by n, the n-row. We next look for a column which repeats exactly the elements heading the rows. This column is headed by n, the n-column. Evidently the element n serves as the *identity element* for $\boxdot$ in this system.

Thus:

$$p \boxdot n = n \boxdot p = p \qquad k \boxdot n = n \boxdot k = k$$
$$m \boxdot n = n \boxdot m = m \qquad n \boxdot n = n$$

Next, does each element possess a $\boxdot$-*inverse?* For example, is there an x such that $p \boxdot x = n$ (the identity element)? Look in the p-row for n. It occurs in the m-column. Hence $p \boxdot m = n$, and m is the $\boxdot$-inverse of p. From the commutative property we know that p is the $\boxdot$-inverse for m. There is one and only one identity element, n, in each row and each column. Hence, each element has a unique $\boxdot$-inverse.

With the inclusion of the equivalence relation, $=$, the requirements for a mathematical system have been fulfilled.

Example 4 A set of motions $\{I, H, R, V\}$ performed on a cardboard in the shape of a rectangle, such that one motion "follows" another, forms a mathematical system.

The elements of $\{I, H, R, V\}$ are illustrated below.

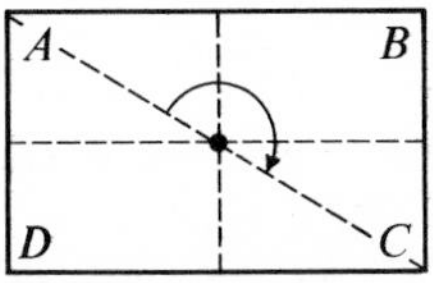

Figure 14–1

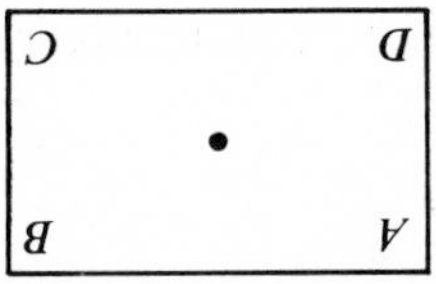

Figure 14–2

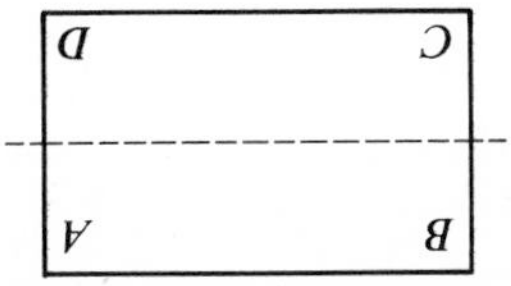

Figure 14–3

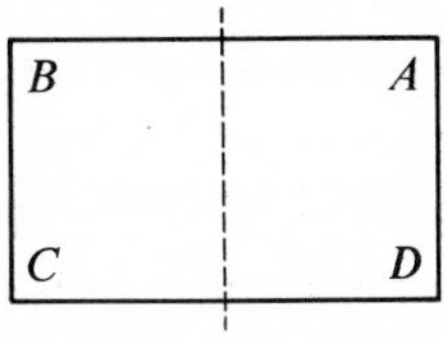

Figure 14–4

Figure 14–1 shows the original position called I. Figure 14–2 shows the result of rotating I one-half turn in a clockwise direction, which is R. Figure 14–3 shows I flipped along the horizontal axis, which gives H. Figure 14–4 shows I flipped along the vertical axis and is called V.

In order to understand the way "follows" (denoted by $\circ$) is used, select H by flipping I along the horizontal. Follow H with R, and the result is

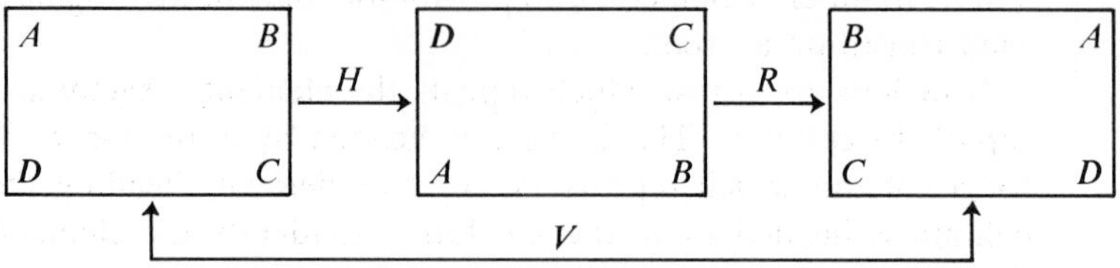

V. A short cut to this would be to begin with I and follow with the motion V. To follow H with R means

$$R \text{ "follows" } H = V.$$

Using the symbol for "follows,"

$$R \circ H = V.$$

This is a "follows" ($\circ$) grid for the set of motions $\{I, V, H, R\}$.

$\circ$	I	V	H	R
I	I	V	H	R
V	V	I	R	H
H	H	R	I	V
R	R	H	V	I

The set of elements $\{I, V, H, R\}$ and $\circ$ form an operational system. Why? Explain why the system is commutative and associative. How do you tell that the identity is I? Does each element have a $\circ$-inverse? Except for associativity, one should be able to detect each of these properties by a careful examination of the grid.

The operational system $(\{I, V, H, R\}, \circ)$, the equivalence relation $=$, and the properties discussed above constitute a mathematical system.

EXERCISE 14.1

1. In Example 1, replace m with x in the grid.
   (a) Does an operational system now exist? Explain.
   (b) Does the system possess an identity element? If so, identify it.

(c) Does each element have a $\odot$-inverse? Name those which do, and give the $\odot$-inverse of each.
(d) Give evidence showing whether associativity holds.
(e) Explain how the grid may be used to identify commutativity. Is this system commutative?
(f) Explain why the change in the grid made possible a mathematical system.

2. Make a list of the pairs of $\boxdot$-inverses in Example 3. What elements in $\{k, p, m, n\}$ are their own $\boxdot$-inverses?
3. Clockwise rotations about point P of the equilateral triangular region ABC are defined as rotations from the original position, r_0.

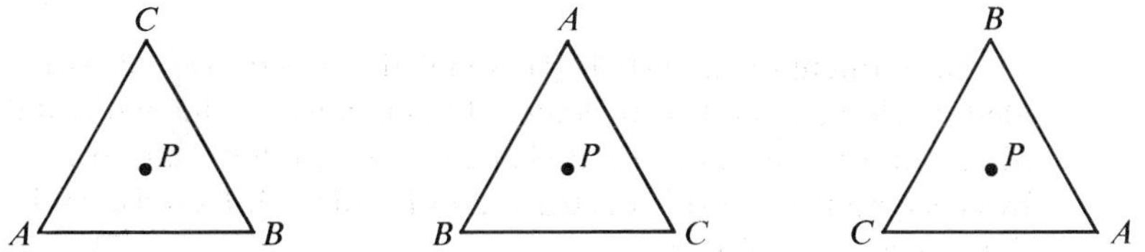

r_0: no turn $\qquad$ r_1: $\frac{1}{3}$ turn $\qquad$ r_2: $\frac{2}{3}$ turn

(a) List the set of elements.
(b) Make a grid which defines "follows" ($\circ$).
(c) Does an operational system exist?
(d) Examine the grid for commutativity and for an identity element.
(e) List the pairs of $\circ$-inverses if there are any.
(f) If associative, cite an instance; if not, cite a counterexample.
(g) Have the requirements for a mathematical system been met? Explain.
(h) Find $r_0 \circ r_1 =$ $\qquad$ $r_0 \circ r_2 =$ $\qquad$ $r_1 \circ r_2 =$
$r_1 \circ r_0 =$ $\qquad$ $r_2 \circ r_0 =$ $\qquad$ $r_2 \circ r_1 =$

4. Repeat the preceding problem using a square region. The clockwise rotations will need to be defined as was done for the triangular region.
5. Consider the set $A = \{0, 2, 4, 6, \ldots\}$, an infinite set, and the operation $*$ defined as

$$x * y = x + 2y, \quad x, y \in A.$$

(a) Find $2 * 4$ and $4 * 2$. Are the results the same? What do you conclude from this information?
(b) Find $(2 * 4) * 6$ and $2 * (4 * 6)$. What conclusion can be drawn from this information?

6. Given $\{0, 1, 2, 3\}$ and $\otimes$ defined as for 5-minute clock numbers:
(a) Construct a $\otimes$ grid.
(b) Examine the grid to determine whether a mathematical system is established.

14.2 A GROUP

Each of the diagrams describes a mathematical system developed in Section 14.1. The first (i) is a diagram of $(C_5, \oplus)$, the second (ii) of $(C_4, \otimes)$. Refer to Exercise 14.1, Problem 6, for $(C_4, \otimes)$.

(i) Operational Systems (ii)

$(C_5, \oplus)$	$(C_4, \otimes)$
Properties	
Commutative	Commutative
Associative	Associative
Identity, 0	Identity, 1
Inverse for each C_5	No inverse for 0 and 2

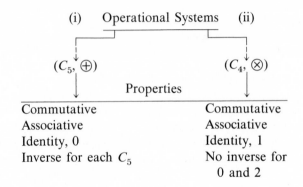

The mathematical system (i) is called a *group*. It is associative, has an identity element, and each element has an inverse. The mathematical system (ii) is not a group, since each element does not have an inverse. Verify this by checking the $\otimes$ grid constructed in Problem 4, Exercise 14.1. The definition for a group follows:

Definition 14.2 A set S of elements (such as $\{a, b, c, \ldots\}$), an operation (such as $\textcircled{\#}$), and an equivalence relation (such as $=$), form a *group* if the following properties hold:

G_1 (Associativity): If a, b, c are any three elements in S, then

$$(a \textcircled{\#} b) \textcircled{\#} c = a \textcircled{\#} (b \textcircled{\#} c).$$

G_2 (Identity element): For any $a \in S$, there exists in S an element i such that

$$a \textcircled{\#} i = i \textcircled{\#} a = a.$$

G_3 (Inverse under #): For any $a \in S$, there exists an a' in S such that

$$a \textcircled{\#} a' = a' \textcircled{\#} a = i.$$

G_4 If also the commutative property holds for every pair of elements, the system is called a *commutative group* or an *Abelian group* (named for Niels Hendrik Abel, a Norwegian mathematician).

By G_4 of the preceding definition, is (i) in the preceding diagram a *commutative group?*

EXERCISE 14.2

1. Determine which of the following are groups. For those which are groups, classify them as commutative or noncommutative. After trying two or three examples, you may *assume* the associative property if you have found no counterexample.
   (a) Problem 1, Exercise 14.1
   (b) Example 3 in Section 14.1

(c) Example 4 in Section 14.1
(d) Problem 3, Exercise 14.1
(e) Problem 4, Exercise 14.1

2. Recall the diagrams of the number systems developed in previous chapters. Examine the two mathematical systems of each separately to determine which are groups and which commutative groups. Justify your answers.
   (a) The system of natural numbers. (HINT: Consider $(N, +)$ and its properties. Is there an additive identity element? Also consider $(N, \times)$ and its properties.)
   (b) The system of whole numbers
   (c) The system of integers
   (d) The system of rational numbers

14.3 MODULAR SYSTEMS

The clock $\otimes$ grids were constructed from results obtained by moving the hands of a clock. Have you discovered how each grid could have been constructed without the use of a clock?

Consider the numbers of a 7-minute clock, $\{0, 1, 2, 3, 4, 5, 6\}$. To find $5 \otimes 6$, proceed as follows. After finding the product of 5×6, "cast out" as many 7's as possible, retaining the remainder, 2, as the result of $5 \otimes 6$. "Casting out" is done by dividing 30 by 7. The result is written,

$$30 = (7 \times 4) + 2.$$

Refer to the Generalization for the Division Algorithm, Chapter 7.

Hence, $5 \otimes 6$: $\quad 5 \times 6 = 30 = (7 \times 4) + 2$

and $5 \otimes 6$ is recorded as 2 in the $\otimes$ grid for the 7-minute clock.

The dividend 30 and the remainder 2 are related in the following way:

$$30 \equiv 2 \text{ (modulo 7)}.$$

This is read: "30 *is congruent to* 2, modulo 7." *Modulo* 7 means with the *modulus* 7.

Example 1 Examine $21 \equiv 1$ (mod 5) to see whether it is a true statement. NOTE: (mod 5) is an abbreviation for modulo 5.

Dividing 21 by 5, the result is written,

$$21 = (5 \times 4) + 1.$$

Since 1 is the remainder, we write,

$$21 \equiv 1 \text{ (mod 5)}.$$

Suppose that we wish to ascertain whether two numbers are congruent. Observe the next example.

Example 2 Is $116 \equiv 71$ (mod 15)?

Dividing 116 and 71 by 15 and applying the Generalization for the

Division Algorithm for recording the results,

(i) $$116 = (15 \times 7) + 11$$

(ii) $$71 = (15 \times 4) + 11.$$

Each having the same remainder, 11, we conclude that

$$116 \equiv 71 \pmod{15}.$$

Subtracting (ii) from (i),

$$\begin{aligned} 116 - 71 &= [(15 \times 7) + 11] - [(15 \times 4) + 11] \\ &= [15 \times (7 - 4)] + (11 - 11). \end{aligned}$$

Therefore,

$$116 - 71 = 15 \times 3.$$

This example leads to the following definition and theorem, which serve as convenient ways of establishing congruence modulo m between two integers.

Definition 14.3 If x, y, and m are integers, $m > 1$, and

$$x - y = mk, \qquad k \text{ an integer},$$

then x *is congruent to* y modulo m, written

$$x \equiv y \pmod{m}.$$

Theorem I *If integers x and y have the same remainder when divided by integer* $m > 1$, *then*

$$x \equiv y \pmod{m}.$$

This may be proved as follows: Let

$$x = mk_1 + r \quad \text{and} \quad y = mk_2 + r.$$

Subtracting, we have

$$\begin{aligned} x - y &= mk_1 - mk_2 \\ &= m(k_1 - k_2). \end{aligned}$$

Then, by Definition 14.3, $x \equiv y \pmod{m}$.

The nonnegative remainder less than m which results when an integer is divided by m is sometimes called the *residue modulo m* of that integer.

The residue modulo m of a positive integer may be found simply by division as above. For negative integers, other methods are used. For example, it is possible to write by inspection

$$^{-}4 = [15 \times {}^{-}1] + 11,$$

and so 11 is the residue modulo 15 of $^{-}4$. That is, $^{-}4 \equiv 11 \pmod{15}$. For negative integers with larger absolute values, the expression in this form may not be immediately evident. One method of approach is illustrated in the following example.

Example 3 Find the residue modulo 15 of $^-109$.

$109 =$	$(15 \times 7) + 4$	
$^-109 =$	$[^-1 \times (15 \times 7)] + {}^-4$	*Why?*
$=$	$(15 \times {}^-7) + ({}^-15 + 15) + {}^-4$	*Adding* 0 *in form* $(^-15 + 15)$
$=$	$(15 \times {}^-7) + (15 \times {}^-1) + (15 + {}^-4)$	*Explain.*
$=$	$[15\,(^-7 + {}^-1)] + 11$	*Explain.*
$=$	$(15 \times {}^-8) + 11$	*Why?*

Thus, 11 is the residue modulo 15 of $^-109$. And,

$$^-109 \equiv 11 \pmod{15}$$

Since 116, 71, $^-4$, and $^-109$ are each congruent to 11 modulo 15, they are congruent to each other.

As an exercise, check the following sentences, called *congruences.*

(1) $21 \equiv 1 \pmod 5$	(5) $21 \equiv {}^-9 \pmod 5$
(2) $154 \equiv 4 \pmod 5$	(6) $^-12 \equiv 23 \pmod 5$
(3) $29 \equiv 14 \pmod 5$	(7) $^-11 \equiv 4 \pmod 5$
(4) $29 \equiv 4 \pmod 5$	(8) $23 \equiv {}^-2 \pmod 5$

It is evident from the sentences above that 154, 29, and $^-11$ (as well as 14 and 4) have a common property—they are congruent to each other modulo 5. It is possible to separate all the integers into five *congruence* or *residue classes modulo* 5 according to their residues. Each integer must belong to one and only one class, since there are exactly five possibilities for residues, namely 0, 1, 2, 3, or 4. Each class has infinitely many elements. We shall use these residues modulo 5 to designate the classes:

Class 0: $5k + 0$, where $k = 0, {}^\pm1, {}^\pm2, {}^\pm3, \ldots$ (that is, all integers congruent to 0 modulo 5)†

Class 1: $5k + 1$, where $k = 0, {}^\pm1, {}^\pm2, \ldots$
Class 2: $5k + 2$, where $k = 0, {}^\pm1, {}^\pm2, \ldots$
Class 3: $5k + 3$, where $k = 0, {}^\pm1, {}^\pm2, \ldots$
Class 4: $5k + 4$, where $k = 0, {}^\pm1, {}^\pm2, \ldots$

Then the set

$$(\text{Class } 0) \cup (\text{Class } 1) \cup (\text{Class } 2) \cup (\text{Class } 3) \cup (\text{Class } 4),$$

a set whose components are disjoint sets of integers, is the entire set of integers I.

The following table displays these classes:

†The symbol, $^\pm1$, is a short way of writing 1, $^-1$.

TABLE OF CONGRUENCE OR RESIDUE CLASSES MODULO 5

Class 0:	...	$^-15$	$^-10$	$^-5$	0	5	10	15	...
Class 1:	...	$^-14$	$^-9$	$^-4$	1	6	11	16	...
Class 2:	...	$^-13$	$^-8$	$^-3$	2	7	12	17	...
Class 3:	...	$^-12$	$^-7$	$^-2$	3	8	13	18	...
Class 4:	...	$^-11$	$^-6$	$^-1$	4	9	14	19	...

Any two integers selected from the same row are congruent modulo 5, while any two selected from different rows are not congruent.

For example, $32 \equiv 2 \pmod 5$, and so 32 would appear in the third line, Class 2, of the table together with 2, 7, 12, 17, ... Where would $^-94$ appear? Now

$$^-94 = (^-18 \times 5) + {^-4} = (^-19 \times 5) + 1.$$

Hence, $^-94 \equiv 1 \pmod 5$ and should appear in the second line, Class 1, along with 1, $^-4$, $^-9$, $^-14$, ...

Test your understanding of the table by placing the integers 20, . . . , 49 and $^-30$, . . . , $^-16$ in their proper places. It should be clear that no more rows can be added. The table will be extended right and left.

We have seen that the modulus 5 separates the set of integers into five disjoint subsets, and that the union of the 5 subsets is the integer set. It is customary to designate such subsets or congruence classes by typical numbers, one to each class, either the smallest numerically, $\{^-2, ^-1, 0, 1, 2\}$, or the least nonnegative numbers, $\{0, 1, 2, 3, 4\}$. We have used the latter set.

Recall that in $x = (m \times q) + r$, with the modulus m and q the greatest possible quotient, the class into which x falls is determined by r. The condition imposed on m throughout the preceding discussion has been, $m > 1$.

Suppose that we examine what would occur if we disregarded this condition, and let $m = 0$. Consider these examples where x takes on different values.

$$x = (0 \times q) + r$$

$$\begin{aligned} x = 7{:}\quad & 7 = (0 \times q) + 7 \\ x = 8{:}\quad & 8 = (0 \times q) + 8 \\ x = 9{:}\quad & 9 = (0 \times q) + 9 \end{aligned}$$

Hence, $m = 0$ leads to infinitely many classes with only one element in each class.

Now, suppose m is allowed to be 1. Then:

$$x = (1 \times q) + r$$

$$\begin{aligned} x = 7{:}\quad & 7 = (1 \times 7) + 0 \\ x = 8{:}\quad & 8 = (1 \times 8) + 0 \end{aligned}$$

As one can see, there will be only one class (0) with infinitely many elements belonging in it.

Let $m = 2$, then:

$$x = (2 \times q) + r$$

$$\begin{aligned} x = 7: &\quad 7 = (2 \times 3) + 1 \\ x = 8: &\quad 8 = (2 \times 4) + 0 \\ x = 9: &\quad 9 = (2 \times 4) + 1 \\ x = 10: &\quad 10 = (2 \times 5) + 0 \end{aligned}$$

Does it seem reasonable to predict that if $m = 2$, there will be only two classes (0 and 1) with infinitely many elements belonging in each?

For $m = 3$, how many classes will there be? Make a generalization about the number of classes for each $m \geq 0$.

We shall now see how it is possible to find at once which class contains the sum or the product of numbers picked at random from two classes. For example, find the class (mod 5) to which ($^-4 + 18$) belongs without actually performing the addition. Referring to the table, we see that $^-4$ is in the 1-class and 18 is in the 3-class. Then their sum should be congruent to $1 + 3$, or 4, (mod 5). Now perform the addition ($^-4 + 18 = 14$) and see that $14 \equiv 4$ (mod 5).

Again, without actually multiplying, show that $12 \times 19 \equiv 3$ (mod 5). This follows because $12 \equiv 2$ (mod 5), $19 \equiv 4$ (mod 5), and their product is congruent to 2×4 (mod 5); $8 \equiv 3$ (mod 5). Check this by finding the product 12×19, dividing it by 5, and noting the remainder. It should be 3.

Make a grid for "addition" modulo 5. Note the similarity between this and the grid for 5-minute clock addition in Section 14.1. However, the elements used in this grid were the numbers 5 (0), 1, 2, 3, 4, while those used here represent classes of numbers modulo 5. Investigate whether the (mod 5) system employing {0, 1, 2, 3, 4} and the operation $\oplus$ has the group properties G_1, G_2, and G_3. If so, is it a commutative group?

The grid for "multiplication" modulo 5 is given below:

"MULTIPLICATION" GRID MODULO 5

$\otimes$	0	1	2	3	4
0	0	0	0	0	0
1	0	1	2	3	4
2	0	2	4	1	3
3	0	3	1	4	2
4	0	4	3	2	1

Notice from the grid that the (mod 5) system made up of {0, 1, 2, 3, 4} and $\otimes$ has commutativity and associativity. The element 1 serves as an identity. However, the element 0 has no inverse under multiplication. All the other elements do have inverses. For example, $2 \otimes b = 1$ has the solution set {3}. Therefore 3 is the multiplicative inverse of 2 in the modulo 5 system. But

$0 \otimes b = 1$ has no solution. Because of this deficiency, this system is not a group. However, verify that the *reduced system,* $\{1, 2, 3, 4\}$ (mod 5) under "multiplication," does form a group. A *reduced system* (mod m) consists of the elements corresponding to numbers relatively prime to m.

Below are tables showing "addition" and "multiplication" facts where the elements represent equivalence classes modulo 6:

"ADDITION" GRID MODULO 6

$\oplus$	0	1	2	3	4	5
0	0	1	2	3	4	5
1	1	2	3	4	5	0
2	2	3	4	5	0	1
3	3	4	5	0	1	2
4	4	5	0	1	2	3
5	5	0	1	2	3	4

"MULTIPLICATION" GRID MODULO 6

$\otimes$	0	1	2	3	4	5
0	0	0	0	0	0	0
1	0	1	2	3	4	5
2	0	2	4	0	2	4
3	0	3	0	3	0	3
4	0	4	2	0	4	2
5	0	5	4	3	2	1

Check the properties of the (mod 6) system under $\oplus$ and also the system under $\otimes$. Do all the nonzero elements have multiplicative inverses (mod 6)? Remember that this same question has just been answered for (mod 5). Compare the answers.

We shall now examine the *congruence relation* to see whether it is an *equivalence relation.* To be an equivalence relation, congruence must obey the reflexive, symmetric, and transitive properties.

Reflexive property: $a \equiv a \pmod{m}$

By Definition 14.3, $\quad a - a = m \times 0$, and $a \equiv a \pmod{m}$.

Symmetric property: If $a \equiv b \pmod{m}$, then $b \equiv a \pmod{m}$.

By Definition 14.3, $\quad a - b = m \times k$.

Multiplying by $^-1$, $\quad -a - (-b) = m \times (-k)$,

which may be written $\quad b - a = m \times (-k)$, and $b \equiv a \pmod{m}$.

Transitive property: If $a \equiv b \pmod{m}$ and $b \equiv c \pmod{m}$, then, $a \equiv c \pmod{m}$.

Since $a = sm + b$ and $b = tm + c$, we have $a = sm + (tm + c)$.
Hence, $a = (s + t)m + c$ and $a \equiv c \pmod{m}$.

Since these three properties hold, the congruence relation is an equivalence relation. Thus, the congruence or residue classes are often called *equivalence classes.* In each instance the modulus is a part of the name, although it may be omitted when the discussion leaves no possibility of misunderstanding.

EXERCISE 14.3

1. Supply the symbol $\equiv$ or $\not\equiv$ in the following to make true statements.

   (a) 5746 _____ 4 (mod 9) (e) 1005 _____ 729 (mod 6)

   (b) 925 _____ ⁻32 (mod 3) (f) 1005 _____ 729 (mod 9)

   (c) 1005 _____ 729 (mod 2) (g) ⁻95 _____ 7 (mod 17)

   (d) 1005 _____ 729 (mod 3) (h) 3 _____ 5 (mod 2)

2. Each of the following open sentences has one solution from the point of view of residue classes. Find four values of x in the same residue class and satisfying the indicated congruences.

   (a) $x \equiv 19 \pmod 6$ (c) $x + 7 \equiv 6 \pmod 4$

   (b) $x \equiv 11 \pmod 5$ (d) $x \equiv {}^-15 \pmod 7$

3. Does the reduced system modulo 6 consisting of those elements corresponding to numbers relatively prime to 6 form a group under "multiplication"?

⊗	1	5
1	1	5
5	5	1

4. Separate the integers from ⁻15 to 40 into equivalence classes:

   (a) mod 3 (b). mod 7

5. Using the integers of Problem 4(a), verify in four different cases that the sum of elements in any two given classes falls in one specific class which is dependent upon the two classes selected, regardless of which elements have been chosen. Do the same verification for products.
6. Repeat Problem 5, using the integers of Problem 4(b).
7. Construct tables of "addition" and "multiplication" facts for the following moduli. Save copies of these tables for use later in this chapter.

   (a) mod 2 (b) mod 3 (c) mod 4 (d) mod 7

14.4 APPLICATIONS OF CONGRUENCES

Congruences may be used to establish rules for divisibility needed in determining factors of composite numbers. They may also be used as the basis for checking computation.

The following theorem is needed in justifying divisibility rules and checking procedures.

Theorem II *Given integers a, b, c, d, k, and m, m > 1, then if*

$$a \equiv b \pmod m \text{ and } c \equiv d \pmod m,$$

it follows that:

(i) $a + c \equiv b + d \ (mod\ m)$

(ii) $ac \equiv bd \ (mod\ m)$

(iii) $ka \equiv kb \ (mod\ m)$

The conclusions of this theorem were suggested in Section 14.3 as part of the discussion of equivalence classes. The reader should prove these statements, as an exercise, by using the definition of the congruence relation.

Example 1 Show that $3765 \equiv 3 + 7 + 6 + 5 \pmod 9$.

In expanded form we have

$$3765 = 3 \cdot 10^3 + 7 \cdot 10^2 + 6 \cdot 10^1 + 5 \cdot 10^0.\dagger$$

Applying Theorem II,

$10^3 \equiv 1 \pmod 9$ and $3 \cdot 10^3 \equiv 3 \pmod 9$ *By part* (iii) *of the theorem*

$10^2 \equiv 1 \pmod 9$ and $7 \cdot 10^2 \equiv 7 \pmod 9$ *By part* (iii) *of the theorem*

$10^1 \equiv 1 \pmod 9$ and $6 \cdot 10^1 \equiv 6 \pmod 9$ *By part* (iii) *of the theorem*

$10^0 \equiv 1 \pmod 9$ and $5 \cdot 10^0 \equiv 5 \pmod 9$ *By part* (iii) *of the theorem*

Thus

$$3 \cdot 10^3 + 7 \cdot 10^2 + 6 \cdot 10^1 + 5 \cdot 10^0 \equiv 3 + 7 + 6 + 5 \pmod 9$$

by part (i) of the theorem.

Then $3765 \equiv 3 + 7 + 6 + 5 \pmod 9$;

but $3 + 7 + 6 + 5 = 21$ and $21 \equiv 3 \pmod 9$;

so, $3765 \equiv 3 \pmod 9$.

That is, if 3765 is divided by 9, the remainder is congruent to the sum of the digits modulo 9.

Example 2 Determine whether 5931 is divisible by 9.

From the preceding example r may be determined.

$$\begin{aligned} 5931 &= (9 \times q) + r \\ r &\equiv 5 + 9 + 3 + 1 \pmod 9 \\ &\equiv 18 \pmod 9 \\ &\equiv 0 \pmod 9 \end{aligned}$$

Hence, 5931 is divisible by 9.

The results from these two examples are used to formulate the two rules which follow:

†Here "·" is used for "×" to indicate multiplication.

Rule 1 *When an integer is divided by 9, the remainder is congruent modulo 9 to the sum of the digits of that integer written in base 10; thus, the integer is congruent modulo 9 to the sum of its digits (base 10).*

Rule 2 *An integer is divisible by 9 if and only if the sum of its digits (base 10) is divisible by 9, that is, congruent to 0 (mod 9).*

With this information, we now proceed to a device for checking addition or multiplication, which is commonly called *casting out nines,* that is, considering the remainders after division by 9. This is done by working with the digits as in Rule 1.

Addition is checked by applying part (i) of the theorem. Again, examples with some comments serve to illustrate the method.

Example 3 Find the sum of the following and check the result by casting out nines.

$$\begin{array}{r} 327 \\ 41 \\ \underline{876} \\ 1244 \end{array} \qquad \begin{array}{rl} 327 \equiv 12 & [12 \equiv 3] \pmod 9 \\ 41 \equiv 5 & \pmod 9 \\ 876 \equiv 21 & [21 \equiv 3] \pmod 9 \end{array}$$

Find the sum $[3 + 5 + 3 = 11]$ of the residues modulo 9 of the addends. Now $11 \equiv 2 \pmod 9$. This is the same as the residue of the sum of the addends. That is,

$$1244 \equiv 11 \quad [11 \equiv 2] \pmod 9.$$

Notice that casting out nines in the addends, adding the residues, and casting out nines again leads to the same result as casting out nines in the sum. In simpler form we have:

$$\left.\begin{array}{r} 327 \equiv 3 \pmod 9 \\ 41 \equiv 5 \pmod 9 \\ \underline{876 \equiv 3 \pmod 9} \end{array}\right\} \rightarrow 3 + 5 + 3 \equiv 2 \pmod 9$$
$$1244 \equiv 2 \pmod 9 \leftarrow$$

Note, however, that while the fact that the sum of the residues is congruent to the residue of the sum modulo 9 is some assurance of the correctness of the sum, it is no guarantee. If, for instance, two digits in the sum had been interchanged, the result of the test would still be favorable, even though the answer were incorrect. If the results do not agree, one should look for errors.

The test can be shortened considerably, once one understands the reasons for its use.

Example 4 Add:

$$\left.\begin{array}{r} 2341 \to 1 \\ 924 \to 6 \\ 7865 \to 8 \\ \underline{121} \to 4 \end{array}\right\} \to 1$$
$$11251 \to 1 \longleftarrow$$

A similar test can be used in multiplication. Refer to part (ii) of the theorem and Rule 1 to justify the procedure.

Example 5 Find the product of 862 and 87, and check by casting out nines.

$$\left.\begin{array}{r} 862 \to 7 \\ \underline{87} \to 6 \end{array}\right\} 7 \cdot 6 = 42 \to 6$$
$$\begin{array}{r} 6034 \\ \underline{6896} \\ 74994 \to 6 \longleftarrow \end{array}$$

A device less commonly used is called *casting out elevens*. It is based on the following congruences.

$$\begin{aligned} 10 &\equiv {}^{-}1 \pmod{11} \\ 10^2 &\equiv 1 \pmod{11} \\ 10^3 &\equiv {}^{-}1 \pmod{11} \end{aligned}$$

In general:

$$10^k \equiv \begin{cases} 1 \pmod{11} \text{ if } k \text{ is even} \\ {}^{-}1 \pmod{11} \text{ if } k \text{ is odd} \end{cases}$$

Rule 3 *An integer is congruent modulo* 11 *to the sum of the digits in the odd-numbered places, counting from the right, and the negatives of the digits in the even-numbered places* (*base* 10).

Example 6 Check the product in Example 5 by casting out elevens.

$$\left.\begin{array}{r} 862 \equiv 2 + {}^{-}6 + 8 \equiv 4 \pmod{11} \\ 87 \equiv 7 + {}^{-}8 \equiv {}^{-}1 \pmod{11} \end{array}\right\} (4 \cdot {}^{-}1) \equiv 7 \pmod{11}$$

The product from Example 5 is:

$$74{,}994 \equiv 4 + {}^{-}9 + 9 + {}^{-}4 + 7 \equiv 7 \pmod{11}$$

We also have a divisibility rule for 11.

Rule 4 *An integer is divisible by* 11 *if and only if, when written in base* 10, *the sum of the digits in the odd-numbered places, counting from the right, and the negatives of the digits in the even-numbered places is congruent to* 0 (mod 11).

EXERCISE 14.4

1. Without actually dividing, determine the remainder when each of the following is divided by 9.

   (a) 73,285 (c) 58,914 (e) 264,858

   (b) 25,837 (d) 195,147 (f) 3,195

2. Determine the remainder when each of the numbers given in Problem 1 is divided by 11.

3. Find the following sums and check by casting out nines.

   (a) $\begin{array}{r} 374 \\ \underline{29} \end{array}$ (b) $\begin{array}{r} 975 \\ \underline{124} \end{array}$ (c) $\begin{array}{r} 1872 \\ \underline{273} \end{array}$

4. Consider the parts of Problem 3 as subtraction problems. Devise a method of checking by casting out nines.

5. Consider the parts of Problem 3 as multiplication problems. Solve and check by casting out nines.

6. Perform the following divisions and devise a method of checking by casting out nines. Remember the relation between dividend, divisor, quotient, and remainder.

   (a) $37\overline{)328}$ (b) $53\overline{)8025}$ (c) $139\overline{)7694}$

7. Check the results of the operations in Problems 3–6 by casting out elevens.

8. Notice that $10^n \equiv 1 \pmod 3$. With this information show that if a number is a multiple of 3, then the sum of its digits is a multiple of 3.

14.5 RINGS, INTEGRAL DOMAINS, AND FIELDS

This section completes the development of the real number system which has been the basic study throughout this textbook. The purpose of the deliberate examination of the structure of each number system (N, W, I, Q) was to prepare for the culmination in which the real number system is recognized as a *field.* To attain this, we shall review the subsystems of the real number system and relate them to an appropriate category—namely, the *group,* the *ring,* the *integral domain,* and the *field.*

The *group,* studied in an earlier section of this chapter, is a mathematical system that consists of *one* operational system, an equivalence relation, and certain properties. See Definition 14.2. Many mathematical systems that have been studied consist of *two* operational systems. A property, familiar to you by now, that connects the two operations in these systems is the *distributive property.*

We shall now present some definitions accompanied by examples designed to help in understanding the definitions. Note that the symbols $\oplus$ and $\otimes$ may represent *any* operations as defined in the system.

Definition 14.5(a) A *ring* is a mathematical system consisting of two binary operational systems, $(S, \oplus)$ and $(S, \otimes)$, an equivalence relation, and these properties:

R_0: S and the operation $\oplus$ form an Abelian group (Definition 14.2).
R_1: The operation $\otimes$ is associative.
R_2: The operation $\otimes$ is distributive over $\oplus$.

Notice the properties which *may* be lacking in a ring. The second operation, $\otimes$, is not necessarily commutative. If it were, then the ring would be a *commutative ring*. Since there need not be an identity for $\otimes$, then $\otimes$-inverses are not required.

Example 1 Let S be $\{w, z\}$ with operations $\oplus$ and $\otimes$ and relation $=$ defined by these tables:

$\oplus$	w	z
w	w	z
z	z	w

$\otimes$	w	z
w	w	w
z	w	z

Establish that $(S, \oplus)$ constitute an Abelian group, that operation $\otimes$ is associative, and that $\otimes$ is distributive over $\oplus$. This means that the system is a ring. It satisfies other conditions not necessary for a ring, namely, commutativity for $\otimes$ and the existence of the $\otimes$-identity, z. However, show that S is not a group under $\otimes$. Does the element w have an inverse under $\otimes$?

Example 2 Given the set of even integers

$$S = \{\ldots, {}^{-}6, {}^{-}4, {}^{-}2, 0, 2, 4, 6, \ldots\},$$

the *usual* operations addition and multiplication, and the *usual* equals relation. Then the following hold for $(S, +)$ and $(S, \times)$:

R_0: $\begin{cases} G_1\text{: Associativity for } + \\ G_2\text{: Identity for } + \\ G_3\text{: Inverse under } + \text{ for each element} \\ G_4\text{: Commutativity for } + \end{cases}$

R_1: Associativity for $\times$
R_2: Distributivity for $\times$ over $+$
R_3: If, as in this particular case, $\times$ is commutative, then the system is called a *commutative ring*.

Notice that there is no identity for multiplication, and hence no possibility of multiplicative inverses.

Let us now refer to the grids for $\oplus$ and $\otimes$ modulo 6 on the set

$$S = \{0, 1, 2, 3, 4, 5\}.$$

We see that this system is a commutative ring. In fact, we may suspect that all modular systems are commutative rings. However, this system has a characteristic that has not heretofore been mentioned. Notice that $2 \otimes 3 \equiv 0 \pmod 6$, while neither 2 nor 3 is zero. In such a situation 2 and 3 are called *proper divisors of zero*.

At this point the reader should recall that elements 2 and 3 represent class 2 and class 3 (mod 6). They represent infinitely many numbers—all the elements in their respective classes.

Notice also that while the system possesses a $\otimes$-identity, 1, this identity does not occur in each row and column. Hence some elements, including zero, lack $\otimes$-inverses. What elements, other than class 2 and class 3, are proper divisors of zero in this (mod 6) system?

Definition 14.5(b) An *integral domain* is a mathematical system consisting of two operational systems. $(S, \oplus)$ and $(S, \otimes)$, an equivalence relation, and the following properties:

$$\text{ID}_0 \begin{cases} G_1\text{: Associative for } \oplus \\ G_2\text{: Identity for } \oplus \\ G_3\text{: } \oplus\text{-inverses} \\ G_4\text{: Commutative for } \oplus \end{cases}$$

$$\text{ID}_1 \begin{cases} R_1\text{: Associative for } \otimes \\ R_2\text{: Distributive for } \otimes \text{ over } \oplus \\ R_3\text{: Commutative for } \otimes \end{cases}$$

ID_2 Identity for $\otimes$

ID_3 No proper divisors of identity for $\oplus$

Notice that, of the modular systems encountered so far, those with composite numbers for moduli (mod 4) and (mod 6) have had proper divisors of zero. Those with prime moduli possessed an identity for $\otimes$, 1, and no zero divisors. A likely generalization would be: *Modular* systems with prime moduli are integral domains, while those with composite moduli are simply rings.

Example 3 Show that the system of integers is an integral domain.

This is a mathematical system consisting of two operational systems $(I, +)$ and $(I, \times)$, the equivalence relation, $=$, and a set of properties, which we shall examine to see whether the requirements for an integral domain are met.

First, we shall check to see if the requirements for a commutative ring have been met. The system is associative for both $+$ and $\times$. There is an identity for $+$, and each element of the set has an additive inverse. There is a distributive property for multiplication over addition. There is commutativity for both $+$ and $\times$. Thus, the requirements for a commutative ring have been met.

Another requirement for an integral domain is that there exist an identity for $\times$. This is the element 1 in this system.

The last requirement is that there be no proper divisors of 0. Since there are no nonzero numbers whose product is 0, this requirement has been met.

We are prepared to say that the *system of integers is an integral domain.*

Definition 14.5(c) A *field* is a mathematical system consisting of two operational systems, $(S, \oplus)$ and $(S, \otimes)$, an equivalence relation, and the following properties:

F_0: Those properties necessary for an integral domain
F_1: $\otimes$-inverses except for the identity for $\oplus$

Example 4 Is the system of rational numbers a field?

The system of rational numbers is a mathematical system consisting of two operational systems $(Q, +)$ and $(Q, \times)$, the equivalence relation, $=$, and a set of properties. Does this set meet the requirements for a field?

Examine the set of properties to see whether the requirements for an integral domain are met. The only other requirement to be satisfied is that each element except 0, the identity for addition, has a $\times$-inverse. You should be satisfied that the *system of rational numbers is a field.*

Example 5 Is the modulo 2 system a field? Given $S = \{0, 1\}$ and operations $\oplus$ and $\otimes$, and $=$ relation defined by these tables:

$\oplus$	0	1
0	0	1
1	1	0

$\otimes$	0	1
0	0	0
1	0	1

Check the conditions in the following list:

For S and $\oplus$ to be a *commutative group,* it is necessary for the following to hold.

G_1: Associativity for $\oplus$
G_2: Identity for $\oplus$
G_3: $\oplus$-inverses
G_4: Commutativity for $\oplus$

For a *commutative ring* it is necessary to have G_1–G_4 and

R_1: Associativity for $\otimes$
R_2: Distributivity for $\otimes$ over $\oplus$
R_3: Commutativity for $\otimes$

For an *integral domain,* all of the above conditions and
ID_2: Identity for $\otimes$
ID_3: No proper divisors of identity for $\oplus$, 0.

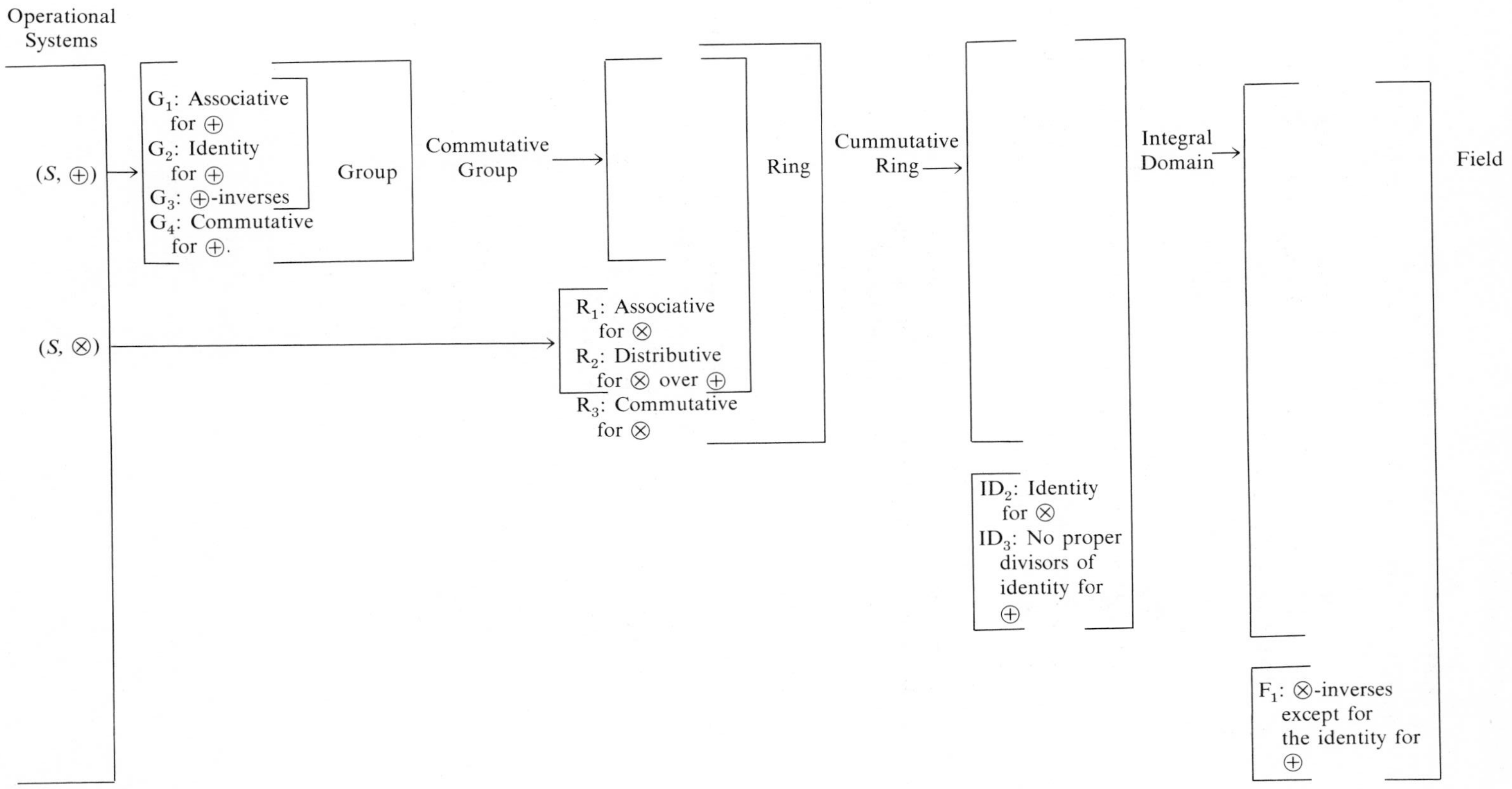
Operational Systems
(S, ⊕)
(S, ⊗)
G₁: Associative for ⊕
G₂: Identity for ⊕
G₃: ⊕-inverses
G₄: Commutative for ⊕.
Group
Commutative Group
R₁: Associative for ⊗
R₂: Distributive for ⊗ over ⊕
R₃: Commutative for ⊗
Ring
Cummutative Ring
ID₂: Identity for ⊗
ID₃: No proper divisors of identity for ⊕
Integral Domain
F₁: ⊗-inverses except for the identity for ⊕
Field

For a *field,* all the above conditions and

F_1: Each element except 0 has an inverse under $\otimes$

The accompanying chart shows the requirements that must be met for a group, a ring, an integral domain, and a field. It is designed to illustrate the relationships that exist.

EXERCISE 14.5

1. Refer to the grid for $\otimes$ modulo 5 (Section 14.3). Also, construct one for $\oplus$ modulo 5. Is the system having these two operations a ring? Is it a commutative ring?
2. Consider (i) the system of natural numbers and (ii) the system of whole numbers.
   (a) Examine (i) for ring, commutative ring, integral domain, and field requirements.
   (b) Do the same for (ii).
3. Examine the following mathematical systems to see whether they satisfy the requirements for ring, integral domain, and field.
   (a) The set of odd numbers and the addition and multiplication operations.
   (b) The set of even numbers and the addition and multiplication operations.
   (c) The operational systems $(\{0, 1\}, +)$ and $(\{0, 1\}, \times)$ (mod 2).
4. Examine the mathematical systems modulo 3 and modulo 4 having operations $\oplus$ and $\otimes$. Is either a field? Discuss.
5. Using C_6 to be the set of numbers in a modulus 6 system and $\otimes$ an operation defined as in Section 14.3, discuss whether $(C_6, \otimes)$ is a group.
6. Discuss whether Problem 3, Exercise 14.1, having $(\{r_0, r_1, r_2\}, \circ)$ is a group. Is it a commutative group. Is it a ring?
7. Is the system given in Example 1 of Section 14.5 $(\{w, z\}, \oplus, \otimes)$ a field? Discuss.
8. Determine if $\{0, 2, 4, 6\}$ (mod 7) is a commutative group with $\oplus$. (HINT: Make up a grid with these elements.)
9. Determine if $\{0, 2, 4, 6\}$ (mod 8) is a commutative group with $\oplus$.
10. Does the reduced system modulo 8 form a group under $\otimes$?
11. Without using the notions of group or ring, list the requirements for each of the following.
    (a) Integral domain (b) Field
12. Is a field an integral domain? Is it a ring? Is it a group?

15

An Introduction to Probability

15.1 INTRODUCTION

Games of chance have long fascinated the young as well as the old. A major concern of the player of a game is his chance of winning. Not only games of chance but much of today's thinking involves probabilistic situations. Few people get through a day without becoming engaged in situations of this type. It may be important to some concerning the probability of rain on August 25, the chance of a Bowl game, or the chance of winning a major prize in a magazine promotional scheme.

Probability theory is increasingly applied in business, industry, and government agencies. Also, it is being used currently in many fields of study to which previously it was thought not to be applicable. Techniques based on probability are used by manufacturers of articles when it is impossible or impractical to test the quality of every item; by pharmaceutical companies to determine the effectiveness of medicines they produce; and by life insurance companies to establish reliable mortality tables in order that sound retirement income plans may be developed.

Because of the role of probability theory in all segments of our society, one should have some knowledge of the way the theory functions. It is not the intent that this chapter will present a highly structured, sophisticated development of the concept of probability. Rather, from the study of this chapter, you should acquire a useful and correct knowledge of this concept at a somewhat elementary level.

15.2 PROBABILITY

We begin our study of probability by returning to the player of a game, a penny-tosser. He considers it reasonable to expect that if a perfectly balanced coin with a head on one side and a tail on the other is tossed, then it is just as likely to fall heads as tails. He estimates that he has a 50–50 chance of getting a head. This means that the chance of getting either *outcome,* a head or a tail, is *equally likely*. When a head appears we say that a head occurs. Now, suppose that the penny-tosser wishes to verify that $\frac{1}{2}$ of the outcomes of an experiment he performs will be heads.

Experiment 1

A 1970 penny is selected from a piggy bank. It is tossed 300 times. A record is kept which shows that heads occurred 125 times. From this data, the player is able to say that $\frac{5}{12}$ of the 300 tosses were heads.

The assignment of the number $\frac{5}{12}$ as the measure of the "chance" of getting heads is called the *probability* of heads occurring with this particular coin. This number is determined by dividing the number of heads by the number of tosses: $\frac{125}{300} = \frac{5}{12}$. The result of this division is the *relative frequency* of getting heads. The probability of heads occurring is written

$$P(H) = \frac{5}{12}.$$

Observe that the penny-tosser's expectation of getting $\frac{1}{2}$ heads and the result of his experiment do not completely agree. However, continuing the experiment over a long sequence of tosses, the probability, $\frac{5}{12}$, determined by the experiment comes closer and closer to the "true" probability $\frac{1}{2}$ based on equally likely outcomes. Hence, the relative frequency over a large number of tosses of a coin is a reliable estimate of the probability.

We now turn our attention to another player, a die-roller. It is important in this discussion to know that consideration is given only to a "true" die; that is, one that is balanced and symmetrical with each face marked in the usual manner with a set of dots ranging in number from 1 to 6 inclusive. When a rolled die (plural, dice) lands with a 3-dot face up, we say that a 3 was rolled or that a 3 occurred. When a true die is rolled, one face is just as likely to occur as another. That is, the probability of getting any one of the outcomes, 1, 2, 3, 4, 5, or 6, is *equally likely*. This means that the probability of each of the outcomes is the same.

The die-roller predicts that the chance of rolling a 4 is one out of six ($\frac{1}{6}$). He attempts to verify this by performing an experiment.

Experiment 2

A certain die is rolled 36 times. The frequency of each outcome is recorded in Table 15.1. The relative frequency for each outcome is found and recorded. Recall that the relative frequency of an outcome is determined by dividing the frequency with which the outcome occurred by the total number of trials.

Table 15.1

Outcomes	*Frequency*	*Relative Frequency*
1	8	$\frac{8}{36}$
2	9	$\frac{9}{36}$
3	2	$\frac{2}{36}$
4	5	$\frac{5}{36}$
5	7	$\frac{7}{36}$
6	5	$\frac{5}{36}$

The table shows that the outcome 4 occurred 5 times in 36 rolls of the die, and the relative frequency is $\frac{5}{36}$. The number $\frac{5}{36}$ is the measure assigned to the probability of 4 occurring in 36 rolls. It is written

$$P(4) = \tfrac{5}{36}.$$

Extending the experiment to include more and more rolls of the die, the relative frequency will come closer and closer to the predicted probability based on equally likely outcomes.

Example 1 What is the sum of the relative frequencies of an experiment? In Experiment 2, the sum of the relative frequencies is

$$\frac{8}{36} + \frac{9}{36} + \frac{2}{36} + \frac{5}{36} + \frac{7}{36} + \frac{5}{36} = \frac{36}{36} = 1.$$

In Experiment 1, the frequency of heads in 300 tosses is 125. The frequency of tails is 175.

$$\begin{aligned} \text{Relative frequency of heads} &= \tfrac{125}{300} \\ \text{Relative frequency of tails} &= \tfrac{175}{300} \\ \text{Total} &= \tfrac{300}{300} = 1 \end{aligned}$$

If each of the experiments were continued over a longer number of trials, would the sum of the relative frequencies of each of the possible outcomes continue to be 1? Would the sum ever be greater than 1? Explain.

At this point some terms necessary in discussing probability will be identified. Some have already been mentioned, while others are introduced for the first time.

We have used the term *experiment* to describe the repeated tossing of a coin and the repeated rolling of a die. Experiments 1 and 2 and all others of this chapter are confined to a finite number of possible outcomes that are equally likely.

In Experiment 1, the possible outcomes consist of $\{H, T\}$, where H and T stand for head and tail respectively. The possible outcomes of Experiment 2 consist of $\{1, 2, 3, 4, 5, 6\}$. These sets are identified as the *sample spaces* of the experiments cited.

Definition 1 The set of all possible outcomes of an experiment is called a *sample space* of that experiment.

Expressing a sample space in set notation we write,

$$S = \{s_1, s_2, \ldots, s_n\},$$

where the elements of set S consist of all possible outcomes of an experiment.

Probability is a *measure* assigned to the chance of a successful outcome in a certain experiment. If an experiment is performed having t equally likely outcomes, and if the number of successful outcomes is s, then the *probability* of success is the measure $\frac{s}{t}$. That is,

$$P(\text{success}) = \tfrac{s}{t}.$$

If the probability of each element of the sample space is the same, then the outcomes are said to be *equally likely*. That is,

$$P(s_1) = P(s_2) = \cdots = P(s_n) = \tfrac{1}{n},$$

where $s_1, s_2, \ldots, s_n$ are the elements of sample space S and n is the number of elements of S. To describe a coin, a die, or any object used in an experiment as "true" means that the outcomes of the sample space of that experiment are equally likely.

Example 2 In tossing a true coin, what is the probability of getting tails?

Since there are two equally likely outcomes, head and tail, and only one considered to be successful, the probability is the number (measure) $\frac{1}{2}$. Symbolically, this is written

$$P(T) = \tfrac{1}{2}.$$

Example 3 Identical cubes, except for color, have been placed in a box. There are 5 white and 7 black ones. What is the probability of drawing a white cube on the first draw? After returning the drawn cube to the box, what is the probability of drawing a black one?

There are 12 cubes in the box, 5 of which are white. The probability of selecting a white cube is determined by

$$\frac{\text{N}\{\text{white cubes}\}}{\text{N}\{\text{cubes in box}\}} = \frac{5}{12}.$$

The number $\frac{5}{12}$ is the probability of drawing a white cube from the box on the first draw:

$$P(W) = \tfrac{5}{12}$$

The drawn cube is returned to the box. The probability of drawing a black cube is

$$\frac{\text{N}\{\text{black cubes}\}}{\text{N}\{\text{cubes in box}\}} = \frac{7}{12}.$$

That is,

$$P(B) = \frac{7}{12}.$$

Observe that the sum of the two probabilities,

$$P(W) + P(B) = \tfrac{5}{12} + \tfrac{7}{12} = \tfrac{12}{12} = 1.$$

Example 4 Eleven tickets placed in a box for drawing have been numbered from 1 through 11. What is the probability of drawing an even number? An odd number?

The sample space consists of $\{1, 2, 3, 4, 5, 6, 7, 8, 9, 10, 11\}$. Since the number of even numbers in the sample space is 5,

$$P(\text{even}) = \tfrac{5}{11}.$$

The probability of drawing an odd number is

$$P(\text{odd}) = \tfrac{6}{11}.$$

This may be found by dividing the number of odd-numbered tickets by the number of tickets, or by

$$P(\text{odd}) = 1 - P(\text{even}).$$

Can you explain why the latter method works?

The experiment described in Example 4 suggests the meaning of an *event*. In speaking of the probability of an even-numbered outcome in this experiment is meant the probability of any one of the elements of $\{2, 4, 6\}$ occurring. This set is a subset of the sample space $S = \{1, 2, 3, 4, 5, 6\}$ and is called an *event*. This event, $\{2, 4, 6\}$, may be described as the set of all even-numbered outcomes of S. Describe the event, $\{1, 3, 5\} \subset S$.

Definition 15.2(a) An *event* is a subset of the sample space of an experiment. A *simple event* is an event consisting of a single element.

Example 5 In a die-rolling experiment, describe event $E = \{1, 6\}$. E is a subset of $S = \{1, 2, 3, 4, 5, 6\}$. By considering the outcomes of the experiment ordered as

$$1, 2, 3, 4, 5, 6,$$

set E could be described as the set consisting of the first and the last outcomes.

Example 6 In a die-rolling experiment having $S = \{1, 2, 3, 4, 5, 6\}$, write in roster notation the event E such that the outcomes are greater than 2 *and* less than 6.

The set of elements belonging to S which are greater than 2 is

$$A = \{3, 4, 5, 6\}.$$

The set of elements belonging to S which are less than 6 is

$$B = \{1, 2, 3, 4, 5\}.$$

The set of elements belonging to S which are greater than 2 *and* less than 6 is the intersection of sets A and B.

$$A \cap B = \{3, 4, 5\}.$$

Exploring the concept of probability in several of the preceding examples suggests that the *probability of an event* be defined as follows:

Definition 15.2(b) The *probability of an event E* is the number determined by the number of elements in E divided by the number of elements in S, where S is the finite sample space of equally likely outcomes of an experiment and E is a set, such that $E \subset S$. Symbolically,

$$P(E) = \frac{\mathrm{N}(E)}{\mathrm{N}(S)}, \qquad E \subset S.$$

Example 7 In a die-rolling experiment, what is the probability of event $\{1, 3, 5\}$? By Definition 15.2(b),

$$P(\{1, 3, 5\}) = \frac{\mathrm{N}\{1, 3, 5\}}{\mathrm{N}\{1, 2, 3, 4, 5, 6\}}$$
$$= \frac{3}{6} \quad \text{or} \quad \frac{1}{2}.$$

Example 8 Fifteen identical red marbles have been placed in a bag. (a) What is the probability of drawing a red marble? (b) What is the probability of drawing a blue marble?

(a) $$P(\text{red}) = \frac{\mathrm{N}\{\text{red marbles}\}}{\mathrm{N}\{\text{marbles in bag}\}} = \frac{15}{15} = 1$$

(b) $$P(\text{blue}) = \frac{\mathrm{N}\{\text{blue marbles}\}}{\mathrm{N}\{\text{marbles in bag}\}} = \frac{0}{15} = 0$$

If all outcomes bring success, $P(\text{success}) = 1$, then success is *certain;* if there are no outcomes that bring success, $P(\text{success}) = 0$, then success is *impossible.*

Results of examples prior to this indicate that the measure of probability is a number greater than 0 and less than 1. This example shows that the probability may be 1 or 0. It follows that

$$0 \leq P(\text{success}) \leq 1.$$

EXERCISE 15.2

1. Given the sample space $S = \{1, 2, 3, 4, 5, 6\}$ of a die-rolling experiment. Express in roster notation the following events.
   (a) $\{x \in S : x \text{ is less than } 2\}$
   (b) $\{x \in S : x \text{ is greater than } 4\}$
   (c) $\{x \in S : x \text{ is less than or equal to } 4 \text{ or greater than or equal to } 5\}$
   (d) $\{x \in S : x \leq 1 \text{ or } x \geq 6\}$
   (e) $\{x \in S : x \text{ is greater than } 3 \text{ and less than } 5\}$
   (f) $\{x \in S : 3 \leq x \leq 5\}$
   (g) $\{x \in S : x \text{ is less than } 3 \text{ and greater than } 4\}$
2. Find the probability of each event described in Problem 1.
   (a) List those events whose probability falls between 0 and 1.
   (b) List those events whose probability is 0; is 1. Interpret a probability of 0 and of 1.
3. A drawing for a pony is being conducted at a neighborhood grocery store. Customers have written their names on cards and dropped them in a box. At the time of the drawing, there were 2500 names in the box. Assume that the cards were thoroughly shuffled before the drawing began. The first name drawn wins the pony.
   (a) Stanley failed to get his name in the box. What are his chances of winning the pony?
   (b) On the other hand, Stephen, anxious to win the pony, had put his name in the box 15 times. What is the probability that his name will be the first one drawn?
4. The sample space of a die-rolling experiment is $S = \{1, 2, 3, 4, 5, 6\}$.
   (a) What is the probability of rolling an even number?
   What is the probability of rolling an odd number?
   Find the sum, $P(\text{even}) + P(\text{odd})$.
   (b) What is the probability of rolling a number less than 3? a number greater than 3? Find the sum, $P(\text{less than } 3) + P(\text{greater than } 3)$. Compare this sum with that of (a) above. Why are they not the same?
   (c) What is the probability of rolling 1? 2? 1 or 2?
   (d) What is the probability of rolling a number less than 1?
5. A conventional, thoroughly shuffled 52-card bridge deck is used for the following drawings.
   (a) Identify the sample space.
   (b) What is the probability of drawing an ace of hearts?
   (c) Find the probability of drawing a king. How many elements are in the event kings?
   (d) What is the probability of drawing a spade?
   (e) What is the probability of drawing a red card?
6. In the game of Honest John, a blindfolded person selects a letter of a previously chosen word by placing the point of a pin on it. "Budget" has been chosen for this particular game.
   (a) Identify the sample space.
   (b) What is the probability that a vowel will be selected?
   (c) What is the probability that a consonant will be selected?

(d) Find $P(\text{vowel}) + P(\text{consonant})$.
(e) What is the probability of the event $\{b, u, d, g, e, t\}$?
(f) What is the probability of selecting the letter r?

15.3 SAMPLE SPACES OF OTHER EXPERIMENTS

The experiments of this discussion differ from those of the preceding section. In Experiment 1, a single coin was tossed; while in Experiment 2, a single die was rolled. Each outcome was represented by a single element. The experiments that follow involve two or more distinguishable objects, such as the tossing of a coin followed by the rolling of a die. The outcomes are represented by ordered pairs or by ordered triples depending on the number of objects used in the experiment.

Experiment 3

In this experiment, a nickel is tossed first and that is followed by a toss of a dime.

Since two objects are to be used in the experiment and the order in which they are to be tossed is established, then (H, T) will be interpreted to be that the nickel fell heads and the dime fell tails. We are now interested in the different ways that the two coins may fall. A simple way of recording the results is shown in the chart.

		Dime	
		Heads	*Tails*
Nickel	*Heads*	(H, H)	(H, T)
	Tails	(T, H)	(T, T)

The set consisting of all the ways that these two coins may fall in the prescribed order is a sample space of the experiment. It is

$$S = \{(H, H), (H, T), (T, H), (T, T)\}.$$

A *tree diagram* is an effective method that may be used to identify and count all the possible outcomes when an experiment involves two or more distinguishable objects in two or more activities.

A tree diagram for this experiment is not a very complicated one. The two figures of the diagram are made to show how the tree grows from the tossing of a nickel (Figure 15–1) which is then followed with the tossing of a dime. By tracing each path of the diagram of Figure 15–2, an outcome is determined.

All possible outcomes may be determined by following each of the paths. Observe that this diagram yields the same outcomes as were recorded in the preceding chart.

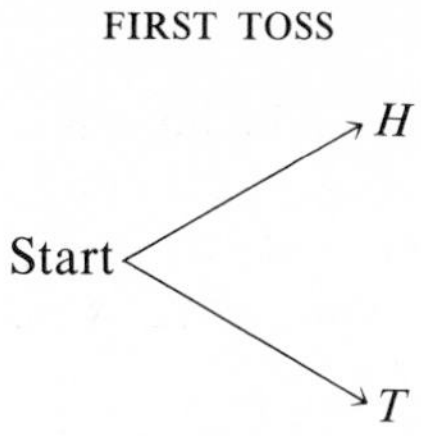

Figure 15–1

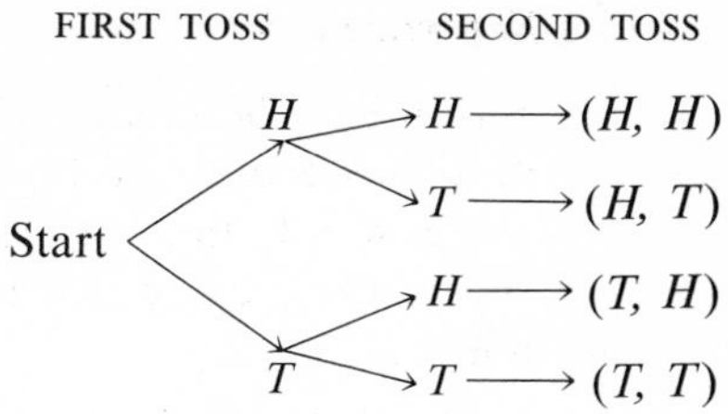

Figure 15–2

The sample space

$$S = \{(H, H), (H, T), (T, H), (T, T)\}$$

was referred to as *a* sample space. This seems to imply that it is not the only one. This is true.

Since there are more sample spaces than one in an experiment such as those discussed in this section, what basis is used in making the selection? It is based on the use that is to be made of it. For instance, suppose one wishes to consider whether the tossed coins fall differently (one head and one tail) or alike (both heads or both tails). Those falling unlike may be represented by U, those falling alike by A. Thus, another sample space of this experiment is

$$S' = \{U, A\}.$$

Observe that each outcome of the experiment corresponds to one and only one element of S'. That is, (H, T) and (T, H) correspond to U; while (H, H) and (T, T) correspond to A. Does each element of S' correspond to exactly one outcome of the experiment? The answer is "no." Obviously, S imparts more information concerning each outcome of the experiment. Because of this, S will be considered the more *basic* sample space. Can you construct other sample spaces of this experiment? If the number of heads and the number of tails of all the outcomes are recorded, another sample space would be

$$S'' = \{(2, 0), (1, 1), (0, 2)\}.$$

Can you explain how these ordered pairs were constructed?

The definition of a sample space will now be extended from the one given in Section 15.2 to the following.

Definition 15.3 A *sample space* of an experiment is a set of elements such that each of the possible outcomes of the experiment is assigned to one and only one element of the set.

Experiment 4

A coin is to be tossed, followed by a roll of a die. The outcomes possible in tossing a coin are heads (H) and tails (T); those for rolling a die are 1, 2, 3, 4, 5, 6. In this experiment, a coin is to be tossed, then a die is rolled. Hence, the ordered pair $(H, 3)$ denotes that the coin fell head and the die rolled 3.

All possible outcomes are shown in the chart and also in the tree diagram that follows.

DIE

		1	*2*	*3*	*4*	*5*	*6*
Coin	*Heads*	$(H, 1)$	$(H, 2)$	$(H, 3)$	$(H, 4)$	$(H, 5)$	$(H, 6)$
	Tails	$(T, 1)$	$(T, 2)$	$(T, 3)$	$(T, 4)$	$(T, 5)$	$(T, 6)$

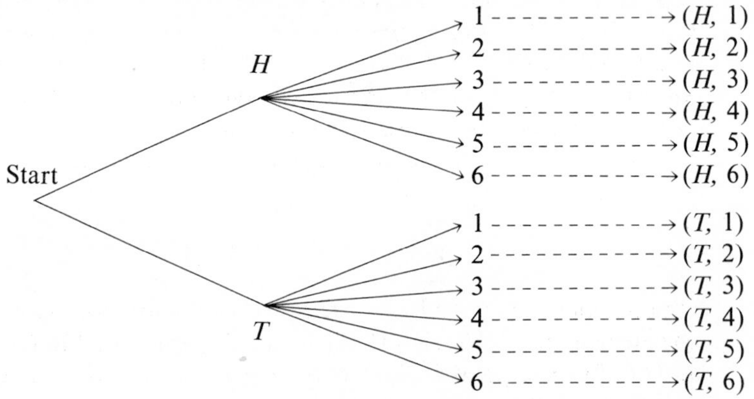

The chart and the tree diagram yield the following basic sample space:

$$Q = \{(H, 1), (H, 2), (H, 3), (H, 4), (H, 5), (H, 6)$$
$$(T, 1), (T, 2), (T, 3), (T, 4), (T, 5), (T, 6)\}$$

Are there other sample spaces for this experiment? Let E represent an even number and D an odd number of $\{1, 2, 3, 4, 5, 6\}$. The following would be a sample space of this experiment:

$$Q' = \{(H, E), (H, D), (T, E), (T, D)\}.$$

Is it possible to assign each outcome of the experiment to exactly one element of Q'?

Experiment 5

A farmer observed that one of his ewes had given birth to three lambs during the past three lambing seasons in the following order: male, female, female.

He wonders how many other kinds of "three-lamb" families would have been possible during three lambing seasons. He makes the following tree diagram:

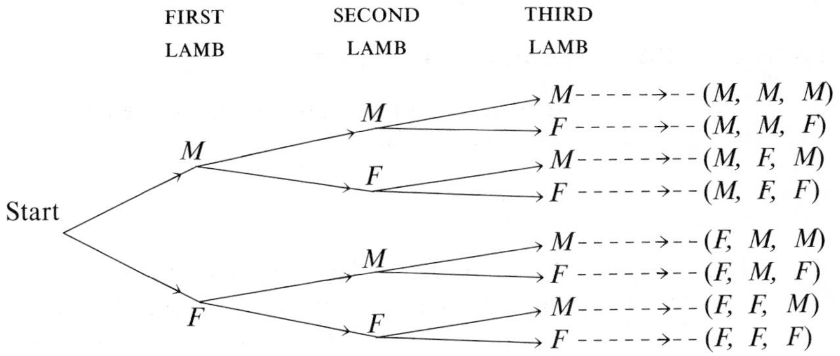

All of the possible outcomes are found in the right column of the tree diagram. The basic sample space is

$$S = \{(M, M, M), (M, M, F), (M, F, M), (M, F, F) \\ (F, M, M), (F, M, F), (F, F, M), (F, F, F)\}.$$

Suppose that the farmer was interested in the number of females in each of these families. An appropriate sample space is

$$S' = \{0, 1, 2, 3\}.$$

Explain what the elements of this set represent. Does each outcome of the experiment correspond to exactly one element of S'?

The exercise set that follows relates to the three preceding experiments.

EXERCISE 15.3

1. In Experiment 3,
   (a) What is the probability of getting exactly 2 tails?
   (b) What is the probability that one coin shows heads and the other tails?
   (c) What is the probability that both coins show the same (both heads or both tails)?
2. In Experiment 4,
   (a) What is the probability that the coin will be heads?
   (b) What is the probability that the die will be an even number?
   (c) What is the probability that the coin will be tails and the die odd?
   (d) What is the probability that the die will be greater than 2?
   (e) What is the probability that the coin will be heads and the die less than 4?
3. In Experiment 5,
   (a) How many elements of the basic sample space consist of the same sex?
   (b) What is the probability that there will be a family of the same sex?
   (c) What is the probability that the second lamb will be a female?
   (d) What is the probability that the first two lambs born will be the same sex?

(e) What is the probability that the last two born will be of opposite sex?
(f) What is the probability that if the first lamb born is male and the second is female that the third one will be male?

15.4 INTERSECTION AND UNION OF EVENTS

The events discussed in this section are subsets of the basic sample space of a two-dice rolling experiment.

Experiment 6

Two dice are rolled, a red one followed by a white one. Each outcome is recorded as an ordered pair of numbers, a subset of $\{1, 2, 3, 4, 5, 6\}$ X $\{1, 2, 3, 4, 5, 6\}$. The order in which the dice are rolled prescribes the order of the number pair. That is, if the red falls a 3 and the white falls a 2, the outcome is recorded $(3, 2)$. The outcome of each roll is recorded (R, W), where R represents the number on the red die and W the number on the white.

The ordered pairs shown in Figure 15–3 are all possible outcomes of a two-dice experiment. The set consisting of all of these outcomes will be referred to as sample space S.

White (W) Die							
	6	(1, 6)	(2, 6)	(3, 6)	(4, 6)	(5, 6)	(6, 6)
	5	(1, 5)	(2, 5)	(3, 5)	(4, 5)	(5, 5)	(6, 5)
	4	(1, 4)	(2, 4)	(3, 4)	(4, 4)	(5, 4)	(6, 4)
	3	(1, 3)	(2, 3)	(3, 3)	(4, 3)	(5, 3)	(6, 3)
	2	(1, 2)	(2, 2)	(3, 2)	(4, 2)	(5, 2)	(6, 2)
	1	(1, 1)	(2, 1)	(3, 1)	(4, 1)	(5, 1)	(6, 1)
		1	2	3	4	5	6

Red (R) Die

Figure 15–3

In the examples of this experiment, which follow, the number plane has an important role. The subset of the number plane R X R that we shall use is described by

(i) $$\{(R, W) \in N \text{X} N : 1 \leq R \leq 6 \text{ and } 1 \leq W \leq 6\}.$$

Figure 15–4 shows the elements of S assigned to points of the number plane described in (i).

Example 1 List the elements of the event, a subset of S, such that (a) R (red) is 2; (b) W (white) is 5.

White (W)	6	(1, 6)	(2, 6)	(3, 6)	(4, 6)	(5, 6)	(6, 6)
	5	(1, 5)	(2, 5)	(3, 5)	(4, 5)	(5, 5)	(6, 5)
	4	(1, 4)	(2, 4)	(3, 4)	(4, 4)	(5, 4)	(6, 4)
	3	(1, 3)	(2, 3)	(3, 3)	(4, 3)	(5, 3)	(6, 3)
	2	(1, 2)	(2, 2)	(3, 2)	(4, 2)	(5, 2)	(6, 2)
	1	(1, 1)	(2, 1)	(3, 1)	(4, 1)	(5, 1)	(6, 1)
		1	2	3	4	5	6

Red (R)

Figure 15–4

The event described in (a) is

$$\{(2, 1), (2, 2), (2, 3), (2, 4), (2, 5), (2, 6)\}.$$

The event described in (b) is

$$\{(1, 5), (2, 5), (3, 5), (4, 5), (5, 5), (6, 5)\}.$$

Example 2 Make a graph of the event, a subset of S, such that $R \leq 2$. The event consists of

$$\{(1, 1), (1, 2), (1, 3), \ldots, (1, 6), (2, 1), (2, 2), \ldots, (2, 6)\}.$$

The graph of this event is shown by encircling the dots representing the ordered pairs of the event.

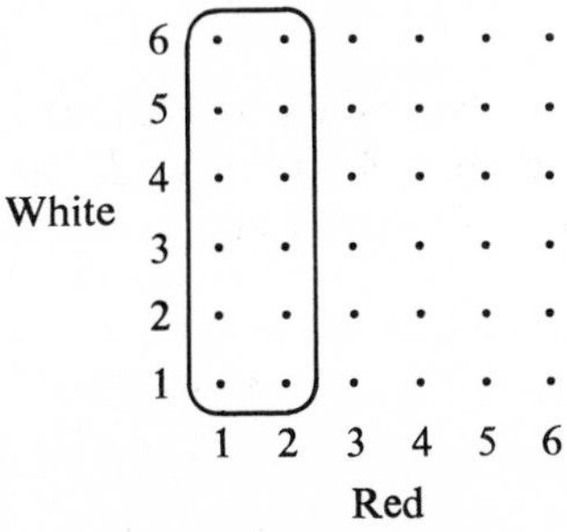

How many elements belong to S? to this event? Is the probability of this event $\frac{12}{36}$? Explain.

Example 3 Consider two events, each a subset of S, such that:

(i) $R = 4$;
(ii) $W = 3$.

Construct a graph of each event.

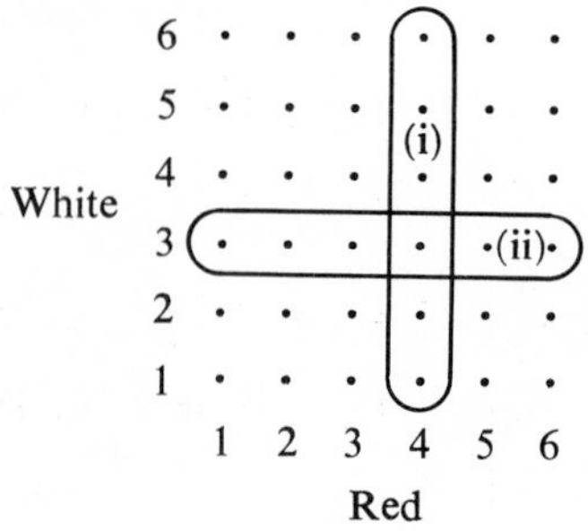

The graphs of (i) and (ii) are shown. By using the graphs, you should see that the set of ordered pairs belonging to the event $R = 4$ *or* $W = 3$ is

$$\{(1, 3), (2, 3), (3, 3), (5, 3), (6, 3), (4, 1), (4, 2), (4, 3), (4, 4), (4, 5), (4, 6)\}.$$

This event may be described in set builder notation as

$$\{(R, W) \in S : R = 4 \text{ or } W = 3\}.$$

Clearly, this is the *union* of the two events (i) and (ii).

What ordered pairs belong to (i) $R = 4$ *and* (ii) $W = 3$? The graph shows that a single ordered pair belongs to the *intersection* of the two sets. This event may be described as

$$\{(R, W) \in S : R = 4 \text{ and } W = 3\}.$$

In roster notation, this is

$$\{(4, 3)\}.$$

Example 4 Event A consists of the ordered pairs in S such that the sum of the components of each is 4. In event B the sum of the components is 11.

Are the graphs of event A and event B correct?

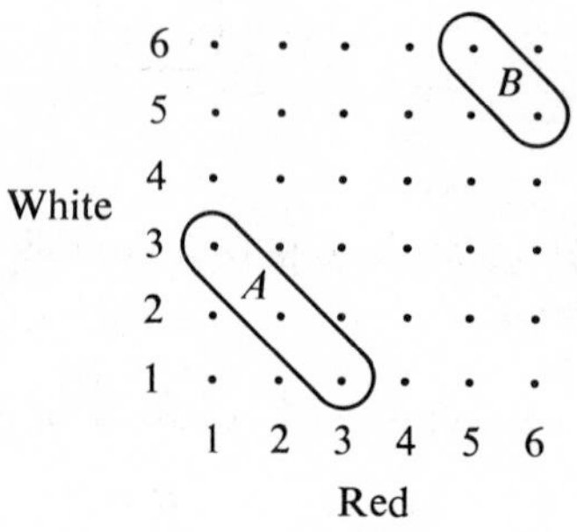

The elements of event A may be determined from the grid. They form the following set:

$$\{(3, 1), (2, 2), (1, 3)\}$$

This event may be described in set builder notation as

$$A = \{(R, W) \in S : R + W = 4\}.$$

The elements of event B form the following set:

$$\{(6, 5), (5, 6)\}.$$

Use set builder notation to describe event B.

List the elements in $A \cup B$ and $A \cap B$.

$$A \cup B = \{(1, 3), (2, 2), (3, 1), (5, 6), (6, 5)\}$$
$$A \cap B = \emptyset$$

Example 5 Event T, a subset of S, is such that $W > 3$; event Q, a subset of S, is such that $W < 4$. Make a graph of the two events.

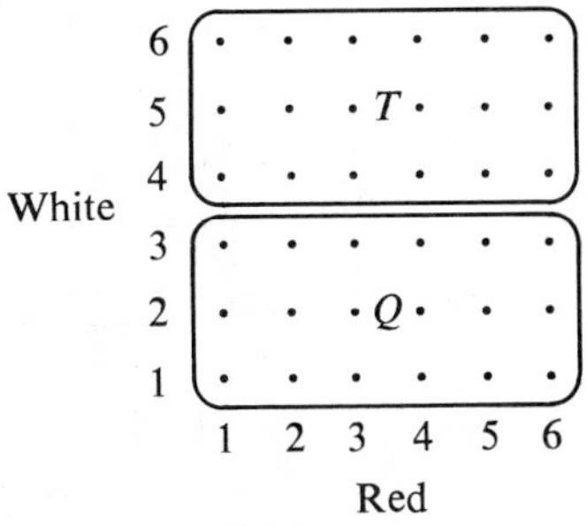

Observations one should make from the graph are that

$$T \cup Q = S,$$
$$T \cap Q = \emptyset.$$

What is the probability of getting an outcome in $T \cup Q$? Explain why it is 1. What is the probability of getting an outcome in $T \cap Q$? Why is it 0?

Example 6 Make a graph of each of the following events described in set builder notation.

(i) $\{(R, W) \in N X N : 1 \leq R \leq 6 \text{ and } 1 \leq W \leq 6 \text{ and } R < W\}$.

(ii) $\{(R, W) \in N X N : 1 \leq R \leq 6 \text{ and } 1 \leq W \leq 6 \text{ and } R = W\}$.

(iii) $\{(R, W) \in N X N : 1 \leq R \leq 6 \text{ and } 1 \leq W \leq 6 \text{ and } R > W\}$.

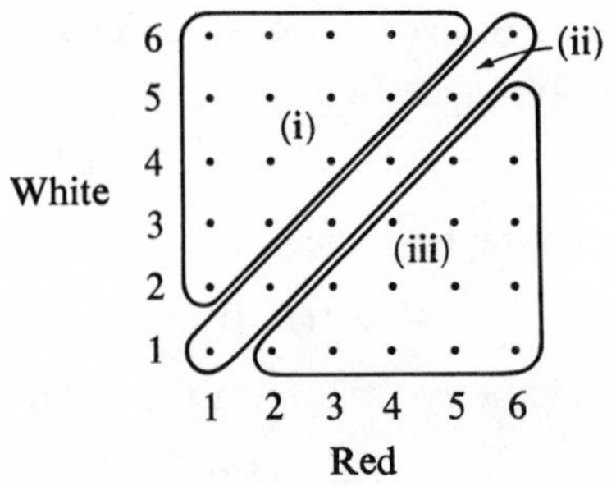

The graph of each event is shown. Do you agree with it? Describe the union of the three events. Are they disjoint?

Example 7 Make a graph of each of the events, a subset of S, such that

$$2 < R < 5 \text{ is event } E$$

and

$$2 \leq W < 5 \text{ is event } F.$$

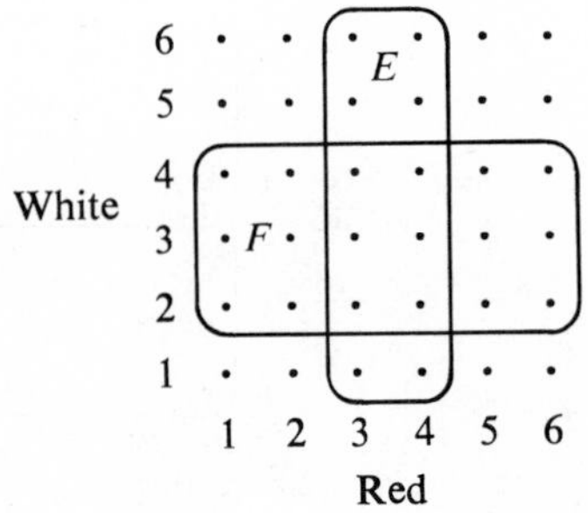

Check to see if you agree with the graph of each event.

List the elements in the intersection of the two events. Are the events disjoint? How many elements are in the union of these two events?

What is the probability of an outcome in the intersection of the two events? How many outcomes are in the intersection? Is the probability $\frac{6}{36}$?

EXERCISE 15.4

1. Event A and event B are subsets of S in a two-dice rolling experiment. Event A consists of all the pairs of doubles. If the sum of the components is 8, those ordered pairs form event B.
   (a) Make a graph of event A and of event B on the same grid.
   (b) Encircle that part of the array of dots which represents $A \cup B$.
   (c) List the elements, if any, for $A \cap B$.
   (d) What is the probability of getting (i) event A? (ii) event B? (iii) event A *or* B? (iv) event A *and* B?

2. Two boys are engaged in a two-dice rolling game. The outcomes of six rolls for Jim is {(6, 4), (2, 2), (3, 5), (2, 5), (1, 2), (5, 5)}; while six rolls for Bill resulted in {(4, 6), (5, 3), (5, 5), (2, 1), (5, 2), (2, 2)}. Make a graph of each, identifying Jim's with □ and Bill's with ×. That is, the location of (6, 4) and (5, 2) will be marked on the grid as shown so as to distinguish the two events.

□	×
(6, 4)	(5, 2)

   (a) List the set of ordered pairs that are marked with □ and ×.
   (b) List the set of ordered pairs that have one or the other, or both marks.
   (c) Let J represent Jim's set of outcomes and B represent Bill's. Find (i) $N(J \cap B)$ and (ii) $N(J \cup B)$.
   (d) What is the probability of getting an outcome in the intersection of J and B? in the union of J and B?

3. In event A the red die is either 2 or 3; in event B the white die is either 2 or 5. Make a graph of the two events, using □ to mark elements in event A and × for those in event B.
   (a) Find $A \cap B$.
   (b) How many ordered pairs belong to event A? to event B? to $A \cup B$?
   (c) Are the two sets disjoint?
   (d) What is the probability of event A? of event B? of $A \cup B$?

15.5 MUTUALLY EXCLUSIVE EVENTS

The events in each of Examples 4 and 5 of Section 15.4 have no elements in common. The sets are disjoint. The events in each example are said to be *mutually exclusive.*

Definition 15.5 Two events A and B are said to be *mutually exclusive,* if $A \cap B = \varnothing$.

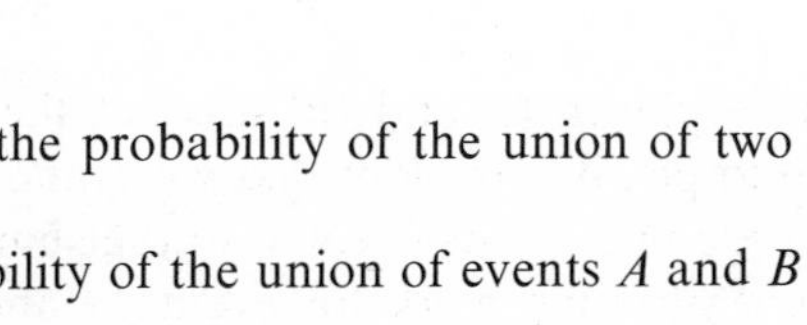

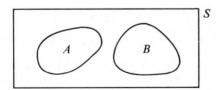

We now discuss the probability of the union of two events.

Example 1 Find the probability of the union of events A and B of Example 4, Section 15.4.

Consider the union of event A and event B:

$$A \cup B = \{(1, 3), (2, 2), (3, 1), (5, 6), (6, 5)\}$$

then

$$N(A \cup B) = 5.$$

The probability of the union of these two events is

(i) $$\frac{N(A \cup B)}{N(S)} = \frac{5}{36}.$$

Next consider the probability of each of the events A and B:

$$P(A) = \frac{3}{36} \quad \text{and} \quad P(B) = \frac{2}{36}$$

Hence,

(ii) $$P(A) + P(B) = \frac{3}{36} + \frac{2}{36} = \frac{5}{36}.$$

We have found the same number for (i) as for (ii). This seems to imply that the probability of the union of two events is the same measure as the sum of the two probabilities. Before accepting this, study the next example.

Example 2 Find the probability of the union of the events E and F in Example 7, Section 15.4.

The union of events E and F is

$$\begin{aligned} E \cup F = \{&(3, 1), (3, 2), (3, 3), (3, 4), (3, 5), (3, 6)\\ &(4, 1), (4, 2), (4, 3), (4, 4), (4, 5), (4, 6)\\ &(1, 2), (1, 3), (1, 4)\\ &(2, 2), (2, 3), (2, 4)\\ &(5, 2), (5, 3), (5, 4)\\ &(6, 2), (6, 3), (6, 4)\} \end{aligned}$$

Then $N(E \cup F) = 24$, and the probability of $N(E \cup F)$ is

(i) $$P(E \cup F) = \tfrac{24}{36}.$$

The probability of each of the events E and F is

$$P(E) = \tfrac{12}{36} \qquad \text{and} \qquad P(F) = \tfrac{18}{36}.$$

From this, we have

(ii) $$P(E) + P(F) = \tfrac{12}{36} + \tfrac{18}{36} = \tfrac{30}{36}.$$

Since the results in (i) and (ii) are not the same as they were in the preceding example, it is evident that we cannot say that the probability of the union of the two events is equal to the sum of their probabilities. Why does this generalization work for events A and B in Example 1 and not for events E and F in this example? Since A and B are mutually exclusive and E and F are not, it appears that if events are not mutually exclusive, the generalization does not apply.

The question now is: How does one find the probability of the union of two nonmutually exclusive events?

Turning again to the graph of the events E and F in Example 7, Section 15.4, by counting we find that

$$N(E \cap F) = 6.$$

Then, the probability of the intersection set is

$$P(E \cap F) = \tfrac{6}{36}.$$

This measure, $\frac{6}{36}$, is the excess of (ii) over (i). Then, it appears that the probability of the union of the two events is equal to the sum of their probabilities minus the probability of their intersection.

We are prepared to state two theorems as a result of these two examples (1 and 2).

Theorem I *If A and B are events belonging to a finite sample space S, then*

$$P(A \cup B) = P(A) + P(B) - P(A \cap B).$$

Proof By Definition 15.2(b),

(i)
$$P(A \cup B) = \frac{N(A \cup B)}{N(S)}.$$

If $A \cap B = \varnothing$, then by Definition 5.2,

$$N(A \cup B) = N(A) + N(B).$$

But, if $A \cap B \neq \varnothing$, then $N(A) + N(B)$ counts the common elements twice. Thus, $N(A \cap B)$ must be subtracted, and

$$N(A \cup B) = N(A) + N(B) - N(A \cap B).$$

Hence, (i) becomes

$$P(A \cup B) = \frac{N(A) + N(B) - N(A \cap B)}{N(S)}$$
$$= \frac{N(A)}{N(S)} + \frac{N(B)}{N(S)} - \frac{N(A \cap B)}{N(S)}.$$

By Definition 15.2(b),

$$P(A \cup B) = P(A) + P(B) - P(A \cap B).$$

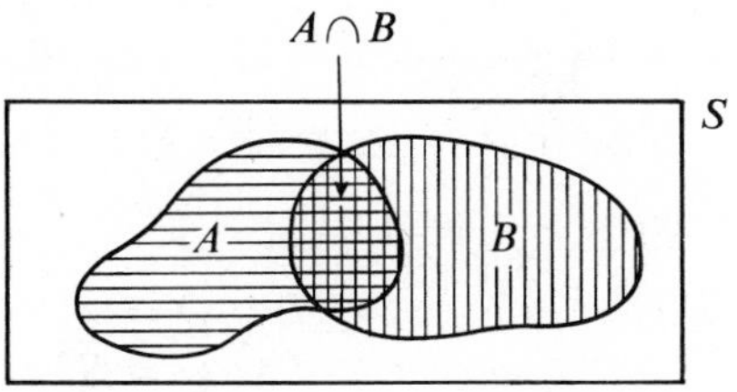

The next theorem follows.

Theorem II *If A and B are events belonging to a finite sample space S and* $A \cap B = \varnothing$, *then*

$$P(A \cup B) = P(A) + P(B).$$

Proof By Theorem I,

$$\text{(i)} \qquad P(A \cup B) = P(A) + P(B) - P(A \cap B).$$

Since $A \cap B = \varnothing$, then $N(A \cap B) = 0$.
By Definition 15.2(b),

$$P(A \cap B) = \frac{N(A \cap B)}{N(S)} = \frac{0}{N(S)} = 0.$$

Therefore (i) becomes

$$P(A \cup B) = P(A) + P(B) - 0$$
$$P(A \cup B) = P(A) + P(B).$$

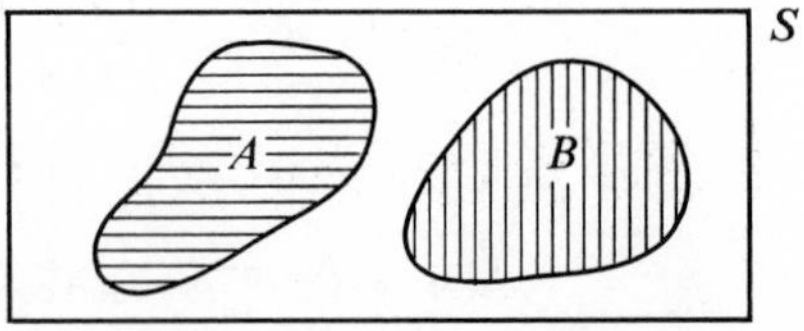

For more than two mutually exclusive events, $E_1, E_2, E_3, \ldots, E_n$, we have the following theorem. You should be able to give a proof of it.

Theorem III *If mutually exclusive sets* $E_1, E_2, E_3, \ldots, E_n$ *belong to the finite sample space S, then*

$$P(E_1 \cup E_2 \cup E_3 \cup \ldots \cup E_n) = P(E_1) + P(E_2) + P(E_3) + \cdots + P(E_n).$$

Example 3 Find the probability of the union of the events T and Q in Example 5, Section 15.4.

By Theorem I,

$$P(T \cup Q) = P(T) + P(Q) - P(T \cap Q).$$

Counting the elements in the graph of events T, Q, and $T \cap Q$, then applying Definition 15.2(b), we have

$$P(T \cup Q) = \tfrac{18}{36} + \tfrac{18}{36} - \tfrac{0}{36} = \tfrac{36}{36} = 1.$$

From the graph, we see that

$$\text{(i)} \qquad T \cap Q = \varnothing$$
$$\text{(ii)} \qquad T \cup Q = S.$$

These two statements are the conditions in Definition 2.5(d) for complementary sets. Since events are sets, T and Q are *complementary events*. That is, Q is the complement of T and T is the complement of Q. Using the symbol introduced in Section 2.5, we shall write $\tilde{T}$ for Q, the complement of T.

The next theorem follows from this example.

Theorem IV *If T and $\tilde{T}$ are complementary events in the sample space S, then*

$$P(T) + P(\tilde{T}) = 1.$$

Proof By Definition 15.2(b),

$$\begin{aligned} P(T) + P(\tilde{T}) &= \frac{N(T)}{N(S)} + \frac{N(\tilde{T})}{N(S)} \\ &= \frac{N(T) + N(\tilde{T})}{N(S)} \\ &= \frac{N(T \cup \tilde{T})}{N(S)} && \textit{Definition 5.2; } T \textit{ and } \tilde{T} \textit{ are disjoint.} \\ &= \frac{N(S)}{N(S)} && T \cup \tilde{T} = S \\ &= 1 \end{aligned}$$

Therefore, $P(T) + P(\tilde{T}) = 1$. It follows that,

$$P(T) = 1 - P(\tilde{T})$$

and

$$P(\tilde{T}) = 1 - P(T).$$

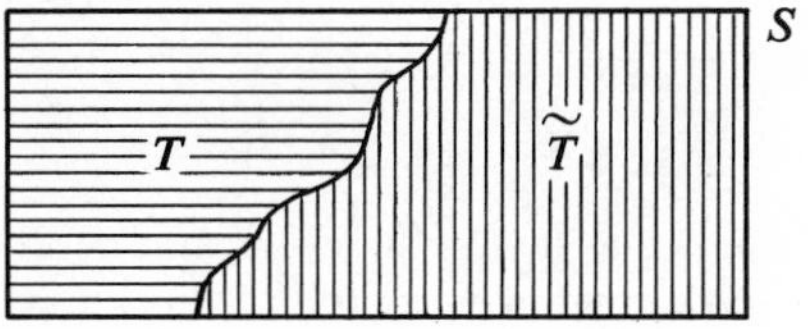

The diagram that follows shows sample space S completely subdivided into 5 mutually exclusive events. S is said to be *partitioned* into n subsets, if and only if

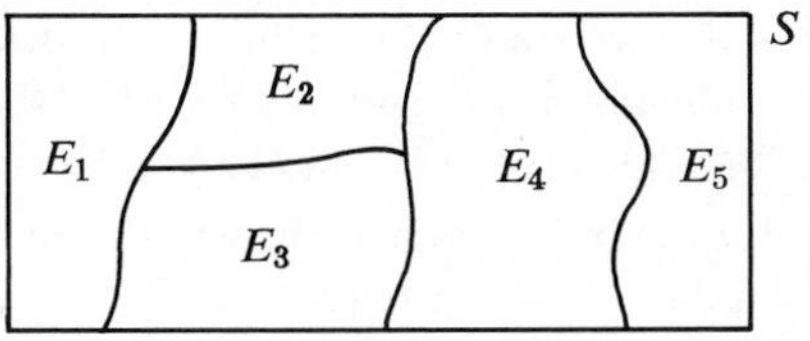

$$E_1 \cup E_2 \cup E_3 \cup \ldots \cup E_n = S.$$

Is S partitioned by the complementary events T and $\tilde{T}$?

You should be able to prove the following theorem.

Theorem V *If sample space S is partitioned into subsets E_1, E_2, $E_3, \ldots E_n$, then*

$$P(E_1) + P(E_2) + P(E_3) + \cdots + P(E_n) = 1.$$

Example 5 A box contains 5 white marbles and 3 black ones.

(a) What is the probability of selecting a white marble on the first draw? ($\frac{5}{8}$)

(b) What is the probability of drawing a nonwhite marble on the first draw? ($\frac{3}{8}$, or $\frac{(8-5)}{8}$)

The *relative probability* of the events white marble and nonwhite marble, complementary events, is

$$\frac{\frac{5}{8}}{\frac{8-5}{8}} = \frac{5}{3}.$$

This relative probability is called the *odds in favor of* the event, a white marble. That is,

$$\text{Odds in favor of } W = \frac{P(W)}{P(\tilde{W})}.$$

$$\frac{P(W)}{P(\tilde{W})} = \frac{\frac{\text{N}(W)}{\text{N}(S)}}{\frac{\text{N}(S-W)}{\text{N}(S)}}$$

$$= \frac{\text{N}(W)}{\text{N}(S-W)}$$

EXERCISE 15.5

1. Bob rolls a fair die. If A is the event that an odd number occurs, and B is the event that an even number occurs, what is the probability of an even or an odd? Are the events mutually exclusive? complementary?
2. What is the probability that in Problem 1 Bob would roll a 4, a 5, or a 6?
3. What is the probability of drawing a jack, a king, a queen, or an ace from a well-shuffled bridge deck of 52 cards?
4. Two fair coins are tossed.
   (a) What is the probability that at least one coin will fall heads?
   (b) What is the probability that neither coin will fall heads?
   (c) What are the odds in favor of at least one coin falling heads?
5. What are the odds in favor of drawing a 5 from a well-shuffled bridge deck?
6. In a two-dice rolling experiment, a red (R) die is rolled first, then followed by a white (W) die. Two events are described as

$$A\colon R < W \qquad \text{and} \qquad B\colon R + W = 7.$$

(a) Make a graph of each event.
(b) Find $A \cap B$. Are the events mutually exclusive? complementary?
(c) Find $P(A \cup B)$.

7. In the experiment of Problem 6, events Q and T are described as,

$$Q\text{: doubles occur} \qquad \text{and} \qquad T\text{: } R > W.$$

(a) Make a graph of the two events.
(b) Find $Q \cap T$. Are the events mutually exclusive? complementary?
(c) Find $P(Q \cup T)$.

8. The following is a list of theorems to be proved:

Theorem VI *The probability of event E, a subset of a finite sample space S, is a nonnegative number.*
(Show that $P(E) \geq 0$.)

Theorem VII *If an event E, a subset of a finite sample space S, is certain, the probability of that event is* 1.
(Since $E = S$, show $P(S) = 1$.)

Theorem VIII *If an event E, a subset of a finite sample space S, is impossible, the probability of that event is* 0.
(Since $E = \varnothing$, show $P(\varnothing) = 0$.)

Theorem IX *The probability of the union of the sample space S and the empty set, ∅, is the probability of S.*
(Establish S and $\varnothing$ as mutually exclusive, then show $P(S \cup \varnothing) = P(S)$.)

Theorem X *If events E and F are subsets of the finite sample space S, and E is a subset of F, then* $P(E) \leq P(F)$.

Theorem XI *The probability of event E, a subset of a finite sample space S, is greater than or equal to* 0 *and less than or equal to* 1.
(Show that $0 \leq P(E) \leq 1$.)

15.6 INDEPENDENT EVENTS

If a die is rolled and a coin is tossed, one of the six occurrences of the die in no way affects one of the two occurrences of the coin, and we say that the occurrences are *independent* of each other.

Example 1 The possible outcomes of rolling a die, then tossing a coin gives the sample space S that follows:

$$\{(1, T), (2, T), (3, T), (4, T), (5, T), (6, T)$$
$$(1, H), (2, H), (3, H), (4, H), (5, H), (6, H)\}$$

The number of outcomes may be counted or may be determined by multiplying the number of rows of the array by the number of columns, (2×6). There are 12 possible outcomes.

Consider event E_1 as those outcomes having numbers less than 3, and event E_2 as those having tails.

$$E_1 = \{(1, T), (2, T), (1, H), (2, H)\}$$
$$E_2 = \{(1, T), (2, T), (3, T), (4, T), (5, T), (6, T)\}$$

Suppose that we are interested in the probability of the event having a number less than 3 *and* having tails, in other words, the probability of $E_1 \cap E_2$. There are two outcomes which meet these requirements, namely, $(1, T)$ and $(2, T)$. Hence,

$$P(E_1 \cap E_2) = \tfrac{2}{12} = \tfrac{1}{6}.$$

Observe the probability of each event separately.

$$P(E_1) = \tfrac{4}{12} = \tfrac{1}{3} \quad \text{and} \quad P(E_2) = \tfrac{6}{12} = \tfrac{1}{2}$$

The product of these two probabilities

$$P(E_1) \cdot P(E_2) = \tfrac{1}{3} \cdot \tfrac{1}{2} = \tfrac{1}{6}.$$

This product is the same as $P(E_1 \cap E_2)$. Both are $\tfrac{1}{6}$. From this example, we may say,

$$P(E_1 \cap E_2) = P(E_1) \cdot P(E_2).$$

Example 2 Two fair coins are tossed. The probability of each of the following events is to be determined, also the probability of their intersection. Event E_3 consists of those outcomes having exactly one head, and event E_4 of those having no double heads.

The sample space S of tossing the coins is

$$S = \{(H, H), (H, T), (T, H), (T, T)\}.$$

The two events consist of

$$E_3 = \{(H, T), (T, H)\} \quad \text{and} \quad E_4 = \{(H, T), (T, H), (T, T)\}.$$

The probability of E_3 and E_4 is given for each.

$$P(E_3) = \tfrac{2}{4} = \tfrac{1}{2} \quad \text{and} \quad P(E_4) = \tfrac{3}{4}$$

Since $E_3 \cap E_4 = \{(H, T), (T, H)\}$, which consists of 2 elements,

$$P(E_3 \cap E_4) = \tfrac{2}{4} = \tfrac{1}{2}.$$

The product of $P(E_3)$ and $P(E_4)$ is found to be $\tfrac{1}{2} \cdot \tfrac{3}{4} = \tfrac{3}{8}$, which is not the same as the probability of $(E_3 \cap E_4)$. Hence, for this example,

$$P(E_3 \cap E_4) \neq P(E_3) \cdot P(E_4).$$

The results of Example 1 and Example 2 say something to us about the events described in each. The events of Example 1 are called *independent events;* those of Example 2 are called *dependent events.* We now give a formal definition for independent events.

Definition 15.6 Events A and B are *independent* if and only if

$$P(A \cap B) = P(A) \cdot P(B).$$

The probability of two events occurring simultaneously may be found by multiplying the probability of one event by that of the other, provided the events are independent.

EXERCISE 15.6

1. In a two-dice experiment consisting of rolling a red (R) die then a white (W) die, are the following events independent?

$$A = \{(R, W) \in S : R < 3\}$$
$$B = \{(R, W) \in S : R + W = 5\}.$$

   (a) Make a graph of each of the events, marking the elements of A with $\times$ and B with $\square$.
   (b) Find $P(A)$; $P(B)$; $P(A \cap B)$.

2. Use the same experiment as in Problem 1 and the events

$$E_1 = \{(R, W) \in S : 3 \leq R < 6\}$$
$$E_2 = \{(R, W) \in S : W > 4\}$$

   (a) Make a graph of each of the events, marking E_1 with $\times$ and B with $\square$.
   (b) Find $P(E_1)$; $P(E_2)$; $P(E_1 \cap E_2)$.
   (c) Are events E_1 and E_2 independent? Explain.

3. A coin is tossed twice. What is the probability that the first coin will fall tails and the second one will fall heads?

4. Three coins are tossed. What is the probability that the first two fall heads and the first and third fall differently?
   (a) Make a tree diagram to determine the sample space.
   (b) List the elements of each of the events.
   (c) Determine whether the events are independent.
   (d) Show whether a 2-coins-tail event and a 3-coins-tail event are dependent.

5. Use the two-dice experiment described in Problem 1 and the events

$$A = \{(R, W) \in S : R \geq 5\}$$
$$B = \{(R, W) \in S : W \geq 3\}.$$

   (a) Make a graph of each of the events A and B.
   (b) Use the graph to determine $P(A)$, $P(B)$, and $P(A \cap B)$. Are A and B independent?
   (c) What is the probability of a red die falling 5 or greater and the white falling 3 or greater? Apply Theorem I.

6. Two cards are drawn from a conventional bridge deck that has been well shuffled. What is the probability of drawing an honor card that is also from a red suit?

7. Give a convincing argument that if any event E is a subset of sample space S then E and S are independent events. Test E if it is empty, if it is nonempty, and if it is the same as S.

Suggested Development of Solutions for Selected Exercises

Chapter 2

Exercise 2.2 Page 10

1. Flock of sheep, class of students, fleet of trucks, etc.
2. (f) Does not describe a set. (g) Describes a set.
3. (c) Since twenty-four expressed in Roman numerals is XXIV, the set of symbols used is {X, I, V}.
   (d) The counting numbers between 5 and 13 are 6, 7, 8, 9, 10, 11, 12. Hence the set of squares of all such numbers is {36, 49, 64, 81, 100, 121, 144}.
4. (a) $V = \{a, e, i, o, u\}$ (b) $e \in V$ (c) $k \notin V$.
5. (a) {68, 69, 70, . . . , 2999}.

Exercise 2.3 Page 14

1. (a) (2) {Jo}, {Mary}, {Jane}, {Jo, Mary}, {Jo, Jane}, {Mary, Jane}, {Jo, Mary, Jane}, $\varnothing$.
   (b) There are four subsets in part (1); eight subsets in part (2); and sixteen subsets in part (3).
   (c) $2^2 = 4, 2^3 = 8, 2^4 = 16$. Perhaps this should suggest 2 raised to a power equal to the number of elements in the set. That is, from a set of N elements the number of possible subsets is 2^N.
   (d) (5) A set with one element, for example $\{a\}$, has two subsets: $\{a\}$, $\varnothing$. This agrees with the number 2^1 suggested in (c) above.
   (6) The set with no elements (0 elements) is $\varnothing$. The only subset is $\varnothing$ itself. $2^0 = 1$, and the pattern holds.
3. (b) $\{\varnothing\}$ is a set with *one* element, $\varnothing$. Hence $\{\varnothing\}$ cannot be the empty set, $\varnothing$. This reasoning should help with (h) and (i).

4. (c) $U = \{a, b, c, \ldots, z\}$ $\{x, y, r\} \not\subset \{s, t\}$
$A = \{x, y, r\}$ $\{s, t\} \subset \{r, s, t, w, z\}$
$B = \{s, t\}$
$C = \{r, s, t, w, z\}$

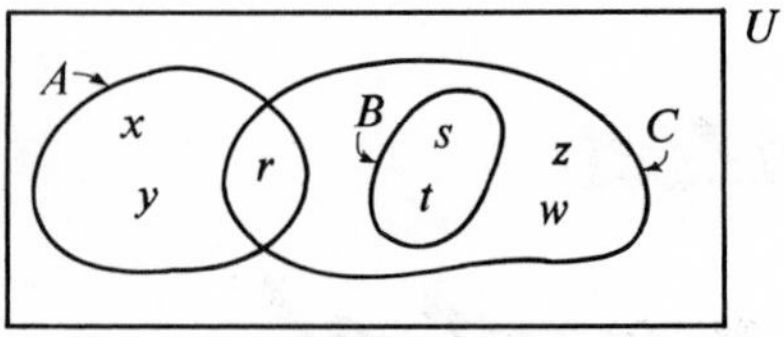

5. (e) The Venn diagram should make it clear that the subset relation, $\subset$, holds here.

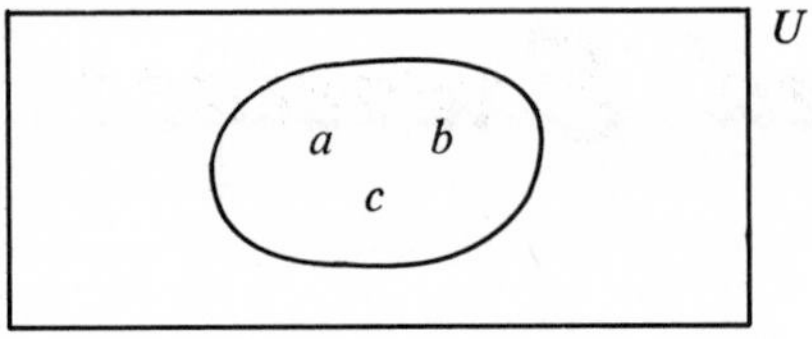

Exercise 2.4 Page 18

1. (d) Let U be the set of toys in the family playroom.
$\{t \in U: t \text{ belongs to Jim}\}$

2. (a) $W_1 = \{$Alabama, Alaska, Arizona, Arkansas, California, Colorado, Connecticut$\}$
(b) $W_2 = \{$Alabama, Connecticut$\}$

4. (a) 5 quarters = \$1.25 is not an element of U, since it is not equal in value to any one of the coins in U.
5 dimes = \$.50 is in U, since it makes a half-dollar, a coin belonging to U.
5 half-dollars = \$2.50 is not an element of U.
5 pennies = nickel is in U.
5 nickels = \$0.25, a quarter, is in U.
Hence, $W_6 = \{$dime, penny, nickel$\}$.

6. $U =$ the set of all counting numbers $= \{1, 2, 3, \ldots\}$.
(c) $C = \{x \in U: (x + 1) < 8\} = \{1, 2, 3, 4, 5, 6\}$.

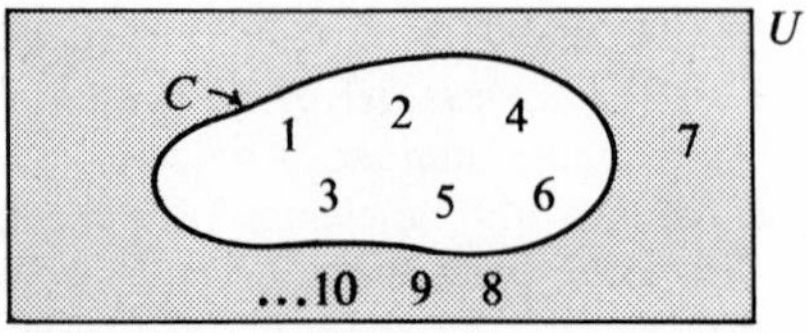

Exercise 2.5 Page 27

3. (c) $\{a, c, d, f\} \cup (\{d, e, a\} \cap \{a, b, c, d\}) \stackrel{?}{=} (\{a, c, d, f\} \cup \{d, e, a\})$
$\cap (\{a, c, d, f\} \cup \{a, c, d\})$

1st step $\{a, c, d, f\} \cup \{a, d\} \stackrel{?}{=} \{a, c, d, e, f\} \cap \{a, c, d, f\}$
2nd step $\{a, c, d, f\} \stackrel{?}{=} \{a, c, d, f\}$ Yes.
Notice that the solutions for the parts of Problem 3 should be done in two steps, first performing the operations enclosed by the parentheses.

8. (a) You may discover certain patterns that speed your work. For example, the union of A with each of the other sets determines the first row and the first column of the $\cup$-grid. Also, what do you notice about the union of H with another set? If you know $B \cup C$, do you know $C \cup B$?
   (b) HINT: Refer to the discussion of subsets in Section 2.3.
14. (a) $\tilde{B} = \{1, 3, 5, 7\}$
    (b) $\widetilde{A \cup C} = \widetilde{\{1, 3, 4, 5, 6, 7\}} = \{2, 8\}$
    $\tilde{A} = \{2, 4, 6, 8\}, \tilde{C} = \{1, 2, 3, 8\}$
    $\tilde{A} \cup \tilde{C} = \{1, 2, 3, 4, 6, 8\}$
    $\widetilde{\tilde{A} \cup \tilde{C}} = \{5, 7\}$
    $\widetilde{\widetilde{\tilde{A} \cup \tilde{C}}} = \{1, 2, 3, 4, 6, 8\}$
    (c) $\tilde{B} \cap \tilde{C} = \{1, 3, 5, 7\} \cap \{1, 2, 3, 8\} = \{1, 3\}$
16. (c) See Definition 2.5(c).

Exercise 2.7 Page 34

2. (d) $A \cap (B \cup C)$

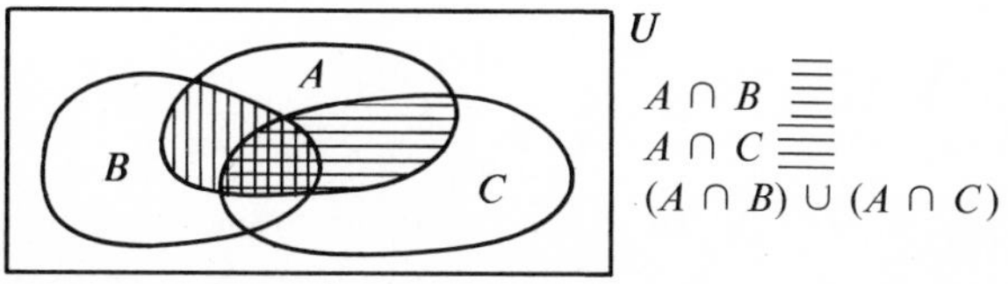

Property: Intersection distributive over union.

4. (b) $(\widetilde{X \cap Y})$ $\tilde{X} \cup \tilde{Y}$

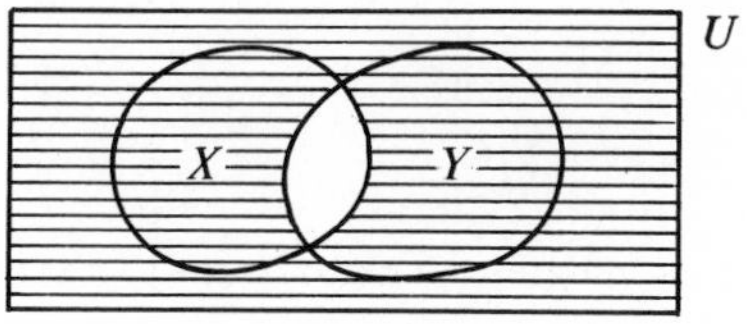

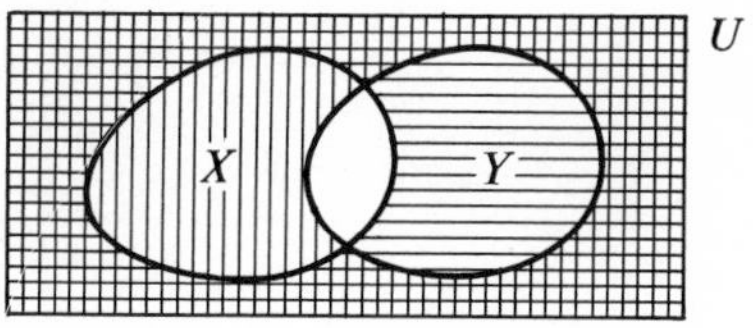

Exercise 2.8 Page 39

1. (b) Onto. Why?
   (c) Try reversing the arrows.
2. (c) Are 1 and 8 members of the range?

Exercise 2.10 Page 46

1. (b) N(A), N$\{1, 2, 3\}$; 3
   (c) N(A) $<$ N(B) or N$\{1, 2, 3\} <$ N$\{0, 2, 4, 6, 8\}$
   N(C) $=$ N(D) or N$\{a, b, c\} =$ N$\{7, 9, 4\}$
2. (a) (6), (9), (13), (17), (18), (19) are some of the statements that are false. You should be able to explain why. Some of the others are true and some are false. Decide about them.

(c) N(A) > N(D) > N(B) > N(G) > N(H).

3. Let the universal set, U, be all the people in town. Let M be the set of Michigan-born people in the block; S be the set of people in school living in the block. Let x = the number of people who are Michigan-born and attend school. In the Venn diagram the people who both attend school and were Michigan-born overlapped and were counted twice. Hence,

$$\begin{aligned} N(M) + N(S) - x &= 43 \\ 27 + 18 - x &= 43 \\ x &= 2. \end{aligned}$$

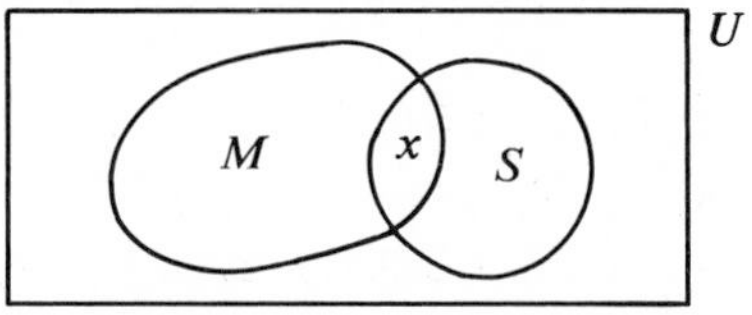

4. (a) 95 here is an ordinal number. What kind of number would 95 be if you were instructed to read 95 pages?
   (g) 5 tells the number of brothers and hence is used as a cardinal number. Number 4 gives his place in the order of birth and hence is an ordinal number.

Chapter 3 *Exercise 3.2 Page 51*

4. Try to think as an Egyptian might have thought. Start with the number of strokes necessary to make a heelbone. This starts the machinery in motion and with no more additions you now have 5 scrolls. What must be added to these 5 scrolls to make a lotus flower?

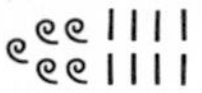

5. (a) The strokes combine into 1 heelbone, the heelbones into 1 scroll, and the scrolls into 1 lotus flower.

   (b) Notice that this requires changing 1 heelbone into 10 strokes and then 1 scroll into 10 heelbones.

Exercise 3.3 Page 54

1. (c) $(2 \times 60^3) + (11 \times 60^2) + (12 \times 60) + 20 = 472{,}340$
   (e) $(1 \times 60^3) + (20 \times 60^2) + (0 \times 60) + 1 = 288{,}001$
   (f) $(1 \times 60^2) + (2 \times 60) + 3 = 3723$
2. Some of them would be:

9 × 8:  10 × 8: 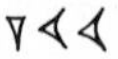 11 × 8:  12 × 8: 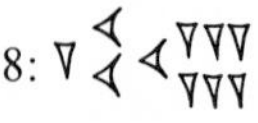

3. (c) 425: $7 \times 60 + 5$: 

   (d) 122: $2 \times 60 + 2$: 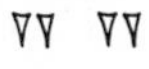

4. $7(60)^2 + 21(60) + 56$. Verify this and then finish the problem.
5. Can you interpret this as 4 quarts per day?

Exercise 3.4 Page 55

1. (a) CMV = 905 (c) MDXCV = 1,595
2. (a) 946 = CMXLVI or DCCCCXXXXVI
   (b) 1,095 = MXCV or MLXXXXV
   (f) 1,944 = MCMXLIV or MDCCCCXXXXIIII
3. (c) As in example, change to longer form.

CC	XX	VII
	XXX	IIII
CL	X	VI
CCCLLXXVII		
or CCCCXXVII		
or CDXXVII		

   (f)

	CLXXXXXXXXXXVIIIIII	
(−)C	XXXXXX	II
	LXXX	VIIII
or	LXXXIX	

Exercise 3.5 Page 59

1. The answers here are yes, no, no. The student should be able to cite, as reasons, the principles involved in the different numeration systems.
2. (b) 375 = 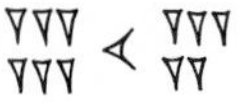 (c) 573 = DLXXIII
3. (b) Since $307 = 5 \times 60 + 7$, a space or a dot must be used to indicate the absence of any symbols representing 10 and to serve to separate the multiples of 60 on the left.

Exercise 3.8 Page 65

2.

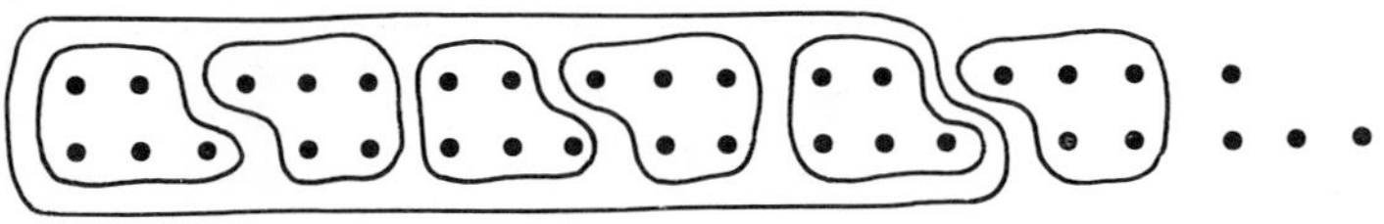

114_{five}

4. Three quarters is 3×5^2 pennies or 300_{five} pennies. Eight nickels is $1 \times 5^2 + 3 \times 5$ pennies or 130_{five} pennies. Six pennies is $1 \times 5 + 1$ pennies or 11_{five} pennies. Their sum is 441_{five} pennies.
6. All powers of 5 are odd numbers. The sum of an odd number of odd numbers is odd, while the sum of an even number of odd numbers is even. Having assured yourself of the truth of these statements, use them to determine whether the numbers given are even or odd without expanding them to base 10.
   Hence, 1110_{five} is odd while 1210_{five} is even.

Exercise 3.9 Page 67

1. Twelve. Because 10 is written in terms of digits already in use. In base twelve 10 means 1 twelve and 0 ones; 11 means 1 twelve and 1 one. Other symbols must be used, since 10 and 11 already have different meanings in the base 12 system.
3. 18 is 16_{twelve}; 15 gross is 1300_{twelve}; 9 great gross is 9000_{twelve}. The sum is $T316_{\text{twelve}}$.
4. (b) $E09000_{\text{twelve}}$

Exercise 3.10 Page 68

2. 0_{two}—off—even
   1_{two}—on—odd
   10_{two}—off—even
   11_{two}—on—odd
   100_{two}—off—even
   101_{two}—on—odd
   110_{two}—off—even

   Notice that the presence of 1 in the units place makes the number odd. If 0 is in the units place, no matter how many 1's are used to the left, signifying the addition of powers of 2 (all even numbers), then the number is even. Hence 0 or 1 in the units place is the key to the situation.
3. (c) $1000_{\text{two}} = (1 \times 2^3) + (0 \times 2^2) + (0 \times 2^1) + (0 \times 2^0)$
4. Refer to the discussion for Problem 6 of Exercise 3.8, remembering that you are now dealing with powers of 2 rather than powers of 5. All powers of 2 except 2^0 are even numbers.
5. (a) $1_{\text{two}} + 1_{\text{two}} = 10_{\text{two}}$
6. (c) $100_{\text{two}} - 1_{\text{two}} = 11_{\text{two}}$

Exercise 3.11 Page 70

3. (d) $4{,}060 = (4 \times 10^3) + (0 \times 10^2) + (6 \times 10^1) + (0 \times 10^0)$
4. (b) In expanded notation for 27,500, 7 means (7×10^3) or 7,000.
5.

$(1 \times 10^2) + (2 \times 10^1) + (3 \times 10^0)$

6. Remember that the addition of 1 to your answer will make the result require 6 digits. It is understood that the digits are not necessarily different.
7. There are 6 ways to arrange the digits to display numerals representing counting numbers.

Exercise 3.12 Page 73

1. The algorithm in the preceding section should be helpful.
   (e) $125 = 11122_{\text{three}}$
2. Since the numerals in each group are related, it is possible to find the others

after one has been determined by the algorithm described in the preceding article.

(c) Since by the algorithm $258 = 10002_{\text{four}}$, then by subtraction in base 4, $256 = 10000_{\text{four}}$, $254 = 3332_{\text{four}}$, $252 = 3330_{\text{four}}$.

3. It will be necessary in each case to convert to base ten and then change to the required base.

(b) $356_{\text{seven}} = 188 = 2330_{\text{four}}$

4. $2a + 5 = 21$

$2a = 16$

$a = 8$ That is: $25_{\text{eight}} = 21$

5. $(1 \times b) + 3 + 5 = 2b + 1$

$7 = b$ That is: $13_{\text{seven}} + 5_{\text{seven}} = 21_{\text{seven}}$

6. (b) $23_b = 13$

$2b + 3 = 13$

$2b = 10$

$b = 5$

$23_{\text{five}} = 13$

Exercise 3.13 Page 75

1.

+	0	1
0	0	1
1	1	10

(b) $1011_{\text{two}} + 111_{\text{two}} = 10010_{\text{two}}$.

$$\begin{array}{r} 1011_{\text{two}} \\ 111_{\text{two}} \\ \hline 10 \\ 10 \\ 1 \\ 1 \\ \hline 10010_{\text{two}} \end{array}$$

2. (b)

$$\begin{array}{rll} 1011_{\text{two}} & = 2^3 + 0 + 2 + 1 & = 11 \\ 111_{\text{two}} & = 2^2 + 2^1 + 1 & = 7 \\ \hline 10010_{\text{two}} & = 2^4 + 0 + 0 + 2^1 + 0 & = 18 \end{array}$$

3. Where necessary change each numeral to base ten. However, a decision may be made in certain cases by inspection. For example, $3411_{\text{five}} < 3411_{\text{seven}}$ since powers of 5 are less than like powers of 7.

4. A number expressed in base 3 employs only numerals 0, 1, 2. These may be represented thus:

Let N{ } be represented by the symbol *
N{a} be represented by the symbol ?
N{x, y} be represented by the symbol $

(c) $9_{\text{ten}} = 100_{\text{three}} =$?** (e) $12_{\text{five}} = 7_{\text{ten}} = 21_{\text{three}} =$ $?

(f) $12_{\text{twelve}} = 14_{\text{ten}} = 112_{\text{three}} =$??$

6. What about: (1) the number of digits required to write the numerals that represent the same number for each of the bases; (2) the number of basic addition and multiplication facts to be learned for each base? Can you think of others?

7.

Base: Ten	Twelve	Eight	Seven	Five	Three	Two
1	1			1		1
2		2			2	10
3	3		3		10	11
⋮	⋮	⋮	⋮	⋮	⋮	⋮
12	10	14	15	22	110	1100
⋮	⋮	⋮	⋮	⋮	⋮	⋮
24	20	30	33	44	220	11000
25	21	31	34	100	221	11001
26	22	32	35	101	222	11010
27	23	33	36	102	1000	11011
28	24	34	40	103	1001	11100
⋮	⋮	⋮	⋮	⋮	⋮	⋮
49	41	61	100	144	1211	110001
50	42	62	101	200	1212	110010

(a) Yes. 1 is the same for all bases; 2 is the same for all bases of three or more; 3 is the same for all bases of four or more; in general, if $n \in N$ then n is the same for all bases of $(n + 1)$ or greater.

(b) EE_{twelve}; 77_{eight}; 66_{seven}; 44_{five}; 22_{three}; 11_{two}

8. A numeral with an even base is even if the ones digit is even, and is odd if the ones digit is odd. Can you justify the above statement? If the numeral has an odd base then probably the safest method is to change the numeral to base ten to check its evenness or oddness, particularly if the numeral contains digits other than ones and zeros.

9. (a) $12_{five} + 21_{three} = 12_{five} + 12_{five} = 24_{five}$

or, $= 21_{three} + 21_{three} = 112_{three}$

(d) $T48_{twelve} + 167_{eight} = T48_{twelve} + 119_{ten}$

$= T48_{twelve} + 9E_{twelve} = E27_{twelve}$

or, $= 2730_{eight} + 167_{eight} = 3117_{eight}$

10.

×	0	1	2	3	4
0	0	0	0	0	0
1	0	1	2	3	4
2	0	2	4	11	13
3	0	3	11	14	22
4	0	4	13	22	31

(a) 42_{five}
(b) 441_{five}
(c) 41_{five}
(d) 434_{five}
(e) 2242_{five}

(f)
$$\begin{array}{rl} 104_{five} & \\ \underline{33_{five}} & \\ 22 & \ldots . 3 \times 4 = 22_{five} \\ 300 & \ldots . 3 \times (1 \times 100) = 300_{five} \\ 220 & \ldots . (3 \times 10) \times 4 = 220_{five} \\ \underline{3000} & \ldots . (3 \times 10) \times (1 \times 100) = 3000_{five} \\ 4042_{five} & \end{array}$$

Chapter 4 *Exercise 4.3 Page 81*

4. (a) $\overleftrightarrow{AX}, \overleftrightarrow{AB}, \overleftrightarrow{XI}, \overleftrightarrow{XA}, \overleftrightarrow{BA}, \overleftrightarrow{BX}$ (b) 6 (c) 2, 12, 20

HINT: In making a generalization the data may be arranged in a grid. The grid is filled in by matching the elements along the vertical with those along the horizontal. See Figure A–1, then the completed Figure A–2.

is matched with	*A*	*X*	*B*
A			
X			
B			

Figure A–1

is matched with	*A*	*X*	*B*
A	*AA*	*AX*	*AB*
X	*XA*	*XX*	*XB*
B	*BA*	*BX*	*BB*

Figure A–2

One could say that the number of different names is $n \times n$, where n is the number of points named. However, names as *AA*, *XX*, and *BB* found in the grid have no meaning as names for lines. Hence, the number of elements along the diagonal must be subtracted from the total number of elements in the grid. So the generalization becomes

$$(n \times n) - n \quad \text{which may be written} \quad n \times (n - 1).$$

Exercise 4.4 Page 84

4. (e)

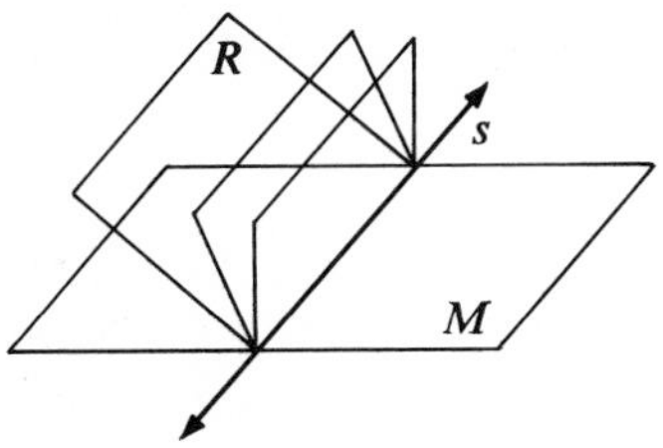

5. (a) The intersection is the plane, an infinite set.
   (b) The intersection is the line, an infinite set.
6. All planes either intersect or they do not intersect. The intersection of planes that intersect is described as nonempty; the intersection of planes that do not intersect is described as empty.

Exercise 4.5 Page 89

1. (b)

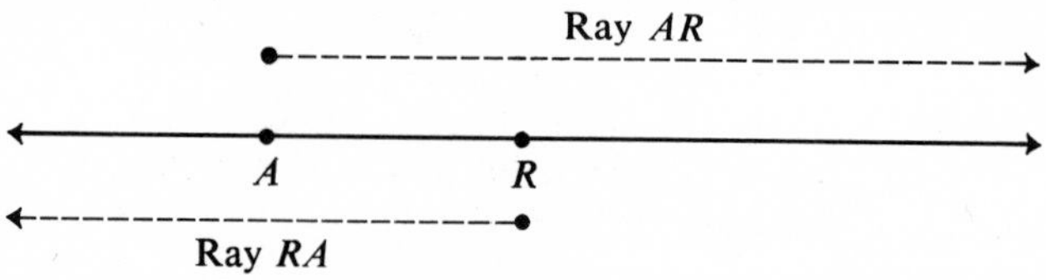

It is clear from the figure that $\overrightarrow{AR}$ and $\overrightarrow{RA}$ do not name the same set of points.

(e) Some of the points of $\overline{AR}$—i.e., those of $\overline{AB}$—are not contained in the set of points in $\overrightarrow{BQ}$.

(i) False. What should replace $\overline{QR}$?

2. (e) Refer to part (c) of this problem and investigate the evenness or oddness of the number of trips.

3. (c) The intersection of $\overline{AD}$ and $\overleftrightarrow{XY}$ is empty; $\overline{AD}$ and $\overleftrightarrow{XY}$ do not intersect.

(e) A and C are on opposite sides of $\overleftrightarrow{XY}$ since $\overline{AC} \cap \overleftrightarrow{XY}$ is nonempty; $\overleftrightarrow{XY}$ separates plane M into two half planes, one containing point A, the other containing point C.

4. (a) The names of the rays may vary and in some cases other rays may be selected.

(3) $\overrightarrow{RQ} \cap \overrightarrow{RT}$

(5) $\overrightarrow{PR} \cup \overrightarrow{QT}$ or $\overrightarrow{QP} \cup \overrightarrow{RT}$

(b) The line segments selected will vary.

(2) $\overline{PR} \cap \overline{QT} = \overline{QR}$

5. (g) Half line on the A-side of P.

Notice this is not $\overrightarrow{PA}$ since the point P would be included in $\overrightarrow{PB}$.

Exercise 4.6 Page 93

1. $\angle APB$ is the same as $\angle BPA$. There are three others. Definition 4.6 does not admit straight angles.

2. See the definitions of exterior and interior of an angle.

3.

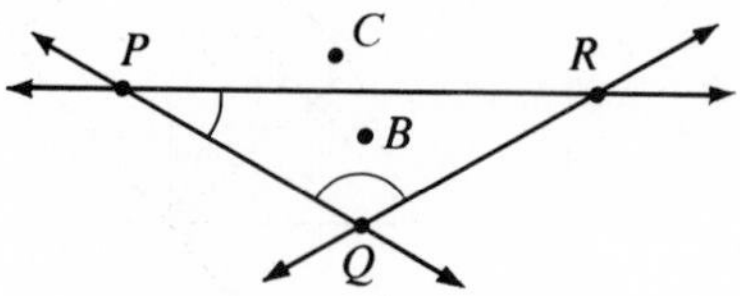

4. (c) $\angle ABC \cap \angle RST$ is the set consisting of the points P and Q.

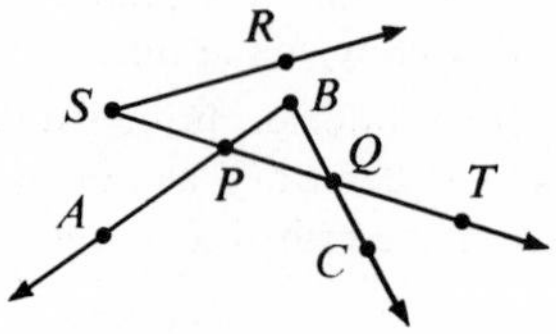

(d) $\angle ABC \cap \angle CPB$ is the set of points consisting of B, D, and C.

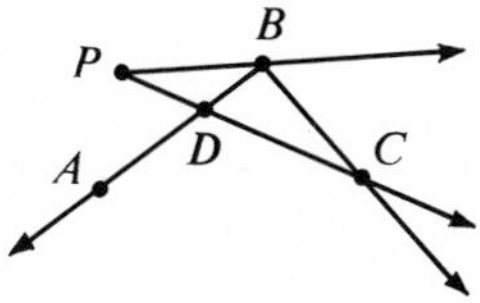

(g) $\angle ABC \cap \angle DCB$ is $\overline{BC}$.

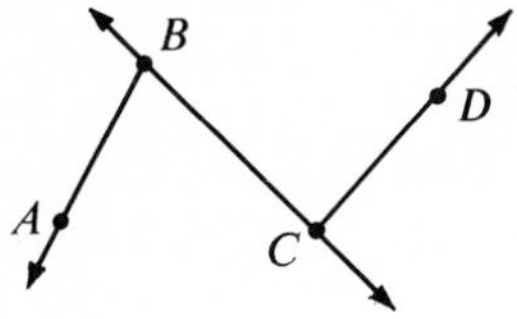

5. (d) $\overline{BC} \cup \angle RBC = \angle RBC$
   (f) $\overline{BC} \cap \angle RBC = \overline{BC}$
   (i) By definition this is not an angle.
   (m) $\varnothing$

6. (d) Half line on the C-side of Y, which is a subset of $\overrightarrow{BY}$.
   (e) The simple closed curve BXY.

Exercise 4.7 Page 96

2. The shaded portion represents the intersection of the exterior of curve a and the interior of curve b.
3. (d) $\varnothing$. Why not point B?
4. (a) In the first drawing, the curves n and p are identical.
   (b)

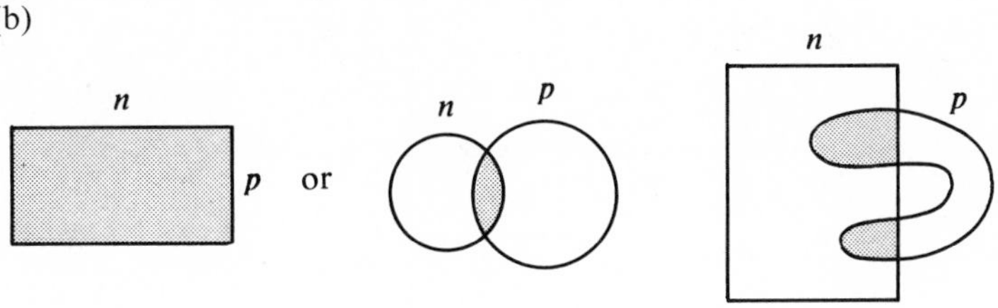

5. 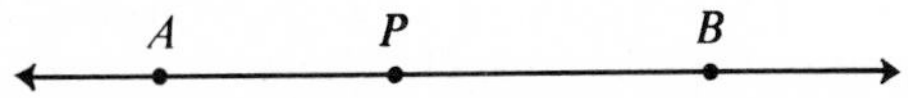

6. (a) Six
   (g) Exterior of curve ABC

Exercise 4.8 Page 105

1. (a) Any point in t, the line of symmetry.
   (b) Point Q itself.
   (c) Point R itself.
   (d) There is no such point. Under translation every point of the plane has one and only one image.

2. (a) Z
(b) E
(c) O

3.

Any point of the line will serve.

4. Any line perpendicular to XY may be used.
5. Any point P in the plane may be used for symmetry in P and the point P will be its own image. Rotation about P also leaves P unchanged.
6. A translation, a rotation about a point, and symmetry with respect to a line each map a set of points onto a congruent set of points.
9. (a) True
(b) False. This would be true only for rotational symmetry of $\frac{1}{2}$ turn.
(c) True
(d) True

10. (a) (b) (d) (g)

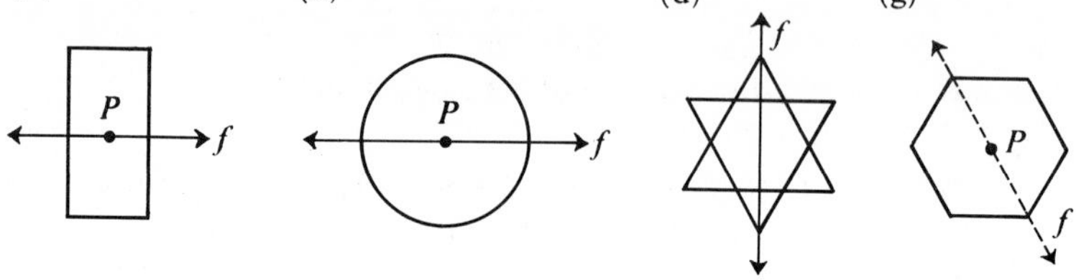

11. To make these problems clearer, cut out a paper model of triangle XYZ and try rotating and reflecting in order to achieve the desired position.

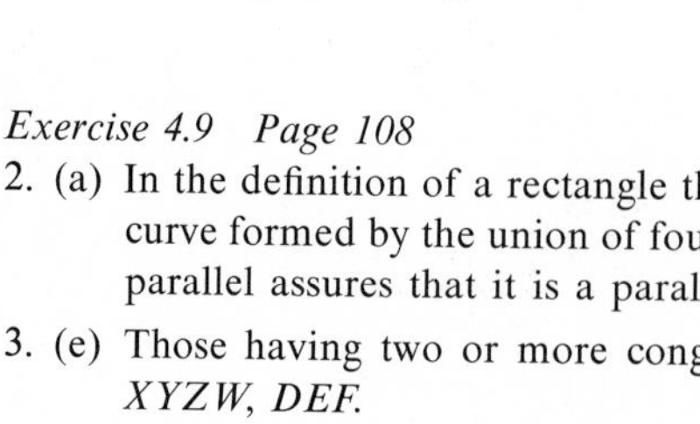

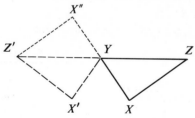

For Figure 4–22:
Step 1: Rotate about Y, $\frac{1}{2}$ turn
Step 2: Reflect through side YZ'.

Exercise 4.9 Page 108

2. (a) In the definition of a rectangle the conditions that it is a simple closed curve formed by the union of four line segments with the opposite sides parallel assures that it is a parallelogram.
3. (e) Those having two or more congruent sides are $ABCD$, XYZ, $PQRS$, $XYZW$, DEF.
5. Line f, the line of symmetry, bisects sides $\overline{AD}$ and $\overline{BC}$. This may be used to show that $\overline{AB} \cong \overline{DC}$.

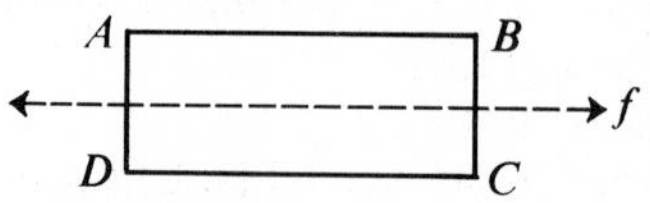

The line f is an extension of $\overline{BM}$ where M is the midpoint of $\overline{AC}$. This line of symmetry may be used to show that $\overline{AB} \cong \overline{BC}$ and also $\angle A \cong \angle C$.

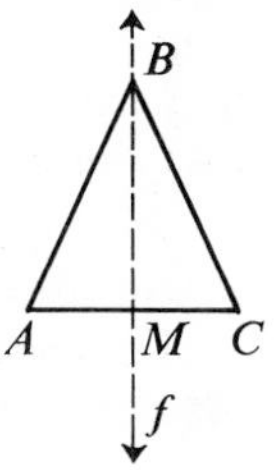

6. Try the midpoint of $\overline{AB}$.
7. The point of rotation, P, is the intersection of the three diagonals that divide the hexagon into six congruent triangles. Rotation through $\frac{1}{6}$ (360°) or 60° will show the sides congruent.

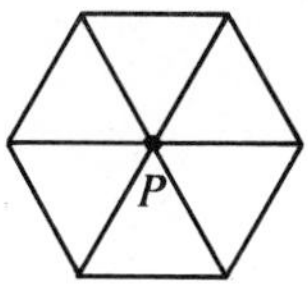

Exercise 4.10 Page 111

2. (a) This statement is false since a triangle is not a convex set, while its region is a convex set. If the statement were changed to read: A triangle and its region *is* a convex set, it would have been true.
4. (a) A ——————— B, a line segment.
   (b) A  B, a semicircle
5. Yes. The intersection contains points that belong to both sets. A line joining any two points of their intersection is a subset of each set by hypothesis. Therefore, this line is a subset of their intersection, and their intersection is convex.
6. Not always. The explanation is left to the student.

Chapter 5 *Exercise 5.2 Page 116*

3. Notice in Definition 5.2 the requirement, $A \cap B = \varnothing$. Has this condition been satisfied here?
5. Refer to your answer in Problem 3.
6. This is an onto mapping. It is not reversible. The domain is the set of ordered pairs of natural numbers. The set of images is the set of natural numbers.

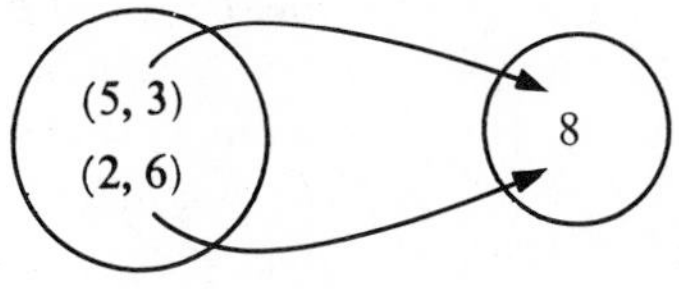

Exercise 5.3 Page 117

1. (f) $(7, 6) \xrightarrow{+} 13$ (g) $(2, 8) \xrightarrow{+} 10$ (h) $(8, 2) \xrightarrow{+} 10$
2. The numbers that match the ordered pair (b, b), $b < 10$, are located on the diagonal from the upper left-hand corner to the lower right-hand corner.
3. The numbers that match the ordered pair $(b, b + 1)$, $b < 9$, are found immediately to the right of the numbers matched with the ordered pair (b, b), $b < 9$.
4. (a) (2, 9), (3, 8), (4, 7), (5, 6), (6, 5), (7, 4), (8, 3), (9, 2).
6. $(b, a) \xrightarrow{+} n$. Refer to Section 4.8 for a discussion of symmetry with respect to a line.

Exercise 5.4 Page 121

1. (b) Associative property for addition.
   (c) Commutative property for addition.
2. 45. There are 9 numbers in the first row. By the commutative property for addition $1 + 2 = 2 + 1$. Therefore, delete this number in the second row; there are now 8 number facts to be learned in the second row. Continue in this manner to get $9 + 8 + 7 + 6 + 5 + 4 + 3 + 2 + 1 = 45$. (Is there an easy way to add these numbers?)
3. (c)

$$
\begin{aligned}
7 + 6 &= (3 + 4) + 6 && \textit{Renaming } 7 \textit{ from the } + \textit{ grid} \\
&= 3 + (4 + 6) && \textit{Associative property for addition} \\
&= 3 + 10 && \textit{Renaming } (4 + 6) \textit{ from the } + \textit{ grid} \\
&= 13 && \textit{Renaming } 3 + 10 \textit{ by decimal notation}
\end{aligned}
$$

   Notice that in these problems it is possible, by using the properties already discussed, to rearrange so that 10 appears as one of the addends. This makes addition much easier.
4. The average of two numbers is one-half their sum. For example, $\frac{1}{2}(3 + 5) = \frac{1}{2}(8) = 4$. The average of a pair of odd numbers is an even number. There are no even numbers in *P*. Notice the requirements for an operation as listed in Section 5.4.

Exercise 5.5 Page 123

1. (a)

	s	t
n	(n, s)	(n, t)
p	(p, s)	(p, t)
r	(r, s)	(r, t)

   (d) Yes. The common elements in the Cartesian product of $A \times B$ and $B \times A$ occur if and only if the sets A and B are not disjoint. The student should verify this statement with examples.
2. (a) $A \times N = \{(a, 1), (a, 2), (a, 3), (b, 1), (b, 2), (b, 3), (c, 1), (c, 2), (c, 3)\}$
   (b) $N \times A = \{(1, a), (1, b), (1, c), (2, a), (2, b), (2, c), (3, a), (3, b), (3, c)\}$
   (c) Yes. Reverse the components of the ordered pairs.

4. (a) 1. The element is (a, x).
   (b) 6. The elements are $(a, x), (a, y), (a, z), (b, x), (b, y), (b, z)$.
   (c) 12. You name them.
   (d) Yes. Find the product of the number of elements in each of the sets.

Exercise 5.6 Page 125

1. (c) No. The components of the ordered pairs are reversed.
   (d) Yes. From parts (a) and (b) above, the number of elements in $A \text{ X } B$ and $B \text{ X } A$ is the same.

3. (a) $N(R \text{ X } V) \times N(P) = 6 \times 3 = 18$
   $N(R) \times N(V \text{ X } P) = 2 \times 9 = 18$

4. 12 ways.

	yellow	*white*	*blue*	*pink*
vanilla	•	•	•	•
chocolate	•	•	•	•
strawberry	•	•	•	•

6. HINT: Refer to your solution of Problem 6 in Exercise 5.2.

Exercise 5.7 Page 126

The questions in this exercise have a close relationship to those in Exercise 5.3. The student should clarify the similarities and differences in the properties of the two operations, addition and multiplication.

Exercise 5.8 Page 131

1. (a) Commutative property for multiplication
   (b) Identity element for multiplication
   (c) Associative property for addition
   (d) Associative property for multiplication
   (g) Distributive property for multiplication over addition
   (o) Commutative property for addition

2. (c) 14×3

$(10 + 4) \times 3$	*Renaming* 14
$(10 \times 3) + (4 \times 3)$	*Distributive property for multiplication over addition*
$(3 \times 10) + (4 \times 3)$	*Commutative property for multiplication*
$30 + (4 \times 3)$	*Renaming* (3×10)
$30 + 12$	*Renaming* (4×3)
42	*Renaming* $(30 + 12)$

   (f) $(50 \times 57) \times 2$

$(57 \times 50) \times 2$	*Commutative property for multiplication*
$57 \times (50 \times 2)$	*Associative property for multiplication*
57×100	*Renaming* (50×2)
5700	*Renaming* (57×100) *by place value notation*

(g) $55 + (92 + 45)$
$55 + (45 + 92)$ — *Commutative property for addition*
$(55 + 45) + 92$ — *Associative property for addition*
$100 + 92$ — *Renaming* $(55 + 45)$
192 — *Renaming* $(100 + 92)$ *by place value notation*

4. (a) Let $A = \{a, b, c\}$ and $B = \{d\}$

$$A \cup B = \{a, b, c\} \cup \{d\} = \{a, b, c, d\}$$
$$\mathrm{N}\{a, b, c\} + \mathrm{N}\{d\} = \mathrm{N}\{a, b, c, d\}$$
$$3 + 1 = 4$$

6. (a) See Problem 2, Exercise 5.4.
   (b) See Problem 5, Exercise 5.3.

8. Closure with respect to the named operation will be assumed if a counterexample cannot be found.
   (a) $\{1, 2, 3\}$ is not closed with respect to addition since $2 + 3 = 5$, and 5 is not an element in the set. $\{1, 2, 3\}$ is not closed with respect to multiplication since $2 \times 3 = 6$, and 6 is not in the set.
   (b) $\{1, 3, 5, 7, \ldots\}$ is not closed with respect to addition. Why? $\{1, 3, 5, 7, \ldots\}$ is closed with respect to multiplication.
   (c) This set is closed with respect to both addition and multiplication.
   (d) Closed with respect to multiplication.
   (e) Not closed with respect to addition or multiplication.
   (f) Closed with respect to multiplication.

10. (a) $B = \{2, 4, 6, 8, \ldots, 2k, \ldots\}, k \in N$
    Let any two elements of B be $2p$ and $2s$. Then $(2p)(2s) = 2[p(2s)]$ by the associative property for multiplication. But $2s$ is an element of N and p is an element of N. Therefore, since N has closure with respect to multiplication, $p(2s)$ is an element of N. Let $p(2s) = t$. Then $(2p)(2s) = 2t$ is an element of B. Thus, the product of any two elements of B is an element of B, and hence, set B has closure for multiplication.
    (b) $C = \{3, 5, 7, \ldots\}$ $\quad 3 \in C$ and $5 \in C$ $\quad 3 + 5 = 8$ $\quad 8 \notin C$
    Therefore, from proof by counterexample the set C does not have closure for addition.

11. The student is familiar with examples for the property: "multiplication distributive over addition." With this as a model write an example to be interpreted as "addition distributive over multiplication." If the latter example is not true, then this constitutes a counterexample.

Chapter 6 *Exercise 6.4 Page 139*

3. (a) $(3, b) \xrightarrow{+} 5$. If you have filled in the grid in Section 6.3, you may look in row 3 until you find 5. This occurs in the column headed by 2. Hence, $(3, 2) \xrightarrow{+} 5$ and $b = 2$.
   (e) and (f) are not possible in W.

Exercise 6.5 Page 141

1. (a) (7) $2 + z = 14$ $\qquad$ (8) $10 + x = 9$
   (b) (7) $z = 14 - 2$ has the solution $z = 12$.
   (8) $x = 9 - 10$ has no solution in W, since $10 \not\leq 9$.
2. (a) (3) If $b + 4 = 9$, then $9 - 4 = b$.

(b) To solve the subtraction problem, $9 - 4 = b$, look in the addition grid in row 4 for the sum, 9. Then look at the head of this column for the number that added to 4 equals 9. The answer is 5.

Exercise 6.8 Page 144

3. $A = \{r, s, t\}$, $B = \{7, 9\}$, $C = \{k, l\}$

$N[(A \text{ X } B) \text{ X } C] \stackrel{?}{=} N[A \text{ X } (B \text{ X } C)]$

It is not suggested that the associative property holds for the Cartesian products. In fact, it does not. It is the *numbers* of the sets which are involved. $N(A) = 3$, $N(B) = 2$, $N(C) = 2$. Use Definition 7.6 to establish that:

$$\begin{aligned} N(A \text{ X } B) &= N(A) \times N(B) \\ &= 3 \times 2 \\ &= 6 \end{aligned}$$

And
$$\begin{aligned} N[(A \text{ X } B) \text{ X } C] &= N(A \text{ X } B) \times N(C) \\ &= 6 \times 2 \\ &= 12 \end{aligned}$$

Likewise,
$$\begin{aligned} N(B \text{ X } C) &= N(B) \times N(C) \\ &= 2 \times 2 \\ &= 4 \end{aligned}$$

and
$$\begin{aligned} N[\text{A X } (B \text{ X } C)] &= N(A) \times N(B \text{ X } C) \\ &= 3 \times 4 \\ &= 12 \end{aligned}$$

Since the results are equal, 12 in each case, we say that the associative property for multiplication has been verified in this particular instance.

It might be of interest for the student to carry out the Cartesian product where three sets are involved. He should notice that the resulting set has elements that are ordered pairs, one of whose components is itself an ordered pair. For example, one element might be $[(r, 7), k]$.

5. $7 \times 7 = 40$

$7 \times 8 = 7 \times (7 + 1)$	*Renaming* 8
$= (7 \times 7) + (7 \times 1)$	*Distributive property for multiplication over addition*
$= (7 \times 7) + 7$	*Multiplicative identity*
$= 49 + 7$	*Multiplication fact already known*
$= 56$	*Renaming* (49 + 7)

6. (c) Distributive property for multiplication over addition.

8. (c) No number in *W*. (d) Any number in *W*.
   (h) No number in *W*.

Exercise 6.9 Page 147

1. (a) (2) $5 \times n = 35$ (7) $12 \times p = 2$
   (9) $0 \times q = 15$
   (b) (2) $n = 7$ (7) Has no solution in *W*.
   (9) Has no solution in *W*.

2. (a) (3) Because multiplication in *W* is commutative, it is possible to rewrite the problem to conform to Definition 6.9 and say:

$$\text{If } 6 \times b = 42, \text{ then } 42 \div 6 = b.$$

However, it is also correct to say:

$$\text{If } b \times 6 = 42, \text{ then } 42 \div b = 6.$$

In other words, either factor can be used as the divisor and the other as the quotient.

(4) If $3 \times 4 = a$, then $a \div 3 = 4$.

Exercise 6.11 *Page 152*

2. (f) $17 \nless 25$ is false. The symbol should be $<$.

(n) $(5 + 7) \leq (3 \times 4)$ is true. The symbol is for "less than or equal to" and in this case "equal to" is the proper statement.

(o) $(3 \times 0) \ngtr (10 \times 0)$. This is true. They are equal.

3. (c) Since $6 < 8$, then

$$6 \times 2 < 8 \times 2.$$

(d) $6 \times b < 8 \times b$, where b is a whole number, holds for all whole numbers except $b = 0$.

$$6 \times 0 = 8 \times 0.$$

5. Given $7 \ngtr 7$. Hence the reflexive property fails. $7 > 4$, but $4 \ngtr 7$. Hence the symmetric property fails. $7 > 4$ and $4 > 2$ leads to $7 > 2$, and the transitive property holds.

The $>$ relation is neither reflexive nor symmetric, and hence is not an equivalence relation.

Exercise 6.13 *Page 157*

1. (c) $(6\square + 3) \leq 15$

2. (d) Some number is greater than 5 and less than 8.

3. (b) $a < 6, a \in N$: $\{1, 2, 3, 4, 5\}$

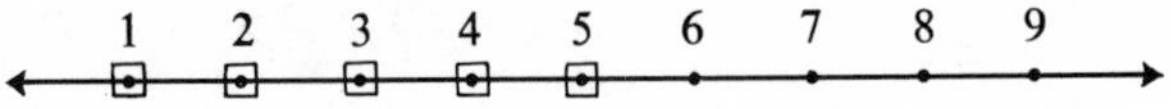

(c) $a \leq 6, a \in W$: $\{0, 1, 2, 3, 4, 5, 6\}$

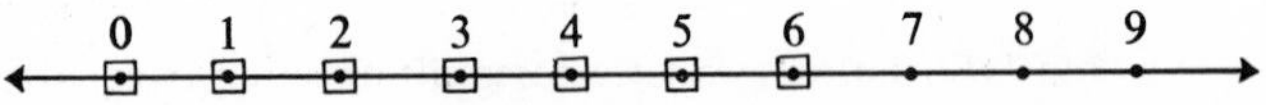

Contrast (b) and (c) and account for the difference in the results.

(h) $\{0, 1, 2, 3, \ldots\}$

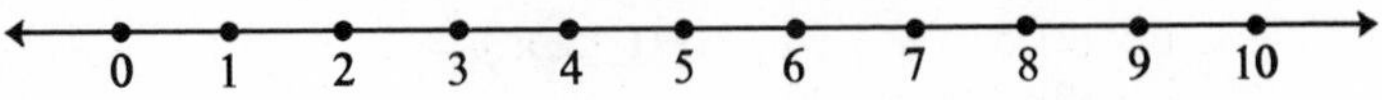

(i) $\{\ \}$

(j) $\{7\}$

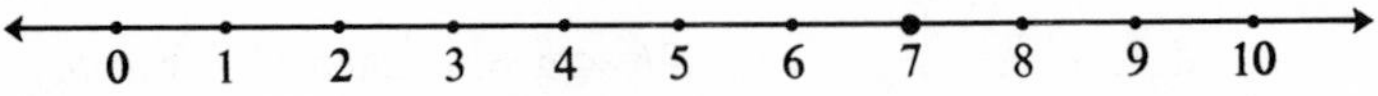

(m) $\{4, 3, 2, 1, 0\} \cap \{3, 4, 5, 6, \ldots\} = \{3, 4\}$

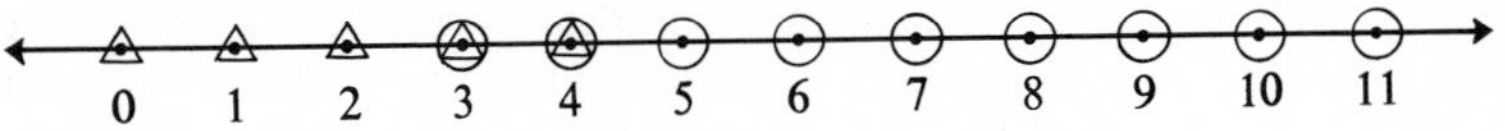

5. (a) The solution sets of (a) and (b) are subsets of the solution set of (c), and the solution set of (b) is a subset of the solution set of (a).
   (e) No. Why?

Chapter 7

Exercise 7.1 Page 160

1. (d) Consider 1 as a possibility.
2. There are six factors.
4. (b) $M_2 = \{0, 2, 4, 6, \ldots\}$
6. (a) $M_{10} = \{0, 10, 20, 30, 40\}$
   (b) $F_{10} = \{1, 2, 5, 10\}$
   (c) There are four factors of 10, but the number of multiples of 10 is infinite.
7. (a) $F_{24} = \{1, 2, 3, 4, 6, 8, 12, 24\}$
   $F_{30} = \{1, 2, 3, 5, 6, 10, 15, 30\}$
   $F_{35} = \{1, 5, 7, 35\}$
   (b) $F_{24} \cap F_{30} = \{1, 2, 3, 6\}$; $F_{30} \cap F_{35} = \{1, 5\}$
   (c) 6 is the greatest element in $F_{24} \cap F_{30}$. 5 is the greatest element in $F_{30} \cap F_{35}$. 1 is the least element in both intersections.
   (d) The elements belong to (or are common to) both of the sets.
8. (b) $M_2 \cap M_3 = \{0, 6, 12, \ldots\}$, which is the same as M_6.
9. (a) $F_{28} = \{1, 2, 4, 7, 14, 28\}$ (b) $1 + 2 + 4 + 7 + 14 = 28$; yes.
   (c) $1 + 2 + 3 = 6$ implies that 6 is a perfect number.
   (d) There is no other perfect number less than 50. A formula for finding perfect numbers is $2^{p-1}(2^p - 1)$, where p and $(2^p - 1)$ are prime numbers. Some students may wish to discover even perfect numbers by using the above formula. It is of interest that no odd perfect numbers have been discovered. This is an unsolved problem in mathematics.

Exercise 7.2 Page 163

1. Extend the sieve in Section 7.2 to include 99.
   (a) There are 25 primes less than 100.
   (b) There are 8 such pairs. Two of them are (11, 13) and (59, 61).
2. There are 3 elements in the set of prime factors of 30 and 4 in the set of composite factors. But in Problem 7(a) of Exercise 7.1, F_{30} listed 8 elements. What has been omitted and why?
3. (a) Two is the only prime number in the set; all others are composite, except 0. See Definition 7.2.
   (e) Find the sum of a number of whole numbers in order to discover the generalization.
   (1) The sum of two even numbers is an even number.
   (2) The sum of two odd numbers is an even number.
   (3) The sum of an even and an odd number is an odd number.
   (f) Proof for (3)
   Given: An even number $2a$ and an odd number $2b + 1$, $a, b \in W$.

Prove: $2a + (2b + 1)$ is an odd number.
Proof:

$2a + (2b + 1) = (2a + 2b) + 1$	*Associative property for addition*
$= 2(a + b) + 1$	*Distributive property*
$(a + b) \in W$	*Closure for addition in W*
$2(a + b) + 1$ is an odd number.	*Problem* 3(d)

Hence, $2a + (2b + 1)$ is an odd number.

5. (b) What about the even number 2?

Exercise 7.3 Page 167

1. (a)
   42
   6×7
   $2 \times 3 \times 7$

   $42 = 2 \times 3 \times 7$ is the prime factorization of 42.
   $F_{42} = \{1, 2, 3, 7, (2 \times 3), (2 \times 7), (3 \times 7), (2 \times 3 \times 7)\}$
   $= \{1, 2, 3, 7, 6, 14, 21, 42\}$

   Notice that 1 is an element in the set of all factors of every number.

   (f) $120 = 2^3 \times 3 \times 5$
   $F_{120} = \{1, 2, 3, 4, 5, 6, 8, 10, 12, 15, 20, 24, 30, 40, 60, 120\}$
3. (e) 31_{five} represents an even number, since $31_{five} = (3 \times 5^1) + (1 \times 5^0) = 15 + 1 = 16_{ten}$.
   (n) 122_{three} represents an odd number, since $122_{three} = (1 \times 3^2) + (2 \times 3^1) + 2 = 9 + 6 + 2 = 17_{ten}$.
4. $334_{six} = (3 \times 6^2) + (3 \times 6^1) + (4 \times 6^0)$
   $= 108 + 18 + 4 = 130_{ten}$
   $334_{six} = 130_{ten} = 2_{ten} \times 5_{ten} \times 13_{ten}$
   $= 2_{six} \times 5_{six} \times 21_{six}$
   $F_{130} = \{1, 2, 5, 10, 13, 26, 65, 130\}$
   $F_{334_{six}} = \{1_{six}, 2_{six}, 5_{six}, 14_{six}, 21_{six}, 42_{six}, 145_{six}, 334_{six}\}$
8. HINT: Only the prime numbers have exactly two factors.
9. (a) $b = 2^5 = 32$; $F_{32} = \{1, 2, 2^2, 2^3, 2^4, 2^5\} = \{1, 2, 4, 8, 16, 32\}$
   (b) $b = p^4$; $F_{p^4} = \{1, p, p^2, p^3, p^4\}$; p^4 has $(4 + 1)$ factors.
10. (a) For all primes greater than 2, the sum of two primes is always an even number.
    (f) False; 1 is a natural number, but is neither prime nor composite. This statement should begin: "All natural numbers greater than 1. . . ." The selection of many true examples does not prove that a statement is true. It takes only one counterexample to prove a statement is not true, however.

Exercise 7.4 Page 169

1. (a) $F_8 = \{1, 2, 4, 8\}$; $F_{12} = \{1, 2, 3, 4, 6, 12\}$; $F_{16} = \{1, 2, 4, 8, 16\}$
   (b) $F_8 \cap F_{12} \cap F_{16} = \{1, 2, 4\}$
   (c) gcf $\{8, 12, 16\} = 4$
2. (c) $630 = 2 \times 3^2 \times 5 \times 7$; $60 = 2^2 \times 3 \times 5$; $600 = 2^3 \times 3 \times 5^2$; $720 = 2^4 \times 2^4 \times 5 \times 3^2$; gcf $\{630, 60, 600, 720\} = 2 \times 3 \times 5 = 30$
   (d) $65 = 5 \times 13$; $84 = 2^2 \times 3 \times 7$; gcf $\{65, 84\} = 1$. Since there are no prime factors common to 65 and 84, they are said to be relatively prime.
4. gcf $\{1, m\}$, $m \in W$, is 1. By Definition 7.4, the greatest common factor is defined only for (whole) numbers greater than 0. Hence, the gcf $\{1, 0\}$ is undefined.

Exercise 7.5 Page 171

1. (a) M_2 less than 50 = {0, 2, 4, 6, 8, 10, . . . , 24, 26, . . . , 46, 48}
M_3 less than 50 = {0, 3, 6, 9, 12, 15, . . . , 42, 45, 48}
M_5 less than 50 = {0, 5, 10, 15, 20, 25, 30, 35, 40, 45}
M_6 less than 50 = {0, 6, 12, 18, 24, 30, 36, 42, 48}
M_{10} less than 50 = {0, 10, 20, 30, 40}

(b) $M_2 \cap M_3 \cap M_6 = \{0, 6, 12, 18, 24, 30, 36, 42, 48\}$

$\text{lcm}\ \{2, 3, 6\} = 6$
$M_2 \cap M_5 \cap M_{10} = \{0, 10, 20, 30, 40\}$
$\text{lcm}\ \{2, 5, 10\} = 10$

(c) The least common multiple of prime numbers is the product of the prime numbers.

2. (e) $15 = 3 \times 5$, $12 = 2^2 \times 3$; $\text{lcm}\ \{15, 12\} = 3 \times 5 \times 2^2 = 60$.
(h) $35 = 5 \times 7$, $32 = 2^5$; $\text{lcm}\ \{35, 32\} = 5 \times 7 \times 2^5 = 1120$.
(i) $45 = 3^2 \times 5$, $15 = 3 \times 5$, $10 = 2 \times 5$;

$$\text{lcm}\ \{45, 15, 10\} = 3^2 \times 5 \times 2 = 90.$$

3. (a) Yes, provided the primes are each distinct prime numbers—that is, no one prime occurs more than once.
(e) True; the set of multiples of b is a subset of the set of multiples of 1 (a set consisting of all whole numbers). Hence, the least element in the intersection is b.
(f) False; $\text{lcm}\ \{3, 5\} = 15$, which is neither 3 nor 5.

6. Yes. The proof follows:

Let $m = a(\text{gcf}\ \{m, n\})$; $n = b(\text{gcf}\ \{m, n\})$ for $a, b \in W$. Then,

$m \times n = a(\text{gcf}\ \{m, n\}) \times b(\text{gcf}\ \{m, n\})$ — *Renaming m and n*
$= a \times b \times (\text{gcf}\ \{m, n\}) \times (\text{gcf}\ \{m, n\})$ — *Commutative property for multiplication*

But $\text{lcm}\ \{m, n\} = a \times b \times (\text{gcf}\ \{m, n\})$ — *Definition of* lcm
Thus, $m \times n = (\text{lcm}\ \{m, n\}) \times (\text{gcf}\ \{m, n\})$ — *Renaming* $[a \times b \times (\text{gcf}\ \{m, n\})]$

Exercise 7.6 Page 187

2. The sequence of steps carried out in justifying each of the following may vary.

(a) 56×7
$= [(5 \times 10) + (6 \times 1)] \times 7$ — *Renaming 56 by expanded notation*
$= [(5 \times 10) \times 7] + [(6 \times 1) \times 7]$ — *Distributive property*
$= [7 \times (5 \times 10)] + [7 \times (6 \times 1)]$ — *Commutative property for multiplication*
$= [(7 \times 5) \times 10] + [7 \times (6 \times 1)]$ — *Associative property for multiplication*
$= (35 \times 10) + (7 \times 6)$ — *Renaming (7 × 5) and the multiplicative identity element*
$= [(30 + 5) \times 10] + (7 \times 6)$ — *Renaming 35 by expanded notation*

$= [(30 + 5) \times 10] + 42$	*Renaming* (7×6)
$= (30 \times 10) + (5 \times 10) + (4 \times 10) + (2 \times 1)$	*Distributive property and renaming* 42 *by expanded notation*
$= (300) + (5 \times 10) + (4 \times 10) + (2 \times 1)$	*Renaming* (30×10)
$= 300 + [(5 + 4) \times 10] + (2 \times 1)$	*Distributive property*
$= (3 \times 100) + (9 \times 10) + (2 \times 1)$	*Renaming* 300 *and* $(5 + 4)$
$= 392$	*Renaming by place value notation*

(b) $35 + 28$

$= [(3 \times 10) + (5 \times 1)] + [(2 \times 10) + (8 \times 1)]$	*Renaming by expanded notation*
$= [(3 \times 10) + (2 \times 10)] + [(5 \times 1) + (8 \times 1)]$	*Associative and commutative properties for addition*
$= [(3 + 2) \times 10] + [(5 + 8) \times 1]$	*Distributive property*
$= [5 \times 10] + [13 \times 1]$	*Renaming* $(3 + 2)$ *and* $(5 + 8)$
$= 50 + 13$	*Renaming* (5×10) *and* (13×1)
$= 50 + (10 + 3)$	*Renaming* 13 *by expanded notation*
$= (50 + 10) + 3$	*Associative property for addition*
$= 60 + 3$	*Renaming* $(50 + 10)$
$= 63$	*Renaming* $(60 + 3)$ *by place value notation*

(c) $79 - 23$

$= (70 + 9) - (20 + 3)$	*Renaming* 79 *and* 23
$= (70 - 20) + (9 - 3)$	*Generalization for subtraction*
$= [(7 \times 10) - (2 \times 10)] + (9 - 3)$	*Renaming* 70 *and* 20
$= [(7 - 2) \times 10] + (9 - 3)$	*Distributive property for multiplication over subtraction*
$= [5 \times 10] + 6$	*Renaming* $(7 - 2)$ *and* $(9 - 3)$
$= 50 + 6$	*Renaming* (5×10)
$= 56$	*Renaming* $(50 + 6)$ *by place value notation*

(d) $249 + 364$

$= [(2 \times 100) + (4 \times 10) + (9 \times 1)] + [(3 \times 100) + (6 \times 10) + (4 \times 1)]$	*Renaming* 249 *and* 364 *by expanded notation*

$= [(2 \times 100) + (3 \times 100)] + [(4 \times 10) + (6 \times 10)] + [(9 \times 1) + (4 \times 1)]$	*Commutative and associative properties for addition*
$= [(2 + 3) \times 100] + [(4 + 6) \times 10] + [(9 + 4) \times 1]$	*Distributive property*
$= (5 \times 100) + (10 \times 10) + (13 \times 1)$	*Renaming* (2 + 3), (4 + 6), *and* (9 + 4)
$= 500 + 100 + 13$	*Renaming what?*
$= 600 + 13$	*Renaming what?*
$= 600 + 10 + 3$	*Renaming what?*
$= 613$	*Renaming by place value notation*

(e) 83×24

$= [(8 \times 10) + (3 \times 1)] \times 24$	*Renaming* 83
$= [(8 \times 10) \times 24] + [(3 \times 1) \times 24]$	*Distributive property*
$= [80 \times (20 + 4)] + [3 \times (20 + 4)]$	*Renaming* 24, (8 × 10), (3 × 1)
$= (80 \times 20) + (80 \times 4) + (3 \times 20) + (3 \times 4)$	*Distributive property*
$= [(8 \times 2) \times (10 \times 10)] + [(8 \times 4) \times 10] + [(3 \times 2) \times 10] + (3 \times 4)$	*Renaming, associative and commutative properties*
$= (16 \times 100) + (32 \times 10) + (6 \times 10) + (3 \times 4)$	*Renaming*
$= (16 \times 100) + [(32 + 6) \times 10] + (3 \times 4)$	*Associative for addition and the distributive property*
$= (16 \times 100) + (38 \times 10) + (10 + 2)$	*Renaming*
$= 1600 + 380 + 10 + 2$	*Renaming*
$= 1992$	*See Problem* 2(d) *above*

4. (a) $4 + 5 = 9$
   (b) $2 \times 3 = 6$
   (c) $n - 5 = 8$ or $n - 8 = 5$
   (e) $n = 46 \div 23$

5. (c)
$$\begin{aligned} 43 - 27 &= (30 + 13) - (20 + 7) \\ &= (30 - 20) + (13 - 7) \\ &= 10 + 6 \\ &= 16 \end{aligned}$$
   (Note that in (c) the generalization requires $d \leq b$; hence 43 is renamed 30 + 13 instead of 40 + 3.)

6. Suggestion: Evaluate the expressions on each side of the = symbol to determine whether the following sentence is true:

$$\begin{aligned} (18 + 6) \div 2 &= (18 \div 2) + (6 \div 2) \\ 24 \div 2 &= 9 + 3 \\ 12 &= 12 \end{aligned}$$

To emphasize that the property must be a "right-hand" distributive property, compare the truth of these sentences:

$$(15 + 10) \div 5 = (15 \div 5) + (10 \div 5)$$
$$25 \div 5 = 3 + 2$$
$$5 = 5$$

$$5 \div (15 + 10) \neq (5 \div 15) + (5 \div 10)$$
$$5 \div 25 \neq (5 \div 15) + (5 \div 10)$$

The generalization that distributes division over addition is a "right-hand" distributive property expressed by:

For each $a, b, c \in W$, $(a + b) \div c = (a \div c) + (b \div c)$.

Using this to find the quotient of $36 \div 3$:

$$\begin{aligned} 36 \div 3 &= (30 + 6) \div 3 \\ &= (30 \div 3) + (6 \div 3) \\ &= 10 + 2 \\ &= 12 \end{aligned}$$

Observe that unless c divides a and b this property is of little use when dividing in the set of whole numbers.

7. Begin with the step in Problem 2(e):

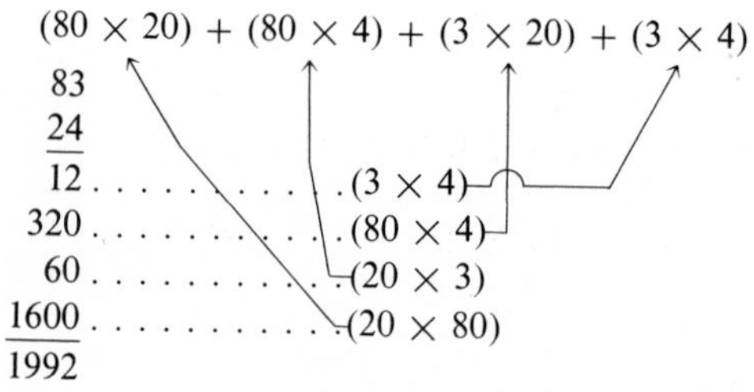

Other steps might be selected. However, this one is probably more comparable to the steps in the usual multiplication algorithm than some of the others.

8. (a) $16 \times (4 \times 10^2) \leq 7001 < 16 \times (5 \times 10^2)$
$16 \times (3 \times 10^1) \leq 601 < 16 \times (4 \times 10^1)$
$16 \times (7 \times 10^0) \leq 121 < 16 \times (8 \times 10^0)$
(d) $7001 = (16 \times 437) + 9$

10. (e) 152) 2584

1520	10
1064	
1064	7
0	17

$2584 = (152 \times 17) + 0$

(f) 203) 307465

203000	1000
104465	
101500	500
2965	
2030	10
935	
812	4
123	1514

$307{,}465 = (203 \times 1514) + 123$

Exercise 7.7 Page 192

1. Refer to the conventional shapes shown near the beginning of this section.
2. Here one must distinguish between input, operation, and output information given in the problem. A suggested sequence is given. Can you construct another one?

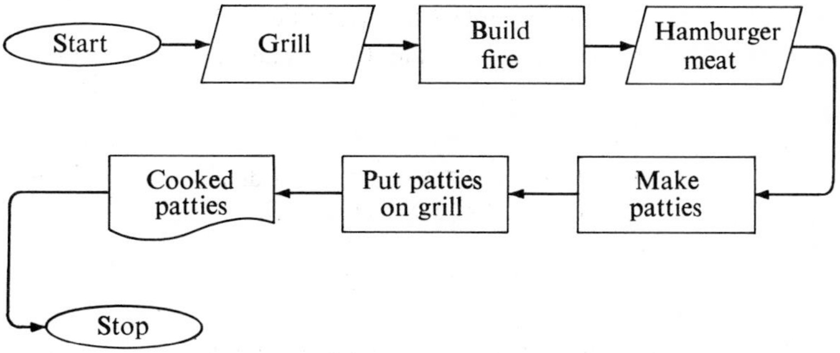

4. (a)

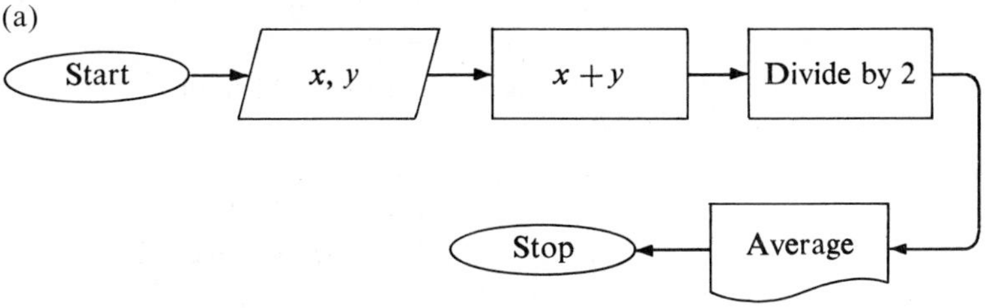

   (b) Alter the preceding flow chart to accommodate the three numbers 96, 81, and 90.

5. (b) The flow chart in Example 7 is given for the addition of two 3-digit numbers. Observe the decision box, "Is $x \geq 2$?" Should the decision box contain "Is $x \geq 3$?" for the addition of two 4-digit numbers?
6. Recall that a trapezoid has exactly two parallel sides, while a parallelogram has two pairs of parallel sides. The decision question should now be obvious.

Chapter 8 *Exercise 8.1 Page 198*

2. (a), (b), (c), (f), (g), (j), (k), and (l). You should be prepared to explain why these are included and others are not.
3. (f) 12 (g) 0
4. Only one ordered pair from W can be matched with 0: the ordered pair (0, 0). There are infinitely many such pairs in I. You should be able to name some.
5. Look at the diagram in the preceding discussion.
6. (e) 5 (f) 0 (h) $^{-}4$

Exercise 8.2 Page 201

4. Follow Example 1 and Definition 8.2(a).

$^{-}21 + 27 = \square$	
$^{-}21 + (21 + 6) = \square$	*Renaming the whole number 27*
$(^{-}21 + 21) + 6 = \square$	*Associative property for addition*
$0 + 6 = \square$	*Additive inverse*
$6 = \square$	*Additive identity*

Hence, $^{-}21 + 27 = 6$.

9. Follow Example 2 and Definition 8.2(b).

23. Follow Example 2 and then Example 3 and Definitions 8.2(b) and 8.2(c).

Exercise 8.3 Page 207

1. (h) $(8 + {}^{-}10) < -({}^{-}6 + {}^{-}3)$. Use definitions in Section 8.2 to rename. They become $^{-}2$ and 9, and $^{-}2 < 9$.
2. (g) $20 > {}^{-}92$
   (l) $({}^{-}2 + {}^{-}6) = {}^{-}(10 + {}^{-}2)$
4. (g) $- - |{}^{-}1| = |{}^{-}1| = 1$
   (k) Since a, b are nonnegative and $b > a$, then $(a + -b) = -(b - a)$. Therefore, $|-(b - a)| = b - a$.
5. (c) $\{{}^{-}4, {}^{-}3, {}^{-}2, {}^{-}1, 0, 1, 2, 3, 4\}$
   (e) Part (c) was the only part for which all the elements could be listed.
6. (a) $\{{}^{-}2, {}^{-}1, 0, 1, 2\}$
   (b) There is no least element in Q.
   (c) $P \cup Q = \{\ldots, {}^{-}4, {}^{-}3, {}^{-}2, {}^{-}1, 0, 1, 2\}$
7. (b) There is no greatest and no least element in set S.
   (d) $R \cup S =$ Set of integers, I.
8. $A = \{\ \}$, $|a|$ must be nonnegative and no nonnegative integer is less than $^{-}2$.

Exercise 8.4 Page 209

2. (c) $9 - 15 = {}^{-}6$, since $15 + {}^{-}6 = 9$.
   (i) $^{-}3 - {}^{-}12 = 9$, since $^{-}12 + 9 = {}^{-}3$.
   (p) $^{-}1 - 18 = {}^{-}19$, since $18 + {}^{-}19 = {}^{-}1$.
3. (a) $5 - 3 = 2$ and $5 + {}^{-}3 = 2$
   (d) $2 - {}^{-}6 = 8$ and $2 + 6 = 8$
   For a and b, nonnegative integers, can you make statements connecting subtraction and addition?

Exercise 8.6 Page 212

1. (c) $15 \times {}^{-}3$
   Use the method of Example 1.

$$\begin{aligned} 15 \times (3 + {}^{-}3) &= (15 \times 3) + (15 \times {}^{-}3) \\ 15 \times 0 &= (15 \times 3) + (15 \times {}^{-}3) \\ 0 &= (15 \times 3) + (15 \times {}^{-}3) \\ 0 &= 45 + (15 \times {}^{-}3) \end{aligned}$$

Then $(15 \times {}^{-}3)$ is the additive inverse of 45. Hence $15 \times {}^{-}3 = {}^{-}45$.

(e) $^-6 \times (8 - 10) = {^-6} \times (8 + {^-10})$ — *Generalization* 8.4 *developed in Problem* 3, *Exercise* 8.4

$= ({^-6} \times 8) + ({^-6} \times {^-10})$ — *Distributive property*

$= {^-48} + 60$ — *Renaming* $({^-6} \times 8)$ *and* $({^-6} \times {^-10})$

$= 12$

Another way:
Rename $(8 - 10)$ as $^-2$ by Definition 8.4 and Definition 8.6(a). Then $^-6 \times (8 - 10)$ becomes $({^-6} \times {^-2})$ or 12.

(f) $({^-6} \times 4) \times 5 = ({^-24} \times 5)$ — *Renaming* $({^-6} \times 4)$

$= {^-120}$ — *Renaming* $({^-24} \times 5)$

Authority for both of the above steps is Definition 8.6(a).

2. (a) Use the procedure in Example 1.
$b \times (1 + -1) = (b \times 1) + (b \times -1)$ — *Distributive property*
Continue as in the example.
(b) See Definition 8.6(b).
(c) Use parts (a) and (b).

3. (f) 6 (g) 20 (i) 0 (j) $^-12$

Exercise 8.8 Page 219

1. $|m| < 4, m \in I.$
If $|m| = 4$, $m \in I$, the graph will be

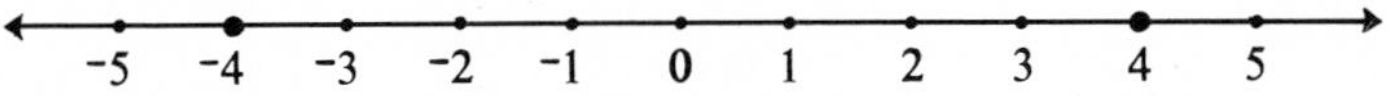

and $m = {^-4}, 4$.
If $|m| < 4$, then $m > {^-4}$ and $m < 4$. This may be represented by:

$$\{m \in I : m > {^-4}\} \cap \{m \in I : m < 4\}$$

Given in roster notation this is:

$$\{{^-3}, {^-2}, {^-1}, 0, 1, 2, 3\}$$

4. No integers satisfy both these conditions and the solution is { }.
5. Look carefully at Definition 8.3(e).
6. $|x + 4| \geq 6, x \in I$ means

$x + 4 \geq 6$ or $x + 4 \leq {^-6}$

$x + 4 \geq 2 + 4$ or $x + 4 \leq {^-10} + 4$ — *Renaming* 6 *and* $^-6$

$x \geq 2$ or $x \leq {^-10}$ — *Cancellation property for addition*

Solution set is $\{x \in I : x \geq 2\} \cup \{x \in I : x \leq {^-10}\}$

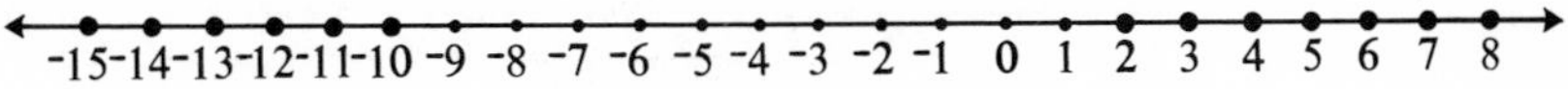

9. Notice that this involves the solution sets of both Problem 6 and Problem 8.

10. $|x| > 3$ means $x > 3$ or $-x > 3$ (that is, $x < {}^-3$). This is written

$$\{x \in I: x > 3\} \cup \{x \in I: x < {}^-3\}.$$

Call this A. The second part is analyzed similarly, that is: $\{x \in I: x > 5\} \cup \{x \in I: x < {}^-5\}$. Call this B. The solution set for the total problem is the intersection of these two solution sets. Showing the two sets on the number line may help you to see that $B \subset A$. Therefore $A \cap B = B$.

11. $\{\ldots, {}^-8, {}^-7, {}^-6, {}^-2, {}^-1, 0, 1, 2, 6, 7, 8, \ldots\}$

Chapter 9 *Exercise 9.2 Page 229*

3. (a) $18 \in I$ and $I \subset Q$.
   (b) $\frac{1}{18}$ is the multiplicative inverse of 18.
   (e) $\frac{{}^-5}{2} = {}^-5 \times \frac{1}{2}$ *Definition* 9.1(b)

4. Follow the patterns set by Examples 1, 2, and 3.
   (c) $8 \times \frac{1}{11} = \frac{8}{11}$ *Definition* 9.1(b)
   (f) Follow Example 1.

Exercise 9.3 Page 234

2. (d)

$$\frac{95}{{}^-7} \div \frac{15}{{}^-21} = \frac{95}{{}^-7} \times \frac{{}^-21}{15} \qquad \textit{Theorem V(b) for multiplicative inverse}$$

$$= \frac{95 \times {}^-21}{{}^-7 \times 15} \qquad \textit{Theorem I}$$

$$= \frac{(19 \times 5) \times ({}^-21)}{({}^-7) \times (3 \times 5)}$$

$$= \frac{(19) \times (5 \times {}^-21)}{({}^-7 \times 3) \times 5}$$

$$= \frac{19}{1} \times \frac{5 \times {}^-21}{({}^-7 \times 3) \times 5}$$

$$= \frac{19}{1} \times 1$$

$$= 19$$

3. (c)

$$\frac{\frac{{}^-19}{15}}{\frac{{}^-2}{3}} = \frac{{}^-19}{15} \div \frac{{}^-2}{3}$$

$$= \frac{{}^-19}{15} \times \frac{3}{{}^-2} \qquad \textit{Theorem V}$$

$$= \frac{({}^-19) \times 3}{15 \times ({}^-2)} \qquad \textit{Theorem I}$$

$$= \frac{({}^-19) \times 3}{(5 \times 3) \times {}^-2}$$

$$= \frac{{}^-19}{5 \times {}^-2} \times \frac{3}{3}$$

$$= \frac{{}^-19}{{}^-10} \times 1$$

$$= \frac{19}{10}$$

4. (c) $\dfrac{\frac{^-19}{15}}{\frac{^-2}{3}} = \dfrac{\frac{^-19}{15} \times \frac{3}{^-2}}{\frac{^-2}{3} \times \frac{3}{^-2}}$

$= \dfrac{\dfrac{^-19 \times 3}{15 \times {^-2}}}{1}$

$= \dfrac{^-19 \times 3}{15 \times {^-2}}$

= ? The rest is left for the student.

Exercise 9.4 Page 239

1. (b) $\left\{\frac{2}{1}, \frac{4}{2}, \frac{6}{3}, \frac{^-10}{^-5}, \frac{^-30}{^-15}, \frac{42}{21}\right\}$

(c) $\left\{\frac{^-21}{13}, \frac{^-42}{26}, \frac{^-63}{39}, \frac{21}{^-13}, \frac{^-105}{65}, \frac{42}{^-26}\right\}$

(d) $\left\{\frac{0}{4}, \frac{0}{^-4}, \frac{0}{8}, \frac{0}{28}, \frac{0}{^-8}, 0\right\}$ Why may 0 be included and $\frac{0}{0}$ may not be included?

3. (a) $\frac{14}{21} = \frac{2 \times 7}{3 \times 7}$ *Renaming* 14 *and* 21.

$= \frac{2}{3} \times \frac{7}{7}$ *Theorem* I

$= \frac{2}{3}$ *Multiplicative identity*

Since 2 and 3 are relatively prime, $\frac{2}{3}$ is the simplest form.

(b) $\frac{13}{17}$ is already in the simplest form since 13 and 17 are relatively prime.

6. There are two of these fractions which are equivalent. Find them and check the others.
7. Find the fraction whose numerator and denominator are relatively prime. Use this to find as many elements of the set as you wish.
8. (e) $\frac{x}{2} = \frac{2}{x}$

$x^2 = 4$; the solution set is $\{^-2, 2\}$. Notice both $^-2$ and 2 satisfy $\frac{x}{2} = \frac{2}{x}$.

(f) $\frac{15 + x}{8} = \frac{5}{2}$

$2(15 + x) = (5 \times 8)$ *Condition for equivalence*

$30 + 2x = 40$ *Renaming and DPMA*

$^-30 + 30 + 2x = 40 + {^-30}$

$(^-30 + 30) + 2x = 10$

$0 + 2x = 10$

$2x = 10$

$x = 5$

The solution set: $\{5\}$

9. For $a, k \in I, k \neq 0$,

$a = a \times 1$ *Multiplicative identity*

Since $\frac{k}{k} = 1$ — *Theorem* II

$a = a \times \frac{k}{k}$ — *Replacing* 1 *with the name* $\frac{k}{k}$

But $a = \frac{a}{1}$ — *Theorem* III

Then $a = \frac{a}{1} \times \frac{k}{k}$ — *Replacing a with* $\frac{a}{1}$

$a = \frac{a \times k}{1 \times k}$ — *Theorem* I

$a = \frac{ak}{k}$ — *Multiplicative identity*

Exercise 9.5 Page 240

1\. (c) $^{-}(\frac{3}{5}) = \frac{^{-}3}{5} = \frac{3}{^{-}5}$. This is a negative rational number.

3\. By agreement (Example 2, Section 9.4) the conditions described in Problem 4, immediately following, are the conditions for $\frac{a}{b}$ in simplest form. (c) $-(\frac{^{-}7}{^{-}9}) = \frac{^{-}7}{9}$, which is a negative rational number in simplest form.

4\. (e)
$$\left(\frac{2}{^{-}3} \times \frac{5}{6}\right) \times \frac{1}{^{-}6} = \frac{2 \times 5}{^{-}3 \times 6} \times \frac{1}{^{-}6}$$
$$= \frac{10}{^{-}18} \times \frac{1}{^{-}6}$$
$$= \frac{10 \times 1}{^{-}18 \times {}^{-}6}$$
$$= \frac{(5 \times 2) \times 1}{^{-}18 \times (^{-}3 \times 2)}$$
$$= \frac{5}{^{-}18 \times {}^{-}3} \times \frac{2}{2}$$
$$= \frac{5}{54} \times 1$$
$$= \frac{5}{54}$$

5\. (c) $a = 7 > 0, b = {}^{-}3 < 0$

$$\frac{-7}{-(^{-}3)} \text{ represents } \frac{-p}{-q},\ p = 7,\ q = {}^{-}3$$
$$= \frac{^{-}1 \times 7}{^{-}1 \times (^{-}3)}$$
$$= \left(\frac{^{-}1}{^{-}1}\right) \times \frac{7}{^{-}3}$$
$$= 1 \times \left(\frac{7}{^{-}3}\right)$$
$$= \frac{7}{^{-}3} \text{ represents } \frac{p}{q}.$$

Exercise 9.6 Page 243

2. (b) $\frac{^-5}{12} + \frac{7}{16} = \left(\frac{^-5}{12} \times \frac{4}{4}\right) + \left(\frac{7}{16} \times \frac{3}{3}\right)$

$$= \frac{^-20}{48} + \frac{21}{48}$$

$$= \frac{^-20 + 21}{48}$$

$$= \frac{1}{48}$$

(e) $\frac{4}{^-5} + \frac{3}{10} = \left(\frac{4}{^-5} \times \frac{^-2}{^-2}\right) + \frac{3}{10}$

$$= \frac{^-8}{10} + \frac{3}{10}$$

$$= \frac{^-8 + 3}{10}$$

$$= \frac{^-5}{10}$$

$$= \frac{^-1}{2}$$

Exercise 9.7 Page 245

1. Using Definition 9.7,

(c) $\frac{4}{15} - \frac{7}{3} = n$ if $\frac{7}{3} + n = \frac{4}{15}$

$\frac{7}{3} \times \frac{5}{5} + n = \frac{4}{15}$

$\frac{35}{15} + n = \frac{4}{15}$

Since $\frac{35}{15} + \frac{^-31}{15} = \frac{4}{15}$, the solution set is $\frac{^-31}{15}$.

2. Using Generalization 9.7:

(d) $\frac{5}{18} - \frac{^-1}{9} = \frac{5}{18} + -(\frac{^-1}{9})$

$= \frac{5}{18} + \frac{2}{18}$

$= \frac{7}{18}$

Exercise 9.8 Page 248

1. (d) $\frac{^-5}{9} < \frac{^-3}{7}$. The common denominator is 9×7.

$$\frac{^-5}{9} = \frac{^-5 \times 7}{9 \times 7} \quad \text{or} \quad \frac{^-35}{63}$$

$$\frac{^-3}{7} = \frac{^-3 \times 9}{7 \times 9} \quad \text{or} \quad \frac{^-27}{63}$$

$$\frac{^-35}{63} + \frac{8}{63} = \frac{^-27}{63}$$

Hence $\frac{^-5}{9} < \frac{^-3}{7}$.

2. (a) $\frac{4}{5} = \frac{16}{20}$ $\qquad$ $\frac{16}{20} + \frac{1}{20} = \frac{17}{20}$

Hence $\frac{4}{5} < \frac{17}{20}$.

(b) $\frac{^-3}{19} = \frac{^-9}{57}$ $\qquad$ $\frac{^-13}{57} + \frac{4}{57} = \frac{^-9}{57}$

Hence $\frac{^-13}{57} < \frac{^-3}{19}$.

3. (a) $\frac{16}{20} = \frac{32}{40}$ $\qquad$ $\frac{17}{20} = \frac{34}{40}$

$\frac{32}{40} < \frac{33}{40} < \frac{34}{40}$. Hence $\frac{33}{40}$ is a rational number between $\frac{4}{5}$ and $\frac{17}{20}$.

4. (a) $\frac{2}{3} = \frac{2 \times 25}{3 \times 25} = \frac{50}{75}$ $\frac{3}{5} = \frac{3 \times 15}{5 \times 15} = \frac{45}{75}$

$\frac{3}{5}$ or $\frac{45}{75} < \frac{46}{75} < \frac{47}{75} < \frac{48}{75} < \frac{49}{75} < \frac{50}{75}$ or $\frac{2}{3}$

Notice that the possibilities for common denominators were 15, 30, 45, 60, 75. The more numbers desired, the larger the denominator required. An even larger denominator (say 120) would have allowed wider choice.

5. (a) You might try to find one.
   (b) Is zero nonnegative? Is it a rational number?

Chapter 10 *Exercise 10.2 Page 258*

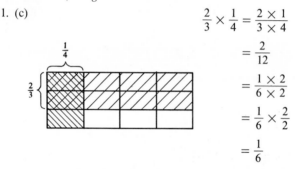

1. (c)

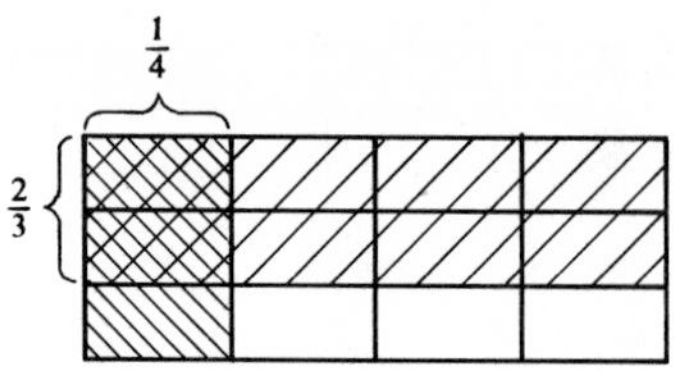

$$\frac{2}{3} \times \frac{1}{4} = \frac{2 \times 1}{3 \times 4} = \frac{2}{12} = \frac{1 \times 2}{6 \times 2} = \frac{1}{6} \times \frac{2}{2} = \frac{1}{6}$$

2. (c) The input is $\frac{3}{7}$. This is "stretched" by 2, giving a result $\frac{6}{7}$. The "shrinker" multiplies by $\frac{1}{5}$, giving a result of $\frac{6}{7} \times \frac{1}{5} = \frac{6}{35}$.

3. (d) $\frac{3}{7} \times \frac{2}{15} = \frac{3 \times 2}{7 \times 15} = \frac{6}{105}$

4. (d) Using the result of part (d) in Problem 3:

$$\frac{3 \times 2}{7 \times 15} = \frac{3 \times 2}{7 \times (3 \times 5)} = \frac{2 \times 3}{(7 \times 5) \times 3} = \frac{2}{7 \times 5} \times \frac{3}{3} = \frac{2}{35} \times 1 = \frac{2}{35}$$

5. (a) $\frac{3}{2} \times \frac{2}{2}$ $\qquad \frac{3}{2} \times \frac{2}{2} = \frac{6}{4}$

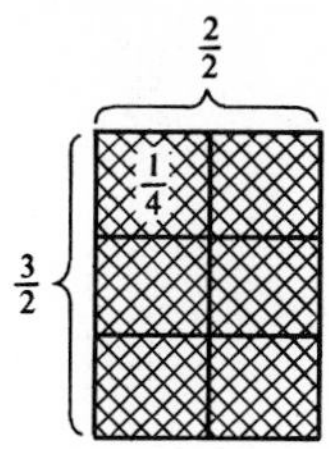

Exercise 10.3 Page 260

1. (b) $\frac{4}{3} \div \frac{7}{2} = \frac{4}{3} \times \frac{2}{7}$ *Multiplicative inverse of* $\frac{7}{2}$ *is* $\frac{2}{7}$.
$= \frac{8}{21}$

2. (d) $$\frac{\frac{7}{8} + \frac{1}{5}}{\frac{1}{10} + 6} \times \frac{40}{40} = \frac{(\frac{7}{8} + \frac{1}{5}) \times 40}{(\frac{1}{10} + 6) \times 40}$$
$$= \frac{(\frac{7}{8} \times 40) + (\frac{1}{5} \times 40)}{(\frac{1}{10} \times 40) + (6 \times 40)}$$
$$= \frac{35 + 8}{4 + 240}$$
$$= \frac{43}{244}$$

Exercise 10.4 Page 264

1. (b) $\frac{7}{13} + \frac{2}{13} = \frac{(7 + 2)}{13}$ *Generalization* 10.4(a)
$= \frac{9}{13}$

(d) $\frac{7}{8} - \frac{2}{8} = n$, if $\frac{2}{8} + n = \frac{7}{8}$ *Definition* 9.7

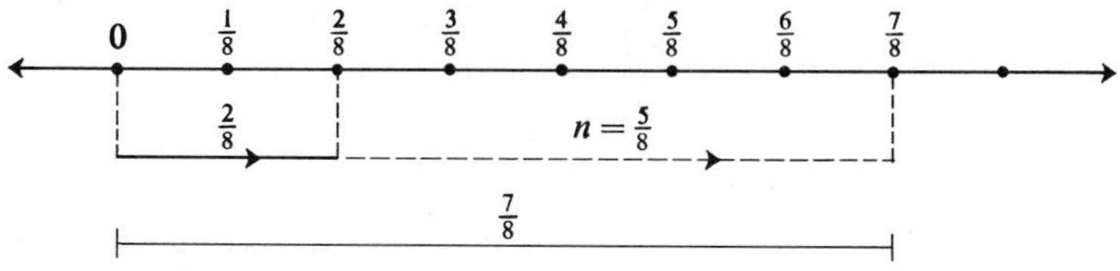

Hence, $\frac{7}{8} - \frac{2}{8} = \frac{5}{8}$.

2. (c) $\frac{3}{10} + \frac{4}{25}$
To find the lcd, find the lcm of 10 and 25.
$10 = 2 \times 5$
$25 = 5 \times 5$
The lcm of 10 and 25 is $2 \times 5 \times 5$.
$\frac{3}{10} \times \frac{5}{5} = \frac{15}{50}$ $\qquad \frac{4}{25} \times \frac{2}{2} = \frac{8}{50}$
$\frac{15}{50} + \frac{8}{50} = \frac{23}{50}$

3. (c) $\frac{6}{14} + \frac{4}{7} = \frac{6}{14} + (\frac{4}{7} \times \frac{2}{2})$
$= \frac{6}{14} + \frac{8}{14}$
$= \frac{14}{14}$
$= 1$

4. (c) $(\frac{1}{2} - \frac{1}{3}) - \frac{1}{5}$ (The lcd is 30.)
$= [(\frac{1}{2} \times \frac{15}{15}) - (\frac{1}{3} \times \frac{10}{10})] - (\frac{1}{5} \times \frac{6}{6})$
$= [\frac{15}{30} - \frac{10}{30}] - \frac{6}{30}$
$= \frac{5}{30} - \frac{6}{30}$. This does not give a nonnegative result.

(d) $\frac{1}{2} - (\frac{1}{3} - \frac{1}{5})$ (Refer to Problem 4(c) above.)
$= \frac{15}{30} - (\frac{10}{30} - \frac{6}{30})$
$= \frac{15}{30} - \frac{4}{30}$
$= \frac{11}{30}$. This is a nonnegative result. Contrast 4(c) and 4(d) and see where they differ.

Exercise 10.5 Page 269

1. (b) $2(10^3) + 0(10^2) + 1\left(\frac{1}{10^1}\right) + 2\left(\frac{1}{10^2}\right)$

Notice that 10^1 and 10^0 are not shown. This means that they are multiplied by 0, just as was (10^2). This number in decimal notation is 2000.12.

(c) $3\left(\frac{1}{10^2}\right) + 4\left(\frac{1}{10^3}\right) + 3\left(\frac{1}{10^4}\right)$

$0\left(\frac{1}{10^1}\right)$ is understood and the number in decimal notation is:

0.0343. Sometimes 0 is written in the ones place. It makes the position of the decimal point clearer.

3. (a) 0.691 (b) 600.091
The student should notice carefully the difference in the way these two numerals are read.

4. (g) $5 \times 1^0 = 5 \times 1 = 5$ (h) $5 \times 1^1 = 5 \times 1 = 5$
(i) $5 \times 1^{-1} = \frac{5}{1} = 5$

Exercise 10.6 Page 273

1. (b) 0.007 $= (0 \times 1) + (0 \times \frac{1}{10}) + (0 \times \frac{1}{100}) + (7 \times \frac{1}{1000})$
26.32 $= (2 \times 10) + (6 \times 1) + (3 \times \frac{1}{10}) + (2 \times \frac{1}{100})$

The sum is: $(2)10 + (0 + 6)1 + (0 + 3)\frac{1}{10} + (0 + 2)\frac{1}{100} + (7)\frac{1}{1000}$
$= (2)10 + (6)1 + (3)\frac{1}{10} + (2)\frac{1}{100} + (7)\frac{1}{1000}$

0.007 + 26.32 = 26.327

2. (b)

```
 0.007
26.32
------
 0.007
 0.02
 0.3
26.
------
26.327
```

Short form:
$$\begin{array}{r} 0.007 \\ \underline{26.32} \\ 26.327 \end{array}$$

3. (d) 21.005×0.014

$$= [(2 \times 10) + (1 \times 1) + (5 \times \tfrac{1}{1000})] \times [(1 \times \tfrac{1}{100}) + (4 \times \tfrac{1}{1000})]$$
$$= (2 \times \tfrac{1}{10} + 8 \times \tfrac{1}{100}) + (\tfrac{1}{100} + 4 \times \tfrac{1}{1000}) + \tfrac{5}{100{,}000} + \tfrac{20}{1{,}000{,}000})$$
$$= \tfrac{2}{10} + \tfrac{8}{100} + \tfrac{1}{100} + \tfrac{4}{1000} + \tfrac{5}{100{,}000} + \tfrac{2 \times 10}{1{,}000{,}000}$$
$$= \tfrac{2}{10} + \tfrac{9}{100} + \tfrac{4}{1000} + \tfrac{7}{100{,}000} = 0.29407$$

4. (e) $86.07 + 7.018$

$$\begin{array}{r} 86.07 \\ \underline{7.018} \\ 80. \\ 13. \\ 0.08 \\ \underline{0.008} \\ 93.088 \end{array} \qquad \text{or} \qquad \begin{array}{r} 86.07 \\ \underline{7.018} \\ 93.088 \end{array}$$

(f) $86.07 - 7.018 = n$ if $n + 7.018 = 86.07$. The algorithm for subtraction was developed in Section 7.6.

$$\begin{array}{rl} 86.070 & \\ \underline{7.018} & \\ 0.002 & \quad 8 + 2 = 10 \text{ or } 10 - 8 = 2 \\ 0.050 & \quad 10 + 50 = 60 \text{ or } 60 - 10 = 50 \\ 9. & \quad 7 + 9 = 16 \text{ or } 16 - 7 = 9 \\ \underline{70.} & \\ 79.052 = n & \end{array}$$

To check 4(f):
$$\begin{array}{r} 79.052 \\ \underline{7.018} \\ 0.010 \\ 0.06 \\ 16. \\ \underline{70.} \\ 86.070 \end{array}$$

Exercise 10.7 Page 284

1. (g) Following Example 1, this is $\frac{134}{10000}$.
2. (d) 3.14 in fraction form is $\frac{314}{100}$.
   In mixed form it is $3\frac{14}{100}$ or $3\frac{7}{50}$.
   It is not possible to write a number in mixed form unless it is greater than 1.
3. (e) Ten and two hundred one thousandths
   (f) One hundred two and one hundredth
   (g) $0.10201 = (0 \times 1) + (1 \times \tfrac{1}{10}) + (0 \times \tfrac{1}{100}) + (2 \times \tfrac{1}{1000}) + (0 \times \tfrac{1}{10000}) + (1 \times \tfrac{1}{100000})$
   $$= 0 + \tfrac{1}{10} + \tfrac{2}{1000} + \tfrac{1}{100000}$$
   $$= 0 + \tfrac{10000}{100000} + \tfrac{200}{100000} + \tfrac{1}{100000}$$
   $$= 0 + \tfrac{10201}{100000}$$
   0.10201 is read, "ten thousand two hundred one hundred-thousandths."

4\. (c) $\frac{17}{25} = \frac{68}{100} = \frac{60}{100} + \frac{8}{100}$

$= \frac{6}{10} + \frac{8}{100} = 0.68$

(i) $\frac{381}{250} \times \frac{4}{4} = \frac{1524}{1000} = 1.524$

5\. The use of the division algorithm in converting to decimal notation has been demonstrated for only (b) and (g).

(b) $\frac{5}{8} = 5 \div 8$

$$\begin{array}{r|l|l}
8\,\overline{)5} & & \\
0 & 0 & (8 \times 0) \leq 5 < (8 \times 1) \\
\hline
5.0 & & \\
4.8 & 0.6 & (8 \times 0.6) \leq 5.0 < (8 \times 0.7) \\
\hline
0.20 & & \\
0.16 & 0.02 & (8 \times 0.02) \leq 0.20 < (8 \times 0.03) \\
\hline
0.040 & & \\
0.040 & 0.005 & (8 \times 0.005) \leq 0.040 < (8 \times 0.006) \\
\hline
 & 0.625 &
\end{array}$$

$\frac{5}{8} = 0.625$

(g) $\frac{1}{9} = 1 \div 9$

$$\begin{array}{r|l|l}
9\,\overline{)1} & & \\
0 & 0 & (9 \times 0) \leq 1 < (9 \times 1) \\
\hline
\frac{10}{10} & & \\
\frac{9}{10} & \frac{1}{10} & (9 \times \frac{1}{10}) \leq \frac{10}{10} < (9 \times \frac{2}{10}) \\
\hline
\frac{10}{100} & & \\
\frac{9}{100} & \frac{1}{100} & (9 \times \frac{1}{100}) \leq \frac{10}{100} < (9 \times \frac{2}{100}) \\
\hline
\frac{10}{1000} & \vdots &
\end{array}$$

$\frac{1}{9} = 0 + \frac{1}{10} + \frac{1}{100} + \ldots = 0.11\ldots = 0.\overline{1}$

Observe that each succeeding remainder is 1, $(1 \times \frac{1}{10})$, $(1 \times \frac{1}{100})$ etc. Hence, the pattern for the repeating decimal may be detected when 1 appears the second time in the above remainder pattern.

The answers to (c) and (d) are:

(c) $0.\overline{142857}$ (d) $3.\overline{571428}$

6\. Remember the test for changing a fraction to a terminating decimal.

(c) $\frac{21}{45} = \frac{7}{(3 \times 5)}$. The factor 3 in the denominator indicates a nonterminating decimal.

(g) $\frac{23}{20} = \frac{23}{(2^2 \times 5)}$. This can be changed to a terminating decimal since its denominator has powers of 2 and 5 and no other factors.

7\. (d)

$$\begin{aligned}
n &= 2.\overline{5} = 2.555\ldots \\
10n &= 25.555\ldots \\
9n + n &= 23 + 2.555\ldots \\
9n &= 23 \\
n &= \tfrac{23}{9}
\end{aligned}$$

(e)

$$\begin{aligned}
n &= 1.\overline{12} \\
100n &= 112.1212\ldots \\
99n + n &= 111 + 1.1212\ldots \\
99n &= 111 \\
n &= \tfrac{111}{99}
\end{aligned}$$

9. (c) 126.8 rounded to nearest ten is 130.0.
126.8 rounded to nearest hundred is 100.0.
(d) 765 rounded to nearest ten is 760 (using the even rule).
765 rounded to nearest hundred is 800.

10. (d) $0.\overline{32} = 32 \times 0.0101\ldots$
$= 32 \times \frac{1}{99}$
$= \frac{32}{99}$
(f) Using the hint in the text, the final result should be $0.3\overline{21} = \frac{53}{165}$.

12. The methods for verifying are varied as shown.
(a) $0.\overline{5} = 0.555\ldots$ Consider $0.\overline{5} + 0.\overline{6} = \frac{5}{9} + \frac{6}{9} = \frac{11}{9} = 1\frac{2}{9}$.
$0.\overline{6} = \underline{0.666\ldots}$ Does $1\frac{2}{9} = 1.\overline{2}$? Explain by considering $1\frac{2}{9} = 1 + \frac{2}{9}$.
$1.221\ldots$
Is it correct to name this as $1.\overline{2}$?
(d) $1.\overline{7} + 1.\overline{6} = 1.777\ldots + 1.666\ldots$
$= 3.\overline{4}$
(e) $0.\overline{23} + 0.\overline{78} = 0.2323\ldots = 0.7878\ldots$
$= 1.\overline{02}$

13. (d) $18 \div .0006 = 180000 \div 6$

14. (b) $7.36 \div 0.21 = 736 \div 21$
$= 35\frac{1}{21}$

Chapter 11 *Exercise 11.3 Page 291*

1. (c) $\dfrac{864}{150} = \dfrac{2^5 \times 3^3}{2 \times 3 \times 5^2}$

Remember that Definition 11.2(b) specifies that a/b be in simplest form. This requires that the common factor 2×3 be removed from numerator and denominator. Then express as a square.

4. $^-64 = {}^-(2^6)$ and the three equal factors required to give a perfect cube are: $(^-2^2)(^-2^2)(^-2^2)$. Can two equal factors ever produce a negative number?

5. (c) Be sure to reduce to simplest form. (d) $^-6$
(e), (f), (g), (h) are introduced to call attention to the difference in the operations multiplication and addition where radicals are involved.

6. (a) The solution set is $\{7, {}^-7\}$.
(d) $x^3 + 27 = 0$
$x^3 + 27 + {}^-27 = {}^-27$ *Uniqueness of addition*
$x^3 = {}^-27$ *Renaming* $(27 + {}^-27)$
$x^3 = (-3)^3$ *Renaming* $^-27$
The solution set is $\{^-3\}$.
(e) Not possible in the set of real numbers. Why?

Exercise 11.5 Page 297

1. (a) $(2.6)^2 = 6.76$ and $(2.7)^2 = 7.29$.
Then $2.6 < \sqrt{7} < 2.7$.
Since 6.76 and 7.29 have missed 7 by nearly the same amounts (.24 and .29), we guess halfway between 2.6 and 2.7, or 2.65. It is easily shown that 2.65 is too large and 2.64 is too small. Therefore, $2.64 < \sqrt{7} < 2.65$

and $\sqrt{7}$ has been estimated between two successive hundredths. The results may be shown as:

$$2.6 < 2.64 < \sqrt{7} < 2.65 < 2.7$$

2. (a) $(2.7)^3 > 19$; $(2.6)^3 < 19$; $2.66 < \sqrt[3]{19} < 2.67$
Distinguish between "correct to two decimal places" and "between successive hundredths." In this case 2.67 is the answer correct to two decimal places, since $(2.67)^3$ is closer to 19 than is $(2.66)^3$.
(c) How is $\sqrt[3]{^-19}$ related to part (a), $\sqrt[3]{19}$?

3. (a) 61.47; 61.474

Exercise 11.6 Page 300

1. (h) π (irrational, real)
(i) $11.\overline{27}$ (rational, real)
(k) $^-\sqrt{16}$ (integer, rational, real)
(l) $\sqrt[4]{-16}$. You should question whether this belongs among the real numbers.

Chapter 12 *Exercise 12.3 Page 305*

3. (b) m $\overline{PQ} = 5$. A portion remains, but Q is nearer to the end of the fifth unit than to the end of the sixth unit.
(c) $5 < \text{m } \overline{PQ} < 6$

Exercise 12.5 Page 311

1. (a) 3; 5; 1; 8. Notice that the first answer here might be in doubt. The answer to Problem 1(b) may help.
2. Since X appears to fall midway between 2 and 3 on the centimeter scale we use the result of Problem 1(b) and say that it falls to the right of this midway position; hence it is nearer 3 than 2.
$Y \rightarrow 1\frac{1}{2}$ in. $\rightarrow$ 4 cm $\quad B \rightarrow 2\frac{3}{4}$ in. $\rightarrow$ 7 cm
$A \rightarrow 2$ in. $\rightarrow$ 5 cm $\quad C \rightarrow 3$ in. $\rightarrow$ 8 cm
$Z \rightarrow 3\frac{3}{4}$ in. $\rightarrow$ 10 cm (Use the same approach suggested for X.)
$R \rightarrow 4$ in. $\rightarrow$ 10 cm
3. $3\frac{3}{4}$ in. $\approx$ 9.52 cm; Z appears midway between 9 and 10 on the centimeter scale.
The matching of the points to the two scales is to be done by inspection. This probably will result in some disagreement.
4. Using 1 cm = .3937 in. and 1 in. = 2.54 cm
(f) m $\overline{YB} \approx 1\frac{1}{4}$ in. $\approx$ 3.18 cm $\approx$ 31.8 mm
7. (f) 29.505 decimeters = 2.9505 meters
= 2 meters 9 decimeters 5 centimeters $\frac{5}{10}$ millimeter.
8. (c) Since 1 cm $\approx$ 0.3937 in.
1 mm $\approx$ 0.03937 in.
3.4 mm $\approx 3.4 \times .03937$ in.
$\approx$ 0.133858 in.
9. 1 mile = 12×5280 inches
Now find 1 mile in cm. You may prefer to convert this to meters and then to kilometers.

12. Can you justify a quick approximation such as: Divide by 5 and multiply by 8? What should be done to convert the posted km per hour to mph?

Exercise 12.6 Page 315

1. (a) 3 cm (i) cm is the unit of measurement.
   (ii) $\frac{1}{2}$ cm is greatest possible error.
   (iii) $\frac{1}{2}/3 = \frac{1}{6} = 0.1\overline{6} = 1\overline{6}\%$ is percent of error.
   (d) 7.148 dm (i) 0.001 dm is unit of measurement.
   (ii) 0.0005 dm is greatest possible error.
   (iii) $\frac{0.0005}{7.148} = \frac{5}{71480} = 0.00007$ or 0.007%
2. 18.7 is more precise.
   0.05 mile ⟶ 2.7%
   $\frac{1}{2}$ mile ⟶ 2.6%
   Hence 19 is the more accurate measurement.

Exercise 12.7 Page 318

1. (d) $m(\text{perimeter}) = [2(5.4 + 8.5) + 5.6 + 10.4]\text{ cm}$
   $= 43.8\text{ cm}$
2. (c) $m(\text{circumference}) = [2 \times (8.3 \times 3.14)]\text{m} = 52.124\text{ m}$

Exercise 12.8 Page 323

1. (b) This figure is called a trapezoid. It has two parallel and two nonparallel sides. It is a special kind of trapezoid with the two nonparallel sides being equal in length.

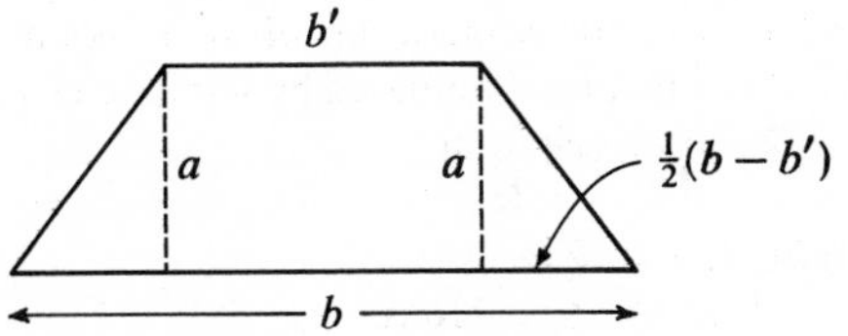

$$A_{\text{trapezoid}} = (a \times b') + 2\left[\frac{1}{2}(b - b') \times \frac{a}{2}\right]$$

$$= (a \times b') + \frac{a}{2}(b - b')$$

$$= (a \times b') + \frac{a \times b}{2} - \frac{a \times b'}{2}$$

$$= \frac{a \times b'}{2} + \frac{a \times b}{2} = \frac{a}{2}(b' + b)$$

Another approach would be to divide the trapezoid into two triangles with bases b and b' and each with altitude a.

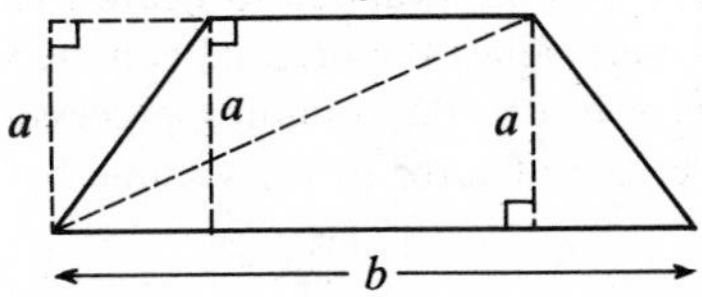

(d) Divide this regular hexagon (six equal sides) into six equal triangles, each with base *s* and altitude *a*.

Exercise 12.9 Page 326

1. (g) 1 dm = 100 mm. Hence 1 sq dm = 10,000 sq mm and 1 sq mm = $\frac{1}{10000}$ sq dm
2. (b) 1 sq ft = 144 sq in. = 144 × 6.45 sq cm
Or using Example 3:

$$1 \text{ sq ft} = 144 \times 6.45 \times 10^2 \text{ sq mm}$$
$$\approx 928.8 \text{ sq cm} = 92880 \text{ sq mm}$$

(d) $1 \text{ sq mm} = \dfrac{1}{144 \times 6.45 \times 100} = \dfrac{1}{92880} \approx .00001$ sq ft. Or using the previously derived formula:

$$1 \text{ sq cm} \approx 0.16 \text{ sq in.}$$
$$1 \text{ sq mm} \approx \tfrac{1}{100} \times 0.16 = 0.0016 \text{ sq in.}$$
$$1 \text{ sq in.} = \tfrac{1}{144} \times 1 \text{ sq ft}$$
$$1 \text{ sq mm} \approx \tfrac{0.0016}{144} \text{ sq ft} = 0.0000\overline{1} \text{ sq ft}$$

3. (a) The dimensions of this rectangle are 17 dm and 53 dm. The area is 17 × 53 sq dm.
4. (a) This area is the sum of the areas bounded by 1 triangle with base 4 in. and altitude 2 in., 1 right triangle with base 4 in. and altitude 3 in., and a rectangle with dimensions 3 in. by 4 in. The total area is 22 sq in.
(d) This figure is made up of areas bounded by a semicircle with radius of $7\frac{1}{5}$ in. plus 4 bounded by semicircles $7\frac{1}{5}$ in. diameter. Area = $\pi(7\frac{1}{5})^2 + 2 \times [\pi(3\frac{3}{5})^2]$ sq in.

Exercise 12.10 Page 331

2. $V = 6 \times 6 \times 6 = 216$ cc
(a) Rounded to the nearest 10, $V = 220$ cc

2. (b)

Measurement	*Unit*	*Greatest Possible Error*	*Percent of Error*
216 cc	1 cc	0.5 cc	$\frac{0.5}{216} = 0.0023 = 0.23\%$
220 cc	10 cc	5 cc	$\frac{5}{220} = 0.023 = 2.3\%$

(c) It is a cube.

4. (iv) $V_{\text{pyramid}} = \frac{1}{3}(4.2)^2\,(3.1)$ cc
$= 18.228$ cc
The volume rounded to nearest tenth = 18.2 cc = 18,200 cubic millimeters. Since the original volume was 18,228 cubic mm, 28 cubic mm were lost in the rounding process.
Percent of error in the volume before rounding is:

$$\frac{0.5}{18228} = \frac{1}{36456} = 0.000024 = 0.0024\%$$

Percent of error after rounding is:

$$\frac{50}{18200} = 0.0027 = 0.27\%$$

Exercise 12.11 Page 336

2. (d) Since measure is a number, a numeral is given here. The use of the degree symbol would be incorrect.
   (5) $m\angle JZE = 75$
   (6) $m\angle CZD = 10$
   (e) (1) $\angle AZB$
   (3) $m\angle AZD = \frac{1}{2}m(\angle GZA)$
3. (d) $18° \, 15' \, 20'' = (18 + \frac{15}{60} + \frac{20}{3600})° = (18\frac{23}{90})°$
4. $18° \, 20' \, 45'' = \left(18 + \dfrac{20}{60} + \dfrac{45}{60 \times 60}\right)^\circ$

   $= (18\frac{83}{240})°.$

   To change this degree measurement to minutes, multiply by 60.
5. (b) $89° \, 16' = 5356'$
   The measure is 1′.
   Possible error is 0.5′.
   $\frac{0.5}{5356} = 0.000093 = 0.0093\%$ is the percent of error.

Chapter 13 *Exercise 13.1 Page 340*

1. (a) In the ordered pair (x, y), x is 1 greater than y, or $x + 1 = y$.
   (c) In the ordered pair (x, y) to the hour x is assigned the temperature y.
2. $\{(3, 1), (5, 1), (5, 3)\}$
3. Using the figure in the text some ordered pairs will be $l_1 \, R \, l_2$, $l_1 \, R \, l_3$, $l_2 \, R \, l_1$.

Exercise 13.2 Page 342

1. (a) For the set of all people:
   Reflexive: John is the brother of John if we define "is the brother of" as having the same parents. If this is accepted, then the relation is reflexive.
   Symmetric: John is the brother of Mary, but Mary is not the brother of John.
   Transitive: If John is the brother of Henry and Henry is the brother of Tom, then John is the brother of Tom.
   Evidently the symmetric property is lacking.
   For the set of all men and boys is this symmetric property present?
   (c) A weighs 120 lbs., B weighs $120\frac{1}{3}$ lbs., while C weighs $120\frac{2}{3}$ lbs. How about A and C? This seems to rule out the transitive property.
2. (b) (5, 1) and (3, 3) are two of them.

Exercise 13.3 Page 350

1. $(3, {}^-4)$ and $({}^-7, 8)$ are two of them; while $({}^-6, 5)$, $(6.7, {}^-7.7)$, and $({}^-1, 0)$ do not belong. (Caution: Remember you are dealing with the *integer* set.)
2. Since the universal set permits real numbers, $(2, \sqrt{21})$ gives $(2)^2 + (\sqrt{21})^2 = 25$. There are two ordered pairs that do not belong. You find them.

3. There are 6 such pairs.
4. 7 and $\sqrt{21}$ are each greater than 4 and hence not included in B. Then $(7, \sqrt{21})$ is not an element of M'.
   The dotted line is used to show the border $(y = x)$ of the region whose points have coordinates satisfying the relation $y > x$. $(2, 2.01)$ is an element of M', but $(2, 2)$ is not.

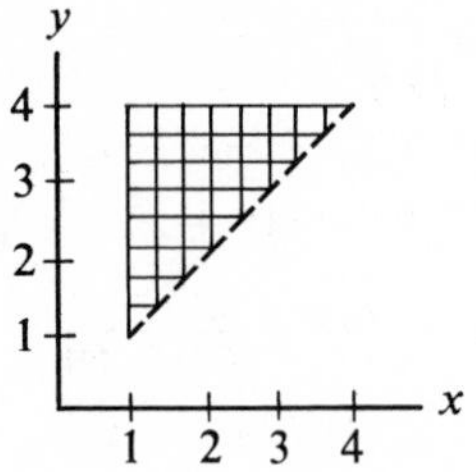

5. First, using permitted values in Problem 3, there are two points satisfying $y = 2x$: $(1, 2)$ and $(2, 4)$. However, when values in R are permitted, there are many elements and the graph is a line segment.
6. (a) $(4, 1)$ (c) $(0, 0)$
7. (b) $U = I$, the set of integers.
   (ii) $P = \{(x, y), x \in U, y \in U: x = {}^{-}2, |y| \leq 3\}$.
   (e) (i) Ordered pairs might be $(.5, \frac{7}{3})$, $(\frac{-3}{2}, \sqrt{17})$.
   (ii) $U = R$, the set of real numbers.
   $P = \{(x, y), x \in U, y \in U: {}^{-}2 < x < 1\}$.
8. $(15, 10)$ and $(4, 12)$ do *not* belong to the new set.

Exercise 13.4 Page 353

1. (a) There are 9 ordered pairs.
   Domain is $\{1, 3, 5\}$; range is $\{1, 3, 5\}$.
   (b) Only one ordered pair.
   (c) Watch out here.
   (d) Notice the domain is $\{2\}$ and the range is $\{4, 5\}$.
2. The student may wish to refer to Section 2.5 for an earlier discussion of the meaning of "and" and "or" conditions.
   (a) $x = 4$ and $y = 3$; $U = N$, the set of natural numbers. The domain is 4 and the range is 3.

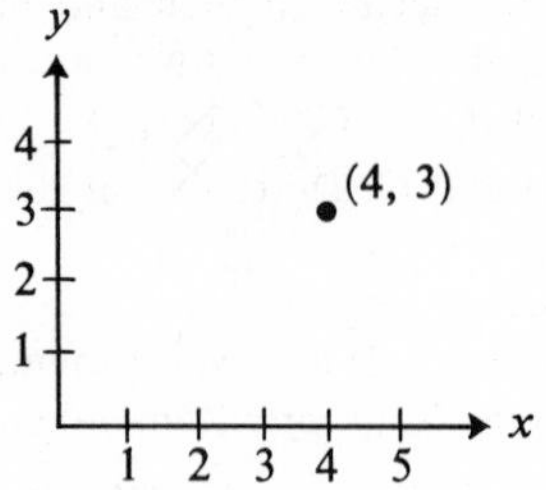

   (b) $x = 4$ or $y = 3$; $U = N$, the set of natural numbers. When $x = 4, y \in N$; or $x \in N$ when $y = 3$.

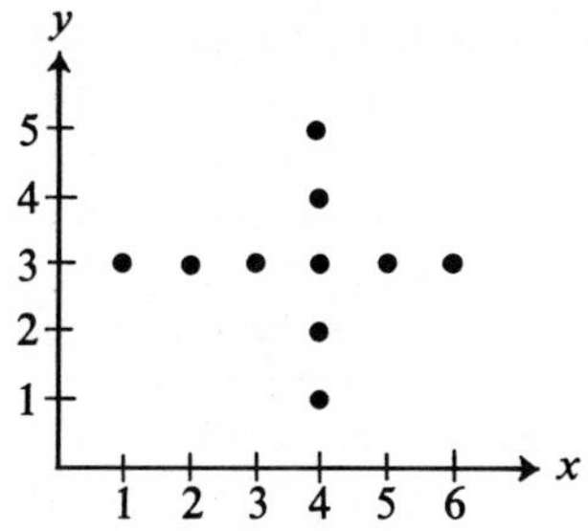

Domain: the natural numbers, N
Range: the natural numbers, N

(c) $x < 4$ and $y < 3$; $U = R$, the set of real numbers. The dotted lines indicate "less than but not equal to." See Problem 4, Exercise 13.3.

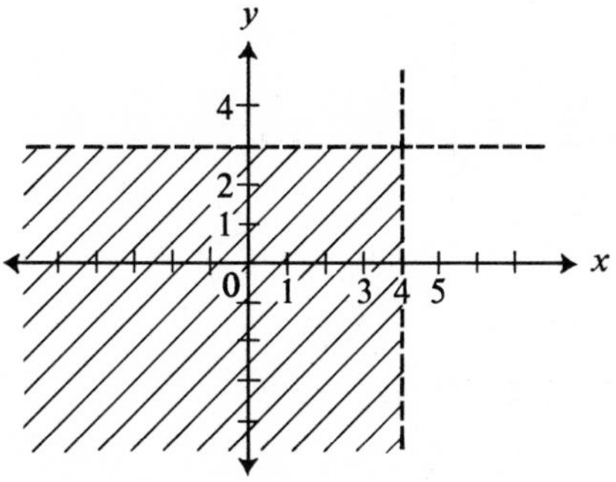

Domain: $x \in R, x < 4$.
Range: $y \in R, y < 3$.

(d) $x < 4$ or $y < 3$, $U = R$, the set of real numbers.

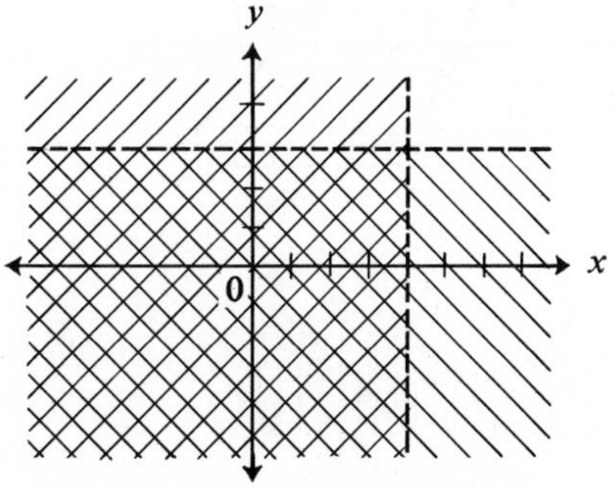

Domain: $x < 4, x \in R$.
Range: $y \in R, y < 3$.

Exercise 13.5 Page 358

1. (a) Domain is $\{7, 4, 3, 2\}$; Range is $\{3, 9, 6, {}^-7, 0, 1, {}^-3\}$; Note $(7, 3)$, $(7, 6)$, $(7, {}^-3)$ prove this relation is not a function. See Definition 13.5.

(c) Domain is what? Range is $\{^-2\}$.
(d) $(2, 3)$ and $(2, \sqrt{9})$ are the same ordered pair. One should be deleted.

2. (a) "y is the uncle of x": The domain is the set of people who have uncles. The range is the set of men and boys who are uncles. Certainly y_1 and y_2 might both be uncles of the same x: (x, y_1), (x, y_2).
(b) Does any triangle have more than one perimeter measure?
(d) "y is the day of the week on which x was the average temperature." Consider the set of ordered pairs:
(90°, Sun.), (95°, Mon.), (92°, Tues.), (90°, Wed.), (89°, Thurs.), (90°, Fri.), (92°, Sat.)

3. A:

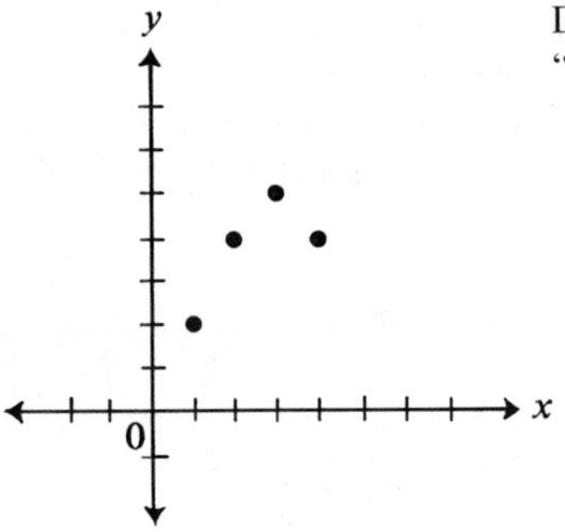

Draw your own conclusion from the "vertical line" test.

4. $M \text{ X } S = \{1, 3, 5\} \text{ X } \{3, 5\}$.
5. Why is (g) a function while (h) is not?
6. (b)

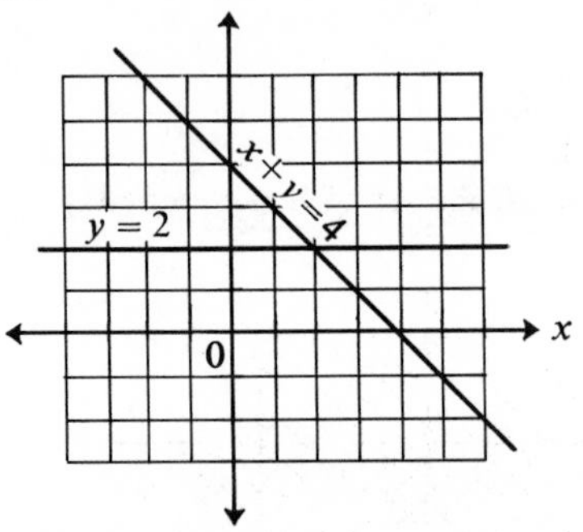

(d)

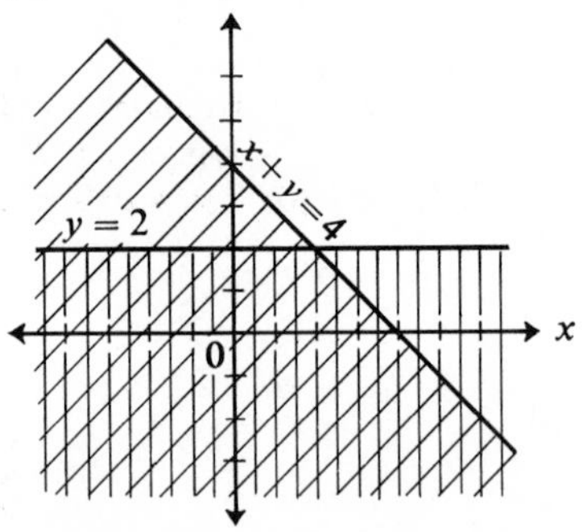

(e)

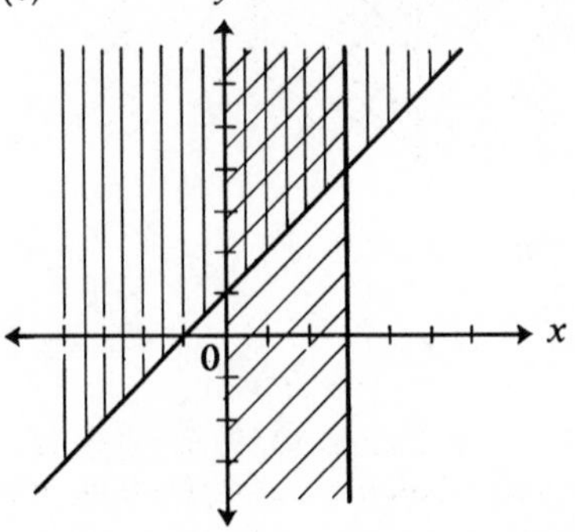

What about (0, 0), (0, 1), (0, 2) and (0, 4)?

Only the crosshatch portion is the graph of this relation.

Exercise 13.6 Page 361

1. The greatest possible range is the set of nonnegative real numbers. Why is "the set of positive real numbers" incorrect?
2. Notice that 27 does not belong in the domain.
4. (b) $g(\frac{1}{3}) = 3$ (c) $g(0)$ is undefined
   (d) $g(\frac{1}{5}) = 5$ (h) No value of x will yield 0 for $\frac{1}{x}$.
   (j) The domain of g is the set of rational numbers with 0 deleted.
5. Is it possible to find any value for $f(x)$ other than 7?
7. $F(x)$ is defined for all real values of x. No value of x makes $F(x)$ zero although very large positive values and their additive inverses make $F(x)$ very small, but still positive. If $x = 0$, $F(x) = 2$. $0 < F(x) \leq 2$ gives the range.

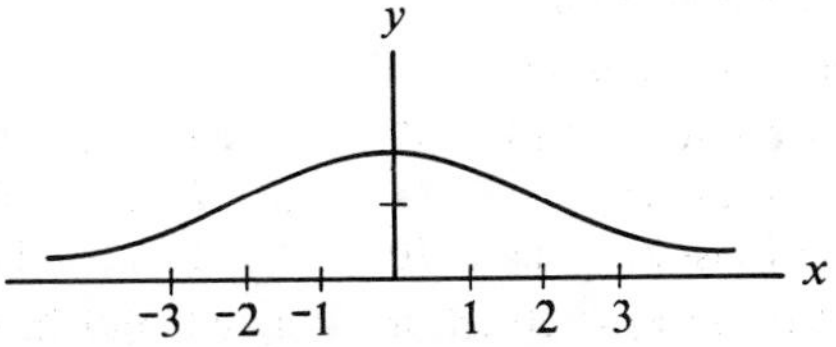

8. Can the range of $F_2(x)$ be greater than 4?

Chapter 14 *Exercise 14.1 Page 370*

1. (a) Yes. Definition 2.5(c)
   (b) Yes. z
   (c) $x \odot x = z$; $y \odot t = z$; $z \odot z = z$.
   (d) $x \odot (y \odot t) = x \odot z = x$
   $(x \odot y) \odot t = t \odot t = x$
   (e) Symmetry with respect to main diagonal. Yes.
   (f) Replacing m by an element in the given set.
3. (a) $\{r_0, r_1, r_2\}$ (b)

$\circ$	r_0	r_1	r_2
r_0	r_0	r_1	r_2
r_1	r_1	r_2	r_0
r_2	r_2	r_0	r_1

4.

r_0: no turn; r_1: $\frac{1}{4}$ turn; etc.

5. (a) $2 * 4 = 2 + 8 = 10$; $4 * 2 = 4 + 4 = 8$

Exercise 14.2 Page 372

2. (a) $(N, +)$. The operation is commutative and associative. There exists no

identity element and no inverse under addition. The system is not a group.
$(N, \times)$. The operation is commutative and associative. There is an identity element, 1. There are no inverses under multiplication for any elements except 1.

2. (b), (c), (d). The student should have no difficulty following this procedure with the other systems. Zero (0) is an element in each of these systems. Does it have an inverse under multiplication $(\times)$?

Exercise 14.3 Page 379

1. (a), (b), (c), (d), (e), all $(\equiv)$; (f) $(\not\equiv)$; (g) $(\equiv)$; (h) $(\equiv)$
Notice that (c) and (d) together make (e) true. This holds because 2 and 3 are relatively prime. You might investigate other examples and formulate a generalization.
2. (d) $\{^-8, {}^-1, 6, 13\}$.
4. (a)

Class 0:	$^-15$	$^-12$	...	$^-3$	...	9	...	18	...	39
Class 1:	$^-14$	$^-11$	...	$^-2$	...	10	...	19	...	40
Class 2:	$^-13$	$^-10$	...	$^-1$	...	11	...	20	...	38

5. $^-2$ (class 1) + 20 (class 2) should fall in class (1 + 2) which is the same as class 0.
$^-2 + 20 = 18$ does fall in class 0.
Also, $^-1$ (class 2) $\times$ 11 (class 2) falls in class (2×2) or class 1.
$^-1 \times 11 = {}^-11$ and $^-11 \equiv 1 \pmod 3$)

7. (a) (mod 2)

⊕	0	1
0	0	0
1	1	0

⊗	0	1
0	0	0
1	0	1

(d) (mod 7)

⊕	0	1	2	3	4	5	6
0		1		3			
1	1						0
2	2		4				
3				6	0		
4						2	
5	5			1			
6							5

(c) (mod 4)

⊗	0	1	2	3
0	0	0	0	0
1	0	1	2	3
2	0	2	0	2
3	0	3	2	1

Exercise 14.4 Page 383

3. (b)
$975 \to 3$
$124 \to 7$ } $\xrightarrow{+} 10 \to 1$
$1099 \to 1 \leftarrow$

4. (b)
$975 \to 3$
(−) $124 \to 7$ } $\xrightarrow{-} {}^-4$
$851 \to 5 \leftarrow$ $\quad {}^-4 \equiv 5 \pmod 9$

5. (b)

$$\left.\begin{array}{r} 975 \rightarrow 3 \\ 124 \rightarrow 7 \end{array}\right\} \xrightarrow{\times} 21 \rightarrow 3$$

$$\begin{array}{r} 975 \\ 124 \\ \hline 3900 \\ 1950 \\ 975 \\ \hline 120900 \rightarrow 3 \end{array}$$

6. (b) $8025 = (53 \times 151) + 22$

   Is $6 \equiv (8 \times 7) + 4 \pmod 9$?

   $\equiv 56 + 4$

   $\equiv 6$ Yes.

7. To check the result of the division in Problem 6(b) by casting out elevens:

$$\begin{aligned} 8025 &\equiv 5 + {}^{-}2 + 0 + {}^{-}8 \equiv {}^{-}5 \pmod{11} \\ 53 &\equiv 3 + {}^{-}5 \equiv {}^{-}2 \pmod{11} \\ 151 &\equiv 1 + {}^{-}5 + 1 \equiv {}^{-}3 \pmod{11} \\ 53 \times 151 &\equiv ({}^{-}2) \times ({}^{-}3) \equiv 6 \pmod{11} \\ 22 &\equiv 0 \pmod{11} \end{aligned}$$

   Substituting these in the results obtained by the division: ${}^{-}5 \equiv 6 + 0 \pmod{11}$. This is a true statement by Theorem II(i), and the check of the division is completed.

8. If $10^n \equiv 1 \pmod 3$, then $k(10^n) \equiv k(1) \pmod 3$, by Theorem II(iii). Any number may be expressed as the sum of such terms where k is a nonnegative integer. Then by Theorem II(i), $k_0 10^n + k_1 10^{n-1} + \cdots + k_n \equiv k_0 + k_1 + k_2 \cdots + k_n$, the sum of the digits of the number.

Warning to Students

Even though skill in carrying out the devices is helpful, it is even more important to know the reasons behind them. Do not accept and use these as checks without knowing why they work.

Exercise 14.5 Page 388

1. Are G_1 to G_4 satisfied? It should be possible to verify this. Check the ⊗ grid to establish that R_1, R_2, R_3 are satisfied. Thus it is a commutative ring.
2. (b) First it will be necessary for the system of whole numbers to be a group under addition. Is G_3 (⊕-inverses) satisfied?
3. (a) The set of odd numbers does not have closure under addition; hence it is not an operational system.

   (b) The set of even numbers does not have ID_2, identity for multiplication. Hence it is a commutative ring but not an integral domain.

   (c) See Example 5 in Section 14.5.
4. Refer to the tables made for Problem 7(b) and 7(c) in Exercise 14.3.
5. Refer to the grid in Section 14.3.

   G_1: Associative for ⊗.

   G_2: Identity for ⊗ is 1.

   G_3; ⊗-inverses? Can you find an inverse for 2, 3, or 4?
6. Can it be a ring with only one operation, "∘"?
7. Does the element w have an inverse under ⊗?

8. It should be evident from the grid that this is not an operational system ($2 \oplus 6 \equiv 1 \pmod 7$).
9. Use a grid to check G_1 to G_4. They should all be satisfied.
10. (mod 8)

$\otimes$	0	2	4	6
0	0	0	0	0
2	0	4	0	4
4	0	0	0	0
6	0	4	0	4

Using this grid, check the requirements for a group. Why is this set not a group under multiplication?

Chapter 15 *Exercise 15.2 Page 395*

1. (d) The event described is a union of $x \leq 1$ and $x \geq 6$. In roster notation, $\{1\} \cup \{6\} = \{1, 6\}$.
   (e) This description is an intersection. $\{4, 5, 6\} \cap \{1, 2, 3, 4\} = \{4\}$.
   (f) Compare this description with that given for (e). Why does this set consist of $\{3, 4, 5\}$, while the preceding set consisted of $\{4\}$?
2. The probability of each of the events (d), (e), and (f) in Problem 1:

Event (d): $\dfrac{N\{1, 6\}}{N\{1, 2, 3, 4, 5, 6\}} = \dfrac{2}{6} = \dfrac{1}{3}$ *Refer to Definition* 15.2(b).

Event (e): $\dfrac{N\{4\}}{N\{1, 2, 3, 4, 5, 6\}} = \dfrac{1}{6}$

Event (f): $\dfrac{N\{3, 4, 5\}}{N\{1, 2, 3, 4, 5, 6\}} = \dfrac{3}{6} = \dfrac{1}{2}$

(a) (a), (b), (d), (f)
(b) The probability of (g) is 0. The event is impossible.
The probability of (c) is 1. The event is certain.

4. (a) The event of rolling an even number is $\{2, 4, 6\}$.
   What is the event of rolling an odd number?

$$P(\text{even}) = \frac{N\{2, 4, 6\}}{N(S)} = \frac{3}{6} = \frac{1}{2}.$$

Show that $P(\text{odd})$ is also $\frac{1}{2}$.

Then, $$P(\text{even}) + P(\text{odd}) = \tfrac{1}{2} + \tfrac{1}{2} = 1$$

(b) The event less than 3 is $\{1, 2\}$.
The event greater than 3 is $\{4, 5, 6\}$.

$$P(\text{less than } 3) = \tfrac{2}{6} \qquad P(\text{greater than } 3) = \tfrac{3}{6}$$

Observe that the sum of these two probabilities is $\frac{5}{6}$, not 1 as for the probabilities in (a).
Consider the element 3 of S which was omitted from the two events. $P(3) = \frac{1}{6}$. Observe this sum:

$P(\text{less than } 3) + P(\text{greater than } 3) + P(3) = \frac{2}{6} + \frac{3}{6} + \frac{1}{6} = \frac{6}{6} = 1$

What does it say to you?

(c) $P(1) = \frac{1}{6}$; $P(2) = \frac{1}{6}$; $P(1 \text{ or } 2) = \frac{2}{6}$

(d) $P(\text{less than } 1) = \dfrac{\text{N}\{\text{number less than } 1\}}{\text{N}(S)} = \dfrac{0}{6} = 0.$

The event is impossible.

5. (a) The sample space consists of 52 cards of 4 suits: clubs, diamonds, hearts, and spades. The clubs and spades are black, while the diamonds and hearts are red. Each suit contains 13 cards beginning with 2 to 10 and the honor cards, jack, queen, king, and ace.
   (b) How many aces of hearts are there? $P(\text{ace of hearts}) = \frac{1}{52}$.
   (c) How many kings are in the deck? $P(\text{kings}) = \frac{4}{52} = \frac{1}{13}$.
   (d) and (e) You should now be able to give answers for these after working through (a), (b), and (c).

Exercise 15.3 Page 399

1. The sample space $S = \{(H, H), (H, T), (T, H), (T, T)\}$
   (a) The event exactly 2 tails is $\{(T, T)\}$.

$$P(\{(T, T)\}) = \frac{\text{N}((T, T))}{\text{N}(S)} = \frac{1}{4}$$

   (b) The event one head and one tail is $\{(H, T), (T, H)\}$. What is the probability of this event?
   (c) The event both heads or both tails is $\{(H, H), (T, T)\}$. What is the probability of this event?
2. Refer to the chart and the tree diagram of this experiment to determine the sample space.
   (a) The event that the coin will be heads is

$$\{(H, 1), (H, 2), (H, 3), (H, 4), (H, 5), (H, 6)\}.$$

   The probability of this event is

$$\frac{6}{\text{N}(S)} = \frac{6}{12} = \frac{1}{2}.$$

   (c) The event of ordered pairs (T, odd) is $\{(T, 1), (T, 3), (T, 5)\}$. What is the probability of this event?
3. Refer to the tree diagram of Experiment 5 to determine the 3-lamb family sample space S.
   (a) The elements of S having the same sex are (M, M, M) and (F, F, F).
   (b) $P(\text{3-lamb families of same sex}) = \dfrac{\text{N(3-lamb families of same sex)}}{\text{N}(S)}$

Exercise 15.4 Page 404

1. Refer to Figures 15–3 and 15–4.
   (a) To distinguish between the elements of the two events, the elements of event A have been marked with □ and those of event B with ×.

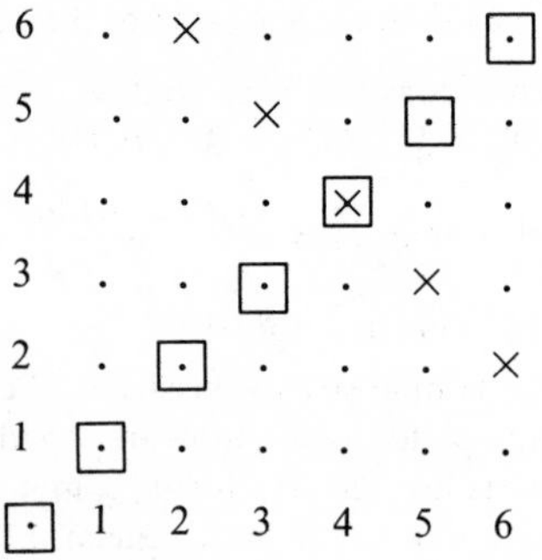

(c) $A \cap B = \{(4, 4)\}$

(d) (i) $P(A) = \dfrac{N(A)}{N(S)} = \dfrac{6}{36}$; (ii) $P(B) = \dfrac{N(B)}{N(S)} = \dfrac{5}{36}$.

(iii) $P(A \text{ or } B) = \dfrac{N(A \text{ or } B)}{N(S)} = \dfrac{10}{36}$. Explain why $N(A \text{ or } B) = 10$.

(iv) $P(A \text{ and } B) = \dfrac{N(A \text{ and } B)}{N(S)} = \dfrac{1}{36}$. Explain why $N(A \text{ and } B) = 1$.

3. Refer to Figures 15–3 and 15–4.

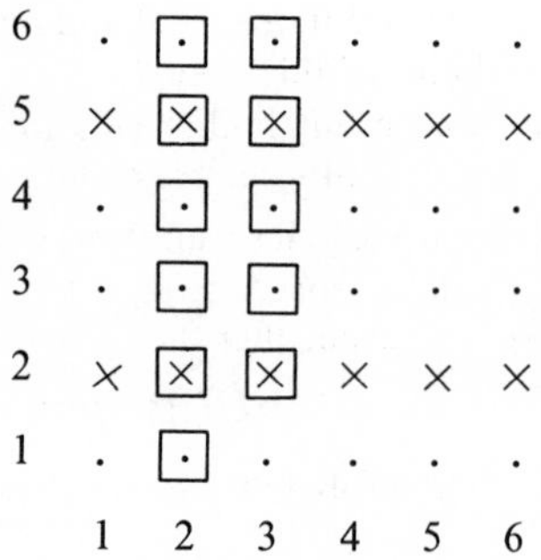

(a) Use the graph to determine

$$A \cap B = \{(2, 2), (2, 5), (3, 2), (3, 5)\}$$

(c) How is it possible to determine whether sets A and B are disjoint?

(d) $P(A) = \frac{12}{36}$; $P(B) = \frac{12}{36}$; $P(A \cup B) = \frac{20}{36}$.
Do you see any explanation for $P(A \cup B) \neq P(A) + P(B)$?

Exercise 15.5 Page 410

1. In a one-die rolling experiment, the sample space $S = \{1, 2, 3, 4, 5, 6\}$. Event $A = \{1, 3, 5\}$; Event $B = \{2, 4, 6\}$

$$P(A \text{ or } B) = \frac{N(A \text{ or } B)}{N(S)} = \frac{N\{1, 3, 5, 2, 4, 6\}}{N\{1, 2, 3, 4, 5, 6\}} = \frac{6}{6} = 1$$

Since $A \cap B = \varnothing$, the events are mutually exclusive. And with $A \cup B = S$, events A and B are complementary.

4. Refer to Experiment 3, Section 15.3. How many elements are in the sample space S?
   (a) In counting the outcomes which have at least one head, should (H, H) be counted?

$$P(\text{at least one head}) = \frac{\text{N}\{(H, H), (H, T), (T, H)\}}{\text{N}(S)} = \frac{3}{4}$$

   (c) Odds in favor of at least one

$$H = \frac{\text{N(at least one } H)}{\text{N}(\overline{\text{at least one } H})} = \frac{3}{1}.$$

   How is $\text{N}(\overline{\text{at least one } H})$ determined? Recall the complement of an event.

6. Refer to Figures 15–3 and 15–4, Section 15.4.

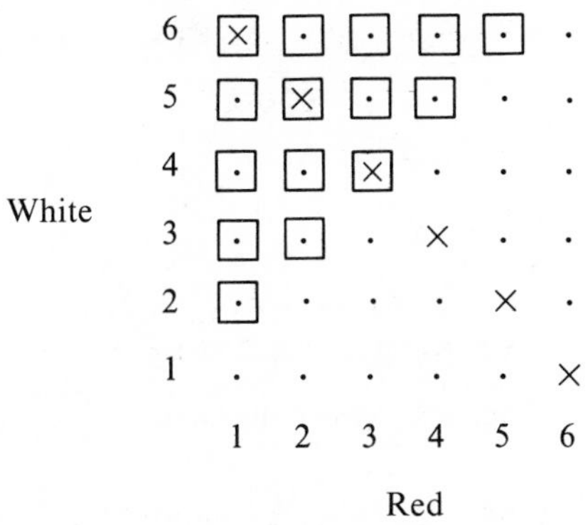

   Elements of event A are designated by □ and those of B by ×.
   (b) Is it possible to tell whether events A and B are mutually exclusive by knowing that $A \cap B = \{(1, 6), (2, 5), (3, 4)\}$? Does this help to determine whether the events are complementary?
   (c) Use the graph to count the number of elements in $(A \cup B)$. Then,

$$P(A \cup B) = \tfrac{18}{36}.$$

   Also, $P(A \cup B)$ may be found by applying Theorem I. Find $P(A)$, $P(B)$, and $P(A \cap B)$.

$$P(A \cup B) = P(A) + P(B) - P(A \cap B) = \tfrac{15}{36} + \tfrac{6}{36} - \tfrac{3}{36} = \tfrac{18}{36}.$$

8. Theorem VI:
   Prove: $P(E)$ is nonnegative, that is, $P(E) \geq 0$.
   Proof: By Definition 15.2(b),

$$P(E) = \frac{\text{N}(E)}{\text{N}(S)}, \text{ where } E \subset S.$$

Since N(E) and N(S) are nonnegative, $\frac{N(E)}{N(S)}$ is nonnegative.

Hence, $P(E)$ is nonnegative.

Theorem VII:
Prove: $P(S) = 1$

Proof: By Definition 15.2(b), $P(S) = \frac{N(S)}{N(S)}$

Let N(S) be s, then $P(S) = \frac{s}{s} = 1$.

Theorem VIII:
Prove: $P(\varnothing) = 0$
Proof: Refer to the proof of Theorem VII.

Theorem IX:
Prove: $P(S \cup \varnothing) = P(S)$

Proof: By Definition 15.2(b), $P(S \cup \varnothing) = \frac{N(S \cup \varnothing)}{N(S)}$

Since $S \cup \varnothing = S$, $P(S \cup \varnothing) = \frac{N(S)}{N(S)}$

But, $P(S) = \frac{N(S)}{N(S)}$. Why?

Hence, $P(S \cup \varnothing) = P(S)$.

Theorem X:
Prove: If $E \subset F$, then $P(E) \leq P(F)$.
Proof: Since $E \subset F$, then $N(E) \leq N(F)$.

Since N(S) is nonnegative, $\frac{N(E)}{N(S)} \leq \frac{N(F)}{N(S)}$.

But, $P(E) = \frac{N(E)}{N(S)}$ and $P(F) = \frac{N(F)}{N(S)}$ by Definition 15.2(b).

Hence, $P(E) \leq P(F)$, if $E \subset F$.
Note that if E is a proper subset of F, $P(E) < P(F)$; if E is a subset, but not a proper subset, then $E = S$ and $P(E) = P(F)$.

Theorem XI:
Prove: $0 \leq P(E) \leq 1$
Proof: Use Theorem VI to state that $P(E) \geq 0$.

Use Definition 15.2(b) to state that $P(E) = \frac{N(E)}{N(S)}$.

Proceed in a manner similar to that done in Theorem X to show that $P(E) \leq 1$. Complete the proof.

Exercise 15.6 Page 413

1. After finding $P(A)$, $P(B)$, and $P(A \cap B)$, you should apply the test (Def. 15.6) to see whether events A and B are independent. You will find that $\frac{1}{18} \neq \frac{1}{3} \cdot \frac{1}{9}$.
3. Make a tree diagram showing a coin tossed twice.

4. A tree diagram for the tossing of three coins is similar to that of Experiment 5. You should construct the diagram.

   (b) The event that the first two are heads is $E_3 = \{(H, H, H), (H, H, T)\}$. The event that the first and last differ is

   $$E_4 = \{(H, H, T), (H, T, T), (T, H, H), (T, T, H)\}.$$

   (c) To determine whether E_3 and E_4 are independent, apply Definition 15.6.

   $$\text{Does} \quad P(E_3 \cap E_4) = P(E_3) \cdot P(E_4)?$$

   Find $P(E_3 \cap E_4)$, $P(E_3)$, and $P(E_4)$. Did you find that you could say $\frac{1}{8} = \frac{2}{8} \cdot \frac{4}{8}$?

   (d) A 2-coins-tail event is $\{(H, T, T), (T, H, T), (T, T, H)\}$
   A 3-coins-tail event is $\{(T, T, T)\}$
   $P(2 \text{ coins}) = \frac{3}{8}$; $P(3 \text{ coins}) = \frac{1}{8}$ $\quad P(2 \text{ coins} \cap 3 \text{ coins}) = P(\varnothing) = 0.$

5. (c) By Theorem I, $P(A \cup B) = P(A) + P(B) - P(A \cap B)$

   $$= \frac{12}{36} + \frac{24}{36} - \frac{8}{36}$$
   $$= \frac{28}{36} = \frac{7}{9}$$

7. If E is $\varnothing$ and $E \subset S$, then $P(\varnothing) = 0$ by Theorem VIII.
   Does $P(\varnothing \cap S) = P(\varnothing) \cdot P(S)$? If so, $\varnothing$ and S are independent.

   $$\begin{array}{ccc} \downarrow & & \downarrow \quad \downarrow \\ P(\varnothing) & & P(\varnothing) \cdot P(S) \\ \downarrow & & \downarrow \quad \downarrow \\ 0 & & 0 \cdot P(S) \\ \downarrow & & \searrow\swarrow \\ 0 & = & 0 \end{array}$$

   Hence Definition 15.6 holds for $\varnothing$ and S.
   Suppose E is nonempty and a proper subset of S.
   Does $P(E \cap S) = P(E) \cdot P(S)$?

   $$\begin{array}{ccc} \downarrow & & \downarrow \quad \downarrow \\ P(E) & & P(E) \cdot \ 1 \qquad \textit{By Theorem } \text{VII}, P(S) = 1 \\ \downarrow & & \downarrow \\ P(E) & = & P(E) \end{array}$$

   Hence Definition 15.6 holds for E and S.
   Suppose E is S.
   $P(S \cap S) = P(S) \cdot P(S)$
   Complete the discussion in a manner similar to the above.

List of Symbols

$\sqrt{5}$	the square root of 5
6^2	the square of 6
$\approx$	is approximately equal to
$=$	is equal to
$\neq$	is not equal to
$\simeq$	is equivalent to
$\not\simeq$	is not equivalent to
$<$	is less than
$\nless$	is not less than
$>$	is greater than
$\ngtr$	is not greater than
$\leq$	is less than or equal to
$\geq$	is greater than or equal to
$\cong$	is congruent to
$\equiv$	is congruent to (modulus)
—	line segment
$\rightarrow$	ray
$\leftrightarrow$	line

$\angle$	angle
$m\overline{AB}$	the measure of $\overline{AB}$
$m\angle ABC$	the measure of $\angle ABC$
$\emptyset$	the empty set
$\in$	is an element of
$\notin$	is not an element of
$\subset$	is a subset of
$\not\subset$	is not a subset of
$\cap$	intersection
$\cup$	union
X	Cartesian product
$\tilde{x}$	the complement of x
(a, b)	the ordered pair a, b
$\{a, b\}$	the set consisting of a, b
{ }	set symbol
( )	parentheses
N{ }	N-notation

NOTE: To negate any relation, a slash mark, /, is drawn through the symbol of the relation.

Index

Index

1 2 3 4 5 6 7 8 9 10